S*pringer* **M***onographs in* **M***athematics*

Kung-Ching Chang

Methods in Nonlinear Analysis

Kung-Ching Chang
School of Mathematical Sciences
Peking University
100871 Beijing
People's Republic of China
E-mail: kcchang@math.pku.edu.cn

Library of Congress Control Number: 2005931137

Mathematics Subject Classification (2000): 47H00, 47J05, 47J07, 47J25, 47J30, 58-01, 58C15, 58E05, 49-01, 49J15, 49J35, 49J45, 49J53, 35-01

ISSN 1439-7382
ISBN-10 3-540-24133-7 Springer Berlin Heidelberg New York
ISBN-13 978-3-540-24133-1 Springer Berlin Heidelberg New York

Springer is a part of Springer Science+Business Media
springeronline.com

Printed in The Netherlands

Typesetting: by the authors and TechBooks using a Springer LaTeX macro package

Cover design: *design & production* GmbH, Heidelberg

Printed on acid-free paper SPIN: 11369295 41/TechBooks 5 4 3 2 1 0

Preface

Nonlinear analysis is a new area that was born and has matured from abundant research developed in studying nonlinear problems. In the past thirty years, nonlinear analysis has undergone rapid growth; it has become part of the mainstream research fields in contemporary mathematical analysis.

Many nonlinear analysis problems have their roots in geometry, astronomy, fluid and elastic mechanics, physics, chemistry, biology, control theory, image processing and economics. The theories and methods in nonlinear analysis stem from many areas of mathematics: Ordinary differential equations, partial differential equations, the calculus of variations, dynamical systems, differential geometry, Lie groups, algebraic topology, linear and nonlinear functional analysis, measure theory, harmonic analysis, convex analysis, game theory, optimization theory, etc. Amidst solving these problems, many branches are intertwined, thereby advancing each other.

The author has been offering a course on nonlinear analysis to graduate students at Peking University and other universities every two or three years over the past two decades. Facing an enormous amount of material, vast numbers of references, diversities of disciplines, and tremendously different backgrounds of students in the audience, the author is always concerned with how much an individual can truly learn, internalize and benefit from a mere semester course in this subject.

The author's approach is to emphasize and to demonstrate the most fundamental principles and methods through important and interesting examples from various problems in different branches of mathematics. However, there are technical difficulties: Not only do most interesting problems require background knowledge in other branches of mathematics, but also, in order to solve these problems, many details in argument and in computation should be included. In this case, we have to get around the real problem, and deal with a simpler one, such that the application of the method is understandable. The author does not always pursue each theory in its broadest generality; instead, he stresses the motivation, the success in applications and its limitations.

The book is the result of many years of revision of the author's lecture notes. Some of the more involved sections were originally used in seminars as introductory parts of some new subjects. However, due to their importance, the materials have been reorganized and supplemented, so that they may be more valuable to the readers.

In addition, there are notes, remarks, and comments at the end of this book, where important references, recent progress and further reading are presented.

The author is indebted to Prof. Wang Zhiqiang at Utah State University, Prof. Zhang Kewei at Sussex University and Prof. Zhou Shulin at Peking University for their careful reading and valuable comments on Chaps. 3, 4 and 5.

Peking University
September, 2003

Kung Ching Chang

Contents

1

Linearization

The first and the easiest step in studying a nonlinear problem is to linearize it. That is, to approximate the initial nonlinear problem by a linear one. Nonlinear differential equations and nonlinear integral equations can be seen as nonlinear equations on certain function spaces. In dealing with their linearizations, we turn to the differential calculus in infinite-dimensional spaces. The implicit function theorem for finite-dimensional space has been proved very useful in all differential theories: Ordinary differential equations, differential geometry, differential topology, Lie groups etc. In this chapter we shall see that its infinite-dimensional version will also be useful in partial differential equations and other fields; in particular, in the local existence, in the stability, in the bifurcation, in the perturbation problem, and in the gluing technique etc. This is the contents of Sects. 1.2 and 1.3. Based on Newton iterations and the smoothing operators, the Nash–Moser iteration, which is motivated by the isometric embeddings of Riemannian manifolds into Euclidean spaces and the KAM theory, is now a very important tool in analysis. Limited in space and time, we restrict ourselves to introducing only the spirit of the method in Sect. 1.4.

1.1 Differential Calculus in Banach Spaces

There are two kinds of derivatives in the differential calculus of several variables, the gradients and the directional derivatives. We shall extend these two to infinite-dimensional spaces.

Let X, Y and Z be Banach spaces, with norms $\|\cdot\|_X$, $\|\cdot\|_Y$, $\|\cdot\|_Z$, respectively. If there is no ambiguity, we omit the subscripts. Let $U \subset X$ be an open set, and let $f : U \to Y$ be a map.

1.1.1 Frechet Derivatives and Gateaux Derivatives

Definition 1.1.1 *(Fréchet derivative) Let $x_0 \in U$; we say that f is Fréchet differentiable (or F-differentiable) at x_0, if $\exists A \subset L(X,Y)$ such that*

$$\| f(x) - f(x_0) - A(x - x_0) \|_Y = \circ(\| x - x_0 \|_X) \, .$$

Let $f'(x_0) = A$, and call it the Fréchet (or F-) derivative of f at x_0.

If f is F-differentiable at every point in U, and if $x \mapsto f'(x)$, as a mapping from U to $L(X,Y)$, is continuous at x_0, then we say that f is continuously differentiable at x_0. If f is continuously differentiable at each point in U, then we say that f is continuously differentiable on U, and denote it by $f \in C^1(U,Y)$.

Parallel to the differential calculus of several variables, by definition, we may prove the following:

1. If f is F-differentiable at x_0, then $f'(x_0)$ is uniquely determined.
2. If f is F-differentiable at x_0, then f must be continuous at x_0.
3. (Chain rule) Assume that $U \subset X, V \subset Y$ are open sets, and that f is F-differentiable at x_0, and g is F-differentiable at $f(x_0)$, where
$$U \xrightarrow{\ f\ } V \xrightarrow{\ g\ } Z$$
Then
$$(g \circ f)'(x_0) = g' \circ f(x_0) \cdot f'(x_0) \, .$$

Definition 1.1.2 *(Gateaux derivative) Let $x_0 \in U$; we say that f is Gateaux differentiable (or G-differentiable) at x_0, if $\forall h \in X, \exists\, df(x_0, h) \subset Y$, such that*

$$\|f(x_0 + th) - f(x_0) - t df(x_0, h)\|_Y = \circ(t) \quad as \quad t \to 0$$

for all $x_0 + th \subset U$. We call $df(x_0, h)$ the Gateaux derivative (or G-derivative) of f at x_0.

We have
$$\frac{d}{dt} f(x_0 + th) \mid_{t=0} = df(x_0, h) \, ,$$
if f is G-differentiable at x_0.

By definition, we have the following properties:

1. If f is G-differentiable at x_0, then $df(x_0, h)$ is uniquely determined.
2. $df(x_0, th) = t df(x_0, h) \ \ \forall t \in \mathbb{R}^1$.
3. If f is G-differentiable at x_0, then $\forall h \in X$, $\forall y^* \in Y^*$, the function $\varphi(t) = \langle y^*, f(x_0 + th)\rangle$ is differentiable at $t = 0$, and $\varphi'(t) = \langle y^*, df(x_0, h)\rangle$.
4. Assume that $f : U \to Y$ is G-differentiable at each point in U, and that the segment $\{x_0 + th \mid t \in [0, 1]\} \subset U$, then
$$\| f(x_0 + h) - f(x_0) \|_Y \leqslant \sup_{0<t<1} \| df(x_0 + th, h) \|_Y$$

Proof. Let

$$\varphi_{y^*}(t) = \langle y^*, f(x_0 + th) \rangle \quad t \in [0,1], \forall y^* \in Y^*$$

$$\begin{aligned} | \langle y^*, f(x_0 + h) - f(x_0) \rangle | &= | \varphi_{y^*}(1) - \varphi_{y^*}(0) | \\ &= | \varphi'_{y^*}(t^*) | \\ &= | \langle y^*, df(x_0 + t^* h, h) \rangle | \end{aligned}$$

for some $t^* \in (0,1)$ depending on y^*. The conclusion follows from the Hahn–Banach theorem. □

5. If f is F-differentiable at x_0, then f is G-differentiable at x_0, with $df(x_0, h) = f'(x_0)h \quad \forall h \in X$.

   Conversely it is not true, but we have:

Theorem 1.1.3 *Suppose that $f : U \to Y$ is G-differentiable, and that $\forall x \in U$, $\exists A(x) \in L(X,Y)$ satisfying*

$$df(x,h) = A(x)h \qquad \forall h \in X .$$

If the mapping $x \mapsto A(x)$ is continuous at x_0, then f is F-differentiable at x_0, with $f'(x_0) = A(x_0)$.

Proof. With no loss of generality, we assume that the segment $\{x_0 + th \mid t \in [0,1]\}$ is in U. According to the Hahn–Banach theorem, $\exists y^* \in Y^*$, with $\| y^* \| = 1$, such that

$$\| f(x_0 + h) - f(x_0) - A(x_0)h \|_Y = \langle y^*, f(x_0 + h) - f(x_0) - A(x_0)h \rangle .$$

Let

$$\varphi(t) = \langle y^*, f(x_0 + th) \rangle .$$

From the mean value theorem, $\exists \xi \in (0,1)$ such that

$$\begin{aligned} | \varphi(1) - \varphi(0) - \langle y^*, A(x_0)h \rangle | &= | \varphi'(\xi) - \langle y^*, A(x_0)h \rangle | \\ &= | \langle y^*, df(x_0 + \xi h, h) - A(x_0)h \rangle | \\ &= | \langle y^*, [A(x_0 + \xi h) - A(x_0)]h \rangle | \\ &= \circ(\| h \|) , \end{aligned}$$

i.e., $f'(x_0) = A(x_0)$. □

The importance of Theorem 1.1.3 lies in the fact that it is not easy to write down the F-derivative for a given map directly, but the computation of G-derivative is reduced to the differential calculus of single variables. The same situation occurs in the differential calculus of several variables: Gradients are reduced to partial derivatives, and partial derivatives are reduced to derivatives of single variables.

Example 1. Let $A \in L(X,Y)$, $f(x) = Ax$. Then $f'(x) = A \quad \forall x$.

Example 2. Let $X = \mathbb{R}^n$, $Y = \mathbb{R}^m$, and let $\varphi_1, \varphi_2 \ldots, \varphi_m \in C^1(\mathbb{R}^n, R^1)$. Set

$$f(x) = \begin{pmatrix} \varphi_1(x) \\ \vdots \\ \varphi_m(x) \end{pmatrix}, \text{ i.e., } \quad f : X \to Y .$$

Then

$$f'(x_0) = \left(\frac{\partial \varphi_i(x_0)}{\partial x_j} \right)_{m \times n} .$$

Example 3. Let $\Omega \subset \mathbb{R}^n$ be an open bounded domain. Denote by $C(\overline{\Omega})$ the continuous function space on $\overline{\Omega}$. Let

$$\varphi : \overline{\Omega} \times \mathbb{R}^1 \longrightarrow \mathbb{R}^1 ,$$

be a C^1 function. Define a mapping $f : C(\overline{\Omega}) \to C(\overline{\Omega})$ by

$$u(x) \mapsto \varphi(x, u(x)) .$$

Then f is F-differentiable, and $\forall u_0 \in C(\overline{\Omega})$,

$$(f'(u_0) \cdot v)(x) = \varphi_u(x, u_0(x)) \cdot v(x) \quad \forall v \in C(\overline{\Omega}) .$$

Proof. $\forall h \in C(\overline{\Omega})$

$$t^{-1}[f(u_0 + th) - f(u_0)](x) = \varphi_u(x, u_0(x) + t\theta(x)h(x))h(x) ,$$

where $\theta(x) \in (0,1)$. $\forall \varepsilon > 0$, $\forall M > 0$, $\exists \delta = \delta(M, \varepsilon) > 0$ such that

$$| \varphi_u(x, \xi) - \varphi_u(x, \xi') | < \varepsilon, \ \forall x \in \overline{\Omega} ,$$

as $|\xi|$, $|\xi'| \leqslant M$ and $|\xi - \xi'| \leqslant \delta$. We choose $M = \| u_0 \| + \| h \|$, then for $|t| < \delta < 1$,

$$|\varphi_u(x, u_0(x) + t\theta(x)h(x)) - \varphi_u(x, u_0(x))| < \varepsilon .$$

It follows that $df(u_0, h)(x) = \varphi_u(x, u_0(x))h(x)$.

Noticing that the multiplication operator $h \mapsto A(u)h = \varphi_u(x, u(x)) \cdot h(x)$ is linear and continuous, and the mapping $u \mapsto A(u)$ from $C(\overline{\Omega})$ into $L(C(\overline{\Omega}), C(\overline{\Omega}))$ is continuous, from Theorem 1.1.3, f is F-differentiable, and

$$(f'(u_0) \cdot v)(x) = \varphi_u(x, u_0(x)) \cdot v(x) \ \forall v \in C(\overline{\Omega}) .$$

$\square$

We investigate nonlinear differential operators on more general spaces. Let $\Omega \subset \mathbb{R}^n$ be a bounded open set, and let m be a nonnegative integer, $\gamma \in (0,1)$. $C^m(\overline{\Omega})$ (and the Hölder space $C^{m,\gamma}(\overline{\Omega})$) is defined to be the function space consisting of C^m functions (with γ-Hölder continuous m-order partial derivatives).

The norms are defined as follows:

$$\| u \|_{C^m} = \max_{x\in\overline{\Omega}} \sum_{|\alpha|\leq m} |\partial^\alpha u(x)| ,$$

and

$$\| u \|_{C^{m,\gamma}} = \| u \|_{C^m} + \max_{x,y\in\overline{\Omega}} \max_{|\alpha|=m} \frac{|\partial^\alpha u(x) - \partial^\alpha u(y)|}{|x-y|^\gamma} ,$$

where $\alpha = (\alpha_1, \alpha_2, \ldots, \alpha_n)$ is a multi-index, $|\alpha| = \alpha_1 + \alpha_2 + \cdots + \alpha_n$, $\partial^\alpha = \partial^{\alpha_1}_{x_1}\partial^{\alpha_2}_{x_2}\cdots\partial^{\alpha_n}_{x_n}$.

We always denote by m^* the number of the index set $\{\alpha = (\alpha_1, \alpha_2, \ldots, \alpha_n) \mid |\alpha| \leqslant m\}$, and $D^m u$ the set $\{\partial^\alpha u \mid |\alpha| \leqslant m\}$.

Suppose that r is a nonnegative integer, and that $\varphi \in C^\infty(\overline{\Omega}\times\mathbb{R}^{r^*})$. Define a differentiable operator of order r:

$$f(u)(x) = \varphi(x, D^r u(x)) .$$

Suppose $m \geqslant r$, then $f : C^m(\overline{\Omega}) \to C^{m-r}(\overline{\Omega})$ (and also $C^{m,\gamma}(\overline{\Omega}) \to C^{m-r,\gamma}(\overline{\Omega})$) is F-differentiable. Furthermore

$$(f'(u_0)h)(x) = \sum_{|\alpha|\leq r} \varphi_\alpha(x, D^r u_0(x)) \cdot \partial^\alpha h(x), \quad \forall h \in C^m(\overline{\Omega}) ,$$

where φ_α is the partial derivative of φ with respect to the variable index α.

The proof is similar to Example 3.

Example 4. Suppose $\varphi \in C^\infty(\overline{\Omega}\times\mathbb{R}^{r^*})$. Define

$$f(u) = \int_\Omega \varphi(x, D^r u(x))dx \quad \forall u \in C^r(\overline{\Omega}) .$$

Then $f : C^r(\overline{\Omega}) \to \mathbb{R}^1$ is F-differentiable. Furthermore

$$\langle f'(u_0), h\rangle = \int_\Omega \sum_{|\alpha|\leq r} \varphi_\alpha(x, D^r u_0(x))\partial^\alpha h(x)dx \quad \forall h \in C^r(\overline{\Omega}) .$$

Proof. Use the chain rule:

$$C^r(\overline{\Omega}) \xrightarrow{\varphi(\cdot, D^r u(\cdot))} C(\overline{\Omega}) \xrightarrow{\int_\Omega} \mathbb{R}^1 ,$$

and combine the results of Examples 1 and 3. □

In particular, the following functional occurs frequently in the calculus of variations ($r = 1, r^* = n+1$). Assume that $\varphi(x, u, p)$ is a function of the form:

$$\varphi(x, u, p) = \frac{1}{2}|p|^2 + \sum_{i=1}^{n} a_i(x)p_i + a_0(x)u ,$$

where $p = (p_1, p_2, \ldots, p_n)$, and $a_i(x)$, $i = 0, 1, \ldots, n$, are in $C(\overline{\Omega})$.
Set

$$f(u) = \int_{\Omega} \left[\frac{1}{2}|\nabla u(x)|^2 + \sum_{i=1}^{n} a_i(x)\partial_{x_i} u + a_0(x)u(x)\right] dx ,$$

we have

$$\langle f'(u), h\rangle = \int_{\Omega} \left[\nabla u(x) \cdot \nabla h(x) + \sum_{i=1}^{n} a_i(x)\partial_{x_i} h(x) + a_0(x)h(x)\right] dx$$
$$\forall h \in C^1(\overline{\Omega}) .$$

Example 5. Let X be a Hilbert space, with inner product $(,)$. Find the F-derivative of the norm $f(x) = \| x \|$, as $x \neq \theta$.

Let $F(x) = \| x \|^2$. Since

$$t^{-1}(\| x + th \|^2 - \| x \|^2) = 2(x, h) + t \| h \|^2 ,$$

we have $dF(x, h) = 2(x, h)$. It is continuous for all x, therefore F is F-differentiable, and

$$F'(x)h = 2(x, h) .$$

Since $f = F^{\frac{1}{2}}$, by the chain rule

$$F'(x) = 2 \| x \| \cdot f'(x) .$$

As $x \neq \theta$,

$$f'(x)h = \left(\frac{x}{\| x \|}, h\right) .$$

In the applications to PDE as well as to the calculus of variations, Sobolev spaces are frequently used. We should extend the above studies to nonlinear operators defined on Sobolev spaces.

$\forall p \geqslant 1$, $\forall$ nonnegative integer m, let

$$W^{m,p}(\Omega) = \{u \in L^p(\Omega) \mid \partial^\alpha u \in L^p(\Omega) \mid |\alpha| \leqslant m\} ,$$

where $\partial^\alpha u$ stands for the α-order generalized derivative of u, i.e., the derivative in the distribution sense. Define the norm

$$\| u \|_{W^{m,p}} = \left(\sum_{|\alpha \leq m} \| \partial^\alpha u \|^p_{L^p(\Omega)}\right)^{\frac{1}{p}} .$$

The Banach space is called the Sobolev space of index $\{m, p\}$.

$W^{m,2}(\Omega)$ is denoted by $H^m(\Omega)$, and the closure of $C_0^\infty(\Omega)$ under this norm is denoted by $H_0^m(\Omega)$.

1.1.2 Nemytscki Operator

On Sobolev spaces, we extend the composition operator $u \mapsto \varphi(x, u(x))$ such that φ may not be continuous in x. The class of operators is sometimes called Nemytski operators.

Definition 1.1.4 *Let (Ω, B, μ) be a measure space. We say that $\varphi : \Omega \times \mathbb{R}^N \to \mathbb{R}^1$ is a Caratheodory function, if*

1. *$\forall$a.e.$x \in \Omega$, $\xi \mapsto \varphi(x, \xi)$ is continuous.*
2. *$\forall \xi \in \mathbb{R}^N$, $x \mapsto \varphi(x, \xi)$ is μ-measurable.*

The motivation in introducing the Caratheodory function is to make the composition function measurable if $u(x)$ is only measurable. Indeed, there exists a sequence of simple functions $\{u_n(x)\}^\infty$, such that $u_n(x) \to u(x)$ a.e., $\varphi(x, u_n(x))$ is measurable according to (2). And from (1), $\varphi(x, u_n(x)) \to \varphi(x, u(x))$ a.e., therefore $\varphi(x, u(x))$ is measurable.

Theorem 1.1.5 *Assume $p_1, p_2 \geqslant 1$, $a > 0$ and $b \in L^{p_2}_{d\mu}(\Omega)$. Suppose that φ is a Caratheodory function satisfying*

$$|\varphi(x, \xi)| \leqslant b(x) + a|\xi|^{\frac{p_1}{p_2}} .$$

Then $f : u(x) \mapsto \varphi(x, u(x))$ is a bounded and continuous mapping from $L^{p_1}_{d\mu}(\Omega, \mathbb{R}^N)$ to $L^{p_2}_{d\mu}(\Omega, \mathbb{R}^N)$.

Proof. The boundedness follows from the Minkowski inequality:

$$\| f(u) \|_{p_2} \leqslant \| b \|_{p_2} + a \| u \|_{p_1}^{\frac{p_1}{p_2}} ,$$

where $\| \cdot \|_p$ is the $L^p_{d\mu}(\Omega, \mathbb{R}^N)$ norm. We turn to proving the continuity. It is sufficient to prove that $\forall \{u_n\}_1^\infty$ if $u_n \to u$ in L^{p_1}, then there is a subsequence $\{u_{n_i}\}$ such that $f(u_{n_i}) \to f(u)$ in L^{p_2}. Indeed one can find a subsequence $\{u_{n_i}\}$ of $\{u_n\}$ which converges a.e. to u, along which $\| u_{n_i} - u_{n_{i-1}} \|_{p_1} < \frac{1}{2^i}$, $i = 2, 3, \ldots$; therefore

$$|u_{n_i}(x)| \leqslant \Phi(x) := |u_{n_1}(x)| + \sum_{i=2}^{\infty} |u_{n_i}(x) - u_{n_{i-1}}(x)| .$$

Since Φ is measurable, and

$$\left(\int_\Omega |\Phi(x)|^{p_1} d\mu \right)^{\frac{1}{p_1}} \leqslant \| u_{n_1} \|_{p_1} + \sum_{i=2}^{\infty} \| u_{n_i} - u_{n_{i-1}} \|_{p_1} < +\infty ,$$

we conclude that $\Phi \in L^{p_1}_{d\mu}(\Omega)$. Noticing

$$f(u_{n_i}) = \varphi(x, u_{n_i}(x)) \to \varphi(x, u(x)) \quad a.e. \ ,$$

and

$$|f(u_{n_i})(x)| \leqslant b(x) + a(\Phi(x))^{\frac{p_1}{p_2}} \in L^{p_2}_{d\mu}(\Omega, R^N) \ ,$$

we have $\| f(u_{n_i}) - f(u) \|_{p_2} \to 0$, according to Lebesgue dominance theorem. This proves the continuity of f. □

Corollary 1.1.6 *Let $\Omega \subset \mathbb{R}^n$ be a smooth bounded domain, and let $1 \leq p_1, p_2 \leq \infty$. Suppose that $\varphi : \Omega \times \mathbb{R}^{m^*} \to \mathbb{R}$ is a Caratheodory function satisfying*

$$|\varphi(x, \xi_0, \ldots, \xi_m)| \leqslant b(x) + a \sum_{j=0}^{m} |\xi_j|^{\frac{\alpha_j}{p_2}} \ ,$$

where ξ_j is a $\#\{\alpha = (\alpha_1, \ldots, \alpha_n) \,|\, |\alpha| = j\}$-vector, $\alpha_j \leqslant (\frac{1}{p_1} - \frac{m-j}{n})^{-1}$, $a > 0$, and $b \in L^{p_2}(\Omega)$. Then $f(u)(x) = \varphi(x, D^m u(x))$ defines a bounded and continuous map from $W^{m,p_1}(\Omega)$ into $L^{p_2}(\Omega)$.

Corollary 1.1.7 *Suppose that $\Omega \subset \mathbb{R}^n$ and that $\varphi : \Omega \times \mathbb{R}^1 \to \mathbb{R}^1$ and $\varphi_\xi(x, \xi)$ are Caratheodory functions. If $|\varphi_\xi(x, \xi)| \leqslant b(x) + a|\xi|^r$, where $b \in L^{\frac{2n}{n+2}}(\Omega)$, $a > 0$, and $r = \frac{n+2}{n-2}$ (if $n \leq 2$, then the restriction is not necessary), then the functional*

$$f(u) = \int_\Omega \varphi(x, u(x)) dx$$

is F-differentiable on $H^1(\Omega)$, with F-derivative

$$\langle f'(u), v \rangle = \int_\Omega \varphi_\xi(x, u(x)) \cdot v(x) dx \ ,$$

where $\langle , \rangle$ is the inner product on $H^1(\Omega)$.

Proof. The Sobolev embedding theorem says that the injection $i : H^1(\Omega) \hookrightarrow L^{\frac{2n}{n-2}}(\Omega)$ is continuous, so is the dual map $i^* : L^{\frac{2n}{n+2}}(\Omega) \hookrightarrow (H^1(\Omega))^*$.

According to Theorem 1.1.5, $\varphi_\xi(\cdot, \cdot) : L^{\frac{2n}{n-2}} \to L^{\frac{2n}{n+2}}$ is continuous. Therefore the Gateaux derivative

$$df(u, v) = \int_\Omega \varphi_\xi(x, u(x)) \cdot v(x) dx \quad \forall v \in H^1(\Omega)$$

is continuous from $H^1(\Omega)$ to $(H^1(\Omega))^*$. Applying Theorem 1.1.3, we conclude that f is F-differentiable on $H^1(\Omega)$. The proof is complete. □

Corollary 1.1.8 *In Corollary 1.1.6, the differential operator*

$$f(u(x)) = \varphi(x, D^m u(x))$$

from $C^{l,\gamma}(\overline{\Omega})$ to $C^{l-m,\gamma}(\overline{\Omega})$, $l \geqslant m$, $0 \leqslant \gamma < 1$, is F-differentiable, with

$$(f'(u_0)h)(x) = \sum_{|\alpha| \leq m} \varphi_\alpha(x, D^m u(x)) \partial^\alpha h(x) \quad \forall h \in C^{l,\gamma}(\overline{\Omega}) \ .$$

1.1.3 High-Order Derivatives

The second-order derivative of f at x_0 is defined to be the derivative of $f'(x)$ at x_0. Since $f' : U \to L(X,Y)$, $f''(x_0)$ should be in $L(X, L(X,Y))$. However, if we identify the space of bounded bilinear mappings with $L(X, L(X,Y))$, and verify that $f''(x_0)$ as a bilinear mapping is symmetric, see Theorem 1.1.9 below, then we can define equivalently the second derivative $f''(x_0)$ as follows: For $f : U \to Y$, $x_0 \in U \subset X$, if there exists a bilinear mapping $f''(x_0)(\cdot,\cdot)$ of $X \times X \to Y$ satisfying

$$\| f(x_0+h) - f(x_0) - f'(x_0)h - \frac{1}{2} f''(x_0)(h,h) \| = \circ(\| h \|^2) \quad \forall h \in X, \text{as } \| h \| \to 0 ,$$

then $f''(x_0)$ is called the second-order derivative of f at x_0.

By the same manner, one defines the mth-order derivatives at x_0 successively: $f^{(m)}(x_0) : X \times \cdots \times X \to Y$ is an m-linear mapping satisfying

$$\left\| f(x_0 + h) - \sum_{j=0}^{m} \frac{f^{(j)}(x_0)(h, \ldots, h)}{j!} \right\| = \circ(\| h \|^m) ,$$

as $\| h \| \to 0$. Then f is called m differentiable at x_0.

Similar to the finite-dimensional vector functions, we have:

Theorem 1.1.9 *Assume that $f : U \to Y$ is m differentiable at $x_0 \in U$. Then for any permutation π of $(1, \ldots, m)$, we have*

$$f^{(m)}(x_0)(h_1, \ldots, h_m) = f^{(m)}(x_0)(h_{\pi(1)}, \ldots, h_{\pi(m)}) .$$

Proof. We only prove this in the case where $m = 2$, i.e.,

$$f''(x_0)(\xi, \eta) = f''(x_0)(\eta, \xi) \quad \forall \xi, \eta \in X .$$

Indeed $\forall y^* \in Y^*$, we consider the function

$$\varphi(t,s) = \langle y^*, f(x_0 + t\xi + s\eta) \rangle .$$

It is twice differentiable at $t = s = 0$; so is

$$\frac{\partial^2}{\partial t \partial s} \varphi(0,0) = \frac{\partial^2}{\partial s \partial t} \varphi(0,0) .$$

Since $f'(x_0 + t\xi + s\eta)$ is continuous as $|t|$, $|s|$ small, one has

$$\frac{\partial}{\partial s} \varphi(t, \cdot)|_{s=0} = \langle y^*, f'(x_0 + t\xi)\eta \rangle ;$$

and then,

$$\frac{\partial^2}{\partial t \partial s} \varphi(t,s)|_{t=s=0} = \langle y^*, f''(x_0)(\xi, \eta) \rangle .$$

Similarly

$$\frac{\partial^2}{\partial s \partial t} \varphi(t,s)|_{t=s=0} = \langle y^*, f''(x_0)(\eta, \xi) \rangle .$$

This proves the conclusion. □

Theorem 1.1.10 *(Taylor formula) Suppose that $f : U \to Y$ is continuously m-differentiable. Assume the segment $\{x_0 + th \,|\, t \in [0,1]\} \subset U$. Then*

$$f(x_0+h) = \sum_{j=0}^{m} \frac{1}{j!} f^{(j)}(x_0)(h,\ldots,h) + \frac{1}{m!}\int_0^1 (1-t)^m f^{(m+1)}(x_0+th)(h,\ldots,h)dt \,.$$

Proof. $\forall y^* \in Y^*$, we consider the function:

$$\varphi(t) = \langle y^*, f(x_0+th)\rangle \,.$$

From the Hahn–Banach theorem and the Taylor formula for single-variable functions:

$$\varphi(1) = \sum_{j=0}^{m} \frac{1}{j!}\varphi^{(j)}(0) + \frac{1}{m!}\int_0^1 (1-t)^m \varphi^{(m+1)}(t)dt \,,$$

we obtain the desired Taylor formula for mappings between B-spaces. □

Example 1. $X = \mathbb{R}^n$, $Y = \mathbb{R}^1$. If $f : X \to Y$ is twice continuously differentiable, then

$$f''(x) = H_f(x) = \left(\frac{\partial^2 f(x)}{\partial x_i \partial x_j}\right)_{i,j=1,\ldots,n} .$$

Example 2. $X = C^1(\overline{\Omega}, \mathbb{R}^N)$, $Y = \mathbb{R}^1$. Suppose that $g \in C^2(\overline{\Omega} \times \mathbb{R}^N, \mathbb{R}^1)$. Define

$$f(u) = \frac{1}{2}\int_\Omega |\nabla u|^2 + \int_\Omega g(x, u(x))$$

as $u \in X$. By definition, we have

$$f'(u)\cdot\varphi = \int_\Omega [\nabla u(x)\nabla\varphi(x) + g'_u(x,u(x))\varphi(x)]dx \,,$$

and

$$f''(u)(\varphi,\psi) = \int_\Omega [\nabla\psi(x)\nabla\varphi(x) + g''_{uu}(x,u(x))\varphi(x)\psi(x)]dx \,.$$

With some additional growth conditions on g''_{uu}:

$$|g''_{uu}(x,u)| \le a(1+|u|^{\frac{4}{n-2}}), \; a>0 \; \forall u \in R^N \,,$$

f is twice differentiable in $H_0^1(\Omega, R^N)$. As an operator from $H_0^1(\Omega, R^N)$ into itself,

$$f''(u) = id + (-\triangle)^{-1} g''_u(\cdot, u(\cdot)) \,.$$

is self-adjoint, or equivalently, the operator $-\Delta + g''_{uu}(x, u(x))\cdot$ defined on L^2 is self-adjoint with domain $H^2 \cap H_0^1(\Omega, R^N)$.

Example 3. Let $X = H_0^1(\Omega, \mathbb{R}^3)$, where Ω is a plane domain. Consider the volume functional

$$Q(u) = \int_\Omega u \cdot (u_x \wedge u_y) \, .$$

One has

$$Q'(u) \cdot \varphi = \int_\Omega \varphi(u_x \wedge u_y) + u \cdot [(\varphi_x \wedge u_y) + (u_x \wedge \varphi_y)] \, ,$$

and

$$\begin{aligned} Q''(u)(\varphi, \psi) = \int_\Omega & \varphi[(\psi_x \wedge u_y) + (u_x \wedge \psi_y)] + \psi[(\varphi_x \wedge u_y) + (u_x \wedge \varphi_y)] \\ & + u[(\varphi_x \wedge \psi_y) + (\psi_x \wedge \varphi_y)] \, , \end{aligned}$$

$\forall \varphi, \psi \in H_0^1(\Omega, \mathbb{R}^3)$.

If further we assume $u \in C^2(\overline{\Omega}, \mathbb{R}^3)$, then from integration by parts and the antisymmetry of the exterior product, we have

$$\begin{aligned} \int_\Omega u(\varphi_x \wedge u_y) &= -\int_\Omega (\varphi \wedge u_y) \cdot u_x + (\varphi \wedge u_{xy}) \cdot u \\ &= \int_\Omega (u_x \wedge u_y)\varphi - \int_\Omega (\varphi \wedge u_{xy}) \cdot u \, , \end{aligned}$$

and

$$\begin{aligned} \int_\Omega u(u_x \wedge \varphi_y) &= -\int_\Omega u_y(u_x \wedge \varphi) + u(u_{xy} \wedge \varphi) \\ &= \int_\Omega \varphi(u_x \wedge u_y) + \int_\Omega (\varphi \wedge u_{xy})u \, . \end{aligned}$$

Therefore,

$$Q'(u) = 3 \int_\Omega \varphi \cdot (u_x \wedge u_y) \, .$$

By the same manner, we obtain

$$Q''(u)(\varphi, \psi) = 3 \int_\Omega u[(\varphi_x \wedge \psi_y) + (\varphi_y \wedge \psi_x)] \, .$$

Geometrically, let $u : \Omega \overset{C^2}{\to} \mathbb{R}^3$ be a parametrized surface in $\mathbb{R}^3$; $Q(U)$ is the volume of the body enclosed by the surface.

As exercises, one computes the first- and second-order differentials of the following functionals:

1. $X = W_0^{1,p}(\Omega, \mathbb{R}^1)$, $Y = \mathbb{R}^1$, $2 < p < \infty$,

$$f(u) = \int_\Omega |\nabla u|^p dx \, .$$

2. $X = C_0^1(\overline{\Omega}, \mathbb{R}^n)$, where $\Omega \subset \mathbb{R}^n$ is a domain,

$$f(u) = \int_\Omega det(\nabla u(x)) dx \, .$$

3. $X = C_0^1(\overline{\Omega}, R^1)$, where $\Omega \subset R^n$ is a domain,

$$f(u) = \int_\Omega \sqrt{1 + |\nabla u|^2} dx \, .$$

1.2 Implicit Function Theorem and Continuity Method

1.2.1 Inverse Function Theorem

It is known that the implicit function theorem for functions of several variables plays important roles in many branches of mathematics (differential manifold, differential geometry, differential topology, etc.). Its extension to infinite-dimensional space is also extremely important in nonlinear analysis, as well as in the study of infinite-dimensional manifolds.

Theorem 1.2.1 *(Implicit function theorem) Let X, Y, Z be Banach spaces, $U \subset X \times Y$ be an open set. Suppose that $f \in C(U, Z)$ has an F-derivative w.r.t. y, and that $f_y \in C(U, L(Y, Z))$. For a point $(x_0, y_0) \in U$, if we have*

$$f(x_0, y_0) = \theta \, ,$$

$$f_y^{-1}(x_0, y_0) \in L(Z, Y) \; ;$$

then $\exists r, r_1 > 0$, $\exists | u \in C(B_r(x_0), B_{r_1}(y_0))$, such that

$$\begin{cases} B_r(x_0) \times B_{r_1}(y_0) \subset U \, , \\ u(x_0) = y_0 \, , \\ f(x, u(x)) = \theta \quad \forall x \in B_r(x_0) \, . \end{cases}$$

Furthermore, if $f \in C^1(U, Z)$, then $u \in C^1(B_r(x_0), Y)$, and

$$u'(x) = -f_y^{-1}(x_0, u(x_0)) \circ f_x(x, u(x)) \quad \forall x \in B_r(x_0) \, . \tag{1.1}$$

Proof. (1) After replacing f by

$$g(x,y) = f_y^{-1}(x_0,y_0) \circ f(x+x_0, y+y_0)\ ,$$

one may assume $x_0 = y_0 = \theta$, $Z = Y$ and $f_y(\theta,\theta) = id_Y$.
(2) We shall find the solution $y = u(x) \in B_{r_1}(\theta)$ of the equation

$$f(x,y) = \theta \quad \forall x \in B_r(\theta)\ .$$

Setting

$$R(x,y) = y - f(x,y)\ ,$$

it is reduced to finding the fixed point of $R(x,\cdot)\quad \forall x \in B_r(\theta)$.

We shall apply the contraction mapping theorem to the mapping $R(x,\cdot)$. Firstly, we have a contraction mapping:

$$\begin{aligned}\| R(x,y_1) - R(x,y_2) \| &= \| y_1 - y_2 - [f(x,y_1) - f(x,y_2)] \| \\ &= \| y_1 - y_2 - \int_0^1 f_y(x, ty_1 + (1-t)y_2)dt \cdot (y_1 - y_2) \| \\ &\leqslant \int_0^1 \| \mathrm{id}_Y - f_y(x, ty_1 + (1-t)y_2) \| dt \cdot \| y_1 - y_2 \|\ .\end{aligned}$$

Since $f_y : U \to L(X,Y)$ is continuous, $\exists r, r_1 > 0$ such that

$$\| R(x,y_1) - R(x,y_2) \| < \frac{1}{2} \| y_1 - y_2 \| \tag{1.2}$$

$\forall (x,y_i) \in B_r(\theta) \times B_{r_1}(\theta)$, $i = 1,2$.

Secondly, we are going to verify $R(x,\cdot) : \overline{B}_{r_1}(\theta) \to \overline{B}_{r_1}(\theta)$. Indeed,

$$\begin{aligned}\| R(x,y) \| &\leqslant \| R(x,\theta) \| + \| R(x,y) - R(x,\theta) \| \\ &\leqslant \| f(x,\theta) \| + \frac{1}{2} \| y \|\ .\end{aligned}$$

For sufficiently small $r > 0$, where

$$\| f(x,\theta) \| < \frac{1}{2} r_1, \quad \forall x \in \overline{B}_r(\theta)\ , \tag{1.3}$$

it follows that $\|R(x,y)\| < r_1$, $\forall (x,y) \in B_r(\theta) \times B_{r_1}(\theta)$. Then, $\forall x \in \overline{B}_r(\theta)$, $\exists | y \in \overline{B}_{r_1}(\theta)$ satisfying $f(x,y) = \theta$. Denote by $u(x)$ the solution y.
(3) We claim that $u \in C(B_r, Y)$. Since

$$\begin{aligned}\| u(x) - u(x') \| &= \| R(x,u(x)) - R(x',u(x')) \| \\ &\leqslant \frac{1}{2} \| u(x) - u(x') \| + \| R(x,u(x)) - R(x',u(x)) \|\ ,\end{aligned}$$

we obtain

$$\| u(x) - u(x') \| \leqslant 2 \| R(x, u(x)) - R(x', u(x)) \| . \tag{1.4}$$

Noticing that $R \in C(U, Y)$, we have

$$u(x') \to u(x) \quad \text{as} \quad x' \to x .$$

(4) If $f \in C^1(U, Y)$, we want to prove $u \in C^1$. First, by (1.2) and (1.4)

$$\begin{aligned} \| u(x) - u(x') \| &\leqslant 2 \| f(x, u(x)) - f(x', u(x')) \| \\ &\leqslant 2 \int_0^1 \| f_x(tx + (1-t)x', u(x)) \| dt \cdot \| x - x' \| . \end{aligned}$$

Therefore

$$\| u(x+h) - u(x) \| = O(\| h \|) \quad \text{for} \quad \| h \| \to 0 .$$

From

$$f(x+h, u(x+h)) = f(x, u(x)) = \theta ,$$

it follows that

$$f(x+h, u(x+h)) - f(x, u(x+h)) + [f(x, u(x+h)) - f(x, u(x))] = \theta ;$$

also

$$f_x(x, u(x+h))h + \circ(\| h \|) + f_y(x, u(x))(u(x+h) - u(x)) + \circ(\| h \|) = \theta .$$

Therefore

$$u(x+h) - u(x) + f_y^{-1}(x, u(x)) \circ f_x(x, u(x+h))h = \circ(\| h \|) ,$$

i.e., $u \in C^1$, and

$$u'(x) = -f_y^{-1}(x, u(x)) \circ f_x(x, u(x)) .$$

□

Remark 1.2.2 *In the first part of Theorem 1.2.1, the space X may be assumed to be a topological space. In fact, neither linear operations nor the properties of the norm were used.*

Theorem 1.2.3 *(Inverse function theorem) Let $V \subset Y$ be an open set, and $g \in C^1(V, X)$. Assume $y_0 \in V$ and $g'(y_0) \in L(X, Y)$. Then there exists $\delta > 0$ such that $B_\delta(y_0) \subset V$ and*

$$g : B_\delta(y_0) \to g(B_\delta(y_0))$$

is a differmorphism. Furthermore

$$(g^{-1})'(x_0) = g'^{-1}(y_0), \ \text{with } x_0 = g(y_0) . \tag{1.5}$$

Proof. Set
$$f(x,y) = x - g(y), \quad f \in C^1(X \times V, X)\,.$$
We use the implicit function theorem (IFT) to f, there exist $r > 0$ and a unique $u \in C^1(B_r(x_0), B_r(y_0))$ satisfying
$$x = g \circ u(x)\,.$$
Since g is continuous, $\exists \delta \in (0,r)$ such that $g(B_\delta(y_0)) \subset B_r(x_0)$, therefore $g : B_\delta(y_0) \to g(B_\delta(y_0))$ is a diffeomorphism. And (1.5) follows from (1.1). □

In the spirit of the IFT, we have a nonlinear version of the Banach open mapping theorem.

Theorem 1.2.4 *(Open mapping) Let X, Y be Banach spaces, and let $\delta > 0$ and $y_0 \in Y$. Suppose that $g \in C^1(B_\delta(y_0), X)$ and that $g'(y_0) : Y \to X$ is an open map, then g is an open map in a neighborhood of y_0.*

Proof. We want to prove that $\exists \delta_1 \in (0,\delta)$ and $r > 0$, such that
$$B_r(g(y_0)) \subset g(B_{\delta_1}(y_0))\,.$$
With no loss of generality, we may assume $y_0 = \theta$ and $g(y_0) = \theta$. Let $A = g'(\theta)$.

Since A is surjective, $\exists C > 0$ such that
$$\inf_{z \in \ker A} \| y - z \|_Y \leqslant C \| Ay \|_X, \; \forall\, y \in Y\,, \tag{1.6}$$
provided by the Banach inverse theorem. One chooses $\delta_1 \in (0,\delta)$ and $r > 0$, satisfying
$$\| g'(y) - A \| \leqslant \frac{1}{2(C+1)} \quad \forall y \in B_{\delta_1}(\theta)\,,$$
and
$$r < \frac{\delta_1}{2(C+1)}\,.$$
Now, $\forall x \in B_r(\theta)$, we are going to find $y \in B_{\delta_1}(\theta)$, satisfying $g(y) = x$. Write
$$R(y) = g(y) - Ay\,.$$
The problem is equivalent to solving the following equation:
$$Ay = x - R(y)\,. \tag{1.7}$$
We solve it by iteration.

Initially, we take $h_0 = \theta$. Suppose that $h_n \in B_{\delta_1}(\theta)$ has been chosen; from (1.6), we can find h_{n+1}, satisfying
$$Ah_{n+1} = x - R(h_n)\,,$$
and

$$\| h_{n+1} - h_n \| \leqslant (C+1) \| A(h_{n+1} - h_n) \| .$$

Thus

$$\begin{aligned} \| h_{n+1} - h_n \| &\leqslant (C+1) \| R(h_n) - R(h_{n-1}) \| \\ &= (C+1) \| \int_0^1 g'(th_n + (1-t)h_{n-1})dt - A \| \cdot \| h_n - h_{n-1} \| \\ &\leq \frac{1}{2} \| h_n - h_{n-1} \| \quad \forall n \geqslant 1 . \end{aligned}$$

Since

$$\| h_1 \| \leqslant (1+C) \| x \| \leqslant \frac{1}{2}\delta_1 ,$$

and

$$\begin{aligned} \| h_{n+1} \| &\leqslant \| h_1 \| + \sum_{j=1}^{n} \| h_{j+1} - h_j \| \\ &\leqslant \left(\frac{1}{2} + \cdots + \frac{1}{2^n} + \frac{1}{2^{n+1}} \right) \delta_1 < \delta_1 , \end{aligned}$$

it follows that $h_{n+1} \in B_{\delta_1}(\theta)$. Then we can proceed inductively.

The sequence h_n has a limit y. Obviously y is the solution of (1.7). □

Essentially, the implicit function theorem is a consequence of the contraction mapping theorem. The continuity assumption of f_y in Theorem 1.2.2 seems too strong in some applications. We have a weakened version.

Theorem 1.2.5 *Let X, Y, Z be Banach spaces, and let $B_r(\theta) \subset Y$ be a closed ball centered at θ with positive radius r. Suppose that $T \in L(Y, Z)$ has a bounded inverse, and that $\eta : X \times B_r \to Z$ satisfies the following Lipschitz condition:*

$$\| \eta(x, y_1) - \eta(x, y_2) \| \leqslant K \| y_1 - y_2 \| \quad \forall y_1, y_2 \in B_r(\theta), \forall x \in X$$

where $K < \| T^{-1} \|^{-1}$. If $\eta(\theta, \theta) = \theta$, and

$$\| \eta(x, \theta) \| \leqslant (\| T^{-1} \|^{-1} - K)r ;$$

then $\forall x \in X$, there exists a unique $u : X \to B_r(\theta)$ satisfying

$$Tu(x) + \eta(x, u(x)) = \theta \quad \forall x \in X .$$

Furthermore, if η is continuous, then so is u.

Proof. $\forall x \in X$ we find the fixed point of the map $-T^{-1}\eta(x, y)$. It is easily verified that $T^{-1} \cdot \eta(x, \cdot) : B_r(\theta) \to B_r(\theta)$ is a contraction mapping. □

1.2.2 Applications

As we mentioned in the beginning of this section, the IFT plays an important role in solving nonlinear equations. However, the IFT is only a local statement. If a problem is local, then it is extremely powerful for local solvability. As to global solvability problems, we first solve them locally, and then extend the solutions by continuation. In this case, the IFT is applied in the first step. However, in this subsection we only present several examples to show how the method works for local problems, and in the next subsection for global problems.

Example 1. (Structural stability for hyperbolic systems)

A matrix $L \in GL(n, R)$ is called hyperbolic if the set of eigenvalues of L, $\sigma(L) \cap iR^1 = \varnothing$. The associate differential system reads as:

$$\dot{x} = Lx, \; x \in C^1(R^1, R^n) .$$

The flow $\phi_t = e^{Lt} \in GL(n, R)$ is also linear. The flow line can be seen on the left of Fig. 1.1. Let $\xi \in C^{0,1}(R^n, R^n)$ be a Lipschitzian (Lip.) map we investigate the hyperbolic system under the nonlinear perturbation:

$$\dot{x} = Lx + \xi(x), \; x \in C^1(R^1, R^n) ,$$

and let ψ_t be the associate flow, the flow line of which is on the right of Fig. 1.1.

What is the relationship between ϕ_t and ψ_t, if ξ is small? One says that the hyperbolic system is structurally stable, which means that the flow lines ϕ_t and ψ_t are topologically equivalent. More precisely, there is a homeomorphism $h : R^n \to R^n$ such that $h \circ \psi_t = \phi_t \circ h$.

We shall show that the hyperbolic system is structurally stable. Let $A = e^L$, then the set of eigenvalues $\sigma(A)$ of A satisfies $\sigma(A) \cap S^1 = \varnothing$. We decompose $\psi_1 = A + f$, where $f \in C^{0,1}(R^n, R^n)$, it is known that the Lipschitzian constant of f is small if that of ξ is.

To the matrix $A \in GL(n, R)$, since $\sigma(A) \cap S^1 = \varnothing$, provided by the Jordan form, we have the decomposition $R^n = E_u \bigoplus E_s$, where E_u, E_s are invariant subspaces, on which the eigenvalues of $A_u := A|_{E_u}$ lie outside the unit circle, and those of $A_s := A|_{E_s}$ lie inside the unit circle. Due to these facts, one has

$$\|A_s\| < 1, \; \|A_u^{-1}\| < 1 .$$

The following notations are used for any Banach spaces X, Y. $C^0(X, Y)$ stands for the space of all bounded and continuous mappings $h : X \to Y$ with norm:

$$\|h\| = \sup_{x \in X} \|h(x)\|_Y .$$

$C^{0,1}(X, Y)$ stands for the space of all bounded Lipschitzian maps from X to Y with norm:

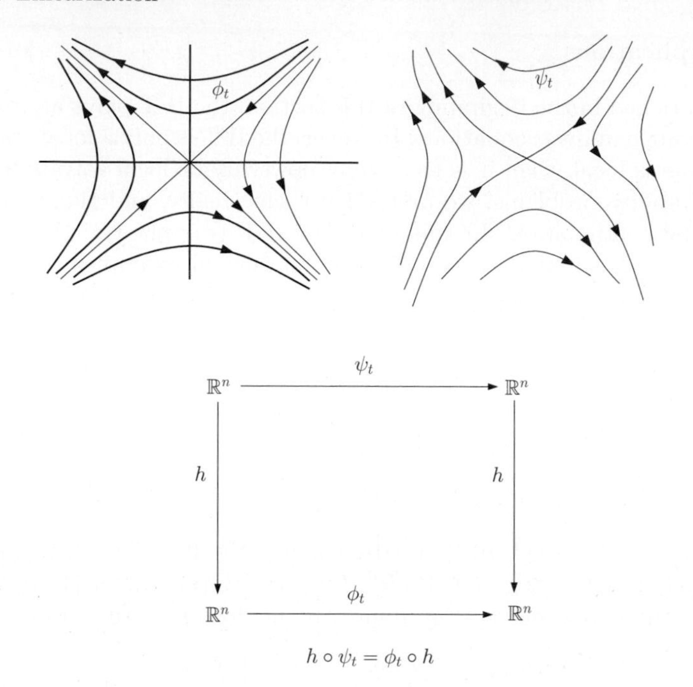

Fig. 1.1.

$$\|h\| = \sup_{x \in X} \|h(x)\|_Y + \sup_{x \in X, y \in X} \frac{\|h(x) - h(y)\|_Y}{\|x - y\|_X} .$$

We shall prove:

Theorem 1.2.6 *(Hartman–Grobman) $\exists \epsilon > 0$, such that there exists a unique homeomorphism $h : R^n \to R^n$ satisfying*

$$\phi_t \circ h = h \circ \psi_t \ \forall t , \tag{1.8}$$

if $\|f\|_{C^{0,1}} < \epsilon$.

Proof. (1) First we prove the case $t = 1$, and would rather consider the problem more generally: assume $g \in C^{0,1}(R^n, R^n)$ with $\|g\|_{C^{0,1}} < \epsilon$. We are going to solve continuous mappings $h, k : R^n \to R^n$ satisfying

$$h \circ (A + f) = (A + g) \circ h \text{ and } k \circ (A + g) = (A + f) \circ k , \tag{1.9}$$

respectively.

Let $0 < \epsilon < \|A^{-1}\|^{-1}$, then both $(A + f)^{-1}$ and $(A + g)^{-1}$ exist. We decompose $R^n = E_u \bigoplus E_s$ and let P_u and P_s be the projections onto E_u and E_s, respectively. Let $r = h - id$, then equation (1.9) for r is equivalent to

$$(A + g) \circ (\text{id} + r) = (\text{id} + r) \circ (A + f) ,$$

i.e.,

$$A \circ r = f - g \circ (\mathrm{id} + r) + r \circ (A + f) ,$$

or

$$\begin{cases} P_u r = A_u^{-1} \circ P_u r \circ (A + f) + A_u^{-1} \circ P_u f - A_u^{-1} \circ P_u g \circ (\mathrm{id} + r) \\ P_s r = A_s \circ P_s r \circ (A + f)^{-1} - P_s f \circ (A + f)^{-1} \\ \qquad + P_s g \circ (\mathrm{id} + r) \circ (A + f)^{-1} . \end{cases} \tag{1.10}$$

We do the same for the equation of k.

(2) From the decomposition $C^0(R^n, R^n) = C^0(R^n, E_u) \bigoplus C^0(R^n, E_s)$, we reduce (1.10) to the form of Theorem 1.2.5. Set $Sr = A_u^{-1} \circ P_u r \circ (A+f) \bigoplus A_s \circ P_s r \circ (A + f)^{-1}$. Then $S \in \mathbf{L}(C^0(R^n, R^n), C^0(R^n, R^n))$, and $\|S\| < 1$. Thus $T := \mathrm{id} - S$ is invertible. By setting

$$\eta(f, g; r) = A_u^{-1} \circ (P_u f - P_u g \circ (\mathrm{id} + r)) \bigoplus (-P_s f + P_s g \circ (\mathrm{id} + r)) \circ (A + f)^{-1} ,$$

Then (1.10) is equivalent to

$$Tr = \eta(f, g; r) .$$

Since $\eta(f, g; r)$ is Lip. with respect to r, with Lip. constant $< \epsilon$, and $\|\eta(f, g; \theta)\| < 2\epsilon$. For sufficiently small $\epsilon > 0$, all conditions of Theorem 1.2.5 are met. There exists a unique continuous map r in a neighborhood of θ in $C^0(R^n, R^n)$, which is continuously dependent on f and g. Similarly, let $q = k - id$, one proves the existence of a unique continuous map $q = q(f, g)$ in the same neighborhood.

(3) Setting $h = \mathrm{id} + r, k = \mathrm{id} + q$, we prove that $h \circ k = k \circ h = \mathrm{id}$. In fact $h \circ k$ and $k \circ h$ satisfy the equations:

$$h \circ k = (A + g)^{-1} \circ (h \circ k) \circ (A + g) ,$$

and

$$k \circ h = (A + f)^{-1} \circ (k \circ h) \circ (A + f) ,$$

respectively. Both the equations have the unique solution id in a neighborhood of θ, and the conclusion is proved.

(4) We now prove the conclusion for arbitrary t. We have

$$\phi_1 \circ (\phi_t \circ h \circ \psi_{-t}) = (\phi_t \circ \phi_1) \circ (h \circ \psi_{-t}) = (\phi_t \circ h \circ \psi_{-t}) \circ \psi_1 .$$

Also as $|t - 1| < \delta$ for small $\delta > 0$, $\phi_t \circ h \circ \psi_{-t} - \mathrm{id} \in C^0(R^n, R^n)$ is in a small neighborhood of θ, and the above equation has a unique solution h there; therefore

$$\phi_t \circ h \circ \psi_{-t} = h .$$

Then we can extend the procedure step by step to all t, i.e. we have $\phi_t \circ h = h \circ \psi_t \ \forall t \in R^1$. □

Example 2. (Local existence of isothermal coordinates)

Given a surface M^2 with a Riemannian metric g, i.e., in local coordinates $x = (x_1, x_2)$,

$$g = E dx_1^2 + 2F dx_1 dx_2 + G dx_2^2 ,$$

where E, F and G are functions of local coordinates $x = (x_1, x_2)$, and $\forall (x_1, x_2)$, $E\xi^2 + 2F\xi\eta + G\eta^2$ is a positive definite quadratic form. Our problem is to find a local coordinate $u(x) = (u_1(x_1, x_2), u_2(x_1, x_2))$ in a neighborhood of $x^0 = (x_1^0, x_2^0)$ such that there exists a function $\lambda = \lambda(x_1, x_2) > 0$ satisfying

$$g = \lambda(x_1, x_2)(du_1^2 + du_2^2) .$$

The local coordinate $u = (u_1, u_2)$ is called an isothermal coordinate.

We shall find $u = (u_1, u_2)$ satisfying

$$\begin{cases} \lambda(x)|\partial_{x_1} u(x)|^2 = E(x) , \\ \lambda(x)\partial_{x_1} u(x) \cdot \partial_{x_2} u(x) = F(x) , \\ \lambda(x)|\partial_{x_2} u(x)|^2 = G(x) . \end{cases} \tag{1.11}$$

We may assume $(x_1^0, x_2^0) = (0, 0)$ and $F(0, 0) \neq 0$ (by translation and rotation of the local coordinates). After eliminating λ, (1.11) is equivalent to:

$$\begin{cases} F|\partial_{x_1} u|^2 = E\partial_{x_1} u \cdot \partial_{x_2} u \\ F|\partial_{x_2} u|^2 = G\partial_{x_1} u \cdot \partial_{x_2} u . \end{cases} \tag{1.12}$$

This is a first-order nonlinear differential system.

In a neighborhood O of $\theta = (0, 0)$, define

$$\varphi(x, \nabla u(x)) = \begin{pmatrix} F(x) \cdot |\partial_{x_1} u(x)|^2 - E(x) \cdot \partial_{x_1} u(x) \cdot \partial_{x_2} u(x) \\ F(x) \cdot |\partial_{x_2} u(x)|^2 - G(x) \cdot \partial_{x_1} u(x) \cdot \partial_{x_2} u(x) , \end{pmatrix} ,$$

$\forall x \in O$, and let f be the map satisfying

$$f(u)(x) = \varphi(x, \nabla u(x)) .$$

We want to solve the equation:

$$f(u) = \theta . \tag{1.13}$$

Let $x = \varepsilon y$, where $y \in D$, the unit disk on the coordinate plane, and $\varepsilon > 0$ is a parameter, and let

$$\varepsilon v(y) = u(x) - \overline{u}(x) . \tag{1.14}$$

where $\overline{u} = (p_1 x_1 + p_2 x_2, q_1 x_1 + q_2 x_2)$, and (p_1, p_2, q_1, q_2) is chosen such that the following conditions are satisfied:

$$\frac{\partial(u_1, u_2)}{\partial(x_1, x_2)} \neq 0 ,$$

and

$$\varphi(0, \nabla \overline{u}) = \theta \; .$$

In fact, this is an algebraic system with four unknowns, and is trivially solvable.

Equation (1.13) is equivalent to

$$\varphi(\varepsilon y, \nabla_y v(y) + \nabla_x \overline{u}(\varepsilon y)) = 0 \; .$$

Let $\dot{C}^{1,\alpha}(\bar{D})$ denote the space $C^{1,\alpha}(\bar{D})$ modulo a constant. Set $F : \mathbb{R}^1 \times (C^{1,\alpha} \cap C_0^0(\overline{D}) \bigoplus \dot{C}^{1,\alpha}(\overline{D})) \to C^\alpha(\overline{D})^2$, for some $\alpha \in (0,1)$, where

$$F(\varepsilon, v) = \varphi(\varepsilon y, \nabla_y v(y) + \nabla_x \overline{u}(\varepsilon y)) \; .$$

Thus

$$F(0, \theta) = \theta \; .$$

It remains to find a pair (ε, v) for small $\varepsilon > 0$, satisfying

$$F(\varepsilon, v) = \theta \; .$$

Then by the transform(1.14), the solution u in a small neighborhood of θ εD is obtained. Note that $(p_1, q_1, p_2, q_2) = (\overline{u}_{1,x_1}, \overline{u}_{2,x_1}, \overline{u}_{1,x_2}, \overline{u}_{2,x_2})$. Let us introduce the notation $(E_0, F_0, G_0) = (E, F, G)|_{(0,0)}$; we have

$$F_v(0, \theta) h = -(\mathbf{A}_1 \nabla h_1 + \mathbf{A}_2 \nabla h_2) \; ,$$

where $h = (h_1, h_2) \in C^{1,\alpha} \cap C_0^0(\overline{D}) \bigoplus \dot{C}^{1,\alpha}(\overline{D})$, and

$$\mathbf{A}_1 = \begin{pmatrix} (E_0 p_2 - 2F_0 p_1) & E_0 p_1 \\ G_0 p_2 & (G_0 p_1 - 2F_0 p_2) \end{pmatrix} ,$$

$$\mathbf{A}_2 = \begin{pmatrix} (E_0 q_2 - 2F_0 q_1) & E_0 q_1 \\ G_0 q_2 & (G_0 q_1 - 2F_0 q_2) \end{pmatrix} .$$

This is a constant coefficient first-order linear differential operator with the 2×2 matrix symbol:

$$L = (\mathbf{A}_1 \omega, \mathbf{A}_2 \omega) \; ,$$

where $\omega = (\xi, \eta)$.

Since

$$\det L = -2F_0(E_0 \xi^2 - 2F_0 \xi \eta + G_0 \eta^2) \frac{\partial(\overline{u}, \overline{v})}{\partial(x, y)} \neq 0$$

$\forall (\xi, \eta) \in \mathbb{R}^2 \backslash \{\theta\}$, L is elliptic.

According to the elliptic theory, $F_v(0, \cdot)$ has a bounded inverse. The IFT is applied to conclude the existence of $\varepsilon_0 > 0$ such that the equation

$$F(\varepsilon, v) = \theta$$

has a unique solution v_ε, $\forall \varepsilon \in (-\varepsilon_0, \varepsilon_0)$.

Remark 1.2.7 *The boundary condition for the first-order system can be deduced from the second-order elliptic theory.* In fact, $h = (h_1, h_2)$ satisfies

$$\mathbf{A}_1 \nabla h_1 + \mathbf{A}_2 \nabla h_2 = f \,, \tag{1.15}$$

where f is given.

Suppose that $\mathbf{A}_2$ is invertible (it is available); we have

$$\nabla h_2 = C \nabla h_1 + \mathbf{A}_2^{-1} f \,, \tag{1.16}$$

where

$$C = -\mathbf{A}_2^{-1} \mathbf{A}_1 \,,$$

Let

$$E = \begin{pmatrix} 0 & 1 \\ -1 & 0 \end{pmatrix} .$$

Then

$$E^{-1} = \begin{pmatrix} 0 & -1 \\ 1 & 0 \end{pmatrix} ,$$

and $E^2 = -I$.

Since

$$h_{2,xy} = h_{2,yx} \,, \tag{1.17}$$

we obtain a second-order equation for h_1, in which the principal symbol of the second-order equation reads as

$$\begin{aligned} -\omega^T E A_2^{-1} A_1 \omega &= -\frac{1}{\det(A_2)} E^2 (A_2 \omega)^T E^{-1} A_1 \omega \\ &= \frac{1}{\det(A_2)} (A_2 \omega)^T E^{-1} A_1 \omega \\ &= \frac{\det(L)}{\det(A_2)} \,. \end{aligned}$$

The later is positive (or negative) definite because the first-order system is elliptic.

Thus the Dirichlet boundary condition for h_1 is well posed, and then h_2 follows from (1.16) modulo a constant.

Remark 1.2.8 *There are many applications of the IFT similar to the above examples, e.g. a necessary and sufficient condition for an almost complex structure being a complex structure (Newlander–Nirenberg theorem, see L. Nirenberg [NN]), prescribing Ricci curvature problem (see DeTurck [DT] etc.). For applications to boundary value problems in ordinary and partial differential equations see S.N. Chow, J. Hale [CH] and J. Mawhin [Maw 3]*

1.2.3 Continuity Method

We have shown the usefulness of the IFT in the existence of solutions for small perturbations of a given equation which has a known solution. As to large perturbations, the IFT is not enough, we have to add new ingredients. The continuity method is a general principle, which can be applied to prove the existence of solutions for a variety of nonlinear equations.

Let X and Y be Banach spaces, and $f : X \to Y$ be C^1. Find the solution of the equation:

$$f(x) = \theta \ .$$

Let us introduce a parameter $t \in [0, 1]$ and a map

$$F : [0, 1] \times X \to Y$$

such that both F and F_x are continuous; in addition,

$$F(1, x) = f(x) \ .$$

Assume that there exists $x_0 \in X$ satisfying $F(0, x_0) = \theta$; we want to extend the solution x_0 of the equation

$$F(0, x) = \theta$$

to a solution of

$$F(1, x) = \theta \ . \tag{1.18}$$

For this purpose, we define a set

$$S = \{t \in [0, 1] \mid \text{such that } F(t, x) = \theta \text{ is solvable}\} \ .$$

What we want to do is to prove:

(1) S is an open set (relative to $[0,\ 1]$). For this purpose, it is sufficient to prove that $\forall t_0 \in S$, $\exists x_{t_0} \in X$, which solves $F(t_0, x_{t_0}) = \theta$ such that $F_x^{-1}(t_0, x_{t_0}) \in L(Y, X)$, provided by the IFT.
(2) S is a closed set. Usually it depends on the a priori estimates for the solution set $\{x \in X \mid \exists t \in S, \text{ such that } F(t, x) = \theta\}$. For most PDE problems it requires special knowledge and features of the equations and techniques in hard analysis.

We present here two major ideas:

(a) If there exist a Banach space X_1, which is compactly embedded in X, and a constant $C > 0$ such that

$$\| x_t \|_{X_1} \leqslant C \quad \forall t \in S \ ,$$

where x_t is a solution of $F(t, x) = \theta$, then S is closed.

In fact, we have $\{t_n\}^\infty \subset S$, which implies $t_n \to t^*$, and

$$\| x_{t_n} \|_{X_1} \leqslant C .$$

Since the embedding $X_1 \hookrightarrow X$ is compact, x_{t_n} subconverges to some point $x^* \in X$ in X. From the continuity, it follows that $F(t^*, x^*) = \theta$. This proves $t^* \in S$, i.e., S is closed.

(b) If $\forall t \in S$, there exists a unique local solution x_t of the equation $F(t, \cdot) = \theta$, and if there exists $C > 0$ such that

$$\| \dot{x}_t \|_X \leqslant C ,$$

where $\dot{x}_t$ is the derivative of x_t, then S is a closed set.

Proof. Let $\{t_n\}_1^\infty$ be a sequence included in an open interval contained in S, with $t_n \uparrow t^*$. Then

$$\| x_{t_n} - x_{t_m} \| \leqslant \int_{t_m}^{t_n} \| \dot{x}_t \| \, dt \leqslant C(t_n - t_m) \to 0 ,$$

as $n \geqslant m \to \infty$. Let x^* be the limit. If the IFT is applicable to (t^*, x^*), then we have $t^* \in S$, i.e., S is closed. □

Once (1) and (2) are proved, S is a nonempty ($0 \in S$) open and closed set. Therefore $S = [0, 1]$, and then the equation $F(1, \cdot) = \theta$ is solvable.

As an application of the continuity method, we have:

Theorem 1.2.9 *(Global implicit function theorem) Let X, Y be Banach spaces, and let $f \in C^1(X, Y)$ with $f'(x)^{-1} \in L(Y, X)$ $\forall x \in X$. If $\exists$ constants $A, B > 0$ such that*

$$\| f'(x)^{-1} \| \leqslant A \| x \| + B \quad \forall x \in X ,$$

then f is a diffeomorphism.

Proof. (1) Surjective. We want to prove that $\forall y \in Y, \exists x \in X$ satisfying

$$f(x) = y .$$

$\forall x_0 \in X$, define $F : [0, 1] \times X \to Y$ as follows:

$$F(t, x) = f(x) - [(1 - t) f(x_0) + ty] .$$

Set $S = \{t \in [0, 1] | \ F(t, \cdot) = \theta \text{ is solvable}\}$. Obviously, $0 \in S$, and since $F_x^{-1}(t, x) = f'(x)^{-1} \in L(Y, X)$, S is open, from the IFT.

It remains to prove the closeness of S. Indeed, in a component (a, b) of S, there is a branch of solutions x_t satisfying

$$F(t, x_t) = \theta \quad \forall t \in (a, b) ,$$

and then

$$f'(x_t)\dot{x}_t = y - f(x_0) .$$

Thus,

$$\| \dot{x}_t \| \leqslant \| f'(x_t)^{-1} \| \| y - f(x_0) \| \leqslant (A \| x_t \| + B) \| y - f(x_0) \| . \tag{1.19}$$

Set $c = \frac{a+b}{2}$, so

$$\| x_t \| \leqslant \| x_c \| + \int_c^t \| y - f(x_0) \| (A \| x_s \| + B) ds \quad \text{as } t > c .$$

Applying Gronwall's inequality, $\exists$ a constant $C > 0$ such that

$$\| x_t \| \leqslant C . \tag{1.20}$$

Substituting (1.20) into (1.19), we have another constant $C_1 > 0$ such that

$$\| \dot{x}_t \| \leqslant C_1, \ \forall t \in (a, b) .$$

This proves the closeness of S, therefore f is surjective.

(2) Injective. We argue by contradiction. If $\exists y \in Y$ and $x_0, x_1 \in X$, satisfying $f(x_i) = y, \ i = 0, 1$. Let $\gamma : [0, 1] \to X$ be the segment connecting these two points:

$$\gamma(s) = (1 - s)x_0 + sx_1 \quad s \in [0, 1] .$$

Thus $f \circ \gamma$ is a loop passing through y. If we could find $x : [0, 1] \to X$ satisfying

$$x(i) = x_i \quad i = 0, 1, \quad \text{and}$$
$$f \circ x(s) = y \ \forall s \in [0, 1] ,$$

then this would contradict with the locally homeomorphism of f.

Define $I = [0, 1]$ and $T : I \times C_0(I, X) \to C_0(I, Y)$ as follows:

$$(t, u(s)) \mapsto f(\gamma(s) + u(s)) - ty - (1 - t)f(\gamma(s)) ,$$

where

$$C_0(I, X) = \{u \in C(I, X) | \ u(0) = u(1) = \theta\} .$$

We want to solve

$$T(t, u) = \theta .$$

Obviously, $T(0, \theta) = \theta$; and if we have $u \in C_0(I, X)$ satisfying $T(1, u(\cdot)) = \theta$, then $x(s) = u(s) + \gamma(s)$ is what we need. Now,

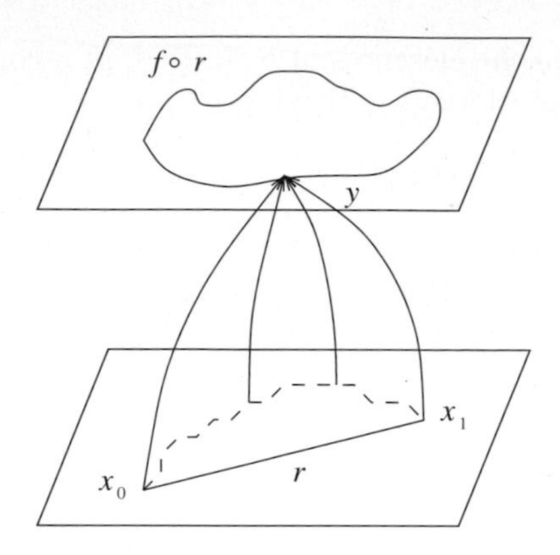

Fig. 1.2.

1.
$$T_u(t,u) = f'(\gamma(\cdot)+u(\cdot)) \in L(C_0(I,X),\ C_0(I,Y))\ ,$$
which has a bounded inverse. Therefore $S = \{t \in [0,1] |\ T(t,u) = \theta$ is solvable$\}$ is open, from the IFT.

2. Let $u_t(s)$ be a solution at $t \in S$. Then
$$f'(\gamma(s)+u_t(s)) \cdot \dot{u}_t(s) = y - f \circ \gamma(s)\ ,$$
where $\dot{u}_t$ denotes the derivative with respect to t. Again we obtain
$$\| \dot{u}_t \|_{C_0(I,X)} \leqslant (A \| u_t \|_{C_0(I,X)} + B_1) \| y - f \circ \gamma \|_{C_0(I,Y)}\ ,$$
where $B_1 > 0$ is another constant depending on B and x_0, x_1 only. As in paragraph (1), $\exists C > 0$ such that
$$\| \dot{u}_t \|_{C_0(I,X)} \leqslant C \quad \forall t \in S\ .$$
Again, by the continuity method, $1 \in S$. This is the contradiction. The injectivity of f is proved. □

Next we shall give an example showing how a priori estimates give the closeness of S.

Theorem 1.2.10 *Suppose $f \in C^1(\overline{\Omega} \times R^1 \times R^n, R^1)$, where $\Omega \subset R^n$ is a bounded domain of smooth boundary. Assume $\exists$ constants $C > 0$, satisfying*

(1) There exists an increasing function $c : R^1_+ \to R^1_+$ such that
$$|f(x,\eta,\xi)| \leq c(|\eta|)(1+|\xi|^2), \forall\, (x,\eta,\xi) \in \overline{\Omega} \times R^1 \times R^n.$$

(2)
$$\frac{\partial f}{\partial \eta}(x,\eta,\xi) \leqslant 0\ .$$

(3) Assume $\exists M > 0$ *such that*

$$f(x,\eta,\theta) = \begin{cases} < 0, & \text{if } \eta > M \\ > 0, & \text{if } \eta < -M \,. \end{cases}$$

Assume $\phi \in C^{2,\gamma}$*, for some* $\gamma \in (0,1)$*. Then the equation*

$$\begin{cases} -\triangle u = f(x,u(x),\nabla u(x)) & x \in \Omega \\ u|_{\partial\Omega} = \phi \end{cases} \tag{1.21}$$

possesses a unique solution in $C^{2,\gamma}$*.*

Lemma 1.2.11 *Under the assumption (3), if* $u \in C^2(\overline{\Omega})$ *is a solution of (1.21), then*

$$\| u \|_{C(\overline{\Omega})} \leqslant \max\left\{ \max_{\partial\Omega} |\phi(x)|, M \right\} .$$

Proof. Assume that $|u(x)|$ attains its maximum at $x_0 \in \overline{\Omega}$. We divide our discussion into two cases.

(1) $x_0 \in \partial\Omega$; the proof is done.

(2) $x_0 \in \overset{\circ}{\Omega}$, then $\nabla u(x_0) = 0$ and $-\triangle u(x_0) = f(x_0, u(x_0), \theta)$.

If $u(x_0) > M$, then LHS $\geqslant 0$, but RHS < 0. It is impossible.

Similarly, if $u(x_0) < -M$, then LHS $\leqslant 0$, but RHS > 0. Again, it is impossible. We have

$$|u(x_0)| \leqslant M \,.$$

□

Lemma 1.2.12 *Assume* $a \in C^{0,\gamma}(\overline{\Omega})$ *and* $\phi \subset C^{2,\gamma}(\partial\Omega)$*. Then the equation*

$$\begin{cases} -\triangle u + u = a(x)(1 + |\nabla u|^2) \\ u|_{\partial\Omega} = \phi \end{cases} \tag{1.22}$$

has a unique solution $u \in C^{2,\gamma}(\overline{\Omega})$*, and*

$$\| u \|_{C^{2,\gamma}} \leqslant C(\| a \|_{C^{0,\gamma}}, \| \phi \|_{C^{2,\gamma}}) \,.$$

Proof. We apply the continuity method to study equation (1.22). Define a map $F : X \times [0,1] \to Y$ as follows:

$$(u,\tau) \mapsto (-\triangle u + u - a(\tau + |\nabla u|^2), u|_{\partial\Omega} - \tau\phi) \,,$$

where $X = C^{2,\overline{\gamma}}(\overline{\Omega}), Y = C^{0,\overline{\gamma}}(\overline{\Omega}) \times C^{2,\overline{\gamma}}(\partial\Omega)$ and $\overline{\gamma} \in (0,\gamma)$.

Noticing that

$$F_u(u,\tau)v = ((-\triangle v + v - a\nabla u \cdot \nabla v), v|_{\partial\Omega}) \,,$$

and from the Schauder estimates, $\forall (u,\tau) \in X \times [0,1]$, $F_u(u,\tau)$ has a bounded inverse. Define the set:

$$S = \{\tau \in [0,1] | \ \exists u_\tau \text{ solving } F(u_\tau,\tau) = \theta\} \ .$$

Equation (1.22) is solvable if $1 \in S$. Since S is open, and $0 \in S$, it remains to verify that the set S is closed. To this end, it is sufficient to prove that there is a constant C, depending on a and ϕ such that $\forall u_\tau \in S$,

$$\| u_\tau \|_{C^{2,\gamma}} \leqslant C \ . \tag{1.23}$$

Since u_τ satisfies $F(u_\tau,\tau) = \theta$, $\dot{u}_\tau = \frac{du_\tau}{d\tau}$ exists, from the IFT, and satisfies:

$$\begin{cases} -\triangle \dot{u}_\tau + \dot{u}_\tau = a + 2a\nabla u_\tau \cdot \nabla \dot{u}_\tau \ , \\ \dot{u}_\tau|_{\partial\Omega} = \phi \ . \end{cases} \tag{1.24}$$

Set $g = a + 2a\nabla u_\tau \cdot \nabla \dot{u}_\tau$; we have

$$\begin{aligned} \| \dot{u}_\tau \|_{W^{2,p}} &\leqslant C(p)(1 + \| g \|_{L^p}) \\ &\leqslant C(p, \| a \|_C)(1 + \| \nabla u_\tau \|_{L^{2p}} \| \nabla \dot{u}_\tau \|_{L^{2p}}) \ , \end{aligned}$$

provided by the L^p estimates.

Since $\dot{u}_\tau$ satisfies (1.24), from Lemma 1.2.11, $\| \dot{u}_\tau \|_{C(\overline{\Omega})}$ is bounded by a constant depending on $\| a \|_{C(\overline{\Omega})}$ and $\|\phi\|$, and from the Gagliardo–Nirenberg inequality

$$\| \nabla \dot{u}_\tau \|_{L^{2p}} \leqslant C_p \| \nabla^2 \dot{u}_\tau \|_{L^p}^{\frac{1}{2}} \| \dot{u}_\tau \|_{L^p}^{\frac{1}{2}} + C \| \dot{u}_\tau \|_{L^\infty} \ ,$$

we obtain:

$$\| \dot{u}_\tau \|_{W^{2,p}} \leqslant C(1 + \| \nabla u_\tau \|_{L^{2p}}^2) \ .$$

Again, by the Gagliardo–Nirenberg inequality, we have

$$\| \nabla u_\tau \|_{L^{2p}} \leqslant C_p \| \nabla^2 u_\tau \|_{L^p}^{\frac{1}{2}} \| u_\tau \|_{L^p}^{\frac{1}{2}} + C \| u_\tau \|_{L^\infty} \ .$$

Repeating the use of Lemma 1.2.11, $\| u_\tau \|_C$ is bounded by a constant depending on $\| \phi \|_C$, and we obtain

$$\| \dot{u}_\tau \|_{W^{2,p}} \leqslant C(\| \phi \|_{C^{2,\gamma}}, \| a \|_C, p)(1 + \| u_\tau \|_{W^{2,p}}) \ .$$

From

$$\frac{d}{d\tau} \| u_\tau \|_{W^{2,p}} \leqslant \| \dot{u}_\tau \|_{W^{2,p}} \ ,$$

and the Gronwall inequality, we obtain

$$\| u_\tau \|_{W^{2,p}} \leqslant C e^{C\tau} \ ,$$

where C depends on p, $\|\phi\|_{C^{2,\gamma}}$ and $\|a\|_C$.

As a consequence of the Sobolev embedding theorem, for $p > \frac{n}{1-\gamma}$, we have

$$\| u_\tau \|_{C^{1,\gamma}} \leqslant C(\| \phi \|_{C^{2,\gamma}}, \| a \|_C, \gamma) . \tag{1.25}$$

Substituting into the equation $F(u_\tau, \tau) = \theta$, we apply the Schauder estimate:

$$\| u_\tau \|_{C^{2,\gamma}} \leqslant C(\| \phi \|_{C^{2,\gamma}}, \| a \|_{C^{0,\gamma}}, \gamma) .$$

The continuity method is applicable; we have a solution u of (1.22).

Indeed, the solution is unique. Let u_1, u_2 be two solutions and set $\omega = u_1 - u_2$, then

$$\begin{cases} -\Delta\omega + \omega = a\nabla(u_1 + u_2) \cdot \nabla\omega & \text{in } \Omega , \\ \omega|_{\partial\Omega} = 0 . \end{cases}$$

If $\max \omega > 0$, then $\exists x_0 \in \overset{\circ}{\Omega}$ such that $\max_\Omega \omega = \omega(x_0)$. Thus, $\nabla\omega(x_0) = 0$, $-\Delta\omega(x_0) \geqslant 0$ and $\omega(x_0) > 0$. This is impossible. Similarly, one proves that $\min \omega < 0$ is impossible. Therefore $\omega \equiv 0$. □

Now we come back to the proof of the theorem.

Proof. Applying the continuity method, we study the equation:

$$\begin{cases} -\Delta u = tf(x, u(x), \nabla u(x)) & t \in [0, 1] \\ u|_{\partial\Omega} = t\phi \end{cases} \tag{1.26}$$

and turn to considering the operator:

$$F : I \times C^{2,\sigma}(\overline{\Omega}) \to C^\sigma(\overline{\Omega}) \times C^{2,\sigma}(\partial\Omega), \quad I = [0, 1] ,$$

$$(t, u) \mapsto (-\Delta u - tf(x, u(x), \nabla u(x)), u|_{\partial\Omega} - t\phi) ,$$

where $\sigma \in (0, \gamma)$.

We want to solve $F(1, u) = \theta$. However, $\forall \overline{u} \in C^{2,\sigma}(\overline{\Omega})$,

$$F_u(t, \overline{u})v = (-\Delta v - tf_\eta(x, \overline{u}(x), \nabla\overline{u}(x))v - tf_\xi(x, \overline{u}(x), \nabla\overline{u}(x))\nabla v, v|_{\partial\Omega}) .$$

By assumption (2) and the maximum principle for linear elliptic equations, $\forall g \in C^\sigma(\overline{\Omega}) \times C^{2,\sigma}(\partial\Omega)$

$$F_u(t, \overline{u})v = g$$

has a unique solution, i.e., $F_u(t, \overline{u}) : C^{2,\sigma} \cap C_0(\overline{\Omega}) \to C^\sigma(\overline{\Omega}) \times C^{2,\sigma}(\partial\Omega)$ has a bounded inverse. Thus the set

$$S = \{t \in I |\ F(t, u) = \theta \text{ is solvable}\}$$

is open, from the IFT.

We prove that S is closed. Noting that if u_t satisfies

$$F(t, u_t) = \theta ,$$

then we set

$$a_t(x) = \frac{tf(x, u_t(x), \nabla u_t(x)) + u_t(x)}{1 + |\nabla u_t(x)|^2} .$$

By assumptions (1) and (3) and Lemma 1.2.11,

$$\|a_t\|_C \leq C(\|u_t\|_C) \leq C(M, \|\phi\|_C) .$$

Since u_t satisfies (1.22), in which a is replaced by a_t, and ϕ by $t\phi$. According to equation (1.25),

$$\|u_t\|_{C^{1,\gamma}} \leq C(\|\phi\|_{C^{2,\gamma}}, M, \gamma) ,$$

and then

$$\|a_t\|_{C^{0,\gamma}} \leq C(\|\phi\|_{C^{2,\gamma}}, M, \gamma, f) .$$

Again by Lemma 1.2.12, we have

$$\|u\|_{C^{2,\gamma}} \leq C(\|\phi\|_{C^{2,\gamma}}, M, \gamma, f) .$$

This is the required estimate. Since the embedding $C^{2,\gamma}(\overline{\Omega}) \hookrightarrow C^{2,\sigma}(\overline{\Omega}) \times C^{2,\sigma}(\partial\Omega)$ is compact, we conclude that S is closed, as we have seen previously.

The existence of the solution for $F(1, u) = \theta$ follows from the continuity method. The uniqueness is a consequence of the maximum principle.

Once we obtain a solution u in $C^{2,\sigma}$, it follows directly by Schauder estimates that $u \in C^{2,\gamma}$ □

Remark 1.2.13 *We shall return to this example by dropping assumption (2) in Chap. 3.*

1.3 Lyapunov–Schmidt Reduction and Bifurcation

1.3.1 Bifurcation

We often meet an equation with a parameter λ:

$$F(x, \lambda) = 0 .$$

The following phenomenon has been observed: a branch of solutions $x(\lambda)$ depending on λ, is either disappeared or split into several branches, as λ attains some critical values. This kind of phenomenon is called bifurcation. For example, a simple algebraic equation:

$$x^3 - \lambda x = 0 \quad \lambda \in \mathbb{R}^1 ,$$

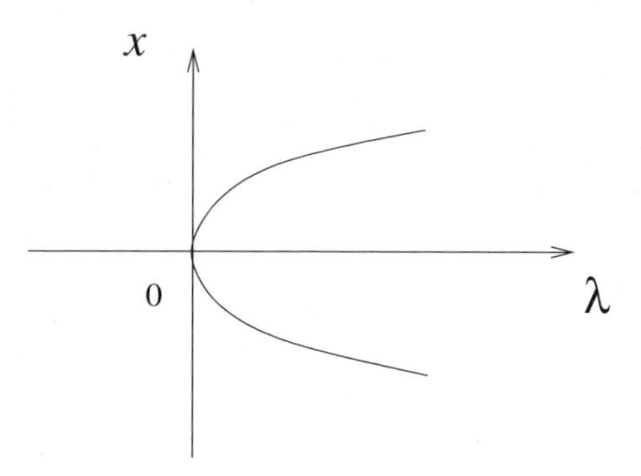

Fig. 1.3.

has a solution $x = 0\ \forall\lambda \in \mathbb{R}^1$. As $\lambda \leqslant 0$, this is the unique solution; but as $\lambda > 0$, we have two more branches of solutions

$$x = \pm\sqrt{\lambda}\,.$$

See Figure 1.3.

Bifurcation phenomena occur extensively in nature. Early in 1744, Euler observed the bending of a rod pressed along the direction of its axis. Let θ be the angle between the real axis and the tangent of the central line of the rod, and let λ be the pressure. The length of the rod is normalized to be π. We obtain the following differential equation with the two free end point conditions:

$$\begin{cases} \ddot{\theta} + \lambda \sin\theta = 0\,, \\ \dot{\theta}(0) = \dot{\theta}(\pi) = 0\,. \end{cases}$$

Obviously, $\theta \equiv 0$ is always a solution of the ODE. Actually the solution is unique, if λ is not large. As λ increasingly passes through a certain value λ_0, it is shown by experiment that there exists a bending solution $\theta \neq 0$.

The same phenomenon occurs in the bending of plates, shells etc. In addition, bifurcation occurs in the study of thermodynamics (Bérnard problem), rotation of fluids, solitary waves, superconductivity and lasers.

Mathematically, we describe the bifurcation by the following:

Definition 1.3.1 *Let X, Y be Banach spaces, and let $\wedge$ be a topological space. Suppose that $F : X \times \wedge \to Y$ is a continuous map. $\forall\lambda \in \wedge$, let*

$$S_\lambda = \{x \in X |\ F(x, \lambda) = \theta\}$$

be the solution set of the equation $F(x, \lambda) = \theta$, where λ is a parameter (Fig. 1.4). Assume $\theta \in S_\lambda, \forall\lambda \in \wedge$. We call (θ, λ_0) a bifurcation point, if for any neighborhood U of (θ, λ_0), there exists $(x, \lambda) \in U$ with $x \in S_\lambda \backslash \{\theta\}$.

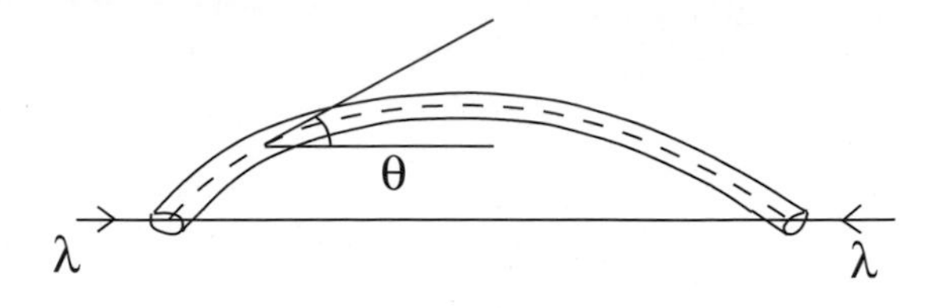

Fig. 1.4.

The following problems are of primary concern:

(1) What is the necessary and sufficient condition for a bifurcation point (θ, λ_0)?
(2) What is the structure of S_λ near $\lambda = \lambda_0$?
(3) How do we compute the solutions near the bifurcation points?
(4) How about the global structure of $\cup_{\lambda \in \wedge} S_\lambda$?
(5) Let $F(x, \lambda) = \theta$ be the steady equation of the evolution equation:

$$\dot{x} = F(x, \lambda) ,$$

we study the stability of solutions in S_λ as λ approaches λ_0.

In this section, we focus our discussions on problems (1) and (2). (4) will be studied in Chap. 3.

For simplicity, we assume that $U \subset X$ is an open neighborhood of the origin θ of the Banach space X, and that $F : U \times \wedge \to Y$ is continuous, and satisfies

$$F(\theta, \lambda) = \theta \quad \forall \lambda \in \wedge .$$

What is the necessary condition for a bifurcation point $\lambda_0 \in \wedge$?

(1) Assume that $F_x(x, \lambda)$ is continuous. If (θ, λ_0) is a bifurcation point, then $F_x(\theta, \lambda_0)$ does not have a bounded inverse.

Proof. By the IFT directly. □

(2) Assume

$$F(x, \lambda) = Lx - \lambda x + N(x, \lambda) ,$$

where $L \in L(X, Y), \lambda \in \mathbb{R}^1$, and that $N : U \times \mathbb{R}^1 \to X$ is continuous with

$$\| N(x, \lambda) \| = \circ(\| x \|) \text{ as } \| x \| \to \theta$$

uniformly for λ in a neighborhood of λ_0. If (θ, λ_0) is a bifurcation point, then $\lambda_0 \in \sigma(L)$, i.e., λ_0 is a spectrum of L.

Proof. If not, $\lambda_0 \in \rho(L)$, the resolvent set of L. Since $\rho(L)$ is open, $\exists \varepsilon > 0$ and $C_\varepsilon > 0$ such that

$$\| (L - \lambda I)^{-1} \| \leqslant C_\varepsilon \text{ as } |\lambda - \lambda_0| < \varepsilon .$$

It follows that

$\| x \| \leqslant \| (L - \lambda I)^{-1} N(x, \lambda) \| = \circ(\| x \|)$ if $|\lambda - \lambda_\varepsilon| < \varepsilon$, and $x \to \theta$, as $x \in S_\lambda$.

Thus $\exists \delta > 0$, such that

$$B_\delta \times (\lambda_0 - \varepsilon, \lambda_0 + \varepsilon) \cap S_\lambda = \{(\theta, \lambda) | \; |\lambda - \lambda_0| < \varepsilon\} ,$$

i.e., (θ, λ_0) is not a bifurcation point. □

(3) The above condition is not sufficient. For example, let $X = \mathbb{R}^2$, let

$$x - \begin{pmatrix} u \\ v \end{pmatrix} ,$$

and let

$$F(x, \lambda) = \begin{pmatrix} u \\ v \end{pmatrix} - \lambda \begin{pmatrix} u \\ v \end{pmatrix} + \begin{pmatrix} -v^3 \\ u^3 \end{pmatrix} .$$

Obviously, $F_x(\theta, \lambda) = (1 - \lambda) Id$. $\lambda = 1$ is in the spectrum, but (θ, λ) is not a bifurcation point, because:

$$F(x, \lambda) = \theta \Leftrightarrow u^4 + v^4 = 0, \text{ i.e., } x = \theta .$$

In order to study the sufficient condition and the local behavior of the solution set near its bifurcation points, we introduce the following:

1.3.2 Lyapunov–Schmidt Reduction

Let X, Y be Banach spaces, and let $\wedge$ be a topological space. Assume that $F : U \times \wedge \to Y$ is continuous, where $U \subset X$ is a neighborhood of θ. We assume that $F_x(\theta, \lambda_0)$ is a Fredholm operator, i.e.,

(1) Im $F_x(\theta, \lambda_0)$ is closed in Y,
(2) $d = \dim \ker F_x(\theta, \lambda_0) < \infty$,
(3) $d^* =$ codim Im $F_x(\theta, \lambda_0) < \infty$.

Set

$$X_1 = \ker F_x(\theta, \lambda_0), \; Y_1 = \text{Im } F_x(\theta, \lambda_0) .$$

Since both dim X_1, and codim Y_1 are finite, we have the direct sum decompositions:

$$X = X_1 \oplus X_2, \; Y = Y_1 \oplus Y_2 ,$$

and the projection operator $P : Y \to Y_1$. $\forall x \in X$, there exists a unique decomposition:

$$x = x_1 + x_2, \; x_i \in X_i, \; i = 1, 2 .$$

Thus

$$F(x, \lambda) = \theta \Leftrightarrow \begin{cases} PF(x_1 + x_2, \lambda) = \theta , \\ (I - P)F(x_1 + x_2, \lambda) = \theta . \end{cases}$$

Now, $PF_x(\theta, \lambda_0) : X_2 \to Y_1$ is a surjection as well as an injection. According to the Banach theorem, it has a bounded inverse. If we already have $F(\theta, \lambda_0) = \theta$, then from the IFT, we have a unique solution

$$u : V_1 \times V \to V_2$$

satisfying

$$PF(x_1 + u(x_1, \lambda), \lambda) = \theta \ ,$$

where V_i is a neighborhood of θ in $U \cap X_i, i = 1, 2$, and V is a neighborhood of λ_0.

It remains to solve the equation:

$$(I - P)F(x_1 + u(x_1, \lambda), \lambda) = \theta$$

on $V_1 \times V$. This is a nonlinear system of d variables and d^* equations.

The above procedure is called the Lyapunov–Schmidt reduction. It reduces an infinite-dimensional problem to a finite-dimensional system. Many applied mathematicians have been doing the same reduction in their concrete problems in their own language.

Before going further, let us study the simple properties of the solution $x_2 = u(x_1, \lambda)$.

Lemma 1.3.2 *Under the assumptions* (1), (2), (3) *of the Lyapunov–Schmidt reduction, if* $F \in C^p(U \times \wedge, Y), p \geqslant 1$, *satisfies* $F(\theta, \lambda) = \theta$, *where* $\wedge$ *is again a Banach space, then we have*

$$u(\theta, \lambda) = \theta,$$
$$u'(\theta, \lambda_0) = \theta \ .$$

If $p = 1$, *then*

$$u(x_1, \lambda) = \circ(\| x_1 \| + |\lambda - \lambda_0|) \ ;$$

and if $p = 2$, *then*

$$u(x_1, \lambda) = O(\| x_1 \|^2 + |\lambda - \lambda_0|^2) \ .$$

Proof. According to the IFT, the solution in $V_1 \times V \to V_2$ is unique. Since $F(\theta, \lambda) = \theta$, we have $u(\theta, \lambda) = \theta$. Again, by the IFT,

$$u'(\theta, \lambda_0)(\overline{x}_1, \overline{\lambda}) = -(PF_x(\theta, \lambda_0))^{-1}(PF_x(\theta, \lambda_0)\overline{x}_1 + PF_\lambda(\theta, \lambda_0)\overline{\lambda})$$

$\forall (\overline{x}_1, \overline{\lambda}) \in X_1 \times \wedge$.

From $F(\theta, \lambda) = \theta$, it follows that $F_\lambda(\theta, \lambda) = 0$. Therefore

$$u'(\theta, \lambda_0) = \theta \ ,$$

provided

$$\overline{x}_1 \in X = \ker F_x(\theta, \lambda_0) \ .$$

The last two conclusions follow from Taylor's formula. □

Next, we turn to the case $d = d^* = 1$.

Theorem 1.3.3 *(Crandall–Rabinowitz) Suppose that $U \subset X$ is an open neighborhood of θ, and that $F \in C^2(U \times \mathbb{R}^1, Y)$ satisfies $F(\theta, \lambda) = \theta$. If $F_x(\theta, \lambda_0)$ is a Fredholm operator with $d = d^* = 1$, and if*

$$F_{x\lambda}(\theta, \lambda_0)u_0 \notin \text{Im } F_x(\theta, \lambda_0) \tag{1.27}$$

for all $u_0 \in \ker F_x(\theta, \lambda_0) \backslash \{\theta\}$, then (θ, λ_0) is a bifurcation point, and there exists a unique C^1 curve $(\lambda, \psi) : (-\delta, \delta) \to \mathbb{R}^1 \times Z$ satisfying

$$\begin{cases} F(su_0 + \psi(s), \lambda(s)) = \theta \\ \lambda(0) = \lambda_0, \psi(0) = \psi'(0) = \theta \,, \end{cases}$$

where $\delta > 0$, and Z is the complement space of span $\{u_0\}$ *in X. Furthermore, there is a neighborhood of (θ, λ_0), in which*

$$F^{-1}(\theta) = \{(\theta, \lambda) | \ \lambda \in \mathbb{R}^1\} \cup \{(su_0 + \psi(s), \lambda(s)) | \ |s| < \delta\} \,.$$

Proof. Decompose the spaces X and Y according to the Lyapunov–Schmidt reduction, and write down the reduction equation. By assumptions,

$$\ker F_x(\theta, \lambda_0) = \text{span}\{u_0\} \,,$$

and $\exists \phi^* \in Y^* \backslash \{\theta\}$, such that $\ker \phi^* = \text{Im} F_x(\theta, \lambda_0)$, from the Hahn–Banach theorem. The reduction equation reads as:

$$g(s, \lambda) = \langle \phi^*, F(su_0 + u(su_0, \lambda), \lambda) \rangle = 0 \,. \tag{1.28}$$

We have the trivial solution $s = 0$, and we look for a nontrivial solution. Noticing

$$g'_s(0, \lambda_0) = \langle \phi^*, F_x(\theta, \lambda_0)(u_0 + u'_s(\theta, \lambda_0)) \rangle \,,$$

and $u_0 \in \ker F_x(\theta, \lambda_0)$, by applying Lemma 1.3.2, we obtain $g'_s(0, \lambda_0) = 0$. Therefore, it is impossible to get a solution $s = s(\lambda)$ directly from the IFT. However, we may consider λ as a function of s, in this case, the only difficulty is that $g(0, \lambda) = 0$ for all λ. Let us introduce a new function:

$$h(s, \lambda) = \begin{cases} \frac{1}{s} g(s, \lambda) & \text{as } s \neq 0 \\ g'_s(0, \lambda) & \text{as } s = 0 \,. \end{cases}$$

When $s \neq 0$, the solutions of $h(s, \lambda) = 0$ are the same as (1.28). Here we define $g'_s(0, \lambda)$ to be the value of h at $s = 0$, in order to make $h \in C^1$.

We assume this conclusion at this moment, and postpone the verification to the end. Since

$$h(0, \lambda_0) = g'_s(0, \lambda_0) = 0 \,,$$

and

$$
\begin{aligned}
h'_\lambda(0,\lambda_0) &= g''_{s\lambda}(0,\lambda_0) \\
&= \langle \phi^*, F_{x\lambda}(\theta,\lambda_0)(u_0 + u'_s(\theta,\lambda_0)) + F_x(\theta,\lambda_0)u''_{s\lambda}(\theta,\lambda_0)\rangle \\
&= \langle \phi^*, F_{x\lambda}(\theta,\lambda_0)u_0\rangle \\
&\neq 0 \,.
\end{aligned}
$$

Again, the IFT is applied, and we obtain a C^1-curve $\lambda = \lambda(s), |s| < \delta$, satisfying

$$
\begin{cases} h(s,\lambda(s)) = 0 \,, \\ \lambda(0) = \lambda_0 \,. \end{cases}
$$

Set

$$
\psi(s) = u(su_0, \lambda(s)) \,;
$$

we have

$$
\begin{aligned}
\psi(0) &= u(0,\lambda_0) = \theta \,, \\
\psi'(0) &= \nabla u(\theta,\lambda_0)(u_0, \lambda'(0)) = \theta \,,
\end{aligned}
$$

and

$$
g(s,\lambda(s)) = \langle \phi^*, F(su_0 + u(su_0,\lambda(s)), \lambda(s))\rangle = 0 \,;
$$

i.e.,

$$
F(su_0 + \psi(s), \lambda(s)) = \theta \,.
$$

This is what we need.

Now we turn to verifying $h \in C^1(B_\eta)$ for some $\eta > 0$, where $B_\eta = \{(s,\lambda) \in \mathbb{R}^2 \mid |s|^2 + |\lambda - \lambda_0|^2 < \eta^2\}$. Indeed, only want to verify that h is C^1 at $s = 0$. By definition,

$$
\lim_{s\to 0} \frac{1}{s} g(s,\lambda) = g'_s(0,\lambda) \,,
$$

which implies the continuity of h. Moreover,

$$
\begin{aligned}
h'_s(0,\lambda) &= \lim_{s\to 0} \frac{1}{s}[h(s,\lambda) - h(0,\lambda)] \\
&= \lim_{s\to 0} \frac{1}{s^2}[g(s,\lambda) - g(0,\lambda) - g'_s(0,\lambda)s] \\
&= \frac{1}{2} g''_{ss}(0,\lambda) \,,
\end{aligned}
$$

therefore

$$
\begin{aligned}
h'_s(s,\lambda) - h'_s(0,\lambda) &= \frac{1}{s^2}\left[g'_s(s,\lambda)s - g(s,\lambda) - \frac{1}{2} g''_{ss}(0,\lambda)s^2\right] \\
&= \circ(1) \quad \text{as } |s| \to 0 \,.
\end{aligned}
$$

Since

$$
\begin{aligned}
h'_\lambda(0,\lambda) &= g''_{s\lambda}(0,\lambda) \,, \\
h'_\lambda(s,\lambda) - h'_\lambda(0,\lambda) &= \frac{1}{s} g'_\lambda(s,\lambda) - g''_{s\lambda}(0,\lambda) \to 0 \quad \text{as } s \to 0 \,.
\end{aligned}
$$

The proof is complete. □

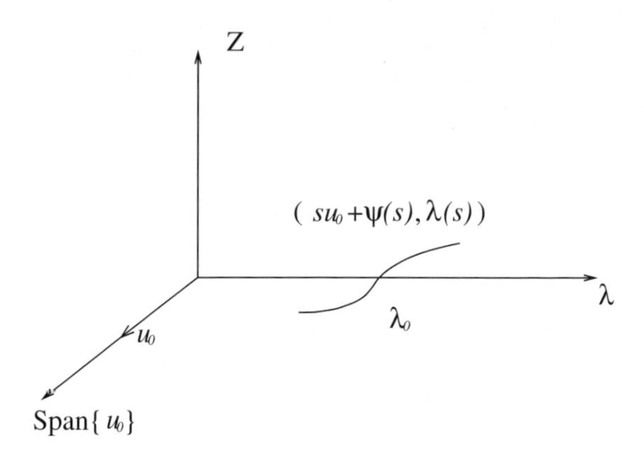

Fig. 1.5.

We present two simple applications.

Example 1. (Euler elastic rod) We study the bending problem raised at the beginning of this section:

$$\begin{cases} \ddot{\varphi} + \lambda \sin \varphi = 0 \quad \text{in } (0, \pi) , \\ \dot{\varphi}(0) = \dot{\varphi}(\pi) = 0 . \end{cases}$$

Let

$$X = \{u \in C^2[0, \pi] |\ \dot{u}(0) = \dot{u}(\pi) = 0)\} ,$$
$$Y = C[0, \pi] ;$$

and let $F : X \times R^1 \to Y$ be the map:

$$(u, \lambda) \mapsto u'' + \lambda \sin u .$$

It is a continuous map satisfying $F(\theta, \lambda) = \theta$. According to the necessary condition, if (θ, λ) is a bifurcation point, then λ is in the spectrum of the linearized operator $(\frac{d}{dt})^2 + \lambda \mathrm{I}$, i.e.,

$$\lambda = n^2, \text{ for some } n = 1, 2, \ldots$$

Since $\ker\left((\frac{d}{dt})^2 + n^2 I\right) = \{s \cos nt |\ s \in \mathbb{R}^1\}$, and since the differential operator $(\frac{d}{dt})^2$ under the free end point condition is self-adjoint, we have

$$\text{coker } F_u(\theta, n^2) = \{s \cos nt |\ s \in \mathbb{R}^1\} .$$

In addition

$$F''_{u\lambda}(\theta, n^2) = \cos nu|_{u=\theta} = \mathrm{I} ,$$

thus

$$F''_{u\lambda}(\theta, n^2) \cos nt = \cos nt \notin \text{Im } F_u(\theta, n^2) .$$

All the assumptions in the Crandall–Rabinowitz theorem are satisfied, and we obtain a family of C^1 curves $(\lambda_n(s), \psi_n(s)) : (-\delta, \delta) \to \mathbb{R}^1 \times Z_n$, where Z_n is the complement of span$\{\cos nt\}$, satisfying

$$\lambda_n(0) = n^2 ,$$
$$\tfrac{d}{ds}\psi_n(0) = \psi_n(0) = \theta ,$$

for $n = 1, 2, 3, \ldots$

If we set

$$\varphi_n(s,t) = s \cos nt + (\psi_n(s))(t) \quad t \in [0, \pi] ,$$

then

$$\left(\frac{\partial}{\partial t}\right)^2 \varphi_n(s,t) + \lambda_n(s) \sin \varphi_n(s,t) = 0 \quad 0 < t < \pi,\ |s| < \delta ,$$
$$\frac{\partial}{\partial t}\varphi_n(s,0) = \frac{\partial}{\partial t}\varphi_n(s,\pi) = 0 .$$

We obtain the bifurcation diagram of Fig. 1.6.

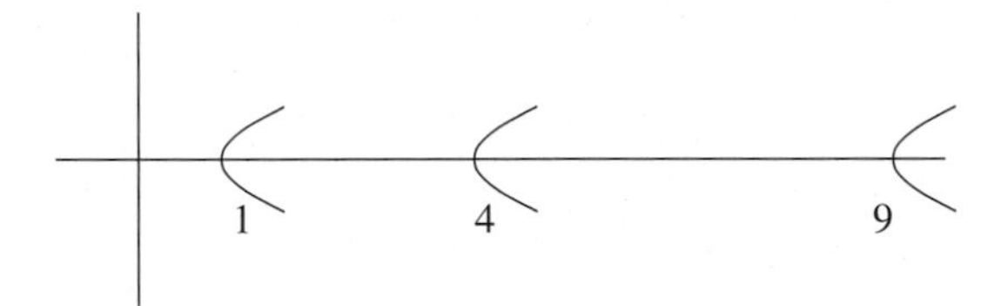

Fig. 1.6.

Example 2. We consider the following elliptic BVP. Let $\Omega \subset \mathbb{R}^n$ be a bounded open domain with smooth boundary $\partial\Omega$, and $p \in (1, \infty)$:

$$\begin{cases} -\Delta u - \lambda u = |u|^{p-1}u & \text{in } \Omega , \\ u|_{\partial\Omega} = 0 . \end{cases}$$

Set

$$X = C^{2,\gamma} \cap C_0(\overline{\Omega}), \text{ and } Y = C^\gamma(\overline{\Omega}) \text{ for some } \gamma \in (0,1) ,$$
$$F : (u, \lambda) \mapsto -\Delta u - \lambda u - |u|^{p-1}u ,$$

we have

$$F_u(\theta, \lambda) = -\Delta - \lambda \mathrm{I} .$$

Thus, for (θ, λ) being a bifurcation point, $-\lambda$ has to be in the spectrum of the Laplacian under the 0-Dirichlet data.

Let us first consider the first eigenvalue λ_1, which is simple, i.e., $\ker -\triangle - \lambda I$ is one dimensional; let φ_1 be the associate eigenfunction. By the same reasoning, coker $F_u(\theta, \lambda_1)$ is also one dimensional, and

$$F_{u\lambda}(\theta, \lambda_1)\varphi_1 = -\varphi_1 \notin \mathrm{Im}\, F_u(\theta, \lambda_1) .$$

Again, we apply the Crandall–Rabinowitz theorem, and conclude that (θ, λ_1) is a bifurcation point, and in its neighborhood, the solution set is the C^1 curve

$$(-\delta, \delta) \mapsto (s\varphi_1(x) + \psi(s)(x), \lambda_1(s))$$

plus the trivial solution set (θ, λ), where $\psi(s) \in Z$, and Z is the complement of span$\{\varphi_1\}$ in $C^{2,\gamma} \cap C_0(\overline{\Omega})$.

We have not studied eigenvalues other than λ_1, because we do not know if they are simple, i.e., if the condition $d = 1$ is satisfied. In fact, if λ is a simple eigenvalue, then we have the similar result.

For some special cases, in which $d \neq 1$, we can extend Theorem 1.3.3 as follows:

Theorem 1.3.4 *Let $\wedge, X$ and Y be Banach spaces. Suppose that $F \in C^2(X \times \wedge, Y)$ has the form*

$$F(x, \lambda) = L(\lambda)x + P(x, \lambda) ,$$

where $L(\lambda) \in L(X, Y)\ \forall \lambda \in \wedge$, and

$$P(\theta, \lambda) = \theta,\ P_x(\theta, \lambda) = \theta,\ P_{x\lambda}(\theta, \lambda_0) = \theta \text{ for some } \lambda_0 \in \wedge .$$

If there exist $u_0 \in \ker L(\lambda_0) \backslash \{\theta\}$ and a closed linear subspace $Z \subset X$ such that $(z, \lambda) \mapsto L(\lambda_0)z + \lambda L'(\lambda_0)u_0 : Z \times \wedge \to Y$ is a linear homeomorphism, then there exist a neighborhood U of (θ, λ_0) in $(\text{span}\{u_0\} \times Z) \times \wedge, \delta > 0$, and a C^1 map: $(-\delta, \delta) \to Z \times \wedge$, defined by $s \mapsto (\varphi(s), \lambda(s))$ satisfying

$$F^{-1}(\theta) \cap U \backslash \{(\theta, \lambda) |\ \lambda \in \wedge\} = \{(s(u_0 + \varphi(s)), \lambda(s)) |\ |s| < \delta\} ,$$

and

$$(\varphi(0), \lambda(0)) = (\theta, \lambda_0) .$$

Proof. Similar to Theorem 1.3.3, we define

$$\Phi(s, z, \lambda) = \begin{cases} \frac{1}{s} F(s(u_0 + z), \lambda) & \text{as } s \neq 0 \\ F_x(\theta, \lambda)(u_0 + z) & \text{as } s = 0 . \end{cases}$$

One wants to verify that $\Phi \in C^1((\mathbb{R}^1 \times Z) \times \wedge, Y)$.

It is sufficient to verify the continuous differentiability at $s = 0$. Since

$$\begin{aligned} \Phi(s, z, \lambda) - \Phi(0, z, \lambda) &= s^{-1}[F(s(u_0 + z), \lambda) - F(\theta, \lambda) - F_x(\theta, \lambda)(s(u_0 + z))] \\ &= s \int_0^1 \int_0^1 F_{xx}(rts(u_0 + z), \lambda) t dt dr \cdot (u_0 + z)^2 , \end{aligned}$$

we have

$$\Phi_s(0,z,\lambda) = \frac{1}{2}F_{xx}(\theta,\lambda)(u_0+z)^2 .$$

Furthermore as $s \to 0$,

$$\begin{aligned}
&\Phi_s(s,z,\lambda) - \Phi_s(0,z,\lambda) \\
&\quad = -s^{-2}\left[F(s(u_0+z),\lambda) - sF_x(\theta,\lambda)(u_0+z) - \frac{s^2}{2}F_{xx}(\theta,\lambda)(u_0+z)^2\right] \\
&\qquad + s^{-1}[F_x(s(u_0+z),\lambda) - F_x(\theta,\lambda)](u_0+z) \\
&\quad = o(1) ,
\end{aligned}$$

and

$$\begin{aligned}
&\Phi_\lambda(s,z,\lambda) - \Phi_\lambda(0,z,\lambda) \\
&\quad = s^{-1}[F_\lambda(s(u_0+z),\lambda) - F_\lambda(\theta,\lambda) - F_{x\lambda}(\theta,\lambda)s(u_0,z)] \\
&\quad = o(1) .
\end{aligned}$$

Thus, $\Phi \in C^1$. Now, we have

$$\begin{aligned}
\Phi(0,z,\lambda) &= L(\lambda)(u_0+z) + P_x(\theta,\lambda)(u_0+z) , \\
\Phi(0,\theta,\lambda_0) &= L(\lambda_0)u_0 + P_x(\theta,\lambda_0)u_0 = \theta ,
\end{aligned}$$

and

$$\begin{aligned}
\Phi_{(z,\lambda)}(0,\theta,\lambda_0)(\overline{z},\overline{\lambda}) &= L(\lambda_0)\overline{z} + L'(\lambda_0)u_0 \cdot \overline{\lambda} + P_{\lambda x}(\theta,\lambda_0)u_0 + P_x(\theta,\lambda_0)\overline{z} , \\
&= L(\lambda_0)\overline{z} + \overline{\lambda}L'(\lambda_0)u_0 ,
\end{aligned}$$

$\forall(\overline{z},\overline{\lambda}) \in Z \times \wedge$. By the assumption, the last linear operator is a homeomorphism; then the IFT is applied. Therefore $\exists$ a neighborhood U of (θ,λ_0), and a unique C^1-curve: $s \mapsto (\varphi(s),\lambda(s)) \in Z \times \wedge$ $\forall |s| < \varepsilon$, satisfying

$$\begin{cases}
(\varphi(0),\lambda(0)) = (\theta,\lambda_0) , \\
\Phi(s,\varphi(s),\lambda(s)) = \theta, \ \ i.e., \ \ F(s(u_0+\varphi(s)),\lambda(s)) = \theta .
\end{cases}$$

The proof is complete. □

The above theorem is applied to Hopf bifurcations in ODE.

Example 3. (Hopf bifurcation)

We study the periodic solutions for a linear ordinary differential system:

$$\dot{x} = Ax , \tag{1.29}$$

where $x \in C^1([0,2\pi],\mathbb{R}^n)$ and $A \in M(n,\mathbb{R})$, the $n \times n$ matrix. Let

$$A = \begin{pmatrix} B & 0 \\ 0 & C \end{pmatrix} ,$$

where

$$B = \begin{pmatrix} 0 & 1 \\ -1 & 0 \end{pmatrix}, \text{ and } C \in M(n-2, \mathbb{R}) .$$

We assume that $(e^{2\pi C} - \mathrm{I})$ is invertible. Obviously the linear system (1.29) admits a family of periodic solutions with two real parameters a and b:

$$x(t) = \begin{pmatrix} a\cos t + b\sin t \\ -a\sin t + b\cos t \\ 0 \\ \vdots \\ 0 \end{pmatrix}, \forall a, b \in \mathbb{R}^1 .$$

Now, we perturb A by introducing a parameter $\mu \in (-1, 1)$, and consider the nonlinear differential system

$$\dot{x} = A(\mu)x + P(x, \mu) , \tag{1.30}$$

where

$$A(\mu) = \begin{pmatrix} B(\mu) & 0 \\ 0 & C(\mu) \end{pmatrix}, \; B(\mu) = \begin{pmatrix} \mu & \beta(\mu) \\ -\beta(\mu) & \mu \end{pmatrix},$$

and $C(\mu) \in M(n-2, \mathbb{R})$, satisfying $e^{2\pi C(\mu)} - \mathrm{I} \in GL(n-2, \mathbb{R})$. We assume that

$$P : \mathbb{R}^n \times (-1, 1) \to \mathbb{R}^n$$

satisfies

$$P(\theta, \mu) = P_x(\theta, 0) = P_{x,\mu}(\theta, 0) = \theta .$$

and that $\beta : (-1, 1) \to \mathbb{R}^1$, satisfies

$$\beta(0) = 1, \text{ and } \beta'(0) \neq 0 .$$

Obviously, $\forall \mu \in (-1, 1),\; x = \theta$ is always a solution. We are interested in nontrivial solutions bifurcating from the branch of trivial solutions.

Theorem 1.3.5 *Suppose that $A(\mu)$ and $P(x, \mu)$ are C^2 functions satisfying the above assumptions. Then $\exists$ positive constants a_0, δ_0 and a C^1 map*

$$(\mu, \omega, x) : (-a_0, a_0) \mapsto \mathbb{R}^1 \times \mathbb{R}^1 \times C^1(\mathbb{R}^1, \mathbb{R}^n) ,$$

satisfying

$$\mu(0) = 0,\; \omega(0) = 1 ,$$

where $x(a)$ is a $2\pi\omega(a)$-periodic function, which satisfies (1.30), and is of the form:

$$(x(a))(t) = \begin{pmatrix} a\sin\omega(a)^{-1}t \\ a\cos\omega(a)^{-1}t \\ 0 \\ \vdots \\ 0 \end{pmatrix} + \circ(|a|) \,.$$

Furthermore, every $2\pi\omega(a)$-periodic solution $y(t)$ of (1.30), satisfying $|y(t)| < \delta_0 \ \forall t$, coincides with $x(a)(t)$ modulating a phase shift when $|\mu| < \alpha_0$ and $|\omega - 1| < \delta_0$.

Proof. Since ω depends on μ, we introduce a new scale τ, and let $t = \omega\tau$; (1.30) is rewritten as

$$\frac{dx}{d\tau} = \omega A(\mu)x + \omega P(x, \mu) \,. \tag{1.31}$$

We find the 2π-periodic solution of (1.31). Set

$$\begin{aligned} &\wedge = \mathbb{R}^2, \ \lambda = (\omega, \mu) \in \wedge \,, \\ &X = C^1_{2\pi}(\mathbb{R}^n) = \{u \in C^1(\mathbb{R}, \mathbb{R}^n) |\ u \text{ is } 2\pi \text{ periodic}\} \,, \\ &Y = C_{2\pi}(\mathbb{R}^n) = \{u \in C(\mathbb{R}, \mathbb{R}^n) |\ u \text{ is } 2\pi \text{ periodic}\} \,, \end{aligned}$$

and set

$$\begin{aligned} &L(\lambda)x = \frac{dx}{d\tau} - \omega A(\mu)x \,, \\ &\mathbb{P}(x, \lambda) = \omega P(x, \mu) \,, \\ &\lambda_0 = (1, 0) \,. \end{aligned}$$

Let

$$u_0 = \begin{pmatrix} \sin\tau \\ \cos\tau \\ 0 \\ \vdots \\ 0 \end{pmatrix}, \ \textit{and } u_1 = \begin{pmatrix} \cos\tau \\ -\sin\tau \\ 0 \\ \vdots \\ 0 \end{pmatrix};$$

then we have

$$\ker L(\lambda_0) = \operatorname{span}\{u_0, u_1\} \,.$$

Define

$$Z = \left\{ z \in X |\ \int_0^{2\pi} z(t)u_j(t)dt = 0, \ j = 0, 1 \right\} \,.$$

It remains to verify the following conditions: $\forall y \in Y$, the linear equation

$$\begin{aligned} &L(\lambda_0)z + \lambda L'(\lambda_0)u_0 \\ &= \frac{dz}{d\tau} - A(0)z - \omega A(0)u_0 - \mu A'(0)u_0 \\ &= y \end{aligned}$$

has a unique solution $(\omega, \mu, z) \in \mathbb{R}^2 \times Z$.

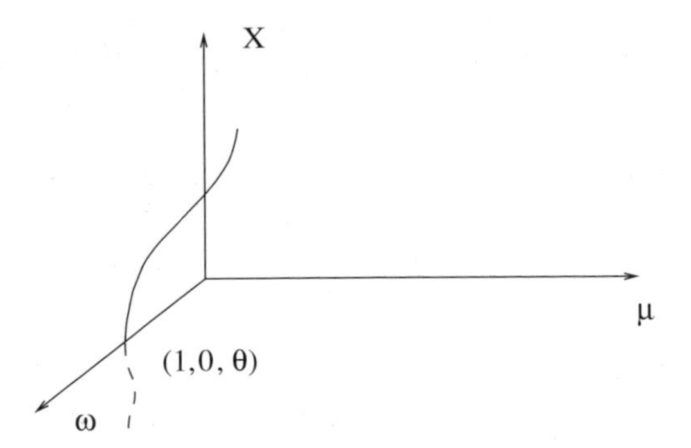

Fig. 1.7. Nontrivial solutions

According to the Fredholm alternative, this is equivalent to saying that $\forall y \in Y, \exists|(\omega,\mu) \in \mathbb{R}^2$ such that

$$y + \omega A(0)u_0 + \mu A'(0)u_0 \in \ker L^*(\lambda_0)^\perp \, .$$

However, $\ker L^*(\lambda_0) = \ker L(\lambda_0)$; this is again equivalent to

$$\det \begin{vmatrix} \int_0^{2\pi}(A(0)u_0(t))u_0(t)dt & \int_0^{2\pi}(A'(0)u_0(t))u_0(t)dt \\ \int_0^{2\pi}(A(0)u_0(t))u_1(t)dt & \int_0^{2\pi}(A'(0)u_0(t))u_1(t)dt \end{vmatrix} \neq 0 \, .$$

Computing the determinant directly, it equals $-4\pi^2\beta'(0)$. Thus all assumptions in Theorem 1.3.4 are satisfied. We obtain our conclusion. □

This kind of bifurcation phenomenon is shown in Fig. 1.7.

1.3.3 A Perturbation Problem

We present here an example of perturbation problems. This is a nonlinear Schrödinger equation (NSE); we look for a nonspreading wave packet solution. Assume that the potential $V \in C^2(R^1)$ is bounded, and the standing wave $\psi = \psi(t,x)$ satisfies the following NSE:

$$ih\frac{\partial\psi}{\partial t} = -\frac{h^2}{2}\frac{\partial^2\psi}{\partial x^2} + V(x)\psi - \psi^3 \, .$$

where $h > 0$ is a small constant, $\psi(t,x) = \exp(-i\lambda t/h)\varphi(x)$ with $\varphi(x) \to 0$ as $|x| \to \infty$, and $\lambda < \inf V$. It is reduced to the following equation:

$$-\frac{h^2}{2}\varphi''(x) + V(x)\varphi(x) = \lambda\varphi(x) + \varphi(x)^3 \, .$$

If V has a nondegenerate critical point x_0, i.e., $V'(x_0) = 0, V''(x_0) \neq 0$, we make a change of variables $y = \frac{x-x_0}{h}$, and the equation is reduced to

$$-\frac{1}{2}\varphi''(y) + (V_h(y) - \lambda)\varphi(y) = \varphi(y)^3 \,, \tag{1.32}$$

where $V_h(y) = V(x_0 + hy)$ and $\varphi(y) \to 0$, as $|y| \to \infty$.

This is a perturbation problem, because $h > 0$ is small. The special case where $h = 0$ reads as:

$$-\frac{1}{2}\varphi'' + E\varphi = \varphi^3 \,, \tag{1.33}$$

where $E = V(x_0) - \lambda$, and $\varphi(y) \to 0$ as $|y| \to \infty$. The equation has a special solution:

$$z(y) = \alpha \text{ sech } (\alpha y), \;\; \alpha = (2E)^{1/2} \,.$$

Let $\tau_\theta\varphi = \varphi(y - \theta)$. Since equation (1.33) is autonomous, it has a one-dimensional solution manifold $Z = \{z_\theta = \tau_\theta z | \; \theta \in R^1\}$. It is easily check: $T_z Z = \text{span}\{e\}$, where $e = z'$, and then $T_{z_\theta} Z = \text{span}\{\tau_\theta e\}$. We write $e_\theta = \tau_\theta e$.

Let us define $X = H^2(R^1), Y = L^2(R^1)$, and

$$F(h,u) = -\frac{1}{2}u'' + (V_h - \lambda)u - u^3.$$

Since V is bounded, $F : R^1 \times X \to Y$.

Thus, $F(0, z_\theta) = 0, \; \forall\theta \in R^1$. Since

$$L_\theta := F_u'(0, z_\theta) = -\frac{1}{2}\frac{d^2}{dt^2} + (E - 3z_\theta^2)\cdot, \text{ with } D(L_\theta) = X \,,$$

L_θ is a self-adjoint operator on Y. We verify that L_θ is a Fredholm operator:

(1) Im (L_θ) is closed.
(2) Ker $(L_\theta) = \text{Coker}(L_\theta) = T_{z_\theta} Z$.

In fact, only (1) remains to be verified; it follows from:

Lemma 1.3.6 *Assume* $\inf(V) > \lambda$. *Then* $\exists\gamma > 0$, *such that* $\|L_\theta v\|_Y \geq \gamma\|v\|_X \;\; \forall v \in (T_{z_\theta} Z)^\perp$.

Proof. We prove it by contradiction. Suppose that $\exists\phi_n \in (T_{z_\theta} Z)^\perp$ such that $\|\phi_n\|_X = 1$, and $\|L_\theta\phi_n\|_Y \to 0$. Then $\phi_n \rightharpoonup \phi$ in X. We claim that $\phi = 0$. In fact,

(1) by weakly convergence, $\phi \in (T_{z_\theta} Z)^\perp$
and
(2) sincc X is dense in $Y, \forall\psi \in X$,

$$\langle L_\theta\phi, \psi\rangle = \langle\phi, L_\theta\psi\rangle = \lim\langle\phi_n, L_\theta\psi\rangle = \lim\langle L_\theta\phi_n, \psi\rangle = 0 \,.$$

Therefore, $L_\theta\phi = 0$. Combining (1) and (2), $\phi = 0$.

By the assumption $E > 0$, let $B = -\frac{1}{2}\frac{d^2}{dt^2} + E$; we have a constant $c > 0$ such that

$$\|Bu\|_Y \geq 2c\|u\|_X.$$

Noticing that $|z(y)| \sim \exp(-\alpha|y|)$ as $|y| \to \infty$, and that $\phi_n \rightharpoonup 0$ in X, in combination with the Sobolev embedding theorem, which implies that $\|\phi_n\|_\infty$ is bounded, and $\phi_n \Rightarrow 0$ uniformly on any bounded interval, we obtain:

$$\|L_\theta \phi_n\|_Y = \|B\phi_n - 3z_\theta^2 \phi_n\|_Y \geq \|B\phi_n\|_Y - \|3z_\theta^2 \phi_n\|_Y \geq 2c\|\phi_n\|_X - c = c\,,$$

as n is large. This is a contradiction. □

We have the following Lyapunov–Schmidt reduction: $X = T_{z_\theta} Z \bigoplus X_1, Y = T_{z_\theta} Z \bigoplus Y_1$, and let $P : Y \to Y_1$ be the orthogonal projection. The equation $F(h, u) = 0$ is equivalent to the system:

$$PF(h, z_\theta, z_\theta + \xi) = 0,\ \langle F(h, z_\theta + \xi), e_\theta \rangle = 0\,.$$

From $F(0, z_\theta) = 0$ and the IFT, for $h > 0$ small, $\exists$ a unique $\xi = \xi(h, \theta)$ solving the first equation, $PF(h, z_\theta + \xi(h, \theta)) = 0$, and satisfying $\|\xi(h, \theta)\| \leq C\|F(h, z_\theta)\|$, for some constant $C > 0$. It remains to solve:

$$w_h(\theta) := \langle F(h, z_\theta + \xi(h, \theta)), e_\theta \rangle = 0\,.$$

Once $\theta = \theta(h)$ is obtained, $u_h = \xi(h, \theta(h)) + z_{\theta(h)}$ solves $F(h, u) = 0$.

We write

$$F(h, u_h) = (F(h, u_h) - F(0, u_h)) + F(0, u_h)\,,$$

and by the Taylor formula,

$$F(0, u_h) = F(0, z_\theta) + L_\theta \xi(h, \theta) - N(h, \theta)\,,$$

where $N(h, \theta) = 3z_\theta \xi^2(h, \theta) + \xi^3(h, \theta)$, and $\theta = \theta(h)$. Since $F(0, z_\theta) = L_\theta e_\theta = 0$, we have

$$|\langle F(0, u_h), e_\theta \rangle| = |\langle N(h, \theta), e_\theta \rangle| \leq C\|\xi(h, \theta(h))\|_Y^2\,,$$

and

$$\begin{aligned} \langle F(h, u_h) - F(0, u_h), e_\theta \rangle| = \int &(V(x_0 + hy) - V(x_0))(z_\theta(h) \\ &+ \xi(h, \theta(h))e_\theta dy = I_1 + I_2\,. \end{aligned}$$

The following estimates hold:

$$\begin{aligned} I_1 &= \langle (V_h - V_0) z_\theta, e_\theta \rangle \\ &= -h\langle z_\theta, V_h' z_\theta \rangle - \langle z_\theta, (V_h - V_0) e_\theta \rangle \\ &= -\frac{1}{2} h \langle z_\theta, V_h' z_\theta \rangle \\ &= -\frac{1}{2} h \int V_h'(y) |z(y - \theta)|^2 dy \\ &= -\frac{1}{2} h \int V'(x_0 + h(y + \theta)) |z(y)|^2 dy\,, \\ I_2 &= |\langle (V_h - V_0) \xi(h, \theta), e_\theta \rangle| \\ &\leq \|(V_h - V_0) e_\theta\| \|\xi(h, \theta)\|\,, \end{aligned}$$

and

$$\begin{aligned}\|F(h,z_\theta)\|_Y^2 &= \|F(h,z_\theta)-F(0,z_\theta)\|_Y\\ &= \int |V_h(y+\theta)-V_0|^2|z(y)|^2dy\\ &\le \int_{B_\rho} |V_h(y+\theta)-V_0|^2|z(y)|^2dy + 4\mathrm{Max}|V|^2\int_{R^1\setminus B_\rho}|z(y)|^2dy\\ &\le \mathrm{Max}_{|y|\le\rho}|V(h(y+\theta))-V_0|^2+4\mathrm{Max}|V|^2e^{-\mu\rho}\\ &\le C_1(h(\rho+|\theta|))^4+e^{-\mu\rho}\end{aligned}$$

for every $\rho > 0$, and for some constant $C_1 > 0$. Let us denote the right-hand side by $M_{h,\theta,\rho}$.

Similarly, we have

$$\|(V_h - V_0)e_\theta\|^2 \le M_{h,\theta,\rho}.$$

Since $z(y)$ is even,

$$\begin{aligned}\left|\frac{1}{h}I_1+\frac{1}{2}V''(x_0)\|z\|^2h\theta\right| &= \frac{1}{2}\left|\int [V''(x_0)h\theta - V'(x_0+h(y+\theta))]|z(y)|^2dy\right|\\ &= \frac{1}{2}\left|\int [V''(x_0)h(y+\theta) - V'(x_0+h(y+\theta))|z(y)|^2dy\right|\\ &\le C_2\int |h(y+\theta)|^2|z(y)|^2dy\\ &\le C_2(h(|\theta|+\rho)|^2+e^{-\mu\rho})\,.\end{aligned}$$

By rescaling $\theta = \frac{s}{h}, v(s) = \frac{1}{h}w_h(\theta), v_0(s) = \frac{1}{2}V''(x_0)\|z\|^2s$, and $v_h(s) = h^{-\nu}v(h^\nu s)$, with $\nu \in (1,2)$, we obtain

$$|v_h(s)-v_0(s)| \le C_3(h^{-\nu}((h^\nu|s|+h\rho)^2+e^{-\mu\rho})+h^{-1-\nu}((h^\nu|s|+h\rho)^4+e^{-\mu\rho}))\,.$$

Choosing $\rho = h^{-\tau}$ with $\tau > 0$, for $|s| \le 1$, we obtain

$$\begin{aligned}|v_h(s)-v_0(s)| \le C_3(&h^{-\nu}((h^\nu+h^{1-\tau})^2+h^{-1-\nu}((h^\nu+h^{1-\tau})^4\\ &+(h^{-\nu}+h^{-1-\nu})e^{-\mu h^{-\tau}}) \to 0\,,\end{aligned}$$

uniformly as $h \to 0$.

Since $v_0(s)$ changes sign at $s = 0$, we have a zero of v_h in $[-1,1]$ as $h > 0$ small. This proves the existence of a zero $s_0 \in [0,1]$ of v_h and then of $w_h(\theta)$. Thus we have proved the following theorem:

Theorem 1.3.7 *Assume that $V \in C^3(R^1)$ is bounded and that x_0 is a nondegenerate critical point of V. If $\lambda < \inf V$, then $\exists h_0 > 0$ such that $\forall h \in (0,h_0)$ $\exists$ a nonzero solution $\phi(x) = z(\frac{x-x_0-s_0}{h}) + \xi(h,s_0(h))(\frac{x-x_0}{h})$ of the equation (1.32), with $s_0 \in [-h^\nu, h^\nu]$ $\nu \in (1,2)$, and $\|\xi(h,s_0(h))\| \to 0$, as $h \to 0$.*

This solution ϕ, which is close to $z(\frac{x-x_0}{h})$, becomes more concentrated about x_0 as $h \to 0$. It is called a nonspreading wave packet with width $\alpha = (2(V(x_0)-\lambda))^{-1/2}$.

1.3.4 Gluing

Gluing is an important technique in nonlinear analysis. It is a method of joining two heteroclinic trajectories to form a new trajectory such that the end point of the first is glued to the starting point of the second, and the new trajectory is closed to the union of the two. For instance, on a Riemannian manifold M, let $f \in C^1(M, \mathbf{R}^1)$. Given $(x, y) \in K \times K$, where K is the critical point set of f, i.e., $K = \{x \in M \mid f'(x) = \theta\}$, let $\mathrm{M}(x, y)$ be the space of all trajectories $c \in H^1(\mathbf{R}^1, M)$ satisfying

$$\dot{c} = -\nabla f \circ c, \quad c(-\infty) = x, \; c(+\infty) = y\,.$$

The importance of the gluing technique is in the study of the compactness of the totality of the trajectory spaces: $\{\mathrm{M}(x, y) \mid (x, y) \in K \times K\}$.

It is proved that the trajectory manifolds $\mathrm{M}(x, y)$ are compact up to the existence of sequences which converge to broken trajectories in the C^∞_{loc} topology. And the "gluing" means to map such broken trajectories equipped with an additional suitable parametrization into the appropriate trajectory space:

$$\mathrm{M}(x, z) \times \mathrm{M}(z, y) \to \mathrm{M}(x, y)\,.$$

This technique is crucial in Floer homology theory (see Hofer and Zehnder [HZ 2], Floer [Fl 2,3], Taubes [Tau]). In order not to involve over specialized knowledge in that theory, it will suffice to introduce the technique by an example of a nonspreading wave packet, which we met in the previous subsection.

We have known the existence of a nonspreading wave packet, if V has a single nondegenerate critical point. However, if V has several nondegenerate critical points $x_1, x_2, \ldots, x_n$, there are at least n nonspreading wave packets with widthes $\alpha_j = (2(V(x_j) - \lambda))^{-1/2}, j = 1, 2, \ldots, n$. As h becomes small, these wave packets are separated. Are there multi-peak wave packets? More precisely, are there new solutions of equation (1.32), which are closed to some of these wave packets on their peak intervals simultaneously? For simplicity, we only consider the case $n = 2$; the result can be extended to any n. Assume that V has only two nondegenerate critical points $x_\pm = \pm R$. Let $\lambda < \inf(V)$. Define $\alpha_\pm = \sqrt{2(V(\pm R) - \lambda)}$, $z_\pm(y) = \alpha_\pm \operatorname{sech}(\alpha_\pm y)$, and $\forall (\theta_+, \theta_-) \in R^2, z(\theta_+, \theta_-) = z_+(\theta_+) + z_-(\theta_-)$, where $z_\pm(\theta_\pm) = \tau_{\pm(R+\theta_\pm)/h} z_\pm$.

Again let

$$F(h, u) = -\frac{1}{2}\frac{d^2}{dy^2}u + (V_h - \lambda)u - u^3\,.$$

We are looking for a solution of the form $u = z(\theta_+, \theta_-) + \phi \in X$ for the equation $F(h, \cdot) = 0$, in which ϕ is small as $h > 0$ is.

Define a two-dimensional manifold: $Z_h = \{(z_+(\theta_+), z_-(\theta_-)) \mid (\theta_+, \theta_-) \in R^2\}$, then $T_{z(\theta_+,\theta_-)} Z_h = \mathrm{span}\{\tau_{(R+\theta_+)/h} e_+, \tau_{-(R+\theta_-)/h} e_-\}$, where $e_\pm = z'_\pm$.

In contrast with the one-peak solution, now $z(\theta_+, \theta_-)$ are not zeroes of $F(0, u)$, therefore the zeroes of $F(h, u)$ cannot be obtained directly by the IFT.

Instead, we appeal to the contraction mapping theorem. However, Lyapunov–Schmidt reduction is again useful. Fixing $(\theta_+.\theta_-) \in R^2$, let $u_0 = z(\theta_+, \theta_-)$, we write down the orthogonal decomposition:

$$X = T_{u_0} Z_h \bigoplus X_1, Y = T_{u_0} Z_h \bigoplus Y_1 ,$$

where the spaces X and Y were introduced in the previous subsection, and let $P : Y \to Y_1$ be the orthogonal projection.

Again,

$$F(h, u) = 0 \Leftrightarrow \begin{cases} PF(h, u_0 + \xi) = 0 \\ (I - P)F(h, u_0 + \xi) = 0 . \end{cases}$$

In fact, by Taylor's formula,

$$F(h, u_0 + \xi) = F(h, u_0) + F_u(h, u_0)\xi + N(h, u_0, \xi) ,$$

where $N(h, u_0, \xi) = 3u_0\xi^2 + \xi^3$. First, fixing $(\theta_+, \theta_-) \in R^2$, we solve the first equation in X_1 for $h > 0$ small. Let $L(\theta_+, \theta_-, h) = PF_u(h, u_0)$. We have the following lemmas:

Lemma 1.3.8 $\exists \gamma > 0, \exists h_0 > 0, \exists \alpha_0 \in (0, 1/2)$ *such that*

$$\|L(\theta_+, \theta_-, h)\|_Y \geq \gamma \|u\|_X \;\; \forall u \in X_1 .$$

as $h \in (0, h_0)$, *and* $|\alpha_\pm| < \alpha_0$.

Lemma 1.3.9 $\forall \rho > 0, \|F(h, u_0)\|_Y \leq C_1(\Sigma_\pm \mathrm{Max}_{|y| \leq \rho h}(V(y) - V(\pm R))^2| \pm R \pm \theta_\pm| + e^{-2\mu\rho} + e^{-\mu R/h})$, *where* $\mu = \mathrm{Min}\{\alpha_+, \alpha_-\}$.

Lemma 1.3.10

$$\|N(h, u_0, \xi)\|_Y \leq C_2 \|\xi\|_X^2,$$
$$\|N(h, u_0, \xi_1) - N(h, u_0, \xi_2)\|_Y \leq C_2 \mathrm{Max}(\|\xi_1\|_X, \|\xi_2\|_X)\|\xi_1 - \xi_2\|_X .$$

Once these lemmas have been proved, according to the contraction mapping theorem, one finds a fixed point ξ of the operator $-L^{-1}(\theta_+, \theta_-, h)N(h, u_0, \xi) - F(h, u_0)$ in a neighborhood of u_0 on X_1. Namely, $\exists h_0 > 0, \exists \alpha_0 > 0, \exists C_1 > 0$ such that $\forall h \in (0, h_0), \forall |\theta_\pm| < \alpha_0, \exists$ a unique $\xi = \xi(\theta_+, \theta_-, h) \in X_1$ such that

$$PF(h, u_0 + \xi) = 0, \;\text{ and }\; \|\xi\| \leq C_1 \|F(h, u_0)\|_Y .$$

Now the second equation is a two system in two variables $(\theta_+, \theta_-) \in [-\alpha_0, \alpha_0]^2$:

$$v_h(\theta_+, \theta_-) := h^{-1}(\langle e_+, F(h, z(\theta_+, \theta_-) + \xi(\theta_+, \theta_-, h)\rangle ,$$
$$\langle e_-, F(h, z(\theta_+, \theta_-) + \xi(\theta_+, \theta_-, h)\rangle) = (0, 0) .$$

Let us define a two-vector function:

$$v_0(\theta_+,\theta_-)=\frac{1}{2}(|z_+|^2V''(R)\theta_+,|z_-|^2V''(-R)\theta_-)\,.$$

Again we need a Lemma:

Lemma 1.3.11 *For $\nu\in(1,2)$ and $0<h<\mathrm{Min}(h_0,\alpha_0)$, we have*

$$h^{-\nu}v_h(h^\nu\theta_+,h^\nu\theta_-)\to v_0(\theta_+,\theta_-)\,,$$

uniformly on $[-1,1]^2$ as $h\to 0$.

The difference between these proofs of the Lemmas 1.3.8 to 1.3.11 and those in the previous subsection lie in the interaction terms (for details refer to [Oh]).

From the Brouwer degree theory (cf. Sect. 3.1),

$$\deg(h^{-\nu}v_h(h^\nu\theta_+,h^\nu\theta_-),[-1,1]^2,(0,0))=\deg(v_0(\theta_+,\theta_-),[-1,1]^2,(0,0))\neq 0\,.$$

This proves the existence of a zero of the above system. Finally we arrive at:

Theorem 1.3.12 *Suppose that $V\in C^3$ is bounded, and $\lambda<\inf(V)$. Then for each pair (x_1,x_2) of nondegenerate critical points of $V,\exists h_0>0$, such that for $h\in(0,h_0)$ equation (1.32) has a nonzero solution u_h of the form $\tau_{(x_1+s_+)/h}z_++\tau_{(x_2+s_-)/h}z_-+\xi(h)$, where $|s_\pm|\le h^{-\nu}$ with $\nu\in(1,2)$, and $\|\xi\|\to 0$ as $h\to 0$.*

1.3.5 Transversality

Transversality is an important notion in differential geometry. It is a condition, induced by the IFT, on a map between two manifolds, under which the preimage of a submanifold is again a submanifold. $f:X\to Y$ is said to be transversal to the submanifold $W\subset Y$ if at every point $x\in f^{-1}(W)$,

$$\mathrm{Im}\,f'(x)+T_{f(x)}(W)=T_{f(x)}(Y)\,,$$

and is denoted by $f\pitchfork W$.

Related notions are regular points and regular values defined as follows.

Definition 1.3.13 *Suppose that X and Y are C^1 Banach manifolds and $f\in C^1(X,Y)$. A point $x\in X$ is called a regular point of f, if $f'(x):T_x(X)\to T_{f(x)}(Y)$ is surjective, and is singular (or critical) if it is not regular. The images of the singular points under f are called singular values (or critical values), and their complement, regular values. If $y\in Y$ is not in the image of f, i.e., $f^{-1}(y)=\emptyset$, then y is a regular value.*

The following Sard theorem reveals the smallness of the set of critical values:

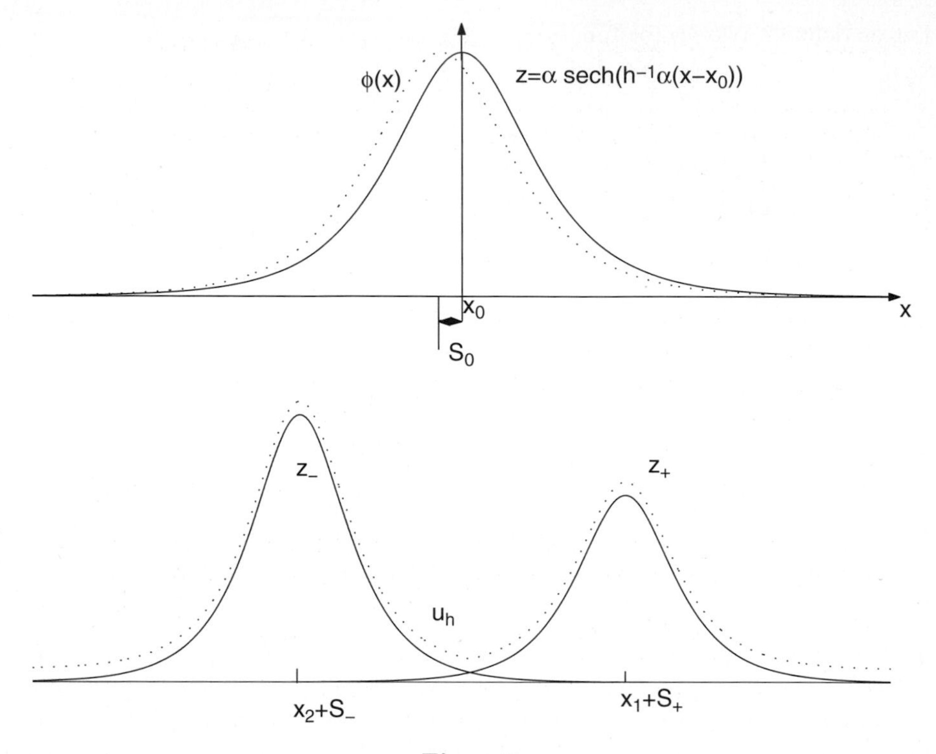

Fig. 1.8.

Theorem 1.3.14 *(Sard) Suppose that X and Y are differential manifolds with dimensions n and m respectively, and that $U \subset X$ is open. If $f \in C^r(X,Y)$, where $r \geq 1$ and $r > \max\{0, \dim X - \dim Y\}$, then the set of critical values of f has measure zero in Y.*

The proof is purely measure theoretic, we shall not give it here. The special case $\dim X = \dim Y < \infty$ will be given in Sect. 3.1, but for the general case refer to Milnor [Mi 2].

We shall extend the study to maps between infinite-dimensional manifolds. Recall a map f between two Banach spaces X and Y is called Fredholm in $U \subset X$, if $f \in C^1(U,Y)$ and $f'(x)$ is a Fredholm operator $\forall x \in U$. The index of f is defined to be

$$\text{ind}\, f'(x) = \dim\ker f'(x) - \dim\operatorname{coker} f'(x)\ .$$

It is known that if U is connected, then $\text{ind}\, f'(x)$ is a constant integer, and then is denoted by $\text{ind}(f)$.

Lemma 1.3.15 *If $f \in C^1(U,Y)$ is a Fredholm map, then the set of critical points is closed.*

Proof. Since $x \mapsto \dim \ker f'(x)$ is u.s.c., and $\text{ind}(f)$ is locally a constant, $\dim \text{coker} f'(x)$ is also u.s.c. Let S be the critical point set of f, then the set

$$S = \{x \in U | f'(x) \text{ is not surjective}\} = \{x \in U | \dim \text{coker } f'(x) \geq 1\}$$

is closed. □

Theorem 1.3.16 *(Sard–Smale) Suppose that X is a separable Banach space and Y is a Banach space. Let $f \in C^r(U, Y)$ be a Fredholm map, where $U \subset X$ is open. If $r > \max(0, \text{ind}(f))$, then the set of critical values is of first category.*

Proof. Since the first category set is a countable union of closed nowhere dense sets, and U is separable, it is sufficient to prove that $\forall x \in U$, there is a neighborhood of $x, V \subset U$ such that the set of critical values of $f|_V, S(f, V)$ is closed and nowhere dense.

By the definition of Fredholm operators, one has the decomposition $X = \ker f'(x) \oplus X_1$. Let Q be the projection of Y onto $\text{Im} f'(x)$; from the IFT, there is a neighborhood $U_0 \times V_0 \subset \ker f'(x) \times X_1$, a neighborhood $W \subset \text{Im} f'(x)$ of $f(x)$, and $h \in C^1(U_0 \times W, V_0)$ such that

$$Qf(u + h(u, w)) = w, \quad \forall (u, w) \in U_0 \times W \ ,$$

and $\forall u \in U_0, h(u, \cdot) : W \to V_0$ is a diffeomorphism. Since U_0 is finite dimensional, it can be chosen compact. We set $V = U_0 \times V_0$,

1. First we show that $S(f, V)$ is closed. Due to Lemma 1.3.15 it is sufficient to show that the map $f|_V$ is closed. In fact, let $x_n = u_n + v_n \in U_0 \times V_0$ be such that $y_n = f(x_n) \to y$. From the compactness of U_0, u_n is subconvergent, and $v_n = h(u_n, Qy_n)$ is convergent, therefore $f|_V$ is closed.

2. Next we show that $S(f, V)$ is a nowhere dense set. Define $H : U_0 \times W \to V$ by $H(u, v) = (u, h(u, w))$. Then H is a diffeomorphism satisfying $Qf \circ H(u, w) = w$. Setting $\tilde{f} = f \circ H$, we have $S(f, V) = S(\tilde{f}, U_0 \times W)$, and $Q\tilde{f}'|_W = id_W$, therefore $\forall w \in W$, $(x, w) \in S(\tilde{f}, U_0 \times W) \Longleftrightarrow x \in S((I - Q)\tilde{f}(\cdot, w), U_0)$. From $r > \text{ind} f'(x) = \dim \ker f'(x) - \dim \text{coker} f'(x)$ and Sard theorem, $S((I - Q)\tilde{f}(\cdot, w), U_0)$ is a null set. Since $S(f, V)$ is also closed, it is a nowhere dense set.

We conclude that the critical set of f is of the first category. □

Corollary 1.3.17 *Let X and Y be Banach spaces, and X be separable. If $f \in C^1(X, Y)$ is Fredholm with negative index, then $f(X)$ does not contain interior points.*

Proof. If not, say $y_0 \in f(X)$ is an interior point, i.e., $\exists$ a neighborhood V of y_0 such that $V \subset f(X)$. According to the Sard–Smale lemma, $\exists y \in V$ such that $f \pitchfork \{y\}$, i.e., $\forall x \in f^{-1}(y), f'(x) : X \to Y$ is surjective, or $\text{codim}\, \text{Im}(f'(x)) = 0$, thus $\text{ind}(f)(x) = \dim \ker f'(x) \geq 0$. A contradiction. □

The following transversality theorem will be often used in the sequel.

Theorem 1.3.18 *Suppose that* X, Z *are Banach spaces, where* X *is separable, and that* S *is a* C^r *Banach manifold. Assume that* $F \in C^r(X \times S, Z)$ *satisfies (1)* $F \pitchfork \{\theta\}$, *and (2)* $\forall s \in S, f_s(u) = F(u, s)$ *is a Fredholm map with index satisfying* $\max\{0, \mathrm{ind}(f_s)\} < r$.

Then $\exists$ *a residual set* $\Sigma \subset S$ *(i.e., the countable intersection of open dense sets) such that* $\forall s \in \Sigma, f_s \pitchfork \{\theta\}$.

Proof. Define the injection: $i : V \longrightarrow X \times S$ and the projections $\pi : X \times S \longrightarrow S, p : X \times S \to X$. Let $V = F^{-1}(\theta)$. In the sequel, both of them are restricted on V. We claim that $\pi \circ i$ is a Fredholm map with $\mathrm{ind}\,(\pi \circ i) = \mathrm{ind}\,(f_s)$.

In fact, $\ker F'(v) = T_v(V) = \mathrm{Im}\,((p \circ i)'(v)) \oplus \mathrm{Im}\,((\pi \circ i)'(v))\ \forall v = (x, s) \in T_v(V)$. Since $F \pitchfork \{\theta\}$, we have a direct sum decomposition: $X \times T_s(S) = Y \oplus T_v(V)$ such that $F'(v) : Y \to Z$ is an isomorphism. By the assumption that $f_s'(x)$ is Fredholm, we have direct sum decompositions: $X = \mathrm{Im}((p \circ i)'(v)) \oplus Y_1, Z = Z_1 \oplus Z_2$, such that $f_s'(x) : Y_1 \to Z_1$. Thus, $Y = Y_1 \oplus Y_2$, where $T_s(S) = \mathrm{Im}(\pi \circ i)'(v) \oplus Y_2$. From $F'(v) = f_s'(x) \oplus \partial_s F(v)$, we have $\partial_s F(v) : Y_2 \to Z_2$ is an isomorphism.

Thus, $\ker f_s'(x) = \mathrm{Im}(p \circ i)'(v) = \ker\,(\pi \circ i)'(v)$, $\mathrm{coker}\, f_s'(x) = Z_2 \simeq Y_2 \simeq \mathrm{coker}(\pi \circ i)'(v)$. Therefore $\mathrm{ind}\,(f_s) = \mathrm{ind}\,(\pi \circ i)$.

According to the Sard–Smale Theorem, the set of regular values of $\pi \circ i$ Σ is a residual set. We are going to prove that for a regular value s, $f_s \pitchfork \{\theta\}$. From $F(x, s) = \theta$, and $F \pitchfork \{\theta\}$, we have $\mathrm{Im}\, F'(x, s) = Z$, i.e., $\forall a \in Z, \exists (\alpha, \beta) \in X \times T_s(S)$ such that $a = (\frac{\partial F}{\partial x}\alpha + \frac{\partial F}{\partial s}\beta)$.

We are going to prove $\mathrm{Im}\, f_s'(x) = Z$, i.e., $\forall$ a $\in Z, \exists y \in X$ such that $a = f_s'(x)y$.

Thus, if $\beta = 0$, we take $y = \alpha$. Otherwise, since π is a projection, $(\pi \circ i)'(x, s) : X \times T_s(S) \longrightarrow T_s(S)$ is a projection. That s is a regular value means that $(\pi \circ i)'(x, s) : T_{(x,s)}(V) \longrightarrow T_s(S)$ is surjective, i.e., to the given $\beta \in T_s(S), \exists w \in X$ such that $(w, \beta) \in T_{(x,s)}(V)$. Since $F^{-1}(\theta) = V$, we have $F'(x, s)(w, \beta) = \theta$.

Setting $y = \alpha - w$, we obtain

$$
\begin{aligned}
f_s'(x)y - a = F'(x, s)[(\alpha, \beta) - (w, \beta)] - a &= \left[\frac{\partial F}{\partial x}\alpha + \frac{\partial F}{\partial s}\beta\right] \\
&- a - F'(x, s)(w, \beta) = \theta\,.
\end{aligned}
$$

□

The same conclusion holds if X and Z are Banach manifolds.

As an application we study the simplicity of eigenvalues of the Laplacian on bounded domains with Dirichlet data. It is well known that all eigenvalues of second-order ODEs on bounded intervals with Dirichlet data are simple, but it is not true for PDE, e.g., the Laplacian on a ball. However, we shall show that for most domains, this is true. What is the meaning of most domains?

Given a bounded open domain $\Omega_0 \subset R^n$ with smooth boundary, we consider the manifold $S = \mathrm{Diff}^3(\Omega_0) := \{g \in C^3(\overline{\Omega}_0, R^n) \,|\, \det(g'(x)) \neq$

$0\ \forall x \in \overline{\Omega}_0\}$. Thus as $g \in S$, the domain $\Omega = g(\Omega_0)$ is C^3. $\forall \Omega$, let $X(\Omega) = H^2(\Omega) \cap H_0^1(\Omega)$ and $Y(\Omega) = L^2(\Omega)$. $\forall g \in S$, define $g^* : X(\Omega_0) \to X(\Omega)$ by

$$(g^*u)(x) = u(g^{-1}(x))\ \ \forall x \in \Omega = g(\Omega_0)\ .$$

One defines a map $F \in C^1((X(\Omega_0)\backslash\{\theta\}) \times S, Y(\Omega_0))$ by

$$F(u, \lambda, g) = g^{*-1}(\Delta + \lambda I)g^*u\ .$$

We want to show that there exists a residual set Σ of S such that $\forall g \in \Sigma$, all eigenvalues of the problem:

$$(\Delta + \lambda)u = 0 \ \text{ on } H_0^1(\Omega)\ ,$$

are simple, where $\Omega = g(\Omega_0)$. Since there are, at most, countable eigenvalues, it is sufficient to show that for each eigenvalue λ after suitable perturbation of the domain, i.e., for after a suitable $g \in S$, it is simple. To this end, let us define $f_g(u, \lambda) = F(u, \lambda, g)$. We claim that it remains to show $f_g \pitchfork \{\theta\}$, or equivalently, $f_g'(u, \lambda) : X(\Omega) \times R^1 \to Y(\Omega)$ is surjective, i.e., $\{(\Delta + \lambda)w + \mu u \,|\, (w, \mu) \in X(\Omega) \times R^1\} = Y(\Omega)\}$, or equivalently, $\text{codim Im}(\Delta + \lambda I) \leq 1$. Since we have assumed that λ is an eigenvalue, i.e., $\text{codim}\,(\text{Im}(\Delta + \lambda I) \geq 1$. Therefore λ is simple.

We know that $f_g'(u, \lambda)$ is a Fredholm map, from Theorem 1.3.18, it is sufficient to verify that $F \pitchfork \{\theta\}$. To this end, $\forall (u_0, \lambda_0, g_0) \in F^{-1}(\theta)$, let $L = F'(u_0, \lambda_0, g_0)$, and $\Omega = g_0(\Omega_0)$. With no loss of generality we may assume $g_0 = Id$. It remains to verify that L is surjective. Since

$$L(w, \mu, h) = (\Delta + \lambda_0)w + \mu u_0 + [h\cdot\nabla, \Delta + \lambda_0]u_0 = (\Delta + \lambda_0)(w - h\cdot\nabla u_0) + \mu u_0\ ,$$

L is Fredholm, where $[A, B] = AB - BA$. Suppose it is not surjective, then $\text{codim Im}(L) > 0$, there exists a nontrivial element $\phi \perp \text{Im}(L)$. First by taking $w = h = \theta$, we have

$$\int_\Omega \phi u_0 = 0\ .$$

Second, we take $h = \theta$; it follows that

$$\int_\Omega \phi(\Delta + \lambda_0 I)w = 0\ \forall w \in X(\Omega)\ .$$

Then $\phi \in C^{2,\alpha}(\bar{\Omega})$, $\alpha \in (0, 1)$ is a solution of the equation:

$$(\Delta + \lambda_0)\phi = 0\ ,$$

with boundary value zero. Then we have

$$\begin{aligned}0 &= \int_{\Omega} \phi(\Delta + \lambda_0)(h \cdot \nabla u_0) \\ &= \int_{\Omega} \phi(\Delta + \lambda_0)(h \cdot \nabla u_0) - (h \cdot \nabla u_0)(\Delta + \lambda_0 I)\phi \\ &= \int_{\partial\Omega} \phi\partial_n(h \cdot \nabla u_0) - \partial_n\phi(h \cdot \nabla u_0) \\ &= -\int_{\partial\Omega} \partial_n\phi(h \cdot \nabla u_0)\end{aligned}$$

for all $h \in C^3(\bar{\Omega}, R^n)$. Thus,

$$\partial_n\phi\nabla u_0 = 0 \text{ on } \partial\Omega \, .$$

According to the uniqueness of the Cauchy problem (see [Hor 1]) for the equations $(\Delta + \lambda_0)\phi = 0$, and $(\Delta + \lambda_0)u_0 = 0$ for $\phi, u_0 \in C^{2,\alpha} \cap C_0(\bar{\Omega})$, this contradicts with $u_0 \neq \theta$ and $\phi \neq \theta$. We have proved a result due to K. Uhlenbeck [Uh]:

Theorem 1.3.19 *There exists a residual set* $\Sigma \subset \text{Diff}^3(\Omega_0)$ *such that* $\forall g \in \Sigma$, *all eigenvalues of the equation*

$$(\Delta + \lambda I)u = 0 \quad on \ H_0^1(g(\Omega_0))$$

are simple.

1.4 Hard Implicit Function Theorem

The inverse function theorem is established for C^1 mappings between Banach spaces, $f : X \to Y$, where X and Y are Banach spaces, under the assumption $f'(x_0)^{-1} \in L(Y, X)$. However, if $f'(x_0)^{-1}$ is not bounded, then the problem will be very complicated. The small divisor problem, arising in the study of the long-time behavior of oscillatory motions in the solar system, is a typical example. The celebrated KAM theory, which states that most (in the sense of measure) quasi-periodic solutions of an integrable smooth Hamiltonian system persist under small perturbations of that Hamiltonian, provided that the Hessian is nondegenerate, is a breakthrough in the progress in this direction. Not only is the result an important conclusion in celestial mechanics, but also the methods developed in this theory have great impact in other nonlinear problems. In fact, Nash–Moser iteration, which was introduced in the study of the isometric embedding problem for Riemann manifolds, see [Na 1,2], and [Mos 1,2], is extensively applied in problems arising in geometry (see for instance Hamilton [Ham]) and in physics (see for instance Hormander [Hor 2]). In recent years, KAM theory has been extended to Hamiltonian systems with infinitely many freedoms (see Poschel [Po], Wayne [Way], Kuksin [Kuk], Bourgain [Bou]etc.)

In avoiding too many complicated computations, we take an approach due to Hormander by appealing to the Leray–Schauder degree theory in the proof of the existence of a solution. Readers may read this subsection after Chap. 3. In fact there are other approaches without using topological arguments (see for instance J. Moser [Mos 1,2,3], Hormander [Hor 2] or J. T. Schwartz [Scw]).

Actually, the boundedness assumption on $f'(x_0)^{-1}$ guarantees the convergence of the simple iteration method in solving the nonlinear equation:

$$y = f(x) \tag{1.34}$$

$$\Leftrightarrow \quad x = x - f'(x_0)^{-1}(y - f(x)) \, .$$

One produces the iteration process as follows:

$$x_{n+1} = x_n - f'(x_0)^{-1}(y - f(x_n)), \quad n = 1, 2, \ldots \, . \tag{1.35}$$

However, if we consider the nonlinear wave equation:

$$\begin{cases} u_{tt} - F(u_{xx}) = f \\ u|_{t=0} = u_t|_{t=0} = 0 \, , \end{cases}$$

where F is a function with $F'(0) = 1$ and $F(0) = 0$, then the above simple iteration process fails, because the linearized equation reads as

$$\begin{cases} u_{tt} - u_{xx} = g \\ u|_{t=0} = u_t|_{t=0} = 0 \, . \end{cases}$$

when $g \in H^s$, with compact support for some $s > 0$. We can at most estimate the boundedness of H^{s+1} norm of u, but not H^{s+2}, as that in the elliptic case. If we use the simple iteration, then we lose derivatives in each step; after a finite number of steps, the iteration cannot go on. This phenomenon is called "loss of derivatives". It occurs in nonelliptic problems.

In the analytic category, occurs phenomenon whereby the inverse maps reduce the convergence radii (e.g. the small divisor problem, see below); this is called "loss of convergence radii". As before, this is also the case that the linearized operator has no bounded inverse.

1.4.1 The Small Divisor Problem

For a given f, analytic in a neighborhood of θ, with $f(0) = 0$, and $f'(0) = \sigma$, find u, analytic in a neighborhood of θ with $u(0) = 0$, and $u'(0) = 1$, satisfying

$$f(u(z)) = u(\sigma z) \, . \tag{1.36}$$

We write it in the IFT form:

Let

$$f(z) = \sigma z + \hat{f}(z) ,$$
$$u(z) = z + \hat{u}(z) .$$

Equation (1.36) is equivalent to

$$\Phi(\hat{f}, \hat{u}) = \hat{f}(z + \hat{u}(z)) + \sigma\hat{u}(z) - \hat{u}(\sigma z) = \theta . \tag{1.37}$$

Obviously, $\Phi(\theta, \theta) = \theta$, and

$$\Phi_{\hat{u}}(\hat{f}, \hat{u})\hat{w} = \hat{f}'(z + \hat{u}(z))\hat{w}(z) + \sigma\hat{w}(z) - \hat{w}(\sigma z) .$$

$\Phi_{\hat{u}}(\theta, \theta)$ is injective, if $\sigma = e^{2\pi i\tau}$, where τ is irrational. In fact, we write

$$\hat{v} = \sum_{j=2}^{\infty} v_j z^j , \quad \hat{w} = \sum_{j=2}^{\infty} w_j z^j .$$

The equation

$$\Phi_{\hat{u}}(\theta, \theta)\hat{w} = \hat{v}$$

has a unique solution $\hat{w}$, in which

$$w_j = \frac{v_j}{\sigma - \sigma^j} \quad j = 2, 3, \cdots .$$

Since the denominator tends to zero for a subsequence, one cannot expect a bounded inverse of $\Phi_{\hat{u}}(\theta, \theta)$.

A real number τ is of type (b, ν), $b > 0, \nu > 2$, if

$$\left|\tau - \frac{p}{q}\right| \geq \frac{b}{q^\nu} \quad \forall p, q \in \mathbb{Z}\backslash\{0\} . \tag{1.38}$$

Claim. For almost all real numbers τ, there exists (b, ν) depending on τ, such that τ is of type (b, ν).

Indeed, for given (b, ν), fixing q, the set of all real numbers $\tau \in [0, 1]$ such that (1.38) does not hold has a measure less than $2bq^{-\nu+1}$. Therefore, the set of all real numbers τ such that (1.38) does not hold has the measure $\leqslant 2b \sum q^{-\nu+1} < \infty$. Since b can be arbitrarily small, the conclusion follows.

Suppose $\sigma = e^{2\pi i\tau}$, where τ is a number of type (b, ν). Then

$$\begin{aligned}
|\sigma - \sigma^j| &= \left|e^{2\pi i(j-1)(\tau - \frac{p}{j-1})} - 1\right| \\
&\geq \left|\sin 2\pi \left(\tau - \frac{p}{j-1}\right)(j-1)\right| \\
&\geq \frac{2}{\pi} \cdot 2\pi(j-1)\left|\tau - \frac{p}{j-1}\right| \\
&\geq 4\frac{b}{(j-1)^{\nu-1}} ,
\end{aligned}$$

therefore

$$\frac{1}{|\sigma - \sigma^j|} \leqslant \frac{j^{\nu-1}}{4b} .$$

In this case, even if $\hat{v}$ has the convergent radius r, $\hat{w}$ can only have a smaller convergent radius. In other words, if we introduce a family of Banach spaces:

$$A(r) = \{\hat{w}(z)| \text{ bounded and analytic in } |z| < r, \hat{w}(0) = \hat{w}'(0) = 0\}$$

with the norm

$$|\hat{w}|_r = \mathrm{Sup}_{|z|<r}|\hat{w}(z)| ,$$

then we have, $\forall \ \hat{v} \in A(r)$

$$\begin{aligned}|\Phi_{\hat{u}}(\theta,\theta)^{-1}\hat{v}|_{r-\delta} &\leqslant \mathrm{Sup}_{|z|<r-\delta}\sum_{j=2}^{\infty}\left|\frac{v_j z^j}{\sigma-\sigma^j}\right| \\ &\leqslant C\sum_{j=2}^{\infty} j^{\nu-1}(r-\delta)^j|v_j| ,\end{aligned}$$

but

$$|v_j| = \frac{1}{2\pi}\left|\int_{|z|=r}\frac{v(z)}{z^{j+1}}dz\right| \leqslant r^{-j}|v|_r \quad ,$$

therefore

$$|\Phi_{\hat{u}}(\theta,\theta)^{-1}\hat{v}|_{r-\delta} \leqslant C|v|_r\sum_{j=2}^{\infty} j^{\nu-1}\left(1-\frac{\delta}{r}\right)^j \tag{1.39}$$

$$\leqslant C\delta^{-\nu}|v|_r \tag{1.40}$$

where $C = C(\nu, r)$ is a constant.

The loss of radius prevent us from simple iterations. One intends to gain back some of the loss by accelerating the convergence. It is known that the Newton method provides more efficient convergence speed. The iteration scheme, in contrast with (1.35), reads as follows:

$$x_{n+1} = x_n - f'(x_n)^{-1}(y - f(x_n)) \tag{1.41}$$

We are led to the computation of $\Phi_{\hat{u}}(\hat{f}, \hat{u})^{-1}$, but the later is too complicated. We intend to find an approximate inverse as a replacement. Of course, the approximation should match the convergence speed.

In doing so, let us perform some computations:

$$(\Phi(\hat{f}, \hat{u}))'(z) = \hat{f}'(z + \hat{u}(z))(1 + \hat{u}'(z)) + [\sigma\hat{u}'(z) - \sigma\hat{u}'(\sigma z)] ,$$

and

$$\Phi_{\hat{u}}(\hat{f}, \hat{u})\hat{v} = \hat{f}'(z + \hat{u}(z))\hat{v(z)} + \sigma\hat{v}(z) - \hat{v}(\sigma z) .$$

Thus

$$\Phi_{\hat{u}}(\hat{f},\hat{u})\hat{v} = \frac{\hat{v}}{1+\hat{u}'}\Phi(\hat{f},\hat{u})' + (1+\hat{u}'\circ\sigma)\Phi_{\hat{u}}(\theta,\theta)\frac{\hat{v}}{1+\hat{u}'} \;.$$

One takes

$$T(\hat{f},\hat{u})\hat{w} = (1+\hat{u}')\Phi_{\hat{u}}(\theta,\theta)^{-1}\left(\frac{\hat{w}}{1+\hat{u}'\circ\sigma}\right) \;, \tag{1.42}$$

as the approximate inverse with $\hat{v}(z) = (1+\hat{u}(z))\hat{w}(z)$. In fact,

$$\Phi_{\hat{u}}(\hat{f},\hat{u})T(\hat{f},\hat{u})\hat{w} - \hat{w} = \Phi(\hat{f},\hat{u})'\Phi_{\hat{u}}(\theta,\theta)^{-1}\left(\frac{\hat{w}}{1+\hat{u}'\circ\sigma}\right) \;,$$

and from the analyticity,

$$|\Phi(\hat{f},\hat{u})'|_{r-\delta} \leqslant C\delta^{-1}|\Phi(\hat{f},\hat{u})|_r \;\;.$$

We take $\gamma \in (0,1)$, and $|\hat{u}'|_r < \gamma$, then $1+\hat{u}'\circ\sigma \neq 0$, and

$$|(\Phi_{\hat{u}}(\hat{f},\hat{u})T(\hat{f},\hat{u}) - Id)\hat{w}|_{r-\delta} \leqslant C\delta^{-(\nu+1)}|\Phi(\hat{f},\hat{u})|_r|\hat{w}|_r \;, \tag{1.43}$$

where $C = C(\gamma,\nu,r)$. Similarly,

$$|T(\hat{f},\hat{u})\hat{w}|_{r-\delta} \leqslant C\delta^{-\nu}|\hat{w}|_r \;\;. \tag{1.44}$$

Let us introduce a new norm:

$$\|\hat{w}\|_r = \mathrm{Sup}_{|z|<r}|\hat{w}'(z)| \;. \tag{1.45}$$

Obviously the following inequality holds:

$$|\hat{w}|_r \leqslant r\|\hat{w}\|_r \;.$$

Moreover, we estimate the error:

$$\begin{aligned}
&\Phi(\hat{f},\hat{u}) - \Phi(\hat{f},\hat{v}) - \Phi_{\hat{u}}(\hat{f},\hat{v})(\hat{u}-\hat{v}) \\
&\quad = \hat{f}(z+\hat{u}(z)) - \hat{f}(z+\hat{v}(z)) - \hat{f}'(z+\hat{v}(z))(\hat{u}(z)-\hat{v}(z)) \\
&\quad = \int_0^1\int_0^1 s\hat{f}''(z+\hat{v}(z)+st(\hat{u}(z)-\hat{v}(z)))dtds(\hat{u}(z)-\hat{v}(z))^2 \;.
\end{aligned}$$

By the Cauchy formula, if $f \in A(r)$ with $|f|_r < \gamma$,

$$|f''(z)| = \frac{1}{\pi}\left|\int_{|\zeta-z|=R}\frac{f(\zeta)}{(\zeta-z)^3}d\zeta\right| \leqslant \frac{2}{R^2}|f|_r$$

for $0 < R \leqslant r-|z|$, we obtain

$$|\Phi(\hat{f},\hat{u}) - \Phi(\hat{f},\hat{v}) - \Phi_{\hat{u}}(\hat{f},\hat{v})(\hat{u}-\hat{v})|_{r-\delta} \leqslant \frac{2\gamma}{\delta^2}\|\hat{u}-\hat{v}\|_r^2 \;. \tag{1.46}$$

The iteration scheme is now modified to

$$\hat{u}_{n+1} = \hat{u}_n - T(\hat{f}, \hat{u}_n)\Phi(\hat{f}, u_n) \ . \tag{1.47}$$

In order to show that $\hat{u}_n$ converges to some point $\hat{u}^*$, one needs to show that $|\Phi(\hat{f}, \hat{u}_n)|_{r_n} \to 0$, while the radius r_n depends on n, very rapidly tends to a certain positive number r^*.

Theorem 1.4.1 *Let X_λ, Y_λ, Z_λ, $\lambda \in (0,1]$ be three families of Banach spaces with nondecreasing norms, i.e., $\|\cdot\|_\mu \leqslant \|\cdot\|_\lambda$ for $\mu \leqslant \lambda$. Let $(\overline{x}, \overline{y}) \in X_1 \times Y_1, r > 0$, and $\Omega_r^\lambda = B_r^\lambda(\overline{x}) \times B_r^\lambda(\overline{y}) \subset X_\lambda \times Y_\lambda$, where $B_r^\lambda(\overline{x})$ is the ball centered at $\overline{x}$, with radius r in X^λ, similarly for the notations $B_r^\lambda(\overline{y})$ etc. Let $\Phi : \Omega_r^\lambda \to Z_\lambda$ $\forall \lambda \in (0,1]$ be C^1 w.r.t. y, and satisfy $\Phi(\overline{x}, \overline{y}) = \theta$. We assume:*

(1) $|\Phi(x,y) - \Phi(x,y') - \Phi_y(x,y')(y-y')|_{\lambda-\delta} \leqslant M\delta^{-2\alpha}\|y-y'\|_\lambda^2, \forall y' \in B_r^\lambda(\overline{y})$,
(2) $\exists T(x,y) \in L(Z_\lambda, Y_{\lambda-\delta})$ such that $|\ T(x,y)z|_{\lambda-\delta} \leqslant M\delta^{-\tau}|z|_\lambda$,
(3) $|(\Phi_y(x,y)T(x,y) - I)z|_{\lambda-\delta} \leqslant M\delta^{-2(\alpha+\tau)}|z|_\lambda|\Phi(x,y)|_\lambda$,

where $M \geq 1, \alpha \geq 0, \tau > 1$ are constants, and $0 < \delta < \lambda$. One concludes that $\exists C = C(M,\alpha,\tau) > 0$ such that $\forall (x,y) \in \Omega_r^{\overline{\lambda}}, \forall \overline{\lambda} \in (0,1]$ with $|\Phi(x,y)|_{\overline{\lambda}} \leqslant C\overline{\lambda}^{-2(\alpha+\tau)}$, we have $y_* = u(x) \in Y_{\frac{\overline{\lambda}}{2}}$ satisfying

$$\Phi(x, u(x)) = \theta \ .$$

Proof. One chooses sequences

$$\lambda_n = \frac{\overline{\lambda}}{2}(1 + 2^{-n}), \quad n = 0, 1, 2, \dots ,$$

and

$$\mu_{n+1} = \frac{1}{2}(\lambda_n + \lambda_{n+1}) = \frac{\lambda}{2}(1 + 3 \cdot 2^{-n-2}), \quad n = 0, 1, 2, \dots .$$

Then $\lambda_0 = \overline{\lambda}$, $\lambda_n \downarrow \frac{\overline{\lambda}}{2}$.

For simplicity, we write $\Phi(y_n) = \Phi(x, y_n)$, $T(y_n) = T(x, y_n)$. The Newton iteration sequence reads as

$$\begin{cases} y_{n+1} = y_n - T(y_n)\Phi(y_n) & n = 0, 1, 2, \dots \\ y_0 = y \ . \end{cases} \tag{1.48}$$

We want to prove:

(1)′ $y_n \in B_r^{\lambda_n}(\overline{y})$,
(2)′ let $c_n = |\Phi(y_n)|_{\lambda_n}$, $\sum_{n=1}^\infty 2^{n\tau} c_n < \infty$,
(3)′ $\|y_{n+1} - y_n\|_{\mu_{n+1}} \leqslant M c_n (2^{(n+3)\tau} \overline{\lambda}^{-\tau})$.

If they are proved, then

$$\sum_{n=1}^{\infty} \|y_{n+1} - y_n\|_{\frac{\overline{\lambda}}{2}} \leqslant M\overline{\lambda}^{-\tau} 2^{3\tau} \sum_{n=1}^{\infty} 2^{n\tau} c_n < \infty .$$

Therefore $\exists y_* = u(x)$ such that $y_n \to y_*$ in $Y_{\frac{\overline{\lambda}}{2}}$, and

$$|\Phi(x, y_*)|_{\frac{\overline{\lambda}}{2}} \leqslant \lim_{n\to\infty} |\Phi(y_n)|_{\lambda_n} = \lim_{n\to\infty} c_n = 0 ,$$

i.e., $\Phi(x, y_*) = \theta$.

First, we prove $(3)'$. In fact, From (1.48) and assumption (2),

$$\begin{aligned} \|y_{n+1} - y_n\|_{\mu_{n+1}} &\leqslant \|T(y_n)\Phi(y_n)\|_{\mu_{n+1}} \\ &\leqslant M(2^{(n+3)\tau}\overline{\lambda}^{-\tau})|\Phi(y_n)|_{\lambda_n} \\ &= Mc_n(2^{(n+3)\tau}\overline{\lambda}^{-\tau}) . \end{aligned}$$

Next, we turn to proving $(2)'$. By (1.48), we have

$$\begin{aligned} \Phi(y_{n+1}) = &[\Phi(y_{n+1}) - \Phi(y_n) - \Phi_y(x, y_n)(y_{n+1} - y_n)] \\ &- [\Phi_y(x, y_n)T(y_n) - I]\Phi(y_n) . \end{aligned} \tag{1.49}$$

Combining $(3)'$, with assumptions (1) and (3), for n large,

$$\begin{aligned} c_{n+1} = |\Phi(y_{n+1})|_{\lambda_{n+1}} &\leqslant M2^{2\alpha(n+3)}\overline{\lambda}^{-2\alpha}\|y_{n+1} - y_n\|^2_{\mu_{n+1}} \\ &+ Mc_n^2(2^{n+3}\overline{\lambda}^{-1})^{2(\alpha+\tau)} \leq aq^n c_n^2 , \end{aligned}$$

where $a = \overline{\lambda}^{-2(\alpha+\tau)} q^3(M + M^3)$ and $q = 4^{\alpha+\tau}$.

Set

$$\alpha_n = aq^n c_n ;$$

we obtain

$$\alpha_{n+1} = aq^{n+1}c_{n+1} \leqslant a^2 q^{2n+1} c_n^2 = q\alpha_n^2 \quad \forall n .$$

One chooses $k \in (1, 2)$, $\epsilon_0 \in (0, 1)$ satisfying

$$q^{\frac{1}{k-1}}\epsilon_0 < 1 ,$$

and $\epsilon_{n+1} = q\epsilon_n^k \ \ \forall n$. Thus

$$\epsilon_{n+1} = q^{1+k+k^2+\cdots+k^n}\epsilon_0^{k^{n+1}} = q^{-\frac{1}{k-1}}(q^{\frac{1}{k-1}}\epsilon_0)^{k^{n+1}} < 1 .$$

For sufficiently small $c_0 = |\Phi(x, y)|_{\overline{\lambda}}$, such that $\alpha_0 = ac_0 \leqslant \epsilon_0$, by mathematical induction, it is easy to show that

$$\alpha_n \leqslant \epsilon_n \quad \forall n .$$

Thus

$$c_{n+1} \leqslant \alpha_n c_n \leqslant \epsilon_n c_n \,. \tag{1.50}$$

It follows $c_{n+1} \leqslant c_n$ and $c_{n+1} \leqslant c_0 \epsilon_n$. However, ϵ_n converges to zero hypergeometrically. That $(2)'$ is proved.

Finally, we verify $(1)'$. From

$$\begin{aligned}\|y_{n+1} - y_0\|_{\mu_{n+1}} &\leqslant \sum_{j=0}^{n} \|y_{j+1} - y_j\|_{\mu_{j+1}} \\ &\leqslant u \sum_{j=0}^{n} c_j q^{\frac{j}{2}} \,,\end{aligned}$$

one sees that $\exists$ a constant $M_1 > 0$, such that

$$\|y_{n+1} - y_0\|_{\mu_{n+1}} \leqslant a c_0 M_1 \,.$$

If c_0 is so small that

$$\|y_{n+1} - y_0\|_{\mu_{n+1}} < r - \|y - \overline{y}\|_{\overline{\lambda}} \quad \forall n \,,$$

then

$$\|y_n - y\|_{\lambda_n} \leqslant \|y_n - y\|_{\mu_n} < r - \|y - \overline{y}\|_{\lambda_n} \,.$$

Therefore

$$\|y_n - \overline{y}\|_{\lambda_n} < r \,.$$

The proof is complete. □

As an application, now we return to the small divisor problem. We have the following theorem.

Theorem 1.4.2 *(Siegel) Suppose that τ is of type $(b, \nu), b > 0, \nu > 2$. Let $\sigma = e^{2\pi i \tau}$. If $f(z) = \sigma z + \hat{f}(z)$ is an analytic function on $|z| < 1$, with $\hat{f}(0) = \hat{f}'(0) = 0$, then there exist γ, $r_0 \in (0, 1)$ and an analytic function u on $|z| < \frac{3r_0}{4(1+\gamma)}$ such that*

$$\sup_{|z|<r_0} |\hat{f}(z)| \leqslant \gamma$$

and $f \circ u = u \circ \sigma$.

Proof. We introduce Banach space families: $\forall r > 0$,

$$A(r) = \{w(z) | \text{analytic and bounded in } |z| < r, \text{ with } w(0) = w'(0) = 0\} \,,$$

with norm

$$|w|_r = \sup_{|z|<r} |w(z)| \,,$$

and
$$\widetilde{A}(r) = \{w(z) \in A(r) \mid w'(z) \text{ is bounded in } |z| < r\} ,$$
with norm
$$\|w\|_r = \sup_{|z|<r} |w'(z)| .$$

Set
$$X_\lambda = A(r_0), \quad Y_\lambda = \widetilde{A}(r_\lambda) \text{ and } Z_\lambda = A(r_\lambda)$$
where $\lambda \in (0,1], r_\lambda = \frac{1+\lambda}{2}\rho,\ \rho = \frac{r_0}{1+\gamma},\ \gamma \in (0,1]$ and $0 < \gamma < 1$ is to be determined. Set $\Phi(\hat{f}, \hat{u}) = \hat{f}(z+\hat{u}(z)) + \sigma\hat{u}(z) - \hat{u}(\sigma z)$ as in (1.37). Set $\overline{\lambda} = 1$. For $\|\hat{f}\|_{A(r_0)} < \gamma,\ \|\hat{u}\|_{\widetilde{A}(\rho)} < \gamma$, we have
$$|\Phi(\hat{f}, \hat{u})|_{A(\rho)} \leqslant \gamma + 2|\hat{u}|_{A(\rho)} \leqslant \gamma(1+2\rho) .$$

If $r_0 > 0$ is small, we may choose $\gamma > 0$ small. Since all assumptions (1)–(3) of Theorem 1.4.1 are satisfied (cf. (1.46), (1.44), and (1.43) respectively), the conclusion follows from Theorem 1.4.1. □

1.4.2 Nash–Moser Iteration

Let us turn to the "loss of derivative" problem. Besides the simple iteration method, there are other ways to solve equation (1.34). Given $c > 0$ and a curve $x(t)$ satisfying
$$f(x(t)) = (1 - e^{-ct})y ,$$
$x(t)$ satisfies the ODE:
$$\dot{x}(t) = (f'(x(t)))^{-1}(y - f(x(t))) .$$

If, for some initial data $x(0) = x_0$, the solution globally exists, and has a limit $x(t) \to x_\infty$, then
$$f(x_\infty) = y .$$

Discretizing the equation, we return to Newton's approximation scheme:
$$x_{n+1} - x_n = cf'(x_n)^{-1}(y - f(x_n)) .$$
This is (1.41) for $c = 1$.

There is enough room to generalize the above method. Let $f(x_0) = y_0$ be a special solution. In order to overcome the problem of "loss of derivatives", we introduce a family of smoothing operators $S_t, t \in [1, \infty)$, which regularizes the function x, and satisfies $S_t \to Id$, as $t \to \infty$. Find a suitable function $g(t)$, which will be described later so that the curve $x(t)$ tends to a solution x_∞ of the equation $f(x) = y$. We intend to solve the ODE:
$$\begin{cases} \dot{x}(t) = f'(v(t))^{-1}g(t) \\ v(t) = S_t x(t) \\ x(0) = x_0 . \end{cases}$$

Since

$$\frac{d}{dt}f(x(t)) = f'(x(t))\dot{x}(t)$$
$$= (f'(x(t)) - f'(v(t)))\dot{x}(t) + g(t) ,$$

we should have

$$f(x_\infty) - f(x_0) = \int_0^\infty e(t)dt + \int_0^\infty g(t)dt ,$$

where $e(t) = (f'(x(t)) - f'(v(t)))\dot{x}(t)$.

Design the iteration scheme:

$$\begin{cases} \delta_n := x_{n+1} - x_n = \triangle_n f'(v_n)^{-1} g_n , \\ v_n = S_{\theta_n} x_n , \\ \triangle_n = \theta_{n+1} - \theta_n , \end{cases} \tag{1.51}$$

where $\theta_1 < \theta_2 < \ldots < \theta_n \to \infty$. Then

$$f(x_{n+1}) - f(x_n) = \triangle_n(e_n + g_n) ,$$

where

$$\triangle_n e_n = f(x_n + \delta_n) - f(x_n) - f'(x_n)\delta_n + (f'(x_n) - f'(v_n))\delta_n .$$

It follows that

$$f(x_{n+1}) - f(x_0) = \sum_{j=0}^{n} \triangle_j(g_j + e_j) .$$

Since we are only interested in the limiting result, sometimes we modify it to be:

$$\sum_{j=0}^{n} \triangle_j g_j + S_{\theta_n} E_n = S_{\theta_n}(y - f(x_0)) ,$$

where $E_n = \sum_{j=0}^{n-1} \triangle_j e_j$, so that g_n can be determined step by step.

$$g_n = \triangle_n^{-1}[(S_{\theta_n} - S_{\theta_{n-1}})(y - f(x_0) - E_{n-1}) - \triangle_{n-1} S_{\theta_n} e_{n-1}], \quad n = 1, 2, \ldots , \tag{1.52}$$

and

$$g_0 = \triangle_0^{-1} S_{\theta_0}(y - f(x_0)) .$$

We start with an abstract framework. $\{E_a\}_{a\geq 0}$ is called a family of Banach spaces with smoothing operators, if $E_b \hookrightarrow E_a$ for $b \geq a$ is an injection, and $\exists C = C(a, b)$ such that

$$\|u\|_a \leqslant C\|u\|_b .$$

Let $E_\infty = \cap_{a\geq 0} E_a$ be endowed with the weakest topology, such that $E_\infty \hookrightarrow E_a$ is continuous. Moreover, we assume that $\exists$ a family of linear operators $S_\theta : E_0 \to E_\infty$, depending on a parameter $\theta \geq 1$, such that

(1) $\|S_\theta u\|_b \leqslant C\|u\|_a \quad b \leqslant a,$
(2) $\|\frac{d}{d\theta}S_\theta u\|_b \leqslant C\theta^{b-a-1}\|u\|_a,$

where C is a constant depending on a and b.

The following inequalities hold:

(1) $\|S_\theta u\|_b \leqslant C\theta^{b-a}\|u\|_a, \forall\, a < b.$
(2) $\|(I - S_\theta)u\|_b \leqslant C\theta^{b-a}\|u\|_a \,\forall\, b < a.$
(3) (Interpolation inequality) $\|u\|_{\lambda a+(1-\lambda)b} \leqslant C\|u\|_a^\lambda \|u\|_b^{1-\lambda} \quad \forall\lambda \in (0,1)\ \forall\, a, b.$

Both (1) and (2) follow from the second inequality of the definition. In fact, if $a > b$, then

$$\begin{aligned}\|u - S_\theta u\|_b &= \left\|\int_\theta^\infty \frac{d}{dt}S_t u\right\|_b \\ &\leqslant C\int_\theta^\infty t^{b-a-1}\|u\|_a dt = C\theta^{b-a}\|u\|_a \quad \text{if } b < a\,.\end{aligned}$$

and if $a < b$, then

$$\begin{aligned}\|S_\theta u\|_b &= \left\|\int_1^\theta \frac{d}{dt}S_t u\right\|_b + \|S_1 u\|_b \\ &\leqslant C\int_1^\theta t^{b-a-1}dt\|u\|_a + C\|u\|_a \leqslant C\theta^{b-a}\|u\|_a\end{aligned}$$

where C denotes various constants.

(3) is a consequence of (1) and (2). In fact, may assume $a < b$. For $c = \lambda a + (1-\lambda)b$,

$$\begin{aligned}\|u\|_c &\leqslant \|S_\theta u\|_c + \|(I - S_\theta)u\|_c \\ &\leqslant C(\theta^{c-a}\|u\|_a + \theta^{c-b}\|u\|_b)\end{aligned}$$

By choosing $\theta = (\frac{C\|u\|_b}{\|u\|_a})^{\frac{1}{b-a}}$, we obtain the desired inequality.

Example. Hölder Spaces with Smoothing Operators

Let Ω be a bounded domain in R^n we write $H^\alpha(\overline{\Omega})$ as the Hölder space, defined as follows: $H^0(\overline{\Omega}) = C(\overline{\Omega})$. If $k \geq 0$ is an integer, $k < a \leqslant k+1$,

$$H^\alpha(\overline{\Omega}) = \begin{cases} C^\alpha(\overline{\Omega}) & \text{if } a < k+1 \\ C^{\alpha-0}(\overline{\Omega}) & \text{if } a = k+1\,. \end{cases}$$

The seminorm is defined to be

$$|u|_a = \sum_{|\alpha|=k} |\partial^\alpha u|_{a-k}\,,$$

and for $0 < a \leqslant 1$,

$$|u|_a = \sup_{x,y\in\overline{\Omega}} \frac{|u(x) - u(y)|}{|x - y|^\alpha} ,$$

and the norm is defined as:

$$\|u\|_a = \|u\|_C + |u|_a .$$

We now come to define the smoothing operators $S_\theta, \theta \geq 1$. For the sake of simplicity, we assume $\Omega = R^n$, the bounded domain case can be modified by standard argument. For a compact set K in $\mathbb{R}^n$, one chooses $\mathcal{X} \in C_0^\infty(\mathbb{R}^n)$ to be 1 in a neighborhood of K, a function $\psi \in C_0^\infty(\mathbb{R}^n)$ to be 1 in a neighborhood of 0, and let φ be the Fourier transform of $\psi, \varphi = \mathcal{F}\psi$, i.e.,

$$\varphi(x) = \int_{R^n} \exp(2\pi i x\xi)\psi(\xi)d\xi .$$

Let

$$\varphi_\theta(x) = \theta^n \varphi(\theta x) \ \theta \geq 1 .$$

If u has support in K, we define

$$S_\theta u = \mathcal{X}(\varphi_\theta * u) \in C_0^\infty(\mathbb{R}^n) .$$

Since $\varphi_\theta(x) = \mathcal{F}\psi(\xi/\theta)$, and $\psi(\frac{\xi}{\theta}) \to 1$ (in the distribution sense), we have $\varphi_\theta \to \delta$, so $S_\theta u \to u$, as $\theta \to +\infty$. The operators S_θ, which approximate the identity and map to smooth functions, are call smoothing operators.

Theorem 1.4.3 *The smoothing operators S_θ have the following properties, for $\theta > 1$ and $u \in H^\alpha$:*

(1) $\|S_\theta u\|_b \leqslant C\|u\|_a \quad b \leqslant a$,
(2) $\|\frac{d}{d\theta} S_\theta u\|_b \leqslant C\theta^{b-a-1}\|u\|_a$.

Proof. (1) following directly from the translation invariance and the convexity of the norms:

$$\|\varphi_\theta * u\|_a \leqslant \|\varphi\|_{L^1}\|u\|_a .$$

We verify (2). Noticing

$$\frac{d}{d\theta} S_\theta u = \mathcal{X} \cdot \mathcal{F}^{-1}(\theta^{-1}\psi_1\left(\frac{\xi}{\theta}\right) \cdot (\mathcal{F}u)(\xi))$$

where

$$\psi_1(\xi) = -\xi \cdot \nabla\psi(\xi) ,$$

and $\mathcal{F}^{-1}$ is the inverse Fourier transform. Again $\psi_1 \in C_0^\infty$, and vanishes in the neighborhood of the origin.

According to (1), (2) holds for $b = a$. We shall only need to verify (2) for $b = 0$ and $b = a + k$, $k \in \mathbb{N}$, because the remaining cases follow from the interpolation inequality.

For $b = a + k$,

$$\left\| \frac{d}{d\theta} S_\theta u \right\|_{a+k} = \sum_{|\alpha|=k} \left\| \partial^\alpha \left(\frac{d}{d\theta} S_\theta u \right) \right\|_a$$
$$= \sum_{|\alpha|=k} \left\| \mathcal{F}^{-1} \left(\theta^{k-1} \left(\frac{\xi^\alpha}{\theta^k} \right) \psi_1 \left(\frac{\xi}{\theta} \right) \cdot (\mathcal{F}u)(\xi) \right) \right\|_a$$
$$\leqslant C\theta^{k-1} \|u\|_a$$

because $\xi^\alpha \psi_1(\xi)$ is again in L^1.

For $b = 0$, we only need to prove for $a = k \in \mathbb{N}$. Let us write

$$\psi_1(\xi) = \sum_{|\alpha|=k} \xi^\alpha \psi_\alpha(\xi) \quad \text{where} \quad \psi_\alpha(\xi) = \frac{\xi^\alpha}{\sum_{|\alpha|=k} |\xi^\alpha|^2} \psi_1(\xi) ,$$

then $\psi_\alpha \in C_0^\infty$ and vanishes in the neighborhood of the origin:

$$\left\| \frac{d}{d\theta} S_\theta u \right\|_0 \leqslant \sum_{|\alpha|=k} \left\| \mathcal{F}^{-1} \left(\psi_\alpha \left(\frac{\xi}{\theta} \right) (\partial^\alpha u)(\xi) \right) \right\|_0 \theta^{-k-1}$$
$$\leqslant \theta^{-k-1} \sum_{|\alpha|=k} \left\| \int [\partial^\alpha u(x-y) - \partial^\alpha u(x)] (\psi_\alpha)_\theta(y) dy \right\|$$
$$= \theta^{-k-1} \cdot \int |y| |(\psi_\alpha)_\theta(y)| dy \|u\|_{k+1}$$
$$= C\theta^{-k-2} \|u\|_{k+1} \ .$$

The proof is complete. □

Now we are going to introduce a family of Banach spaces associated with E_a as follows: Let $\theta_j = 2^j$, $j = 0, 1, 2, \ldots$, $\triangle_j = \theta_{j+1} - \theta_j$, $j \geq 1, \triangle_0 = 1$:

$$R_0 = \triangle_0^{-1} S_{\theta_1},\ R_j = \triangle_j^{-1} (S_{\theta_{j+1}} - S_{\theta_j}),\ j \geq 1$$

According to (2),

$$\|R_j u\|_b \leqslant \triangle_j^{-1} \left\| \int_{\theta_j}^{\theta_{j+1}} \frac{d}{dt} S_t u \right\|_b \leqslant C \frac{\theta_{j+1}^{b-a} - \theta_j^{b-a}}{(b-a)\triangle_j} \|u\|_a \ .$$

We obtain:

(4) $\|R_j u\|_b \leqslant C_{ab} \theta_j^{b-a-1} \|u\|_a$,

and

(5) $\sum_{j=0}^\infty \triangle_j R_j u = u$.

The series is convergent in E_b if $u \in E_a, a > b$.

This is called the Paley–Littlewood decomposition.

Conversely, suppose we have a sequence $\{u_j\} \subset E_a,\ a \in [a_1, a_2]$, satisfying

$$\|u_j\|_{a_i} \leqslant C\theta_j^{a_i-a-1}, \quad i = 1, 2\ \forall j\ ;$$

then, by interpolation inequality,

(6) $\|u_j\|_b \leqslant C\theta_j^{b-a-1} \quad \forall b \in [a_1, a_2],\ \forall j,$

and then the series $\sum_j \triangle_j u_j$ converges in E_b if $b < a$.

Definition 1.4.4 *For $a \in [a_1, a_2]$, we define $E'_a = \{u = \sum_{j=0}^{\infty} \triangle_j u_j \mid \|u_j\|_{a_i} \leqslant C\theta_j^{a_i-a-1} \quad i = 1, 2,\ \forall j\}$ with norm:*

$$\|u\|'_a = \inf_{u=\sum_j \triangle_j u_j} \sup_{i,j} \theta_j^{-a_i+a+1} \|u_j\|_{a_i}\ .$$

Then E'_a is a family of Banach spaces. The following properties hold:

(7) $\|u\|_b \leqslant C\|u\|'_a \leqslant C_1\|u\|_a \quad$ if $b < a$.
To prove the first inequality. $\forall \epsilon > 0\, \forall u \in E'_a\ \exists \{u_j\} \subset E_{a_i}, i = 1, 2$ such that $u = \sum_j \triangle_j u_j$, and

$$\|u_j\|_{a_i} \leqslant C\theta_j^{a_i-a-1}(\|u\|'_a + \epsilon)\ .$$

From (6), $\|u_j\|_b \leqslant C\theta_j^{b-a-1}$. It follows that $\|u\|_b \leqslant C(\|u\|'_a + \epsilon)$.
The second inequality follows from (4) and (5).

(8) The space E'_a does not depend on $a_1,\ a_2$. It follows from the interpolation inequality directly.

(9) $\|(I - S_\theta)u\|_b \leqslant C\theta^{b-a}\|u\|'_a \quad$ for $b < a$. Indeed, $\forall u = E'_a$ one chooses $a_1 < a_2$ such that $b < a_1 < a < a_2$. $\forall \epsilon > 0$, $\exists \{u_j\} \subset E_{a_i}, i = 1, 2$, such that $u = \Sigma \triangle_j u_j$ and $\|u_j\|_{a_i} \leq C\theta_j^{a_i-a-1}(\|u\|'_a + \epsilon)$. Then

$$\|(I - S_\theta)u\|_b \leqslant \sum \triangle_j \|u_j - S_\theta u_j\|_b\ ,$$

and

$$\|(I - S_\theta)u_j\|_b \leqslant C\theta^{b-a_i}\|u_j\|_{a_i} \leqslant C\theta^{b-a_i}\theta_j^{a_i-a-1}(\|u\|'_a + \epsilon),\ i = 1, 2\ ,$$

therefore

$$\begin{aligned} \|(I - S_\theta)u\|_b &\leqslant C(\|u\|'_a + \epsilon) \\ &\times \left(\sum_{\theta_j > \theta} \triangle_j \theta^{b-a_1}\theta_j^{a_1-a-1} + \sum_{\theta_j < \theta} \triangle_j \theta^{b-a_2}\theta_j^{a_2-\theta-1} \right). \end{aligned}$$

Thus,

$$\|(I - S_\theta)u\|_b \leqslant CM\theta^{b-a}\|u\|'_a\ .$$

Theorem 1.4.5 *Suppose that $\{E_a\}$ and $\{F_a\}$ are two families of Banach spaces with smoothing operators, and that the embedding $F_b \hookrightarrow F_a$ is compact when $b > a$. Let $\alpha, \beta, r > 0$, and B_r^α be a ball with radius r centered at the origin in E'_α. Assume that $f : B_r^\alpha \to F_\beta$ is C^2 with $f(\theta) = \theta$, and satisfies*

(1) $f'(v)^{-1}$ exists for $v \in B_r^\alpha \cap E_\infty$, and $\forall g \in F_\infty$, the map $(v, g) \mapsto f'(v)^{-1}g$: $(E_\infty \cap B_r^\alpha) \times F_\infty \to E_{a_2}$ is continuous for some $a_2 > \alpha > a_1 \geqslant 0$, and satisfies

$$\|f'(v)^{-1}g\|_a \le C(\|g\|_{\beta+a-\alpha} + \|g\|_0\|v\|_{\beta+\alpha}) \quad \forall a \in [a_1, a_2]\ ,$$

(2) $\|f''(u)(v,w)\|_{\beta+\delta} \leqslant C\sum_{\max(l-\alpha,0)+\max(m,a_1)+n<2\alpha}(1+\|u\|_l)\|v\|_m\|w\|_n$,

for some $\delta > 0$. Then $\forall y \in F'_\beta$ with $\|y\|'_\beta$ small, $\exists u \in E'_\alpha$ satisfying

$$f(u) = y\ .$$

Proof. One uses the iteration scheme (1.51), so one should determine g first. As we have seen from above, after decomposition, it satisfies a series of recursive equations.

$\forall g \in F'_\beta$ we have the decomposition:

$$g = \sum_j \triangle_j g_j\ ,$$

with

$$\|g_j\|_b \leqslant C_b \theta_j^{b-\beta-1}\|g\|'_\beta\ . \tag{1.53}$$

Define

$$\begin{cases} u_{j+1} = u_j + \triangle_j f'(v_j)^{-1} g_j,\ u_0 = \theta\ , \\ v_j = S_{\theta_j} u_j\ , \\ \delta_j = \triangle_j f'(v_j)^{-1} g_j\ . \end{cases} \tag{1.54}$$

We are going to prove

$$\|f'(v_j)^{-1}g_j\|_a \leqslant C_1\|g\|'_\beta \theta_j^{a-\alpha-1} \quad a \in [a_1, a_2]\ , \tag{1.55}$$

$$\|v_j\|_a \leqslant C_2\|g\|'_\beta \theta_j^{a-\alpha} \quad a \in (\alpha, a_2]\ , \tag{1.56}$$

$$\|u_j - v_j\|_a \leqslant C_3\|g\|'_\beta \theta_j^{a-\alpha} \quad a \leqslant a_2\ , \tag{1.57}$$

inductively.

Suppose (1.56) and (1.57) are true for $j \leqslant k$ and (1.55) holds for $j < k$, we prove (1.55) for $j = k$. By the assumption (1) and (1.53):

$$\begin{aligned} \|f'(v_k)^{-1}g_k\|_a &\leqslant C(\|g_k\|_{\beta+a-\alpha} + \|g_k\|_0\|v_k\|_{\beta+a}) \\ &\leqslant C\big(\theta_k^{a-\alpha-1}\|g\|'_\beta + \theta_k^{-\beta-1}\|g\|'_\beta \cdot \theta_k^{\beta+a-\alpha}\|g\|'_\beta\big) \\ &\leqslant C\theta_k^{a-\alpha-1}\|g\|'_\beta\ , \end{aligned}$$

if $\|g\|'_\beta$ is small.

Now we prove (1.57) for $j = k+1$, from (9), in the case $a < \alpha$,

$$\begin{aligned}\|u_{k+1} - v_{k+1}\|_a &= \|u_{k+1} - S_{\theta_{k+1}} u_{k+1}\|_a \\ &\leqslant C\theta_{k+1}^{a-\alpha} \|u_{k+1}\|'_\alpha .\end{aligned}$$

Since

$$u_{k+1} = \sum_{j=0}^{k} \triangle_j f'(v_j)^{-1} g_j ,$$

and by the definition of F'_α norm, we have

$$\|u_{k+1}\|'_\alpha \leqslant C\|g\|'_\beta, \forall \alpha \in [a_1, a_2] .$$

Thus

$$\|u_{k+1} - v_{k+1}\|_a \leqslant C\theta_{k+1}^{a-\alpha} \|g\|'_\beta .$$

In the case $a = a_2$,

$$\begin{aligned}\|u_{k+1} - v_{k+1}\|_{a_2} &\leqslant C\|u_{k+1}\|_{a_2} \\ &\leqslant C \sum_{j=0}^{k} \|\triangle_j f'(v_j)^{-1} g_j\|_{a_2} \\ &\leqslant C\|g\|'_\beta \sum_{j=0}^{k} \triangle_j \theta_j^{a_2-\alpha-1} \\ &\leqslant C\theta_{k+1}^{a_2-\alpha} \|g\|'_\beta .\end{aligned}$$

The other cases, $\alpha \leqslant a \leqslant a_2$, are verified by the interpolation property. Finally, we prove (1.56) for $j = k+1$:

$$\begin{aligned}\|v_{k+1}\|_a &\leqslant \|u_{k+1}\|_a + \|v_{k+1} - u_{k+1}\|_a \\ &\leqslant 2C\|g\|'_\beta \theta_{k+1}^{a-\alpha} .\end{aligned}$$

Thus the construction of the sequence u_k is possible. And u_k, v_k are all in B_r^α if $\|g\|'_\beta$ is small.

Now

$$\begin{aligned}f(u_{j+1}) - f(u_j) &= (f(u_j + \delta_j) - f(u_j) - f'(u_j)\delta_j) \\ &\quad + (f'(u_j) - f'(v_j))\delta_j + \triangle_j g_j \\ &= \triangle_j (e'_j + e''_j + g_j) \qquad (1.58)\end{aligned}$$

where

$$e'_j = \triangle_j^{-1} (f(u_{j+1}) - f(u_j) - f'(u_j)(u_{j+1} - u_j)) ,$$

and

$$e_j'' = \int_0^1 f''(v_j + t(u_j - v_j))(\triangle_j^{-1}\delta_j, u_j - v_j)dt \,.$$

We obtain from the assumption (2), for $n < \alpha < l$,

$$\|e_j''\|_{\beta+\delta} \leqslant C\sum(1 + \|tu_j + (1-t)v_j\|_l)\|f'(v_j)^{-1}g_j\|_n\|u_j - v_j\|_n \quad (1.59)$$

$$\leqslant C\theta_j^{-1-\epsilon}\|g\|_\beta'^{\,2} \quad (1.60)$$

where $\epsilon = 3\alpha - l - 2n$. Similarly by Taylor's formula,

$$\|e_j'\|_{\beta+\delta} \leqslant \theta_j^{-1-\epsilon}\|g\|_{\beta'}'^{\,2} \quad (1.61)$$

Let $T(g) = \sum_j \triangle_j(e_j' + e_j'')$. Then $\|T(g)\|_{\beta+\delta}' \leqslant C\|g\|_\beta'^{\,2}$. According to (7), and the assumption on the compactness of embedding, $F_b \to F_a$ as $b > a$, we conclude that $T : F_\beta' \to F_\beta'$ is compact. According to the recursive formula (1.52), g is uniquely determined by y, therefore $I + T$ is locally injective. Now one can apply the Leray–Schauder invariance of domain theorem (see Chap. 3, Corollary 3.4.12) to conclude that $\forall y \in F_\beta'$ with small $\|y\|_\beta'$, there exists $g \in F_\beta'$ such that $(I + T)(g) = y$.

Substituting this g into the iteration scheme (1.54), u_n is convergent to some u in E_a', provided by (1.55), and (7). Again, by (1.58), $f(u) = (I + T)(g) = y$. □

2

Fixed-Point Theorems

It is well known that the contraction mapping theorem is one of the most important fixed-point theorems in analysis. It is based on the metric of the underlying space. It is simple and is strongly dependent on the chosen metric, but useful. In particular, the unique solution can be computed by iteration. In fact, the implicit function theorem and then the first three Sects. of Chap. 1 are based on it.

The Brouwer fixed theorem (1911), which says that every continuous self-mapping on a closed ball B^n has a fixed-point, is a fundamental fixed-point theorem in topology. It is based on the notion of retraction. Because of its importance, there are a lot of proofs, roughly speaking, divided into three classes according to methods: (1) Combinatorics (Sperner lemma) by which computing methods are developed. (2) Algebraic topology; the topological degree and other algebraic topological invariants are introduced. (3) Differential topology; the proofs are simple and beautiful (see Dunford and Schwartz [DS], Milnor [Mi 2] etc.). In the study of analysis, we need infinite-dimensional versions of this theorem. A new ingredient – the compactness – is added, while the ball is replaced by its topological equivalent – the convex set. The Schauder fixed-point theorem and its extensions are all based on convexity and compactness. They are widely used in combining with a priori estimates for solutions in differential equations.

In an ordered space, the Bourbaki–Kneser principle is another basic fixed-point theorem with applications in analysis, in case compactness is unavailable.

The chapter is organized as follows: The order method is studied in Sect. 2.1. Several fixed-point theorems based on the Bourbaki–Kneser principle are derived, by which, the sub- and super-solutions method in ordered Banach space are developed with applications in PDEs. The convexity-compactness method is developed in detail in Sect. 2.3. We start with the KKM map and the Ky Fan inequality. All other fixed-point theorems, including the Schauder fixed-point theorem and its generalizations, the Nash equilibrium, and the Von Neumann–Sion saddle point theorem, are derived as consequences. Various applications are studied, in particular, the Ky Fan fixed-point theorem for set valued mappings is applied to free boundary problems. Section 2.4 is devoted to the existence and iteration method for fixed points of nonexpansive maps, which are on the borderline of the contraction mappings. The prototype of the monotone mapping is the subdifferential of a convex function. Due to the special feature of monotonicity, the compactness requirement can be reduced considerably. Since monotone mappings map a Banach space into its dual, one studies the surjectivity of monotone mappings. The main results in this aspect are due to Minty [Min] and Browder [Bd 3,4]. Applications to variational inequalities and quasi-linear elliptic problems are studied as well. We shall show how the monotonicity is applied instead of the compactness in the existence theory. These are the contents of Sects. 2.5 and 2.6. Convex sets and convex functions will be used form time to time; we collect their most important properties in Sect. 2.2.

2.1 Order Method

A kind of nonlinear operator defined on ordered spaces is order preserving; this special feature makes the fixed-point problem easy to handle.

A set E is said to be ordered if a partial order $\leqslant$ is defined by the following axioms: $\forall x, y, z \in E$

(i) $x \leqslant x$,
(ii) $x \leqslant y$, and $y \leqslant x$ imply $x = y$,
(iii) $x \leqslant y$, and $y \leqslant z$ imply $x \leqslant z$.

A chain (or totally ordered set) E is an ordered set, on which $\forall x, y \in E$ either $x \leqslant y$ or $y \leqslant x$.

The following terminologies are introduced: The smallest (greatest) element $\underline{x}$ ($\overline{x}$ resp.): $\underline{x} \leqslant x$ ($x \leqslant \overline{x}$), $\forall x \in E$; the minimum (maximum) element x_* (x^* resp.): if $x \leqslant x_*$ ($x^* \leqslant x$), $x \in E$ then $x_* = x$ ($x^* = x$); the lower (upper) bound of a subset $F \subset E$: $\underline{a} \in E$ ($\overline{a} \in E$) and $\underline{a} \leqslant x$ ($x \leqslant \overline{a}$) $\forall x \in F$; the infimum (supremum) of F $\inf F$ ($\sup F$ respectively) is the greatest (smallest) element of the subset, which consists of all lower (upper) bounds of F.

$\forall a \in E$, we denote

$$S_+(a) = \{x \in E |\ a \leqslant x\} ,$$

and call it the right section of a. Similarly we define the left section $S_-(a)$. Set

$$[a, b] = S_+(a) \cap S_-(b)$$

for $a \leqslant b$, we call it an order interval.

Zorn's lemma, which is equivalent to the Zermelo selection axiom, is our starting point: In a nonempty ordered set $(E, \leq)$, if every chain has an upper bound in E, then the set has a maximum element.

Let $(E, \leqslant)$ and $(F, \leqslant)$ be two ordered sets. A map $f : E \to F$ is called (strict) order preserving, if $x \leqslant (<) y$ implies $f(x) \leqslant (<) f(y)$. Similarly, we define the (strict) order reversing map.

By definition, if $f : E \to F$ is order preserving, and $a \leqslant f(a)$ $(f(a) \leqslant a)$, then $f(S_\pm(a)) \subset S_\pm(a)$. Thus $a \leqslant f(a)$ and $f(b) \leqslant b$ imply $f([a,b]) \subset [a,b]$.

Theorem 2.1.1 *(Bourbaki–Kneser principle) Let $(E, \leqslant)$ be an ordered set, in which every chain has an upper bound. If $f : E \to E$ satisfies $x \leqslant f(x)$ $\forall x \in E$, then f has a fixed point.*

Proof. By Zorn's lemma, E has a maximum element a. Since $a \leqslant f(a)$, and a is maximal, $a = f(a)$. □

To a self-map f on E, an element $x \in E$, satisfying $x \leqslant f(x)$ $(f(x) \leqslant x)$, is called a sub-(or super-) solution of f.

The Bourbaki–Kneser fixed-point theorem is a general principle. Many important fixed point theorems can be derived from it.

Theorem 2.1.2 *(Caristi) Suppose that (E, ρ) is a complete metric space and that $\varphi : E \to \mathbb{R}^1$ is a (l.s.c.) function bounded from below. Assume $f : E \to E$ is a mapping satisfying*

$$\rho(x, f(x)) \leqslant \varphi(x) - \varphi(f(x)) , \tag{2.1}$$

then f has a fixed point.

Proof. (1) An order structure on E is introduced by φ:

$$\forall x, y \in E \quad x \leqslant y \quad \text{iff} \quad \rho(x,y) \leqslant \varphi(x) - \varphi(y) .$$

It is easy to verify that $(E, \leqslant)$ is an ordered set.

(2) It is sufficient to verify that every chain X has an upper bound. By definition, φ is a monotone decreasing function on X: $x \leqslant y \Longrightarrow \varphi(y) \leqslant \varphi(x)$ $\forall x, y \in X$. Let $c = \inf_X \varphi$; we can find an increasing sequence $x_1 \leqslant x_2 \leqslant \cdots \leq x_n \leq \cdots$ in X such that $\varphi(x_n) \to c$.

We conclude that there is at most one $a \in X$ such that $x_n \leqslant a$ $\forall n$. Indeed, if $\exists a_1, a_2 \in X$ satisfy $x_n \leqslant a_i, i = 1, 2$, then

$$c = \varphi(a_1) = \varphi(a_2), \text{ because } c \leqslant \varphi(a_i) \leqslant \varphi(x_n) \ \forall n .$$

But X is a chain, either $a_1 \leqslant a_2$ or $a_2 \leqslant a_1$, it follows that

$$\rho(a_1, a_2) \leqslant |\varphi(a_1) - \varphi(a_2)| = 0 ,$$

i.e., $a_1 = a_2$.

Noticing that $\{x_n\}$ is a Cauchy sequence:

$$\rho(x_{n+m}, x_n) \leqslant \varphi(x_n) - \varphi(x_{n+m}) \quad \forall n, \forall m ,$$

there exists a limit $b \in E$. We conclude that X has an upper bound. In fact, from the l.s.c. of φ, one has $x_n \leqslant b\ \forall n$. Now, $\forall x \in X$, if $\exists n \in \mathbb{N}$, such that $x \leqslant x_n$, then $x \leqslant b$, so b is an upper bound, otherwise, x itself is the unique upper bound in X.

(3) The condition (2.1) on f means $x \leqslant f(x)\ \forall x \in E$. Applying Theorem 2.1.1, f has a fixed point. □

It is worth noting that in the Caristi fixed-point theorem, there is no continuity assumption on the map f. It has many applications.

The following version of the Ekeland variational principle is often used.

Theorem 2.1.3 *(Ekeland) Suppose that* (E, ρ) *is a complete metric space and that* $\varphi : E \to \mathbb{R} \cup \{+\infty\}$ *is l.s.c. and bounded from below.* $\forall \varepsilon > 0$ *if* $x \in E$ *satisfies* $\varphi(x) \leqslant \inf_E \varphi + \varepsilon$*, then* $\forall \lambda > 0\ \exists y_\lambda \in E$ *satisfying*

$$\varphi(y_\lambda) \leqslant \varphi(x) \tag{2.2}$$

$$\rho(x, y_\lambda) \leqslant \frac{1}{\lambda} \tag{2.3}$$

$$\varphi(y) > \varphi(y_\lambda) - \varepsilon\lambda\rho(y, y_\lambda) \quad \forall y \in E \backslash \{y_\lambda\}\ . \tag{2.4}$$

Proof. Define a subset of E:

$$E_1 = \{y \in E |\ \varphi(y) + \lambda\varepsilon\rho(x, y) \leqslant \varphi(x)\}\ ,$$

E_1 is closed (φ is l.s.c.) and nonempty ($x \in E_1$), so is complete. We claim that $\exists y_\lambda \in E_1$ such that

$$\varphi(y) > \varphi(y_\lambda) - \lambda\varepsilon\rho(y, y_\lambda) \quad \forall y \in E_1 \backslash \{y_\lambda\}\ .$$

Indeed, if not, $\exists \lambda > 0, \forall y_\lambda \in E_1, \exists y \in E_1 \backslash \{y_\lambda\}$ such that

$$\varphi(y) \leqslant \varphi(y_\lambda) - \lambda\varepsilon\rho(y, y_\lambda)\ .$$

Define $f(y_\lambda) = y \neq y_\lambda$, then $f : E_1 \to E_1$ satisfying $\lambda\varepsilon\rho(y_\lambda, f(y_\lambda)) \leqslant \varphi(y_\lambda) - \varphi(f(y_\lambda))$. According to the Caristi theorem, f has a fixed point; this is a contradiction.

Since (2.2) and (2.3) are trivially true, it remains to verify (2.4) for $y \in E \backslash E_1$. In fact, if

$$\varphi(y) \leqslant \varphi(y_\lambda) - \lambda\varepsilon\rho(y, y_\lambda) \text{ and } y \in E \backslash E_1\ ,$$

then

$$\begin{aligned} \varphi(y) &\leqslant \varphi(x) - \lambda\varepsilon\rho(x, y_\lambda) - \lambda\varepsilon\rho(y, y_\lambda) \\ &\leqslant \varphi(x) - \lambda\varepsilon\rho(x, y)\ . \end{aligned}$$

It follows that $y \in E_1$. This is a contradiction. □

Theorem 2.1.4 *(Amann) Assume that every chain of $(E, \leqslant)$ has a supremum, and that $f : E \to E$ is order-preserving mapping. If f has a subsolution $a \leqslant f(a)$ then f has a least fixed point in $S_+(a)$.*

Proof. (1) We verify the existence of a fixed point. Define $E_+ = \{x \in E |\ x \leqslant f(x)\} \cap S_+(a)$. According to the Bourbaki–Kneser principle, we only want to show: (i) $E_+ \neq \emptyset$, (ii) $f(E_+) \subset E_+$, (iii) every chain of E_+ has an upper bound in E_+.

Indeed, $a \in E_+$; (i) follows.

Since f is order preserving, $f(E_+) \subset E_+$.

Let X be a chain of E_+. By assumption, it has a supremum $b \in E$. From $x \leqslant f(x) \leqslant f(b)\ \forall x \in X$, we have $b \leqslant f(b)$, and then $b \in E_+$, i.e., b is an upper bound of X in E_+.

(2) Let $\mathrm{Fix}_{S_+(a)}(f)$ be the fixed point set of f in $S_+(a)$. Define

$$G_+ = \{y \in E_+ |\ y \leqslant z\ \forall z \in \mathrm{Fix}_{S_+(a)}(f)\}\ .$$

We claim that $\mathrm{Fix}_{G_+}(f) \neq \emptyset$, then $\forall y_0 \in \mathrm{Fix}_{G_+}(f)$, y_0 is the least fixed point in $S_+(a)$. To this end, from the Bourbaki–Kneser principle, it is sufficient to verify: (i) $G_+ \neq \emptyset$, (ii) $f(G_+) \subset G_+$, (iii) every chain of G_+ has an upper bound in G_+. In fact, $a \in G_+$; (i) follows. $\forall y \in G_+, \forall z \in \mathrm{Fix}_{S_+(a)}(f), f(y) \leqslant f(z) = z$, which implies $f(y) \in G_+$, and then (ii) follows.

Let X be a chain of G_+, by assumption, it has a supremum $b \in E$, so is $b \in E_+$ as in (1) and $b \leqslant z\ \forall z \in \mathrm{Fix}_{S_+(a)}(f)$. i.e., $b \in G_+$. This is an upper bound of X in G_+. □

An ordered set is called chain complete if every chain has an infimum and a supremum.

Corollary 2.1.5 *Let E be a chain complete ordered set, and let $f : E \to E$ be order preserving. Suppose that $\exists$ a pair of sub- and super- solutions $\underline{a} \leqslant \overline{a}$. Then f has at least a least and a greatest fixed point in $[\underline{a}, \overline{a}]$.*

An ordered set E is called a lattice, if for every pair $x, y \in E$, $x \vee y = \sup\{x, y\}$ and $x \wedge y = \inf\{x, y\}$ are in E.

A lattice is called complete if every nonempty subset possesses an infimum and a supremum.

Corollary 2.1.6 *(Birkhoff–Tarski) Let E be a complete lattice and let f be an order-preserving self-map on E, then there exist a smallest and a greatest fixed point of f.*

Proof. Let $\underline{a} = \inf E$ and $\overline{a} = \sup E$, then $\underline{a}$ and $\overline{a}$ are sub- and super- solutions of f respectively, satisfying $\underline{a} \leqslant \overline{a}$. □

We study chain complete ordered sets.

Let X be a Banach space, and let $\mathbb{P} \subset X$ be a nonempty closed convex positive cone with $\mathbb{P} \cap (-\mathbb{P}) = \{\theta\}$. It reduces a partial order structure on X:

$$x \leqslant y \text{ if and only if } y - x \in \mathbb{P}.$$

The order defined above matches the linear structure and the topology on X, but not necessarily the magnitude of the norm.

$$x_i \leqslant y_i, i = 1, 2 \Longrightarrow x_1 + x_2 \leqslant y_1 + y_2 ,$$
$$x \leqslant y, \lambda \geqslant 0 \Longrightarrow \lambda x \leqslant \lambda y ,$$

and

$$x_n \leqslant y_n, x_n \to x, y_n \to y \Longrightarrow x \leqslant y .$$

A Banach space X with a closed convex positive cone $\mathbb{P}$, induces an ordered Banach space (OBS) $(X, \mathbb{P})$. In particular $\forall a \in X,\ S_\pm(a)$ is closed.

Lemma 2.1.7 *Every compact subset E of an OBS is chain complete.*

Proof. For every chain $X \subset E$, we want to show that it has a supremum. In fact $\forall x \in X$, the family of sets: $\{S_+(x) \,|\, x \in X\}$ is a family of closed subsets with finite intersection property, i.e., $\forall x_1, x_2, \ldots, x_n \in X,\ \bigcap_{i=1}^{n} S_+(x_i) = S_+(x^*)$, where $x^* = \sup\{x_1, \ldots, x_n\} \in X$, provided that X is a chain. Since $\overline{X}$ is compact, $\bigcap_{x \in X} S_+(x) \cap \overline{X} \neq \emptyset$. Let $\overline{x}$ be an element in the intersection. Then (1) $x \leqslant \overline{x}\ \forall x \in X$, i.e., $\overline{x}$ is an upper bound of X, and (2) $\forall$ upper bound d of X, from $x \leq d$ and $\bar{x} \in \overline{X}$, it follows $\bar{x} \leqslant d$. Thus $\bar{x}$ is a supremum. Similarly, one proves that it has an infimum. □

A positive cone $\mathbb{P}$ is called normal if every ordered interval $[a, b]$ in X is bounded in the norm.

Lemma 2.1.8 *$\mathbb{P}$ is normal if and only if $\exists$ a constant $C > 0$ such that $0 \leqslant x \leqslant y \Longrightarrow \| x \| \leqslant C \| y \|$.*

Proof. "$\Longrightarrow$" If not, $\exists \theta \leqslant x_n \leqslant y_n$ such that $\| x_n \| \geqslant n^3 \| y_n \|$ $n = 1, 2, \ldots$. Let $z_n = \frac{x_n}{n^2 \|y_n\|}$, then $\| z_n \| \geqslant n$. But $\theta \leqslant z_n \leqslant \frac{y_n}{n^2 \|y_n\|}$. Define $y = \sum_{n=1}^{\infty} \frac{1}{n^2} \frac{y_n}{\|y_n\|}$; we have $\theta \leqslant z_n \leqslant y$, i.e., $z_n \in [\theta, y]$. But z_n is unbounded. A contradiction.

"$\Longleftarrow$" $\forall x \in [a, b]$, we have $\theta \leqslant x - a \leqslant b - a$. Since

$$\| x - a \| \leqslant C \| b - a \| ,$$

it follows that

$$\| x \| \leqslant \| x - a \| + \| a \| \leqslant \| a \| + C \| b - a \| .$$

□

Corollary 2.1.9 *Suppose that $(X, \mathbb{P})$ is a normal OBS, and that $D = [a, b]$ is an order interval. If f is an order-preserving compact self-mapping on D, then it has a smallest and a greatest fixed point.*

Proof. By normality, D is bounded in the norm. Let $E = f(D)$. E is a compact subset of an OBS, provided by the compactness of f. Since f is a self-mapping on D, $a \le f(a), f(b) \le b$, so $a < b$ is a pair of sub- and super-solutions of f. The conclusion follows from Corollary 2.1.5 and Lemma 2.1.7. □

Noticing that a bounded closed convex set of a reflexive Banach space is weakly compact, when we make use of the weak topology, Corollary 2.1.9 can be modified as follows:

Corollary 2.1.10 *Suppose that $(X, \mathbb{P})$ is a reflexive normal OBS, and that $D = [a, b]$ is an order interval. If f is an order-preserving self-mapping on D, then f has a smallest and a greatest fixed point.*

Example 1. Let M be a compact space, and let $X = C(M)$ with norm

$$\| x \| = \max_{\xi \in M} |x(\xi)| .$$

Define $P = \{x \in X | \; x(\xi) \geqslant 0\}$. Then P is normal and (X, P) is an OBS.

Example 2. Let $(\Omega, \mathcal{B}, \mu)$ be a measure space and let $L^p(\Omega, \mathcal{B}, \mu)$, $1 \leqslant p \leqslant \infty$ be the L^p space over $(\Omega, \mathcal{B}, \mu)$. Define $P = \{x \in L^p(\Omega, \mathcal{B}, \mu) | \; x(\xi) \geqslant 0 \text{ a.e. } \xi \in \Omega\}$. Then (X, P) is an OBS, and P is normal.

Example 3. Let $X = C^1([0, 1])$ with norm

$$\| x \| = \max_{t \in [0,1]} |x(t)| + \max_{t \in [0,1]} |x'(t)| ,$$

and $P = \{x \in X | \; x(t) \geqslant 0 \; \forall t \in [0, 1]\}$. Then (X, P) is an OBS, but P is not normal.

Corollaries 2.1.9 and 2.1.10 are the foundation of the sub- and super-solutions method in the theory of differential equations.

Example 4. Consider the following semi-linear elliptic BVP:

$$\begin{cases} -\triangle u(x) = \varphi(x, u(x)) & \text{in } \Omega , \\ u|_{\partial\Omega} = 0 \end{cases} \tag{2.5}$$

where Ω is a bounded domain with smooth boundary in $\mathbb{R}^n$, and $\varphi : \overline{\Omega} \times \mathbb{R}^1 \to \mathbb{R}^1$ is continuous. Assume that $\forall x \in \overline{\Omega}$, the function $t \mapsto \varphi(x, t)$ is nondecreasing. Define $\mathbb{K} = (-\triangle)^{-1}$ with 0-Dirichlet condition; it is known that $\mathbb{K}$ is a positive linear operator from the maximum principle. Let F be the mapping $u(x) \mapsto \varphi(x, u(x))$. Since φ is nondecreasing in u, the map F is order preserving.

Setting $X = C_0(\overline{\Omega})$, we consider the following mappings:

$$C_0(\overline{\Omega}) \xrightarrow{F} C_0(\overline{\Omega}) \xrightarrow{i} L^p(\Omega) \xrightarrow{\mathbb{K}} W_p^2 \cap \overset{\circ}{W}{}_p^1(\Omega) \xrightarrow{j} C_0(\overline{\Omega}) ,$$

where i and j are injections, and $p > \frac{n}{2}$. The later is compact, provided by the Sobolev embedding theorem. Define $f = j \circ \mathbb{K} \circ i \circ F : C_0(\overline{\Omega}) \to C_0(\overline{\Omega})$; the problem (2.5) is equivalent to finding the fixed point of f. Since f is compact and order preserving, it has a fixed point if one can find a sub-solution $\underline{u}$ and super-solution $\overline{u}$ of (2.5) in $W_p^2 \cap \overset{\circ}{W}{}_p^1(\Omega)$. i.e., $\underline{u} \leqslant \overline{u}$ satisfying

$$-\triangle \underline{u} \leqslant \varphi(x, \underline{u}(x)) \text{ and } -\triangle \overline{u} \geqslant \varphi(x, \overline{u}(x)) \text{ a.e. in } \Omega .$$

Namely, we have proved:

Statement 2.1.11 *Suppose that $\varphi : \Omega \times \mathbb{R}^1 \to \mathbb{R}^1$ is continuous and $\forall x \in \Omega$, $t \mapsto \varphi(x,t)$ is nondecreasing. If there is a pair of sub- and super-solutions $\underline{u} \leqslant \overline{u}$ in $W_p^2 \cap \overset{\circ}{W}{}_p^1(\Omega)$, $\frac{n}{2} < p < \infty$, then equation (2.5) has the smallest and the greatest solutions in the order interval $[\underline{u}, \overline{u}] \subset W_p^2 \cap \overset{\circ}{W}{}_p^1(\Omega)$.*

Remark 2.1.12 *The nondecreasing condition on φ can be weakened as follows: $\exists \omega \geqslant 0$, such that $\forall x \in \overline{\Omega}$, the function:*

$$\varphi_\omega(x,t) = \varphi(x,t) + \omega t$$

is nondecreasing in t.

Indeed, instead of $\mathbb{K}$ and F, we introduce the linear positive operator $\mathbb{K}_\omega = (-\triangle + \omega I)^{-1}$ and the order-preserving mapping $F_\omega : u(x) \mapsto \varphi_\omega(x, u(x))$. Then we find the fixed point of $f_\omega = j \circ \mathbb{K}_\omega \circ i \circ F_\omega$.

Moreover, the continuity condition on φ can be replaced by the Caratheodory condition. In this case, we consider the map $f : L^p(\Omega) \to L^p(\Omega)$, with $f = i \circ j \circ \circ K \circ F$ instead, where F is the Nemytski operator.

Remark 2.1.13 *The sub- and super-solutions method has an advantage in numerical analysis, because one can get the smallest and the greatest solutions in the order interval $[\underline{u}, \overline{u}]$ by iteration. Setting*

$$\begin{aligned} &\underline{u}_{i+1} = f(\underline{u}_i), \overline{u}_{i+1} = f(\overline{u}_i) \quad i = 0, 1, 2, \ldots , \\ &\underline{u}_0 = \underline{u}, \qquad \overline{u}_0 = \overline{u} . \end{aligned}$$

Then one has

$$\underline{u}_0 \leqslant \underline{u}_1 \leqslant \cdots \leqslant \underline{u}_n \leqslant \overline{u}_n \leqslant \cdots \leqslant \overline{u}_1 \leqslant \overline{u}_0 .$$

The sequences $\{\underline{u}_n\}$ and $\overline{u}_n$ converge to the least solution u_* and the greatest solution u^* respectively. If $u_* = u^*$, then this is the unique solution in $[\underline{u}, \overline{u}]$; and the two sequences provide lower and upper bounds estimates for $u_* = u^*$.

Remark 2.1.14 *In contrast to Corollaries 2.1.9 and 2.1.10, we consider an order-reversing compact mapping f on a normal OBS $(X, \mathbb{P})$. Let $a \in X$ be a* *sub- (or super-) solution of both f and f^2. Setting $x_{i+1} = f(x_i)$, $i = 0, 1, 2, \ldots$, and $x_0 = a$, we have either*

$$x_0 \leqslant x_2 \leqslant x_4 \leqslant \cdots \quad \cdots \leqslant x_3 \leqslant x_1 \quad \text{if } a \leqslant f^2(a) \leqslant f(a) \, ,$$

or

$$x_1 \leqslant x_3 \leqslant x_5 \leqslant \cdots \quad \cdots \leqslant x_2 \leqslant x_0 \quad \text{if } f(a) \leqslant f^2(a) \leqslant a, \text{respectively} \, .$$

Thus

$$x_{2n} \nearrow \underline{x} \text{ and } x_{2n+1} \searrow \overline{x} \quad \text{if } a \leqslant f(a) \, ,$$

$$(x_{2n+1} \nearrow \underline{x} \text{ and } x_{2n} \searrow \overline{x} \quad \text{if } f(a) \leqslant a \text{ respectively.})$$

If the fixed-point set of f, $\mathrm{Fix}(f) \cap S_+(a)$ (or $\mathrm{Fix}(f) \cap S_-(a)$) is not empty, then it is contained in $[\underline{x}, \overline{x}]$.

Example 5. Let (M, g) be a compact Riemannian manifold without boundary, and let

$$\Delta_M = \frac{1}{\sqrt{\det(g)}} \sum_{i,j=1}^{2} \partial_i (g^{ij} \sqrt{\det(g)} \partial_j)$$

be the Laplace–Beltrami operator with respect to g. Assume the Gaussian curvature $k(x) < 0 \ \forall x \in M$. Given a function $K(x) < 0$ on M, we consider the following equation:

$$\triangle_M u(x) - k(x) + K(x) e^{2u(x)} = 0 \quad x \in M \, . \tag{2.6}$$

This equation arises from geometry. The solution u defines a conformal metric $\widetilde{g} = e^{2u} g$. Under the new metric $\widetilde{g}$, $K(x)$ is the Gaussian curvature of M.

We construct a pair of sub- and super-solutions: Choose $C > 0$ sufficiently large such that

$$-k(x) + K(x) e^{2C} \leqslant 0 \leqslant -k(x) + K(x) e^{-2C} \quad \forall x \in M \, .$$

Then $(-C, C)$ is a pair of sub- and super solutions.

Statement 2.1.15 *Suppose that $K, k \in C(M)$ satisfy $K(x) < 0$, $k(x) < 0 \ \forall x \in M$, then equation (2.6) has a unique solution.*

Proof. Since $\varphi(x, u) = -k(x) + K(x) e^{2u}$ is decreasing in u, one cannot apply Statement 2.1.11 directly. Assume $K = \max_{x \in M}(-K(x))$ and set $\omega = 2K e^{2C}$, then the function $t \mapsto \omega t + K(x) e^{2t}$ is nondecreasing. After Remark 2.1.13, Statement 2.1.11 is applied, and we conclude that there is a smallest and a greatest solution of (2.6) in the order interval $[-C, C]$ of $C(M)$ for any sufficiently large constant $C > 0$.

However, since $K(x) < 0$ and e^{2u} is increasing in u, from the maximum principle, the solution is unique. References can be found in Kazdan [Ka], Kazdan and Warner [KW 1][KW 2] and Aubin [Au 1]. □

Example 6. Consider the following nonlinear heat equation:

$$\begin{cases} \frac{\partial u}{\partial t} - \triangle u = \lambda u - a u^k & \text{in } \Omega \times \mathbb{R}^1_+ \\ u = 0 & \text{on } \partial\Omega \times \mathbb{R}^1_+ \\ u(\cdot, 0) = \varphi & \text{on } \Omega \times \{0\} \end{cases} \tag{2.7}$$

where $\lambda > 0, a > 1, k > 1$ and $\varphi \in W^2_p(\Omega)$ is nonnegative, for some $p > \frac{n}{2}$. Notice that one can find a large constant $M > 0$ such that $\lambda M - a M^k < 0$ and $\max \varphi \leqslant M$. Since $\exists \omega > 0$ such that $(\lambda + \omega)u - a u^k$ is increasing in the interval $[0, M]$, one can use the sub- and super-solution method to prove the existence of a positive solution of (2.7).

In fact, $\forall T > 0$ define a mapping $A : g \mapsto v = Ag$, $L^p(\Omega \times [0, T]) \to W^{2,1}_p(\Omega \times [0, T]) \hookrightarrow C(\overline{\Omega} \times [0, T])$ as follows:

$$\begin{cases} \frac{\partial v}{\partial t} - \triangle v + \omega v = g & \text{in } \Omega \times [0, T] \\ v = 0 & \text{on } \partial\Omega \times [0, T] \\ v(\cdot, 0) = \varphi & \text{on } \Omega \times \{0\} . \end{cases}$$

From the maximum principle for heat equations, A is order preserving. Define $F(u) = (\lambda + \omega)u - a u^k$; again F is order-preserving in the order interval $D = [\theta, M]$ of the OBS $C(\overline{\Omega} \times [0, T])$. Setting $f = A \circ F \circ i$, where i is the injection $C(\overline{\Omega} \times [0, T]) \to L^p(\Omega \times [0, T])$, f is a compact order-preserving map on D. Now $\theta \leqslant f(\theta)$, and $f(M) \leqslant M$. From the parabolic maximum principle, we have a fixed point $u = f(u)$ in $[\theta, M]$. This is a solution of (2.7).

The sub- and super-solutions method is extensively used in the study of nonlinear differential equations. The analytic basis is the maximum principle. As to systems, the method is also applicable if the associated maximum principle is extended to the linearized operator. However, the construction of a pair of sub- and super-solutions is technical, but crucial. It depends on the special feature of the nonlinear term.

2.2 Convex Function and Its Subdifferentials

2.2.1 Convex Functions

Let X be a vector space, a function $f : X \to \mathbb{R} \cup \{+\infty\}$ is convex, if $\forall x, y \in X$, $\forall \lambda \in [0, 1]$, we have

$$f(\lambda x + (1 - \lambda)y) \leqslant \lambda f(x) + (1 - \lambda) f(y) .$$

It is called concave, if $-f$ is convex. f is called affine if f is both convex and concave. The domain, the epigraph and the level set of f are defined as follows:

$$\begin{aligned} \mathrm{dom}(f) &= \{x \in X |\ f(x) < +\infty\}\,, \\ \mathrm{epi}(f) &= \{(x,\lambda) \in X \times \mathbb{R}^1 |\ f(x) \leqslant \lambda\}\,, \\ f_a &= \{x \in X |\ f(x) \leqslant a\} \quad \forall a \in \mathbb{R}^1\,. \end{aligned}$$

f is said to be proper if $f \not\equiv +\infty$.

By definition,

$$f \text{ is convex} \Longleftrightarrow \forall x_1, \ldots, x_n \in X, \forall \lambda_1, \ldots, \lambda_n \geqslant 0, \sum_{i=1}^n \lambda_i = 1\,,$$

$$f\left(\sum_{i=1}^n \lambda_i x_i\right) \leqslant \sum_{i=1}^n \lambda_i f(x_i) \Longleftrightarrow \mathrm{epi}(f) \text{ is convex}\,.$$

Also f is convex $\Longrightarrow \forall a \in \mathbb{R}^1$, f_a is convex, but the converse is not true.

A function f is called quasi convex (quasi concave) if f_a is convex (concave resp.) $\forall a \in \mathbb{R}^1$.

For a subset C of X, the indicator function is defined to be

$$\chi_C(x) = \begin{cases} 0 & \text{if } x \in C \\ +\infty & \text{if } x \notin C\,. \end{cases}$$

Thus

$$C \text{ is convex} \Longleftrightarrow \chi_C \text{ is a convex function.}$$

The following simple propositions hold:

(1) If f and g are convex, then $\forall \alpha, \beta \geqslant 0$ $\alpha f + \beta g$ is convex.
(2) If A is a linear mapping: $X \to Y$ and if $f : Y \to \mathbb{R} \cup \{+\infty\}$ is convex, then $f \circ A : X \to \mathbb{R} \cup \{+\infty\}$ is convex.
(3) If $\{f_\iota |\ \iota \in \wedge\}$ is a set of convex functions, then $\sup\{f_\iota |\ \iota \in \wedge\}$ is convex.

We study the continuity of convex functions on Banach spaces.

Theorem 2.2.1 *Suppose that $f : X \to \mathbb{R} \cup \{+\infty\}$ is a convex function on a Banach space X. If f is bounded from above in a neighborhood of a point x, then f is continuous at x. Moreover, f is locally Lipschitzian in* $\mathrm{int}(\mathrm{dom}(f))$.

Proof. (1) We may assume $x = \theta$ and $f(\theta) = 0$. From the assumption, $\exists \eta > 0, \exists M > 0$ such that

$$f(y) \leqslant M \quad \forall y \in B_\eta(\theta)\,. \tag{2.8}$$

On one hand, $\forall z \in B_\eta(\theta) \backslash \{\theta\}$, we have

$$f(z) = f\left(\frac{\| z \|}{\eta}\left(\frac{\eta z}{\| z \|}\right)\right) \leqslant \frac{\| z \|}{\eta} f\left(\frac{\eta z}{\| z \|}\right) \leqslant \frac{M}{\eta} \| z \|\,. \tag{2.9}$$

One the other hand, $\forall z \in X$

$$0 = f(\theta) = f\left(\frac{\eta}{\eta + \| z \|} z + \frac{\| z \|}{\eta + \| z \|}\left(-\frac{\eta}{\| z \|} z\right)\right)$$
$$\leqslant \frac{\eta}{\eta + \| z \|} f(z) + \frac{M \| z \|}{\eta + \| z \|} ,$$

It follows that

$$f(z) \geqslant -\frac{M}{\eta} \| z \| . \tag{2.10}$$

Combining (2.9) with (2.10), we obtain

$$|f(z)| \leqslant \frac{M}{\eta} \| z \| \quad \forall z \in B_\eta(\theta) .$$

(2) To prove the second assertion, we first show that $\forall y \in \mathrm{int}(\mathrm{dom}(f))$, f is bounded above in a neighborhood of y, i.e., $\exists \eta_1 > 0, \exists M_1 > 0$ such that

$$f(z) \leqslant M_1 \quad \forall z \in B_{\eta_1}(y) . \tag{2.11}$$

To this end one chooses $w = (1+t)y$ such that the segment $\overline{yw} \subset \mathrm{int}(\mathrm{dom}(f))$. Setting $\eta_1 = \frac{t}{1+t}\eta$ and $M_1 = \max\{M, f(w)\}$, we have $f(w) \leqslant M_1$, and

$$\left\| \frac{1+t}{t}(z-y) \right\| \leqslant \eta ,$$

if $z \in B_{\eta_1}(y)$. Since

$$z = \frac{t}{1+t}\left(\frac{1+t}{t}(z-y)\right) + \frac{w}{1+t}$$

it follows from the convexity that

$$f(z) \leqslant \frac{t}{1+t} f\left(\frac{1+t}{t}(z-y)\right) + \frac{1}{1+t} f(w) \leqslant M_1 .$$

Thus (2.11) holds.

(3) We turn to proving that f is locally Lipschitzian in $\mathrm{int}(\mathrm{dom}(f))$. In fact, after (2), we only want to show that

$$|f(z) - f(z')| \leqslant \frac{M}{\eta - \delta} \| z - z' \| \quad \forall z, z' \in B_\delta(\theta), 0 < \delta < \eta .$$

Dividing the segment $\overline{zz'}$ into n equal length subsegments $\overline{y_i y_{i+1}}, i = 1, \ldots, n$ with $y_1 = z, y_{n+1} = z'$, and $n > \frac{\|z-z'\|}{\eta-\delta}$, we have $y_{i+1} \in B_{\eta-\delta}(y_i) \subset B_\eta(\theta)$. Applying (1), it follows that

$$|f(y_{i+1}) - f(y_i)| \leqslant \frac{M}{\eta - \delta} \| y_{i+1} - y_i \| \quad i = 1, 2, \ldots, n .$$

Then

$$|f(z) - f(z')| \leqslant \frac{M}{\eta - \delta} \| z - z' \| .$$

We also need to study lower semi-continuity of functions defined on a topological space X. □

Definition 2.2.2 *A function $f : X \to \mathbb{R}^1 \cup \{+\infty\}$ is said to be lower semi-continuous (l.s.c.), if $\forall\lambda \in \mathbb{R}^1$, the level set $f_\lambda = \{x \in X | \, f(x) \leqslant \lambda\}$ is closed. It is called sequentially lower semi-continuous (s.l.s.c.), if for any sequence $\{x_n\}$, with $x_n \to x \in X$, we have*

$$\underline{\lim} f(x_n) \geqslant f(x) \, .$$

From the definition, it follows directly that

(1) f is l.s.c. $\Longleftrightarrow$ epi(f) is closed in $X \times (\mathbb{R}^1 \cup \{+\infty\})$.
(2) If $\{f_\alpha | \alpha \in \wedge\}$ is a family of l.s.c. functions. Then

$$f(x) = \sup \{f_\alpha(x) | \alpha \in \wedge\}$$

is l.s.c.

(3) If f, g are l.s.c. and $\lambda, \mu \geqslant 0$, then $\lambda f + \mu g$ is l.s.c.

Theorem 2.2.3 *Let X be a Banach space and let $f : X \to \mathbb{R}^1 \cup \{+\infty\}$ be quasi convex. Then*

(1) f is l.s.c. $\Longleftrightarrow$ f is weakly lower semi-continuous (i.e., l.s.c. in weak topology).
(2) If X^* is separable, f is weakly l.s.c. $\Longleftrightarrow$ f is sequentially weakly l.s.c.
(3) If $X = Y^*$, where Y is a separable B-space, then f is w^* l.s.c. $\Longleftrightarrow$ f is sequentially w^* l.s.c.

Proof. (1) "$\Longleftarrow$" is trivial; we prove "$\Longrightarrow$". Since $\forall\lambda \in R^1$, f_λ is convex and closed. By the Hahn–Banach theorem, it is also weakly closed, i.e., f is weakly l.s.c.

(2) In this case, the weak topolopy restricted to a normed bounded set is metrizable.

(3) In this case, the w*-topology restricted to a normed bounded set is metrizable. □

We collect the following theorems on weak compactness, w*-compactness, weak sequential compactness and w*-sequential compactness for further reference.

Theorem 2.2.4 *(Banach–Alaoglu) Let X be a Banach space, and let $E \subset X^*$. Then E is w^*-compact if and only if E is normed bounded and w^*-closed.*

In particular, if further, X is reflexive, then for any $E \subset X$, E is weakly compact if and only if E is normed bounded and weakly closed.

See for instance Larsen [La], pp. 254–257.

Theorem 2.2.5 *Let E be a weakly closed set in a Banach space, then it is weakly compact if and only if it is weakly sequentially compact.*

See for instance Larsen [La], pp. 303–309.

2.2.2 Subdifferentials

Convex functions on Banach spaces are not differentiable in general. However, the notion of subdifferentials is introduced as a replacement of derivatives of differentiable functions.

Definition 2.2.6 *(Subdifferential) Let $f : X \to \mathbb{R} \cup \{+\infty\}$ be a convex function on a vector space X. $\forall x_0 \in \mathrm{dom}(f)$, $x^* \in X^*$ is called a subgradient of f at x_0 if*

$$\langle x^*, x - x_0 \rangle + f(x_0) \leqslant f(x) \quad \forall x \in X .$$

The set of all subgradients at x_0 is called the subdifferential of f at x_0, and is denoted by $\partial f(x_0)$.

Geometrically, $x^* \in \partial f(x_0)$ if and only if the hyperplane

$$y = \langle x^*, x - x_0 \rangle + f(x_0)$$

lies below the epigraph of f, i.e., it is a support of epi(f).

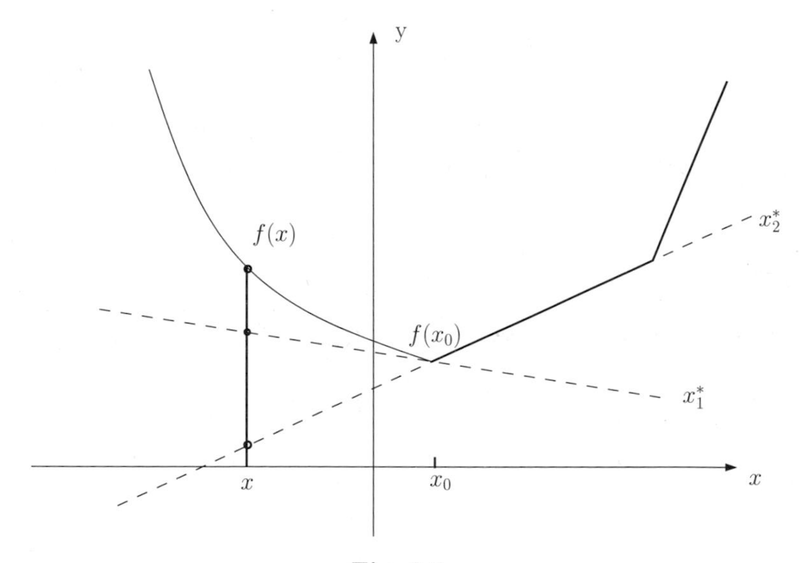

Fig. 2.1.

Obviously, $\partial f(x_0)$ may contain more than one point. The following propositions hold, if X is a Banach space.

(1) $\partial f(x_0)$ is a w*-closed convex set.
(2) If $x_0 \in \mathrm{int}(\mathrm{dom}(f))$, then $\partial f(x_0) \neq \emptyset$.

Proof. We apply the Hahn–Banach separation theorem to the convex set epi(f): $\exists (x^*, \lambda) \in X^* \times \mathbb{R}^1 \backslash \{(\theta, 0)\}$ such that

$$\langle x^*, x_0 \rangle + \lambda f(x_0) \geqslant \langle x^*, x \rangle + \lambda t \quad \forall (x,t) \in \mathrm{epi}(f) \; .$$

Since $(x_0, f(x_0)+1) \in \mathrm{epi}(f)$, it follows that $\lambda \leqslant 0$. However, $\lambda \neq 0$. Otherwise, we would have $\langle x^*, x - x_0 \rangle \leqslant 0 \; \forall x \in \mathrm{dom}(f)$. From $x_0 \in \mathrm{int}(\mathrm{dom}(f))$, we conclude that $x^* = \theta$. It contradicts with $(x^*, \lambda) \neq (\theta, 0)$. Setting $x_0^* = \frac{1}{-\lambda} x^*$, we obtain $x_0^* \in \partial f(x_0)$. □

(3) $\forall \lambda \geqslant 0 \; \partial(\lambda f)(x_0) = \lambda \partial f(x_0)$.
(4) If $f, g : X \to \mathbb{R}^1 \cup \{+\infty\}$ are convex, then $\forall x_0 \in \mathrm{int}(\mathrm{dom}(f) \cap \mathrm{dom}(g))$

$$\partial(f+g)(x_0) = \partial f(x_0) + \partial g(x_0) \; .$$

Proof. "$\supset$" is trivial; we are going to prove "$\subset$". We may assume $x_0 = \theta$, $f(\theta) = g(\theta) = 0$, and $\theta \in \partial(f+g)(\theta)$. We want to show that $\exists x_0^* \in \partial f(\theta)$ such that $-x_0^* \in \partial g(\theta)$. Since the set $C = \{(x,t) \in \mathrm{dom}(g) \times \mathbb{R}^1 | \; t \leqslant -g(x)\}$ is convex, and $C \cap \mathrm{int}(\mathrm{epi}(f)) = (\theta, 0)$, from the fact that $f(x) + g(x) \geqslant 0$. According to the Hahn–Banach separation theorem, $\exists (x^*, \lambda) \in X^* \times \mathbb{R}^1$ such that

$$\langle x^*, x \rangle + \lambda f(x) \geqslant 0 \geqslant \langle x^*, x \rangle + \lambda t \quad \forall (x,t) \in C \; .$$

In the same manner as the proof of (2), we verify $\lambda > 0$. Setting $x_0^* = \frac{-x^*}{\lambda}$, we have

$$\langle x_0^*, x \rangle \leqslant f(x), \; -\langle x_0^*, x \rangle \leqslant g(x) \quad \forall x \in X \; .$$

i.e., $x_0^* \in \partial f(\theta)$, and $-x_0^* \in \partial g(\theta)$. □

(5) If $f : X \to \mathbb{R}^1 \cup \{+\infty\}$ is convex, and is G-differentiable at a point $x_0 \in \mathrm{int}(\mathrm{dom}(f))$, then $\partial f(x_0)$ is a single point x_0^* satisfying $\langle x_0^*, h \rangle = df(x_0, h) \; \forall h \in X$.

Proof. We may assume $x_0 = \theta$ and $f(\theta) = 0$. Define the functional on X:

$$L(h) = df(\theta, h) \; .$$

It is homogeneous. From the convexity of f, we have

$$L(h_1 + h_2) \leqslant L(h_1) + L(h_2), \;\; \forall h_1, h_2 \in X \; .$$

Then L is linear. Again by the convexity:

$$-f(-h) \leqslant L(h) \leqslant f(h) \quad \forall h \in X \; .$$

Combining with Theorem 2.2.1, L is continuous. Therefore $\exists x_0^* \in X^*$ such that

$$L(h) = \langle x_0^*, h \rangle \; .$$

It follows that $x_0^* \in \partial f(\theta)$.
Now, suppose $x^* \in \partial f(\theta)$, i.e., $\langle x^*, h \rangle \leqslant f(h) \; \forall h \in X$, and then $\langle x^*, h \rangle \leqslant \frac{1}{t} f(th) \; \forall h \in X, \; \forall t > 0$. By taking limit $t \to +0$, it follows that $\langle x^*, h \rangle \leqslant \langle x_0^*, h \rangle \; \forall h \in X$. Then $x^* = x_0^*$. □

(6) If $f : X \to \mathbb{R}^1 \cup \{+\infty\}$ is convex and attains its minimum at x_0, then $\theta \in \partial f(x_0)$.

Let us present a few examples in computing the subdifferentials of convex functions.

Example 1. (Normalized duality map) Let X be a real Banach space, and let $f(x) = \frac{1}{2} \| x \|^2$. Then $\partial f(x) = F(x) := \{x^* \in X^* | \| x^* \| = \| x \|, \langle x^*, x \rangle = \| x \|^2\}$. It is called the normalized duality map.

In the particular case where X is a real Hilbert space, it is known that $f'(x) = x$, and indeed, $F(x) = x$. However, for Banach spaces, f may not be differentiable. We verify $\partial f(x) = F(x)$ as follows:

"$\subset$" $\forall x^* \in \partial f(x)$, we have

$$\langle x^*, y - x \rangle \leqslant \frac{1}{2}(\| y \|^2 - \| x \|^2)\,,$$

and then $\forall h \in X$,

$$\langle x^*, h \rangle \leqslant \frac{1}{2t}(\| x + th \|^2 - \| x \|^2) \leqslant \| x \| \| h \| + \frac{t}{2} \| h \|^2\,,$$

as $t > 0$. It follows that $\langle x^*, h \rangle \leqslant \| x \| \| h \|$, and then $\| x^* \| \leqslant \| x \|$. On the other hand, setting $y = \lambda x$, we have $(\lambda - 1)\langle x^*, x \rangle \leqslant \frac{1}{2}(\lambda^2 - 1) \| x \|^2$. Dividing by $(\lambda - 1)$ as $\lambda \in (0, 1)$, and then letting $\lambda = 1$ we obtain $\langle x^*, x \rangle \geq \| x \|^2$; it follows that $\| x \| \leqslant \| x^* \|$. Thus $\|x\| = \|x^*\|$, and $\langle x^*, x \rangle = \|x\|^2$.

"$\supset$" $\forall x^* \in F(x)$, one has

$$\begin{aligned}
\langle x^*, y - x \rangle &= \langle x^*, y \rangle - \| x \|^2 \\
&\leqslant \| x^* \| \| y \| - \| x \|^2 \\
&\leqslant \frac{1}{2}(\| y \|^2 - \| x \|^2) \\
&= f(y) - f(x)
\end{aligned}$$

$\forall y \in X$, i.e., $x^* \in \partial f(x)$.

Example 2. Let C be a convex subset of X. The support function of C is defined to be $S_C(x^*) = \sup_{x \in C} \langle x^*, x \rangle$. Geometrically, $S_C(x^*) = \langle x^*, x_0 \rangle$ if and only if $\{x \in X | \langle x^*, x \rangle = S_C(x^*)\}$ is a support hyperplane of C at x_0.

We consider the subdifferential of the indicator function of C:

$$\partial \chi_C(x_0) = \begin{cases} \{\theta\} & \text{if } x_0 \in \overset{\circ}{C} \\ \{x^* \in X^* | S_C(x^*) = \langle x^*, x_0 \rangle\} & \text{if } x_0 \in \partial C \\ \emptyset & \text{if } x_0 \notin \overline{C}\,. \end{cases}$$

Example 3. Let $\beta : \mathbb{R}^1 \to \mathbb{R}^1$ be a monotone nondecreasing function. Then

$$\varphi(t) = \int_0^t \beta(s)ds$$

is a continuous convex function, and $\partial\varphi(t_0) = [\beta(t_0 - 0), \beta(t_0 + 0)]\ \forall t_0 \in \mathbb{R}^1$.

If further, $\exists p > 1, C_1, C_2 > 0$ such that $|\beta(t)| \leqslant C_1 + C_2|t|^{p-1}$, then the functional

$$J(u) = \int_\Omega \varphi(u(x))dx$$

is a convex functional on $L^p(\Omega)$, where Ω is a measurable set in $\mathbb{R}^n$ with bound measure. We conclude that

$$\begin{aligned} \partial J(u_0) &= [\beta(u_0(x) - 0), \beta(u_0(x) + 0)] \\ &:= \{v \in L^{p'}(\Omega) |\ \beta(u_0(x) - 0) \leqslant v(x) \leqslant \beta(u_0(x) + 0)\ \forall \text{ a.e. } \Omega\} \end{aligned}$$

where $\frac{1}{p'} + \frac{1}{p} = 1$. In fact,

$$\begin{aligned} v \in \partial J(u_0) &\Longleftrightarrow \int_\Omega v(x)(u(x) - u_0(x)) \leqslant \int_\Omega [\varphi(u(x)) - \varphi(u_0(x))]\ \ \forall u \in L^p(\Omega) \\ &\Longleftrightarrow v(x)(u(x) - u_0(x)) \leqslant \varphi(u(x)) - \varphi(u_0(x)) \ \text{ a.e.} \\ &\Longleftrightarrow v(x) \in \partial\varphi(u_0(x)) \ \text{ a.e.} \end{aligned}$$

2.3 Convexity and Compactness

The Schauder fixed-point theorem is one of the most important fixed-point theorems in nonlinear analysis. In this section we shall study a series of fixed-point theorems on compact convex set, including the Schauder fixed-point theorem and its extensions. All these theorems are set up by considering the Ky Fan inequality and KKM mapping from the outset.

Definition 2.3.1 *Suppose that X is a vector space, $E \subset X$ is a subset of X. A set-valued mapping $G : E \to 2^X$ is called a KKM mapping, if $\forall x_1, \ldots, x_n \in E$,* $\text{conv}\{x_1, \ldots, x_n\} \subset \overset{n}{\underset{i=1}{\cup}} G(x_i)$.

Knaster, Kuratowski, and Mazurkiweicz (1929) discovered the following (KKM) theorem: Let $[p_0, \ldots, p_n]$ be an n-simplex generated by $n+1$ points $p_0, p_1, \ldots, p_n$ in a vector topological space, and let $M_0, \ldots, M_n$ be $(n+1)$ closed sets, satisfying $[p_{i_0}, \ldots, p_{i_k}] \subset \overset{k}{\underset{j=0}{\cup}} M_{i_j}\ \forall$ index subset $\{i_0, \ldots, i_k\} \subset \{0, 1, \ldots, n\}$. Then $\overset{n}{\underset{i=0}{\cap}} M_i \neq \emptyset$.

The following version of the KKM theorem is due to Ky Fan.

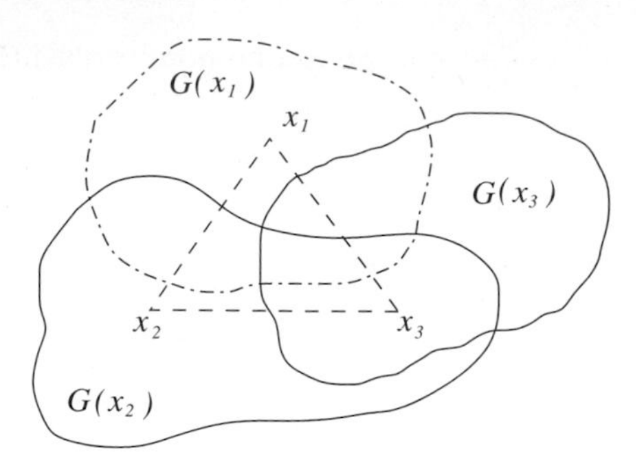

Fig. 2.2.

Theorem 2.3.2 *(FKKM) Suppose that X is a locally convex Hausdorff space (LCS), $E \subset X$, and that $G : E \to 2^X$ is a closed-valued KKM map. If $\exists x_0 \in E$ such that $G(x_0)$ is compact, then $\bigcap_{x \in E} G(x) \neq \emptyset$.*

Proof. Define $F(x) = G(x) \cap G(x_0)$. Then $\{F(x)|\ x \in E\}$ is a family of closed subsets of $G(x_0)$. If one can show the finite intersection property of $\{F(x)|\ x \in E\}$, then our conclusion is proved.

To this end, we only want to verify the finite intersection property for $\{G(x)|\ x \in E\}$. If not, $\exists \{x_1, \ldots, x_n\} \subset E$, such that $\bigcap_{i=1}^{n} G(x_i) = \emptyset$. Let us introduce the Euclidean metric d on $L = \text{span}\{x_1, \ldots, x_n\}$. Since L is finite dimensional, the metric derives the same topology on L, from the following facts: Each finite-dimensional LCS is normable (a consequence of the Kolmogorov normable theorem), and all norms on a finite-dimensional space are equivalent. Then the function

$$\lambda(x) = \sum_{i=1}^{n} d(x, L \cap G(x_i)) > 0 \quad \forall x \in L \,.$$

Define

$$\beta_i(x) = \frac{1}{\lambda(x)} d(x, L \cap G(x_i)) \quad i = 1, \ldots, n \,.$$

$\forall x \in L$, we have

$$\beta_i(x) > 0 \Longleftrightarrow x \notin G(x_i) \quad i = 1, \cdots, n \,.$$

Again, define $C = \text{conv}\{x_1, \ldots, x_n\}$ and a map $\varphi : C \to C$ to be

$$\varphi : x \mapsto \sum_{i=1}^{n} \beta_i(x) x_i \,.$$

According to the Brouwer fixed-point theorem, $\exists x_0 \in C$ such that $x_0 = \varphi(x_0)$.

Letting $I(x_0) = \{i|\ \beta_i(x_0) > 0\}$, we have

$$\varphi(x_0) \in \text{conv}\{x_i|\ i \in I(x_0)\}\ .$$

But $x_0 \notin G(x_i)\ \forall i \in I(x_0)$ implies $x_0 \notin \bigcup_{i \in I(x_0)} G(x_i)$. This contradicts with the KKM mapping. □

Ky Fan's inequality is the foundation of this section, by which all fixed-point theorems are derived.

Theorem 2.3.3 *(Ky Fan's inequality) Suppose that X is an LCS and that $E \subset X$ is a nonempty convex set. Assume $\phi : E \times E \to \mathbb{R}^1$ satisfying*

(1) $\forall y \in E\ x \mapsto \phi(x, y)$ is l.s.c.,
(2) $\forall x \in E\ y \mapsto \phi(x, y)$ is quasi-concave,
(3) $\exists y_0 \in E$ such that

$$\{x \in E|\ \phi(x, y_0) \leqslant \sup_{x \in E} \phi(x, x)\} \text{ is compact}\ .$$

Then $\exists x_0 \in E$ such that

$$\sup_{y \in E} \phi(x_0, y) \leqslant \sup_{x \in E} \phi(x, x)\ .$$

Before going to the proof of Theorem 2.3.3, we first introduce a geometric version of the statement. Given a set E, let $A \subset E \times E$ be a subset. Define the sections:

$$\forall x \in E,\ A_1(x) = \{y \in E|\ (x, y) \in A\}\ ,$$

and

$$\forall y \in E,\ A_2(y) = \{x \in E|\ (x, y) \in A\}\ .$$

Denote the diagonal $\{(x, x)|\ x \in E\}$ by $\triangle$.

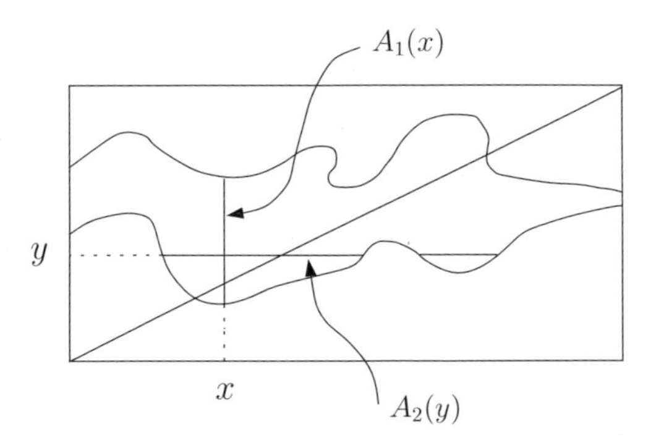

Fig. 2.3.

In this manner, we define two set-valued mappings: $E \to 2^E$,

$$A_1 : x \mapsto A_1(x) \text{ and } A_2 : y \mapsto A_2(y)\ .$$

Theorem 2.3.4 *Suppose that X is an LCS and that $E \subset X$ is a nonempty convex set. Assume that $\Gamma \subset E \times E$ is a subset satisfying*

(1′) $\forall y \in E$, $\Gamma_2(y)$ *is open,*
(2′) $\forall x \in E$, $\Gamma_1(x)$ *is convex,*
(3′) $\triangle \cap \Gamma = \emptyset$,
(4′) $\exists y_0 \in E$ *such that* $E \backslash \Gamma_2(y_0)$ *is compact.*

Then $\exists x_0 \in E$ *such that* $\Gamma_1(x_0) = \emptyset$.

We are going to show that Theorem 2.3.3 $\Longleftrightarrow$ Theorem 2.3.4.

Proof. "$\Longrightarrow$" Define a function on $E \times E$:

$$\phi(x,y) = \begin{cases} 1 & (x,y) \in \Gamma \\ 0 & (x,y) \notin \Gamma \end{cases}$$

$\forall y \in E$, we see

$$\{x \in E |\ \phi(x,y) > \lambda\} = \begin{cases} E & \text{if } \lambda < 0 \\ \Gamma_2(y) & \text{if } 0 \leqslant \lambda < 1 \\ \emptyset & \text{if } 1 \leqslant \lambda\,, \end{cases}$$

then $x \mapsto \phi(x,y)$ is l.s.c., from (1′).

Similarly, $\forall x \in E$

$$\{y \in E |\ \phi(x,y) > \lambda\} = \begin{cases} E & \text{if } \lambda < 0 \\ \Gamma_1(x) & \text{if } 0 \leqslant \lambda < 1 \\ \emptyset & \text{if } 1 \leqslant \lambda\,, \end{cases}$$

then $y \mapsto \phi(x,y)$ is quasi-concave, from (2′).

Since $\triangle \cap \Gamma = \emptyset$, $\phi(x,x) = 0\ \forall x \in E$. The set

$$\{x \in E |\ \phi(x,y_0) \leqslant 0\} = E \backslash \Gamma_2(y_0)$$

is compact, from (4′). Theorem 2.3.3 is applied to conclude that $\exists x_0 \in E$ such that $\phi(x_0,y) \leqslant 0\ \forall y \in E$, i.e., $\Gamma_1(x_0) = \emptyset$.

"$\Longleftarrow$" Define

$$\Gamma = \{(x,y) \in E \times E |\ \phi(x,y) > \mu\}\,,$$

where $\mu = \sup_{x \in E} \phi(x,x)$. Then (1) and (2) imply (1′) and (2′) respectively. By definition $\Gamma \cap \triangle = \emptyset$. Also

$$E \backslash \Gamma_2(y_0) = \{x \in E |\ \phi(x,y_0) \leqslant \mu\}$$

is compact. We then apply Theorem 2.3.4, and conclude that $\exists x_0 \in E$ such that $\Gamma_1(x_0) = \emptyset$, i.e., $\phi(x_0,y) \leqslant \mu\ \forall y \in E$. □

Now we turn to giving a proof of Theorem 2.3.4.

Proof. Define a set-valued mapping

$$G : y \mapsto E \backslash \Gamma_2(y) .$$

From $(1')$, it is a closed set-valued mapping.

$(4')$ implies that $G(y_0)$ is compact. If we can show that G is a KKM map, then the FKKM theorem is applied to conclude that $\bigcap_{y \in E} G(y) \neq \emptyset$, i.e.,

$$\exists x_0 \in \bigcap_{y \in E} G(y) \iff (x_0, y) \in (E \times E) \backslash \Gamma \ \forall y \in E \iff \Gamma_1(x_0) = \emptyset .$$

Verification of the KKM map.

If not, $\exists \{y_1, \ldots, y_n\} \subset E$ and $\exists \lambda_1 \geq 0, \ldots, \lambda_n \geq 0$, with $\Sigma_1^n \lambda_i \leq 1$, such that

$$\begin{aligned} w = \sum_{i=1}^n \lambda_i y_i \notin \bigcup_{i=1}^n G(y_i) &\iff w \in \bigcap_{i=1}^n \Gamma_2(y_i) , \\ &\iff y_i \in \Gamma_1(w) \ \forall i , \\ &\overset{\text{by } (2')}{\Longrightarrow} w \in \Gamma_1(w) , \\ &\Longrightarrow \triangle \cap \Gamma \neq \emptyset . \end{aligned}$$

This contradicts with $(3')$. □

First we mention that this theorem is very useful in game theory. In particular, the Nash equilibrium theorem and Von Neumann–Sion saddle point theorem can be derived from it.

In an n-person game, there are n players $\{1, 2, \ldots, n\}$. The ith person has a strategy set E_i and a payoff function $f_i, i = 1, 2, \ldots, n$. The payoff functions depend on the strategies of all players, i.e., $f_i, i = 1, 2, \ldots, n$ are functions of $x = (x_1, \ldots, x_n)$, where $x_j \in E_j, j = 1, \ldots, n$. Nash's solution of the game is such a strategy $(x_1^*, \ldots, x_n^*) \in E_1 \times \cdots \times E_n$: no player has any incentive to change his strategy as long as his enemies don't change theirs.

Mathematically, let $X_1, \ldots, X_n$ be n-LCS, let $E_i \subset X_i$ be nonempty compact convex sets, $\forall i = 1, \ldots, n$, and let $f_1, \ldots, f_n$ be n continuous functions on $E = E_1 \times \cdots \times E_n$. A point $x^* = (x_1^*, \ldots, x_n^*) \in E$ is called a Nash equilibrium if

$$f_i(x^*) \geqslant f_i(x_1^*, \ldots, y_i, \ldots, x_n^*) \quad \forall i = 1, 2, \ldots, n \ \forall y = (y_1, \ldots, y_n) \in E .$$

Corollary 2.3.5 *(Nash) If the continuous functions*

$$x_i \mapsto f_i(x_1, \ldots, x_i, \ldots, x_n) \quad \forall i = 1, \ldots, n$$

are quasi-concave $\forall (x_1, \ldots, \hat{x}_i, \ldots, x_n) \in \prod_{j \neq i} E_j \ \forall i$. *Then the Nash equilibrium exists.*

In fact, let us define a function on $E \times E$:

$$\phi(x, y) = \sum_{i=1}^{n} (f_i(x_1, \ldots, y_i, \ldots, x_n) - f_i(x_1, \ldots, x_n))$$

It is easy to verify that all the assumptions of the Ky Fan inequality are satisfied (in particular, $y \to \phi(x, y)$ is quasi-concave), and $\phi(x, x) = 0 \; \forall x \in E$. Then $\exists x^* \in E$ satisfying $\phi(x^*, y) \leqslant 0 \; \forall y \in E$. This is the Nash equilibrium.

Corollary 2.3.6 *(Von Neumann–Sion) Suppose that X, Y are reflexive Banach spaces, $E \subset X$ and $F \subset Y$ are nonempty closed convex sets. Assume that $f : E \times F \to \mathbb{R}^1$ satisfies*

(1) $\forall y \in F$, $x \mapsto f(x, y)$ is l.s.c. and quasi-convex.
(2) $\forall x \in E$, $y \mapsto f(x, y)$ is u.s.c. and quasi-concave.
(3) $\exists (x_0, y_0) \in E \times F$ such that $f(x_0, y) \to -\infty$ as $\| y \| \to +\infty$ and $f(x, y_0) \to +\infty$ as $\| x \| \to +\infty$.

Then there exist $(x^*, y^*) \in E \times F$ such that

$$f(x^*, y) \leqslant f(x^*, y^*) \leqslant f(x, y^*) \quad \forall (x, y) \in E \times F \, . \tag{2.12}$$

Proof. Define

$$\phi(u, v) = f(x, w) - f(z, y) \, ,$$

where $u = (x, y), v = (z, w) \in E \times F$. Then we have the following:

From Theorem 2.2.3, $\forall v \in E \times F, u \mapsto \phi(u, v)$ is w.l.s.c.
It is easy to verify directly that $\forall u \in E \times F, v \mapsto \phi(u, v)$ is quasi-concave, $\phi(u, v_0) = f(x, y_0) - f(x_0, y) \to +\infty$ as $\| x \| + \| y \| \to +\infty$, where $v_0 = (x_0, y_0)$.

The set $\{u \in E \times F | \; \phi(u, v_0) \leqslant 0\}$ is weakly compact. We apply Ky Fan's inequality to the space $E \times F$ with weak topology. There exists $u^* = (x^*, y^*)$ such that

$$f(x^*, w) - f(z, y^*) \leqslant 0 \quad \forall (z, w) \in E \times F \, ,$$

which implies

$$f(x^*, y) \leqslant f(x^*, y^*) \leqslant f(x, y^*) \quad \forall (x, y) \in E \times F \, .$$

□

The solution (x^*, y^*) is called the saddle point of f.
As a special case where $\dim Y = 0$, we have:

Corollary 2.3.7 *Suppose that X is a reflexive Banach space. If $f : X \to \mathbb{R}^1 \cup \{+\infty\}$ is an l.s.c. convex function, satisfying $f(x) \to +\infty$ as $\| x \| \to \infty$, then there exists $x_0 \in X$ such that $f(x_0) = \min\{f(x) | \, x \in X\}$.*

In the case where E and F are compact, the Von Neumann–Sion theorem is also rewritten as follows:

$$\min_{x\in E}\max_{y\in F} f(x,y) = \max_{y\in F}\min_{x\in E} f(x,y) \, . \tag{2.13}$$

Indeed, (2.12) $\Longleftrightarrow$ (2.13).

Proof. "$\Longrightarrow$" We always have

$$\beta := \max_{y\in F}\min_{x\in E} f(x,y) \leqslant \min_{x\in E}\max_{y\in F} f(x,y) =: \alpha \, .$$

However, (2.12) implies

$$\alpha \leqslant \max_{y\in F} f(x^*,y) \leqslant f(x^*,y^*) \leqslant \min_{x\in E} f(x,y^*) \leqslant \beta \, .$$

Therefore $\alpha = \beta$, and then (2.13) holds.

"$\Longleftarrow$" Since both E and F are compact and the functions $x \mapsto f(x,y)$ is l.s.c., $y \mapsto f(x,y)$ is u.s.c., the functions $x \to \max_{y\in F} f(x,y)$ is l.c.s, and $y \to \min_{x\in E} f(x,y)$ is u.s.c., we have $(x^*,y^*) \in E \times F$ such that

$$\begin{aligned}\min_{x\in E} f(x,y^*) &= \max_{y\in F}\min_{x\in E} f(x,y)\\ &= \min_{x\in E}\max_{y\in F} f(x,y)\\ &= \max_{y\in F} f(x^*,y) \, .\end{aligned}$$

Since

$$\min_{x\in E} f(x,y^*) \le f(x^*,y^*) \le \max_{y\in F} f(x^*,y) \, ,$$

these three are equal. It (2.12) follows. □

In this sense, the Von Neumann–Sion theorem is called the mini-max theorem, which is the fundamental theorem in two-person game theory, and has many applications in other fields of mathematics (linear programming, convex programming, potential theory, the dual variational theory in mechanics etc.)

Next, we turn to fixed-point theorems.

For a set-valued mapping $\Gamma : E \to 2^F$, where E, F are Hausdorff topological vector spaces, we say Γ is upper (or lower) semi-continuous (u.s.c. or l.s.c., respectively for short), if for any closed (open resp.) set $W \subset F$, the pre-image $\Gamma^{-1}(W) = \{x \in E | \, \Gamma(x) \cap W \neq \emptyset\}$ is closed (open resp.).

For single-valued mappings, both u.s.c. and l.s.c. are all reduced to continuity.

An alternative description of the upper semi-continuity is that for any open set $W \subset F$, the set $\{x \in E | \, \Gamma(x) \subset W\}$ is open. A point $x_0 \in E$ is called a fixed point of Γ if $x_0 \in \Gamma(x_0)$.

Theorem 2.3.8 *(Ky Fan–Glicksberg) Suppose that X is an LCS, and that $E \subset X$ is a nonempty compact convex set. If $\Gamma : E \to 2^E$ is u.s.c., and $\forall x \in E, \Gamma(x)$ is a nonempty closed convex subset of E, then $\exists x_0 \in E$ such that $x_0 \in \Gamma(x_0)$.*

Proof. We prove by contradiction. If $\forall x \in E, x \notin \Gamma(x)$, then from the Hahn–Banach theorem $\exists x^* \in X^*\ \exists t \in \mathbb{R}^1$ such that

$$\langle x^*, y\rangle < t < \langle x^*, x\rangle \quad \forall y \in \Gamma(x) .$$

Since Γ is u.s.c. there exists a neighborhood $U(x)$ of x such that

$$\langle x^*, y\rangle < t < \langle x^*, z\rangle \quad \forall y \in \Gamma(z)\ \forall z \in U(x) .$$

Since E is compact, we have $x_1, \ldots, x_n \in E$ such that $E \subset \bigcup_{i=1}^{n} U(x_i)$. Let $\{x_1^*, \ldots, x_n^*\} \subset X^*$ be the associated elements; one has

$$\langle x_i^*, y\rangle < \langle x_i^*, x\rangle \quad \forall y \in \Gamma(x)\ \forall x \in U(x_i) .$$

We construct a partition of unity of E:

$$\{\beta_i : E \to \mathbb{R}^1, i = 1, \ldots, n \mid \beta_i \geqslant 0, \sum_{i=1}^{n} \beta_i = 1 \text{ on } E \text{ and } \beta_i(x) > 0 \text{ iff } x \in U(x_i)\} .$$

Define a continuous map $f : E \to X^*$ by

$$f(x) = \sum_{i=1}^{n} \beta_i(x) x_i^* ,$$

one has

$$\begin{aligned} \langle f(x), y\rangle &= \sum_{i=1}^{n} \beta_i(x)\langle x_i^*, y\rangle < \sum_{i=1}^{n} \beta_i(x)\langle x_i^*, x\rangle \\ &= \langle f(x), x\rangle \quad \forall x \in E, \forall y \in \Gamma(x) . \end{aligned} \tag{2.14}$$

Define $\phi : E \times E \to \mathbb{R}^1$ by

$$\phi(x, y) = \langle f(x), x - y\rangle ,$$

All the assumptions of Theorem 2.3.3 are satisfied, and we conclude that $\exists x_0 \in E$ such that

$$\phi(x_0, y) \leqslant 0 \quad \forall y \in E ;$$

it follows that

$$\langle f(x_0), x_0\rangle = \min_{y \in E} \langle f(x_0), y\rangle .$$

This contradicts (2.14), because $\Gamma(x) \subset E$. □

This is a very general fixed-point theorem concerning convexity and compactness. The very special case is the Schauder fixed-point theorems, which are stated as follows:

Corollary 2.3.9 *(Schauder–Tichonov) Suppose that X is an LCS and that $E \subset X$ is a nonempty compact convex set. If $f : E \to E$ is continuous, then f has a fixed point.*

Corollary 2.3.10 *(Schauder) Suppose that X is a Banach space and that $E \subset X$ is a nonempty bounded closed convex set. If $f : E \to E$ is a compact map, then f has a fixed point.*

Proof. It is a direct consequence of Corollary 2.3.9. Let us define

$$E_1 = \overline{\text{conv}} f(E) .$$

Then E_1 is a nonempty compact convex subset of E, and $f : E_1 \to E_1$. □

We would rather give a direct proof of the second Schauder fixed-point theorem as follows:

Proof. Define $\phi(x, y) = -\|y - f(x)\|$ on $E_1 \times E_1$, where $E_1 = \overline{\text{conv}} f(E)$ is compact. Applying Ky Fan's inequality, we have $x_0 \in E_1$ such that $\sup_{y \in E_1} \phi(x_0, y) \le \sup_{x \in E_1} \phi(x, x)$. Thus, $\inf_{x \in E_1} \|x - f(x)\| \le \inf_{y \in E_1} \|y - f(x_0)\| = 0$, i.e., $\exists x \in E_1$ such that $x = f(x)$. □

Remark 2.3.11 *Theorem 2.3.8 was obtained by Ky Fan [FK 1] and Glicksberg independently in 1952. The earlier results for $X = \mathbb{R}^n$ and X being a linear normed space are due to Kakutani (1941) [Kak] and Bohnenblust, Karlin (1950) resp.*

Applications

1. Convex programming

Let X be a reflexive Banach space. Given l.s.c. convex functions $f, g_1, \ldots, g_n : X \to \mathbb{R}^1$, assume that the set $C = \{x \in X | g_i(x) \le 0, i = 1, 2, \ldots, n\}$ is nonempty; find $x_0 \in C$ such that

$$f(x_0) = \min \{f(x) | x \in C\} . \tag{2.15}$$

Let us introduce Lagrange multipliers $\{y_i\}_1^n, y_i \geqslant 0, i = 1, \ldots, n$, and the following function:

$$L(x, y) = f(x) + \sum_{i=1}^n y_i g_i(x) \quad (x, y) \in X \times \mathbb{R}_+^n ,$$

which is called the Lagrangian.

Statement 2.3.12 *If* (x_0, y_0) *is a saddle point of* L*:*

$$L(x_0, y) \leqslant L(x_0, y_0) \leqslant L(x, y_0) \quad \forall (x, y) \in X \times \mathbb{R}^n_+ . \tag{2.16}$$

Then x_0 *is a solution of (2.15) and satisfies*

$$y_i^\circ g_i(x_0) = 0, \ i = 1, 2, \ldots, n ,$$

where $(y_1^\circ, \ldots, y_n^\circ) = y_0$*. Conversely, if there exists* $x^* \in X$ *satisfying* $g_i(x^*) < 0$*,* $i = 1, 2, \ldots, n$*, and if* $x_0 \in X$ *solves (2.15), then* $\exists \xi \in \mathbb{R}^n_+$ *such that* (x_0, ξ) *is a saddle point of* L*.*

Proof. "$\Longrightarrow$" From the first inequality of (2.16), we have

$$\sum_{i=1}^n (y_i^\circ - y_i) g_i(x_0) \geqslant 0 .$$

Letting $y_i \to +\infty$ and $y_j = y_j^\circ, j \neq i$, it follows that $g_i(x_0) \leqslant 0$, $i = 1, 2, \ldots, n$. Then letting $y_i = 0$, $i = 1, 2, \ldots, n$, it follows that $\sum_{i=1}^n y_i^\circ g_i(x_0) \geqslant 0$. But for $y_i^\circ \geqslant 0$, we have either $g_i(x_0) = 0$ or $y_i^\circ = 0$, i.e., $y_i^\circ g_i(x_0) = 0$, $i = 1, 2, \ldots, n$.

Returning to the second inequality of (2.16), we obtain

$$f(x_0) \leqslant f(x) + \sum_{i=1}^n y_i^\circ g_i(x) \quad \forall x \in X .$$

Thus

$$f(x_0) \leqslant f(x) ,$$

whenever $g_i(x) \leqslant 0$, $i = 1, 2, \ldots, n$.

"$\Longleftarrow$" We consider two convex sets:

$$E = \{(y^0, \ldots, y^n) \in \mathbb{R}^{n+1} |\ y^0 < f(x_0),\ y^i < 0,\ i = 1, 2, \ldots, n\} ,$$

and

$$F = \{(y^0, \ldots, y^n) \in \mathbb{R}^{n+1} |\ \exists x \in X \text{ such that } f(x) \leqslant y^0,\ g_i(x) \leqslant y^i,\ \forall i\} .$$

Claim: $E \cap F = \emptyset$. If not, $\exists (\overline{y}^0, \ldots, \overline{y}^n) \in E \cap F$, i.e., $\exists x_1 \in X$ satisfying $f(x_1) \leq \overline{y}^0 < f(x_0)$, $g_i(x_1) \leqslant \overline{y}^i < 0$, $i = 1, \ldots, n$. Then x_0 cannot be a solution of (2.15). This is a contradiction.

Applying the Hahn–Banach theorem, $\exists \omega = (\omega_0, \ldots, \omega_{n+1}) \in \mathbb{R}^{n+1} \backslash \{\theta\}$ such that

$$\sum_{i=0}^n \omega_i y^i \leqslant \sum_{i=0}^n \omega_i z^i \tag{2.17}$$

$\forall y = (y^0, \ldots, y^n) \in E$, $\forall z = (z^0, \ldots, z^n) \in F$. By setting $z = (f(x_0), g_1(x_0), \ldots, g_n(x_0))$,

$y = (f(x_0)-\epsilon_0, g_1(x_0)-\epsilon_1, \dots, g_n(x_0)-\epsilon_n), \forall(\epsilon_0, \epsilon_1, \dots, \epsilon_n) \in R_+^{n+1}$, it follows that $\omega_i \geqslant 0,\ i = 0, 1, \dots, n$.

Again, $\forall \varepsilon > 0$, substituting $y^0 = f(x_0) - \varepsilon, z^0 = f(x),\ y^i = -\varepsilon, z^i = g_i(x)\ i = 1, \dots, n$ into (2.17) $\forall x \in X$, and letting $\varepsilon \to 0$, we have

$$\omega_0 f(x_0) \leqslant \sum_{i=1}^{n} \omega_i g_i(x) + \omega_0 f(x) \quad \forall x \in X . \tag{2.18}$$

Claim $\omega_0 > 0$. Otherwise, $\omega_0 = 0$, thus $\sum_{i=1}^{n} \omega_i g_i(x) \geqslant 0$. Since we have assumed $g_i(x^*) < 0$, combining with the fact that $\omega_i \geqslant 0,\ i - 0, 1, \dots, n$, it follows that $\omega_i = 0,\ i = 1, \dots, n$. This is impossible.

Dividing (2.18) by ω_0, we obtain

$$f(x_0) \leqslant \sum_{j=1}^{n} \xi_j g_j(x) + f(x) \quad \forall x \in X ,$$

where $\xi_j = \frac{\omega_j}{\omega_0}$. Then

$$\sum_{j=1}^{n} \xi_j g_j(x_0) = 0 ,$$

it follows that $\xi_j g_j(x_0) = 0\ j = 1, \dots, n$. Thus

$$f(x_0) + \sum_{j=1}^{n} y_j g_j(x_0) \leqslant f(x_0) + \sum_{j=1}^{n} \xi_j g_j(x_0) \leqslant f(x) + \sum_{j=1}^{n} \xi_j g_j(x)$$

$\forall (x, y) \in X \times \mathbb{R}_+^n$, i.e., (x_0, ξ) is a saddle point of L. $\sqcup$

Statement 2.3.13 *In the convex programming problem, assume*

(1) $f(x) \to +\infty$ *as* $\| x \| \to \infty$*,*
(2) $\exists x_0 \in X$ *such that* $g_i(x_0) < 0,\ i = 1, 2, \dots, n$*.*

The problem (2.16) has a solution.

Proof. Consider the Lagrangian on $X \times \mathbb{R}_+^n$,

$$L(x, y) = f(x) + \sum_{i=1}^{n} y_i g_i(x) .$$

Obviously,

$$\forall y \in \mathbb{R}_+^n,\ x \mapsto L(x, y) \text{ is l.s.c. and convex,}$$
$$\forall x \in X,\ y \mapsto L(x, y) \text{ is linear ,}$$

and by (2) and (1),

$$L(x_0, y) \to -\infty \text{ as } \| y \| \to \infty ,$$
$$L(x, \theta) = f(x) \to +\infty \text{ as } \| x \| \to \infty .$$

Applying Corollary 2.3.6, there is a saddle point of L, which is the solution of the problem (2.15), according to Statement 2.3.12. □

2. Periodic solutions of ODE

We study the 2π-periodic solutions of the following ODE:

$$\ddot{x} + g(x) = f(t) \tag{2.19}$$

where $g \in C(\mathbb{R}^1)$ and $f \in L^2([0, 2\pi])$. Assume that $\exists n \in \mathbb{N}$ and $\varepsilon > 0$ such that

$$n^2 + \varepsilon \leqslant \frac{g(x)}{x} \leqslant (n+1)^2 - \varepsilon \quad \text{for } |x| \text{ large} . \tag{2.20}$$

Statement 2.3.14 *The problem (2.19) has a 2π-periodic solution under the assumption (2.20).*

Proof. Let $\gamma = n^2 + n + 1/2$. Then $\exists R > 0$ such that

$$|g(x) - \gamma x| \leqslant \left(n + \frac{1}{2} - \epsilon\right) |x| \quad \text{as } |x| \geqslant R .$$

Rewrite (2.19) as

$$\ddot{x} + \gamma x + g(x) - \gamma x - f(t) = 0 \tag{2.21}$$

and define the linear operator

$$L : x \mapsto -(\ddot{x} + \gamma x)$$

with domain $D(L) = \{x \in H^2(0, 2\pi) |\ x(0) = x(2\pi),\ \dot{x}(0) = \dot{x}(2\pi)\}$. Then L has a bounded inverse: $L^{-1} : L^2(0, 2\pi) \to D(L)$. Since the injection $D(L) \to L^2(0, 2\pi)$ is compact, (2.21) is equivalent to

$$x = L^{-1} \cdot G(x) ,$$

where $G : x \mapsto g(x) - \gamma x - f$ is a bounded continuous mapping on $L^2(0, 2\pi)$. Then $F = L^{-1} \circ G$ is a compact map on $L^2(0, 2\pi)$.

Moreover $\| L^{-1} \| \leqslant \frac{1}{n+1/2}$, $\| Gx \| \leqslant (n + \frac{1}{2} - \varepsilon) \| x \| + C$, where $C = \sqrt{2\pi} \max\limits_{|x| \leqslant R} |g(x) - \gamma x| + \| f \|_2$. Setting $R_1 \geqslant \frac{C}{\varepsilon}$, we have $F : \overline{B}_{R_1}(\theta) \to \overline{B}_{R_1}(\theta)$. The Schauder fixed-point theorem is applied to ensure the existence of $x_0 \in B_{R_1}(\theta)$ satisfying $F(x_0) = x_0$, thus $x_0 \in D(L)$ and satisfies (2.21). □

3. Obstacle problem

Find the equilibrium position u of a membrane with a fixed boundary, acted upon by an external force f, and obstructed by an fixed obstacle ψ.

The problem is posed as a minimizing problem on a closed convex set. Let $\Omega \subset \mathbb{R}^n$ be a bounded domain, $f \in L^2(\Omega)$, $\psi \in H^1(\Omega)$. Find $u \in H_0^1(\Omega)$ such that u is the minimizer of the problem: $\min\{J(v)|\ v \in E\}$, where

$$J(v) = \int_\Omega \left(\frac{1}{2}|\nabla v|^2 - f \cdot v\right) ,$$

and

$$E = \left\{v \in H_0^1(\Omega)|\ v(x) \leqslant \psi(x) \text{ a.e.}\right\} .$$

From $J(u) \leqslant J(u + t(v-u))\ \forall v \in E$ and $t \in (0,1)$, and letting $t \to 0$, we obtain the following variational inequality:

$$\int_\Omega \nabla u \cdot \nabla(v-u) \geqslant \int_\Omega f(v-u) \quad \forall v \in E . \tag{2.22}$$

The existence of a weak solution for (2.22) can be studied by variational method, see Chap. 4. However, we present here a fixed-point approach.

First we reduce the problem to a system of inequalities. One has

$$u \leqslant \psi \text{ a.e. }, \tag{2.23}$$

Denote the positive cone of $H_0^1(\Omega)$ by $C = \{w \in H_0^1(\Omega)|\ w(x) \geqslant 0 \text{ a.e.}\}$. Let $v \in H_0^1(\Omega)$, and $v \le u$, then $v \in E$, and let $w = u - v$, then (2.22) implies that $\int_\Omega \nabla u \nabla w \leqslant \int_\Omega fw\ \forall w \in C$ i.e.,

$$-\triangle u - f \leqslant 0 \text{ a.e. }, \tag{2.24}$$

Substituting $v = \psi$ and $2u - \psi$, respectively into (2.22), we obtain

$$\int_\Omega (-\triangle u - f) \cdot (\psi - u) \ge 0 ,$$

and

$$\int_\Omega (-\triangle u - f) \cdot (u - \psi) \ge 0 ,$$

respectively, therefore $\int_\Omega (-\triangle u - f)(u - \psi) = 0$.

If $u \in W_p^2(\Omega)$ then,

$$(-\triangle u - f)(u - \psi) = 0 \text{ a.e.} \tag{2.25}$$

Conversely, from (2.24), (2.25) and $v \le \psi$, we obtain $\int_\Omega (-\Delta u - f)(v-u) \ge 0$, i.e., (2.22)

Thus, for $u \in W_p^2 \cap H_0^1(\Omega)$, it is a solution of the variational inequality (2.22) if and only if it satisfies the system of inequalities (2.23), (2.24), (2.25).

Second, the system of inequalities is equivalent to a PDE with discontinuous nonlinearity, i.e., for $\psi \in W_p^2(\Omega)$ the system of inequalities (2.23), (2.24), (2.25) is again equivalent to

$$-\triangle u = \begin{cases} \min\{f(x), -\triangle\psi(x)\} & \text{if } u(x) \geqslant \psi(x) \\ f(x) & \text{if } u(x) < \psi(x)\,. \end{cases} \tag{2.26}$$

Verification:

"$\Longrightarrow$" On the set $\{x \in \Omega|\ u(x) < \psi(x)\}$ the equality follows from (2.25). On the complement set $\{x \in \Omega|\ u(x) = \psi(x)\}$ we have $-\triangle\psi(x) = -\Delta u(x) \leqslant f(x)$ a.e., and then $-\Delta u(x) = -\Delta\psi(x) = \min\{f(x), -\triangle\psi(x)\}$.

"$\Longleftarrow$" (2.24) is trivial.

If the set $U = \{x \in \Omega|\ u(x) > \psi\} \neq \emptyset$, (2.26) implies that $-\triangle(u - \psi) \leqslant 0$ on U. This contradicts the maximum principle; then (2.23) holds, and consequently (2.25).

Now, (2.26) is a PDE with discontinuous nonlinearity:

$$-\triangle u = \phi(x, u) \tag{2.27}$$

where

$$\phi(x, u) = \begin{cases} \min\{f(x), -\triangle\psi(x)\} & \text{if } u \geqslant \psi(x) \\ f(x) & \text{if } u < \psi(x)\,. \end{cases}$$

We shall solve the equation by the Ky Fan fixed-point theorem.

First, let us define a set-valued mapping

$$F : C(\overline{\Omega}) \to 2^{L^p(\Omega)} : u \mapsto Fu = [\underline{\phi}(x, u(x)), \overline{\phi}(x, u(x))],\ \text{the order interval in } L^p(\Omega)\,,$$

where $\underline{\phi}(x, t) = \underline{\lim}_{t' \to t} \phi(x, t')$, and $\overline{\phi}(x, t) = \overline{\lim}_{t' \to t} \phi(x, t')$.

We consider the following maps:

$$C(\overline{\Omega}) \xrightarrow{F} 2^{L^p(\Omega)} \xrightarrow{\mathbb{K}} 2^{W^2_p \cap H^1_0(\Omega)} \xrightarrow{j} 2^{C(\overline{\Omega})}\,,$$

where $\mathbb{K} = (-\triangle)^{-1}$ and j is the embedding map.

It is not difficult to verify that $\Gamma := j \circ \mathbb{K} \circ F$ is u.s.c., and $\forall u \in C(\overline{\Omega})$, $\Gamma(u)$ is a nonempty closed convex set of $C(\overline{\Omega})$.

Second, we consider the set $E = \mathbb{K}(\overline{B}_R(\theta))$ of $C(\overline{\Omega})$, where $R = \max\{\| f \|_p, \| \triangle\psi \|_p\}$. It is a nonempty compact convex set in $C(\overline{\Omega})$, and $\Gamma : E \to 2^E$. Ky Fan's fixed-point theorem is applied to ensure the existence of $u_0 \in E$ such that $u_0 \in \Gamma(u_0)$, i.e., $u_0 \in W^2_p \cap H^1_0(\Omega)$ satisfying

$$-\triangle u_0(x) \in \begin{cases} \min\{f(x), -\triangle\psi(x)\} & \text{if } u_0(x) > \psi(x) \\ [\min\{f(x), -\triangle\psi(x)\}, f(x)] & \text{if } u_0(x) = \psi(x) \\ f(x) & \text{if } u_0(x) < \psi(x)\,. \end{cases}$$

Third, however, on the set $U = \{x \in \Omega|\ u_0(x) = \psi(x)\}$, $-\triangle u_0(x) = -\triangle\psi(x)$ a.e. This implies that $-\triangle\psi(x) \leqslant f(x)$ a.e., i.e.,

$$-\triangle u_0(x) = \begin{cases} \min\{f(x), -\triangle\psi(x)\} & \text{if } u_0(x) \geqslant \psi(x) \\ f(x) & \text{if } u_0(x) < \psi(x)\,. \end{cases}$$

This is exactly the solution of the PDE with discontinuous nonlinearity (2.26).

Statement 2.3.15 *If $\psi \in W_p^2(\Omega)$, $f \in L^p(\Omega)$, $p > \frac{n}{2}$, then there exists $u \in W_p^2 \cap \overset{\circ}{W}{}_p^1(\Omega)$ satisfying (2.26).*

The obstacle problem is a free boundary problem for PDEs. The boundary where the membrane detaches from the obstacle is unknown, and is to be determined simultaneously with the position of the membrane. A kind of free boundary problem, having the feature that the boundary values and the normal derivatives of the unknown function coincide on both sides of the unknown free boundary, can be studied via the above approach, i.e., pose the free boundary problem as a PDE with discontinuous nonlinearity, in which the unknown boundary is automatically absorbed into the unknown function. Fixed-point theorems for set-valued mappings can be used to study the existence of the solutions. Examples of the obstacle problem are the seepage surface in dams, the free boundary of confined plasma, water cones in oil reservoirs etc. (see K.C. Chang [Ch 1,2]).

4. Stefan problem

Consider a container Ω with boundaries Γ' and Γ'', filled with ice. Cooling on Γ'' and heating on Γ' (see Fig. 2.4), we study the melting process.

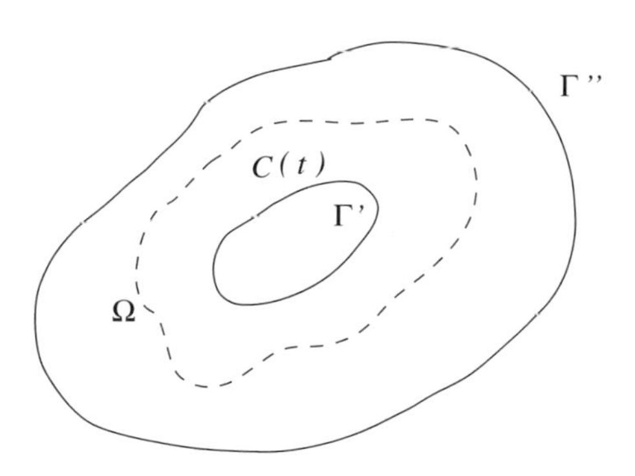

Fig. 2.4.

At time t, the domain occupied by water is denoted by $C(t)$, and let $\theta(x,t) \geqslant 0$ be the temperature at the space–time (x,t) as $x \in C(t)$. Both θ and C are unknown, but $g(x,t) = \theta(x,t)|_{\Gamma' \times [0,T]}$ is given $\forall T > 0$. Let $S(t) = \partial C(t)$ be the unknown boundary. Let us suppose that $S(t)$ is given by an equation $\sigma(x) = t$. The temperature distribution is governed by the following equations:

$$\begin{cases} \frac{\partial}{\partial t}\theta - \triangle\theta = 0 & \forall t \in [0,T],\ \forall x \in C(t), \\ \theta(x,t)|_{\Gamma' \times [0,T]} = g(x,t), & \\ \theta(x,t)|_{S(t)} = 0 & \forall t \in [0,T], \\ \nabla\theta \cdot \nabla\sigma|_{S(t)} = -L, & \\ \theta(x,0) = 0 & \forall x \in \Omega, \end{cases} \tag{2.28}$$

where $L > 0$ is a constant. The boundary conditions on $S(t)$ are derived from the continuity of the temperature and the conservation of the energy, respectively.

We set

$$\widetilde{\theta}(x,t) = \begin{cases} \theta(x,t) & \text{if } x \in C(t) \\ 0 & \text{if } x \notin C(t) \ , \end{cases}$$

and

$$u(x,t) = \int_0^t \widetilde{\theta}(x,s)ds \ .$$

The latter is called the modified Baiocchi transformation of θ. Let

$$\psi(x,t) = \int_0^t g(x,s)ds \ .$$

The problem (2.28) is reduced to a parabolic equation with discontinuous nonlinearity:

$$\begin{cases} \frac{\partial u}{\partial t} - \triangle u = -LH(u) & \text{in } Q_T := \Omega \times (0,T) \ , \\ u|_{\Gamma' \times [0,T]} = \psi \ , & \\ u|_{\Gamma'' \times [0,T]} = 0 \ , & \\ u(x,0) = 0, & \text{on } \Omega \ . \end{cases} \tag{2.29}$$

Again, the advantage of the formulation (2.29) is that the unknown domain $C(t)$ is implicitly involved. The price we paid is the nonlinear term being a discontinuous nonlinear function: the Heaviside function $H(u) = 1$ as $u > 0$, and $= 0$ as $u \leqslant 0$.

The problem can be solved by the Ky Fan fixed-point theorem as before. Let us introduce the anisotropic Sobolev space

$$W_p^{1,2}(Q_T) = \{u \in L^p(Q_T)|\ \partial_t u,\ \nabla_x u,\ \nabla_x^2 u \in L^p(Q_T)\} \ ,$$

with norm:

$$\| u \|_{W_p^{1,2}} = \| u \|_p + \| \partial_t u \|_p + \| \nabla_x^2 u \|_p \ .$$

The boundary values of functions in $W_p^{1,2}(Q_T)$ span a fractional Sobolev space $W_p^{1-\frac{1}{2p},2-\frac{1}{p}}(\Gamma \times (0,T))$, where $\Gamma = \Gamma' \cup \Gamma''$. In particular, $C^{1,2}(\Gamma \times [0,T]) = \{u \in C(\Gamma \times [0,T])|\ \partial_t u,\ \nabla_x u\ \nabla_x^2 u \in C(\Gamma \times [0,T])\} \subset W_p^{1-\frac{1}{2p},2-\frac{1}{p}}(\Gamma \times (0,T))$.

Statement 2.3.16 *Assume* $\psi \in W_p^{1-\frac{1}{2p},2-\frac{1}{p}}(\Gamma \times (0,T))$, $p > n$, *and* $\psi \geqslant 0$; *the equation (2.29) has a unique solution* $u \in W_p^{1,2}(Q_T)$.

Proof. Define a set-valued mapping

$$u \mapsto F(u) = \begin{cases} 1 & u(x,t) > 0 \ , \\ [0,1] & u(x,t) = 0 \ , \\ 0 & u(x,t) < 0 \ , \end{cases}$$

then $F : C(\overline{Q}_T) \to 2^{L^p(Q_T)}$ is u.s.c. Let $\mathbb{K}_\psi : L^p(Q_T) \mapsto W_p^{1,2}(Q_T)$ be the affine operator $f \mapsto u$, where u satisfies the linear parabolic equation:

$$\begin{cases} \frac{\partial u}{\partial t} - \Delta u = f & \text{in } Q_T \ , \\ u|_{\Gamma' \times (0,T)} = \psi \ , \\ u|_{\Gamma'' \times (0,T)} = 0, \\ u(x,0) = 0, & \text{on } \Omega \ . \end{cases}$$

Then the set-valued mapping $\Gamma := \mathbb{K}_\psi \circ (L \circ F) : C(\overline{\Omega}) \to 2^{C(\overline{\Omega})}$ is a u.s.c. closed convex set-valued mapping. Define a compact convex set:

$$E = \{\mathbb{K}_\psi u | \ u \in L^p(Q_T), \ \| u \|_p \leqslant \mathrm{Lm}(\Omega)^{\frac{1}{p}}\} \ .$$

Obviously $\Gamma : E \to 2^E$. According to Ky Fan's fixed-point theorem, one finds $u \in W_p^{1,2}(Q_T)$ satisfying:

$$\begin{cases} \frac{\partial u}{\partial t} - \Delta u \in -LF(u) \\ u|_{\Gamma' \times (0,T)} = \psi, \ u|_{\Gamma'' \times (0,T)} = 0, \ u|_{\Omega \times \{0\}} = 0 \ . \end{cases}$$

On the set $u^{-1}(0) = \{(x,t) \in Q_T | \ u(x,t) = 0\}$, we have $\frac{\partial u}{\partial t} - \Delta u = 0$. This means u automatically satisfies (2.29).

Now, we prove the uniqueness of the solution u of equation (2.29). In fact, let u_1 and u_2 be two solutions of (2.29), and $u = u_1 - u_2$, then

$$\begin{cases} \frac{\partial u}{\partial t} - \Delta u = -L(H(u_1) - H(u_2)) & \text{in } Q_T, \\ u|_{\partial Q_T \setminus (\Omega \times \{T\})} = 0 \ . \end{cases}$$

By the maximum principle for parabolic equations, both sets $\{(x,t) \in Q_T | \ \pm u(x,t) > 0\}$ are null sets. Thus $u_1 = u_2$ a.e.

Returning to the original problem (2.28), we define $C(t) = \{x \in \Omega | \ u(x,t) > 0\}$, $\widetilde{\theta}(x,t) = \frac{\partial}{\partial t} u(x,t)$, and $\theta(\cdot,t) = \widetilde{\theta}|_{C(t)}$ $\forall t \in (0,T)$.

The verification of (2.28) is omitted. □

2.4 Nonexpansive Maps

Let X be a Banach space, and E be a closed subset of X. A map $T : E \to X$ is called a nonexpansive map, if

$$\| T(x) - T(y) \| \leqslant \| x - y \| \quad \forall x, y \in E .$$

In contrast with the fact that a contraction mapping $T : E \to E$, (i.e., $\exists \alpha \in (0, 1)$ such that

$$\| T(x) - T(y) \| \leqslant \alpha \| x - y \|)$$

always has a fixed point, the nonexpansive map does not in general. This can be seen by the following example:

Example 1. $X = c_0$, the space consisting of all sequences $x = \{\xi_i\}_1^\infty$ converging to zero, with the norm

$$\| x \| = \max_{i \geqslant 1} |\xi_i| .$$

Let

$$T(x) = (1 - \| x \|, \xi_1, \xi_2, \ldots) ,$$

and E be the closed unit ball. Then $T : E \to E$ is non-expansive:

$$\| T(x) - T(y) \| = \max \left\{ | \| x \| - \| y \| |, \quad \max_{i \geqslant 1} |\xi_i - \eta_i| \right\} \leqslant \| x - y \| .$$

However, T does not have a fixed point. In fact, if $x = T(x)$, then $\xi_1 = \xi_2 = \ldots = 1 - \| x \|$. This is impossible.

We are going to study the fixed points for a nonexpansive map on a closed convex subset of a Banach space X with a normal structure, which means that for every closed convex subset E of X containing at least two points, there exists a point $x_0 \in E$ such that

$$\sup_{x \in E} \| x - x_0 \| < \mathrm{diam}\,(E) . \tag{2.30}$$

It is easy to see that Hilbert spaces are Banach spaces with normal structure: Assume $a, b \in E$, then $\forall x \in E$, by parallelogram identity,

$$\left\| x - \frac{a+b}{2} \right\|^2 + \left\| \frac{a-b}{2} \right\|^2 = \frac{1}{2}(\| x - a \|^2 + \| x - b \|^2) \leqslant \mathrm{diam}\,(E) .$$

Taking $x_0 = \frac{1}{2}(a + b)$, (2.30), is satisfied.

Now, for any closed bounded convex subset E, one defines a number

$$r_E = \inf_{x \in E} \sup_{y \in E} \| x - y \| .$$

By definition, $r_E \leqslant \mathrm{diam}\,(E)$. If further, X is a Banach space with a normal structure, then $r_E < \mathrm{diam}\,(E)$.

Lemma 2.4.1 *If X is a reflexive Banach space and if E is a bounded closed convex set, then $\exists x_0 \in E$ such that*

$$\sup_{y \in E} \| x_0 - y \| = r_E .$$

Proof. The function $f(x) = \sup_{y \in E} \| x - y \|$ is convex, l.s.c., and coercive. We now apply Corollary 2.3.7; there exists $x_0 \in E$ such that $f(x_0) = \min_{x \in E} f(x)$. □

Setting $G(y) = \{x \in E | \; \| y - x \| \leqslant r_E\} \; \forall y \in E$, we define

$$\mathrm{ct}\,(E) = \bigcap_{y \in E} G(y) .$$

Provided by Lemma 2.4.1, $x_0 \in \mathrm{ct}\,(E)$. Therefore $\mathrm{ct}\,(E)$ is a nonempty closed convex subset of E. It is called the center of E. By definition, we have

$$\sup_{x,y \in ct(E)} \|x - y\| \leq r_E . \tag{2.31}$$

Theorem 2.4.2 *Suppose that E is a bounded closed convex set of a reflexive Banach space with a normal structure, and that $T : E \to E$ is nonexpansive, then the fixed-point set of T $\mathrm{Fix}\,(T)$ is closed and nonempty.*

Proof. (1) Let X be the set of all nonempty closed convex subsets of E. Define the partial order $\leqslant$ on X by inclusion:

$$C_1 \leqslant C_2 \text{ if } C_2 \subset C_1 \quad \forall C_1, C_2 \in \mathsf{X} .$$

Then $(\mathsf{X}, \leqslant)$ is an ordered set, in which every chain X_1 has a supremum: $\bigcap_{C \in \mathsf{X}_1} C$. The intersection is not empty, because E is weakly compact and C is weakly closed.

(2) Define

$$f(C) = \mathrm{ct}\,(\overline{\mathrm{conv}}(T(C))) ,$$

then $f : \mathsf{X} \to \mathsf{X}$ is order preserving and $E \leq f(E)$. Applying the Amann theorem, f has a fixed point $C_0 \in \mathsf{X}$, i.e., $C_0 \subset E$ is a nonempty closed convex T-invariant set satisfying:

$$C_0 = \mathrm{ct}(\overline{\mathrm{conv}}(T(C_0))) .$$

(3) It remains to verify that C_0 consists of a single point. If not, from the fact that $r_C < \mathrm{diam}\,(C)$ for all bounded closed convex sets C containing more than one point, we obtain:

$$\mathrm{diam}(C_0) = \mathrm{diam}(\mathrm{ct}(\overline{\mathrm{conv}}(T(C_0))))$$

$$
\begin{aligned}
&= \sup_{x,y \in \mathrm{ct}(\overline{\mathrm{conv}}(T(C_0)))} \| x - y \| \\
&\le r_{\overline{\mathrm{conv}}(T(C_0))} \\
&< \mathrm{diam}\,(\overline{\mathrm{conv}}(T(C_0))) \\
&= \mathrm{diam}\,(T(C_0)) \\
&\leqslant \mathrm{diam}\,(C_0)\ .
\end{aligned}
$$

This is a contradiction.

Thus, C_0 is a single-point set in E, say $C_0 = \{x_0\}$. It follows that

$$x_0 = f(\{x_0\}) = ct(\overline{\mathrm{conv}}\,(Tx_0)) = Tx_0\ .$$

Obviously Fix (T) is closed. □

Remark 2.4.3 *A Banach space X is called uniformly convex, if* $\forall \varepsilon > 0\ \exists \delta > 0$ *such that* $\forall x, y \in X \quad \| x \|, \ \| y \| \leqslant 1, \ \| x - y \| \geqslant \varepsilon$ *implies that* $\| \frac{x+y}{2} \| \leqslant 1 - \delta$. *Obviously, a Hilbert space is a uniformly convex Banach space. Moreover,* $L^p(\Omega, \mathcal{B}, \mu),\ 1 < p < +\infty$ *are also, but* $C(\overline{\Omega})$ *is not.*

It is known [DS] that a uniformly convex Banach space is a reflexive Banach space with normal structure. One has the following:

Theorem 2.4.4 *(Browder–Göhde) Suppose that X is a uniformly convex Banach space, and that E is a nonempty bounded closed convex set of X. If $T : E \to E$ is nonexpansive, then T has a fixed point.*

A natural question: can we find out a fixed point of a nonexpansive map T by iteration as we did for contraction mappings?

Starting from any point $x_0 \in E$, set $x_i = T^i x_0,\ \forall i$, from

$$\| x_{i+1} - x_i \| \leqslant \| x_i - x_{i-1} \| \leqslant \ldots \leqslant \| x_1 - x_0 \|\ ,$$

we know that $\{\| x_{i+1} - x_i \|\}_{i=0}^{\infty}$ is a nonincreasing sequence. It is obviously true that even if $\{x_i\}$ has a convergent subsequence $x_{n_i} \to x^*$, we still cannot conclude that x^* is a fixed point of T, e.g., $Tx = -x$, and $x_0 \neq \theta$.

To this end, Mann [Man] introduced an iteration method for the following perturbed mappings: $\forall \alpha \in (0, 1)$, set

$$T_\alpha x = \alpha x + (1 - \alpha) T x\ .$$

One easily verifies that if E is convex and $T : E \to E$ is nonexpansive then:

(1) $T_\alpha : E \to E$.
(2) T_α is nonexpansive.
(3) T_α shares the same fixed point set with T.
(4) Let X be a uniformly convex Banach space, and E be a bounded closed convex subset of X. Let p be the fixed point of T. If further, $I - T$ is proper (a map f is called proper if the inverse image of any compact set $C, f^{-1}(C)$ is compact), then $\forall \alpha \in (0,1)$, $\| T_\alpha^i x - p \| \to 0$ as $i \to \infty\ \forall x \in E$.

The proofs of (1)–(3) are trivial. In order to give a proof of (4). Let $z_i = T^i_\alpha x$. Since the sequence $\|z_{i+1} - z_i\|$ is nonincreasing, either (1) $\exists \epsilon > 0$ such that $\|z_{i+1} - z_i\| \geq \epsilon$, or (2) $\|z_{i+1} - z_i\| \to 0$.

But case (1) is impossible. In fact, from the definition of the uniform convexity $\exists \delta_0 = \delta(\alpha, \epsilon) > 0$, such that

$$\begin{aligned}\| z_{i+1} - p \| &= \| \alpha(z_i - p) + (1-\alpha)(Tz_i - Tp) \| \\ &\leqslant (1-\delta_0) \max\{\| z_i - p \|, \| Tz_i - Tp \|\} \\ &= \beta \| z_i - p \| \\ &\leqslant \cdots \\ &\leqslant \beta^i \| x - p \| ,\end{aligned}$$

where $\beta = 1 - \delta_0$. This proves that $z_i \to p$, a contradiction. In case (2), by the assumption that $I - T$ is proper, there exists a convergent subsequence $z_{n_i} \to p$. It follows that $p = Tp$, i.e., $p \in \text{Fix}\,(T) = \text{Fix}\,(T_\alpha)$. Thus $\|z_n - p\| \leq \|z_{n_i} - p\| \to 0$ as $n \geq n_i \to \infty$.

We have thus proved:

Theorem 2.4.5 *Assume that X is a uniformly convex Banach space, and that E is a nonempty bounded closed convex set of X. If $T : E \to E$ is nonexpansive and $I - T$ is proper, $\forall \alpha \in (0,1)$, let $T_\alpha = \alpha I + (1-\alpha)T$, then $\forall x \in E$, the sequence $T^i_\alpha x$ converges to a fixed point of T.*

Applications

(1) The iteration method is an effective computing method in the convex feasibility problem:

Given a family of closed convex sets $\{E_i|\ i = 1, \ldots, n\}$ in a Hilbert space H, find a point $x \in \bigcap_{i=1}^{n} E_i$, if the intersection is nonempty.

In fact, given a closed convex set E in H, $\forall x \in H$, there is a projection onto E: $P_E x \in E$, which is defined by the minimizer of the problem:

$$c = \min_{y \in E} \| x - y \| .$$

The existence of $P_E x$ follows from the parallelogram identity:

$$\| y_n - y_m \|^2 = 2(\| x - y_n \|^2 + \| x - y_m \|^2) - 4 \left\| x - \frac{y_n + y_m}{2} \right\|^2 ,$$

where $\{y_n\}$ is a minimizing sequence: $\| x - y_n \| \to c$. Again by the parallelogram identity, the minimizer is unique. We define the minimizer y^* to be the projection $P_E x$.

The projection can be characterized by the following variational inequality:

$$(x - P_E x, P_E y - P_E x) \leqslant 0 \quad \forall x, y \in H . \tag{2.32}$$

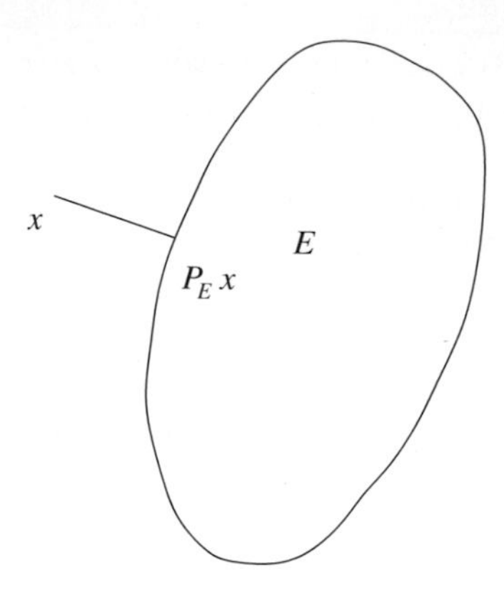

Fig. 2.5.

In fact, $\forall t \in (0,1)$, $tP_E y + (1-t)P_E x \in E$. One has

$$\| x - tP_E y - (1-t)P_E x \|^2 \geqslant \| x - P_E x \|^2 .$$

i.e.,

$$-2t(x - P_E x, P_E y - P_E x) + t^2 \| P_E y - P_E x \|^2 \geqslant 0 .$$

Letting $t \to 0$, we obtain the desired result (2.32).

Statement 2.4.6 *P_E is a nonexpansive map on H.*

Proof. By definition, $\forall x, y \in H$

$$(x - P_E x, P_E y - P_E x) \leqslant 0 ,$$
$$(y - P_E y, P_E x - P_E y) \leqslant 0 .$$

Summing the two inequalities:

$$(x - y + (P_E y - P_E x), P_E y - P_E x) \leqslant 0 ,$$

it follows that

$$\| P_E y - P_E x \|^2 \leqslant (y - x, P_E y - P_E x) \leqslant \| x - y \| \| P_E y - P_E x \| .$$

Therefore P_E is nonexpansive. □

One may use Mann iteration to compute a feasible solution. Various algorithms have been introduced to improve the convergent rate and the computation.

(2) Periodic solution for nonlinear evolution equations:

Let X be a real Hilbert space. Given $\varphi \in X$, and $f : \mathbb{R}^1_+ \times X \to X$. We consider the following nonlinear evolution equation: Find $x \in C^1(\mathbb{R}^1_+, X)$ satisfying

$$\dot{x}(t) = f(t, x(t)) \quad t \in \mathbb{R}^1_+ , \tag{2.33}$$
$$x(0) = \varphi . \tag{2.34}$$

Statement 2.4.7 *Assume $\omega > 0$ and that*

(1) f is ω-periodic in t, i.e., $f(t,x) = f(t+\omega, x)$,
(2) $(f(t,x) - f(t,y), x-y) \leqslant 0 \; \forall t \in \mathbb{R}^1_+, \; \forall x, y \in X$,
(3) $\exists R > 0$ such that $(f(t,x), x) < 0, \; \forall t \in [0,\omega], \; \forall x \in \partial B_R(\theta)$,
(4) $\forall \varphi \in \overline{B}_R(\theta)$, the initial-value problem (2.33) (2.34) has a solution.

Then (2.33) has a unique ω-periodic solution, i.e., $x(t+\omega) = x(t)$.

Proof. From (4), we consider the Poincaré map on X:

$$T : x(0) \mapsto x(\omega) ,$$

and want to show that T has a unique fixed point.

First we show that $T : \overline{B}_R(\theta) \to \overline{B}_R(\theta)$ is a well-defined nonexpansive map:

1. We show that the solution for the initial-value problem is unique, i.e., if $x(t)$, $y(t)$ are solutions with the same initial data φ, then $x(t) = y(t)$. Indeed, from

$$\begin{aligned} \frac{d}{dt} \| x(t) - y(t) \|^2 &= 2(x'(t) - y'(t), x(t) - y(t)) \\ &= 2(f(t, x(t)) - f(t, y(t)), x(t) - y(t)) \\ &\leqslant 0 , \end{aligned}$$

it follows that

$$\| x(t) - y(t) \| \leqslant \| x(0) - y(0) \| . \tag{2.35}$$

Thus $x(t) = y(t)$, and then T is well defined.
2. Moreover, (2.35) implies that T is nonexpansive.
3. We turn to showing that T maps $\overline{B}_R(\theta)$ into itself. Because of assumption (3)

$$\frac{d}{dt} \| x \|^2 = 2(x'(t), x(t)) = 2(f(t, x(t)), x(t)) < 0 ,$$

whenever $x(t) \in \partial B_R(\theta)$. If the conclusion is not true, then $\exists t > 0$ such that $x(t) \notin \overline{B_R(\theta)}$ and then $\exists t_0 \in (0,t)$, such that $x(t_0) \in \partial B_R(\theta)$ and $x(t) \notin \overline{B_R(\theta)}$ as $t_0 < t < t_0 + \delta$ for some $\delta > 0$. But this contradicts the differential inequality.

Now, we apply Theorem 2.4.2; there is a fixed point of T in $\overline{B}_R(\theta)$. Again applying step 1, the ω-periodic solution is also unique. □

2.5 Monotone Mappings

Let us consider the subdifferential of a convex function f on a vector space X: $\forall x, y \in \text{dom}\,(f)$, $\forall x^* \in \partial f(x)$, $\forall y^* \in \partial f(y^*)$, we have

$$\langle x^*, y - x \rangle + f(x) \leqslant f(y) ,$$
$$\langle y^*, x - y \rangle + f(y) \leqslant f(x) .$$

By addition,

$$\langle y^* - x^*, y - x \rangle \geqslant 0 \tag{2.36}$$

This is what we call the monotonicity of ∂f.

Definition 2.5.1 *Suppose that X is a real Banach space and that $E \subset X$ is a nonempty subset. A set-valued mapping $A : E \to 2^{X^*}$ is called monotone if $\forall x, y \in E$, $\forall x^* \in A(x)$, $\forall y^* \in A(y)$, one has*

$$\langle y^* - x^*, y - x \rangle \geqslant 0 .$$

The set $\{x \in E |\ A(x) \neq \emptyset\}$ is called the domain of A, denoted by $D(A)$, and the set $\Gamma_A = \{(x, x^*) \in E \times X^* |\ x \in D(A),\ x^* \in A(x)\}$ is called the graph of A. A single-valued monotone mapping is called a monotone operator.

Example 1. Suppose that X is a real Hilbert space, and that A is a linear positive operator, i.e., $\forall x \in D(A)$, $(Ax, x) \geqslant 0$. Then A is a monotone operator.

Example 2. Suppose that X is a real Hilbert space, and that T is a nonexpansive map, then $A = \mathrm{id} - T$ is a monotone operator:

$$(Ay - Ax, y - x) = \| y - x \|^2 - (Ty - Tx, y - x) \geqslant 0 .$$

Example 3. Let X be a real Banach space, and let $f : X \to \mathbb{R}^1 \cup \{+\infty\}$ be convex, then ∂f is a monotone mapping.

Example 4. (p-Laplacian) Let Ω be a bounded domain of $\mathbb{R}^n$. For $1 < p < \infty$, the operator

$$Au = -\mathrm{div}\left(|\nabla u|^{p-2} \nabla u\right) = -\sum_{i=1}^n \frac{\partial}{\partial x_i} \left(\left[\sum_{j=1}^n \left(\frac{\partial u}{\partial x_j} \right)^2 \right]^{\frac{p-2}{2}} \frac{\partial u}{\partial x_i} \right)$$

defines a map from $\mathring{W}_p^1(\Omega)$ to $W_{p'}^{-1}(\Omega)$, $\frac{1}{p} + \frac{1}{p'} = 1$, as follows:

$$\langle Au, v \rangle = \sum_{i=1}^n \int_\Omega \left[\sum_{j=1}^n \left(\frac{\partial u}{\partial x_j} \right)^2 \right]^{\frac{p-2}{2}} \frac{\partial u}{\partial x_i} \frac{\partial v}{\partial x_j} dx \quad \forall v \in \mathring{W}_p^1(\Omega) .$$

In fact, by the Hölder inequality:

$$\int_\Omega |\nabla u|^{p-1} |\nabla v| \leqslant \left(\int_\Omega |\nabla u|^p \right)^{\frac{1}{p'}} \left(\int_\Omega |\nabla v|^p \right)^{\frac{1}{p}} ,$$

$Au \in (\overset{\circ}{W}{}_p^1(\Omega))^* = W_{p'}^{-1}(\Omega)$.

We verify that A is monotone: According to the elementary inequalities:

$$(|b|^{p-2}b - |a|^{p-2}a)(b-a) \geqslant c_p \begin{cases} |b-a|^p & \text{if } p \geqslant 2\,, \\ (1+|b|+|a|)^{p-2}|b-a|^2 & \text{if } 1<p<2\,, \end{cases}$$

where $c_p > 0$ is a constant, and $a, b \in \mathbb{R}^n$, and the Hölder inequality, we obtain

$$\begin{aligned} &\langle Au - Av, u-v\rangle \\ &\geqslant c_p \begin{cases} \int_\Omega |\nabla u - \nabla v|^p & \text{if } p \geqslant 2\,, \\ (\int_\Omega (1+|\nabla u|+|\nabla v|)^p)^{1-\frac{2}{p}} (\int_\Omega |\nabla u - \nabla v|^p)^{\frac{2}{p}} & \text{if } 1<p<2\,. \end{cases} \end{aligned}$$

The requirement of the continuity for monotone operators is very weak.

Definition 2.5.2 *Let X be a real Banach space and let $E \subset X$ be a nonempty subset. A map $A : E \to X^*$ is called hemi-continuous at $x_0 \in E$, if $\forall y \in X$, $\forall t_n \downarrow 0$ with $x_0 + t_n y \in E$, imply that $A(x_0+t_n y) \overset{*}{\rightharpoonup} A(x_0)$. It is called demi-continuous at $x_0 \in E$, if $\forall \{x_n\} \subset E$, $x_n \to x_0$ implies that $A(x_n) \overset{*}{\rightharpoonup} A(x_0)$, where $* \rightharpoonup$ is the w^* -convergence.*

Obviously,

$$\text{"continuous"} \Longrightarrow \text{"demi-continuous"} \Longrightarrow \text{"hemi-continuous"}\,.$$

The monotone operator in Example 4 is hemi-continuous. In fact, $\forall u, v, w \in \overset{\circ}{W}{}_p^1(\Omega)$, we consider the function $t \mapsto \langle A((1-t)x+ty), w\rangle$, and verify the continuity. Now

$$\begin{aligned} \langle A((1-t)u+tv), w\rangle = \int_\Omega & \left[\sum_{j=1}^n \left((1-t)\frac{\partial u}{\partial x_j} + t\frac{\partial v}{\partial x_j}\right)^2\right]^{\frac{p-2}{2}} \\ & \times \sum_{i=1}^n \left((1-t)\frac{\partial u}{\partial x_i} + t\frac{\partial v}{\partial x_i}\right)\frac{\partial w}{\partial x_i}\,. \end{aligned}$$

Since the integrand on the RHS is dominated by an integrable function $(|\nabla u| + |\nabla v|)^{p-1}|\nabla w|$, the Lebesgue dominance theorem is applied.

An important property for monotone operators reads as:

Lemma 2.5.3 *Let E be a convex subset of a real Banach space X. If $A : E \to X^*$ is hemi-continuous and monotone, then for any sequence $\{x_j\} \subset E$ with $x_j \rightharpoonup x \in E$ and $\overline{\lim}\langle A(x_j), x_j - x\rangle \leqslant 0$, we have*

$$\underline{\lim}\langle A(x_j), x_j - y\rangle \geqslant \langle A(x), x-y\rangle \quad \forall y \in E\,.$$

Proof. First we claim that

$$\lim \langle A(x_j), x_j - x\rangle = 0 \,. \tag{2.37}$$

Indeed, since A is monotone, it follows that

$$0 = \underline{\lim}\langle A(x), x_j - x\rangle \leqslant \underline{\lim}\langle A(x_j), x_j - x\rangle \leqslant \overline{\lim}\langle A(x_j), x_j - x\rangle \leqslant 0 \,.$$

Again by the monotonicity and (2.37), $\forall z \in E$

$$\underline{\lim}\langle A(x_j), x - z\rangle = \underline{\lim}\langle A(x_j), x_j - z\rangle \geqslant \lim \langle A(z), x_j - z\rangle = \langle A(z), x - z\rangle \,. \tag{2.38}$$

Now, $\forall y \in E$, $\forall t_n \downarrow 0$, substituting $z = z_n := (1 - t_n)x + t_n y$ into (2.38), we obtain

$$\underline{\lim}\langle A(x_j), x - y\rangle \geqslant \langle A(z_n), x - y\rangle \,. \tag{2.39}$$

By using the hemi-continuity, the RHS of (2.39) tends to $\langle A(x), x - y\rangle$. Combining (2.37) (2.39) and the last fact, it follows that

$$\underline{\lim}\langle A(x_j), x_j - y\rangle \geqslant \langle A(x), x - y\rangle \,. \tag{2.40}$$

□

Remark 2.5.4 *The deduction argument from (2.37) to (2.40) is called Minty's trick, in which a combination of the monotonicity and the hemi-continuity is applied. This lemma is substantial in the study of monotone operators. The following notion on pseudo-monotonicity is abstracted from it.*

Definition 2.5.5 *Let X be a reflexive Banach space and let $E \subset X$ be a nonempty closed convex subset. An operator $A : E \to X^*$ is called pseudo monotone, if*

(1) $\forall$ finite-dimensional linear subspace $L \subset X$, $A|_{L\cap E} : L \cap E \to X^*$ is demi-continuous.
(2) $\forall$ sequence $\{x_j\} \subset E$ with $x_j \rightharpoonup x \in E$, the condition $\overline{\lim}\langle A(x_j), x_j - x\rangle \leqslant 0$ implies that $\underline{\lim}\langle A(x_j), x_j - y\rangle \geqslant \langle A(x), x - y\rangle$, $\forall y \in E$.

Thus, a hemi-continuous monotone operator is pseudo monotone. Moreover, a completely continuous mapping $A : X \to X^*$ (i.e., for any $x_j \rightharpoonup x$ in X, we have $Ax_j \to Ax$ in X^*) is pseudo monotone.

In contrast with the fixed-point problem for compact maps and non-expansive mappings, we shall study the surjection of pseudo monotone operators, because the latter maps a subset of a Banach space into its dual space.

The following generalized Ky Fan's inequality is the basis of this section.

Theorem 2.5.6 *(Brezis, Nirenberg, Stampacchia) Assume that E is a nonempty convex set of an LCS X. Assume $\Phi : E \times E \to \mathbb{R}^1$ satisfying*

(1) $\forall$ *finite-dimensional linear subspace* L *of* X, $\forall y \in E \cap L$, $x \mapsto \Phi(x,y)|_{L\cap E}$ *is l.s.c.*
(2) $\forall x \in E$, $y \mapsto \Phi(x,y)$ *is quasi concave.*
(3) $\exists$ *a compact set* $K \subset X$, $\exists y_0 \in E$ *such that* $\{x \in E |\ \Phi(x,y_0) \leqslant 0\} \subset K$.
(4) $\sup_{x\in E} \Phi(x,x) = 0$.
(5) $\forall x, y \in E$ *and for any net* $x_\alpha \to x$, $\forall t \in [0,1]$, $\Phi(x_\alpha, (1-t)x + ty) \leqslant 0$ *implies that* $\Phi(x,y) \leqslant 0$.

Then $\exists x_0 \in E$ *such that* $\sup_{y\in E} \Phi(x_0, y) \leqslant 0$.

Proof. Again we transform the problem into its geometric version by setting $\Gamma = \{(x,y) \in E \times E |\ \Phi(x,y) > 0\}$. Then (1)–(5) are transformed into:

(1′) $\forall L$, letting $E_L = E \cap L$ and $\Gamma^L = \Gamma \cap (L \times L)$, $\forall y \in E_L$ $\Gamma_2^L(y)$ is open in E.
(2′) $\forall x \in E$, $\Gamma_1(x)$ is convex.
(3′) $\exists y_0 \in E$ such that $E \backslash \Gamma_2(y_0) \subset K$.
(4′) $\triangle \cap \Gamma = \emptyset$, where $\triangle$ is the diagonal of $E \times E$.
(5′) Let $\overline{xy}$ be the segment connecting x and y, then $\Gamma_1(x_\alpha) \cap \overline{xy} = \emptyset$ implies $y \notin \Gamma_1(x)$ $\forall y \in E$ for any net $x_\alpha \to x$.

Now we set $\mathsf{X} = \{L |$ finite-dimensional linear subspace, with $y_0 \in L$ and $E_L \neq \emptyset\}$. $\forall L \in \mathsf{X}$, applying Ky Fan's inequality, $\exists x_L \in E_L$ such that $\Gamma_1^L(x_L) = \emptyset$. From (3′), $x_L \in K$ as $L \in \mathsf{X}$. $\forall L \in \mathsf{X}$, setting $N^L = \{x \in K |\ \Gamma_1^L(x) = \emptyset\}$, from the above discussion, it is nonempty. We claim that the family $\{\overline{N}^L |\ L \in \mathsf{X}\}$ has the finite intersection property: Indeed, $\forall$ $L_1, \ldots, L_n \in \mathsf{X}$, setting $L = \text{span}\{L_1, \ldots, L_n\}$ we have $L \in \mathsf{X}$. Then $\Gamma_1^L(x_L) = \bigcup_{i=1}^n \Gamma_1^{L_i}(x_L)$ and $N^L = \bigcap_{i=1}^n N^{L_i}$. Since K is compact, $Z := \bigcap_{L\in\mathsf{X}} \overline{N}^L \neq \emptyset$.

We shall prove that $\forall x \in Z$, $\Gamma_1(x) = \emptyset$. To this end, $\forall y \in E$, we choose $L \in \mathsf{X}$ such that the segment $\overline{xy} \subset L$. Since $x \in \overline{N}^L$, $\exists$ a net $x_\alpha^L \in N^L$, such that $\Gamma_1(x_\alpha^L) \cap \overline{xy} = \emptyset$, and $x_\alpha^L \to x$. From assumption (5′), $y \notin \Gamma_1(x)$. Since $y \in E$ is arbitrary, we obtain $\Gamma_1(x) = \emptyset$, Therefore there exists $x_0 \in Z$ such that $\sup_{y\in E} \Phi(x_0, y) \leqslant 0$. □

In many of the applications, the LCS is taken to be a reflexive Banach space with its weak topology. Since then for weakly closed set it is weakly compact if and only if it is sequentially weakly compact (cf. Theorem 2.2.5), the net convergence in condition (5) of the above theorem can be replaced by the sequential convergence.

The following theorem is the foundation of the theory of variational inequalities, which arise in free boundary problems for partial differential equations.

Theorem 2.5.7 *(Hartman–Stampacchia) Suppose that X is a real reflexive Banach space, and that $E \subset X$ is a nonempty closed convex set. Assume*

(1) $A : E \to X^*$ is pseudo monotone,
(2) $\varphi : E \to \mathbb{R}^1$ is convex and l.s.c.,
(3) $\exists y_0 \in E\ \exists R_0 > 0$ such that

$$\langle Ax, x - y_0 \rangle + \varphi(x) - \varphi(y_0) > 0 \quad \text{as } \| x \| \geqslant R_0 \text{ and } x \in E\ .$$

Then $\exists x_0 \in X$ such that

$$\langle Ax_0, x_0 - y \rangle + \varphi(x_0) - \varphi(y) \leqslant 0, \quad \forall y \in E\ .$$

Proof. (1) Assign the weak topology on X, and define

$$\Phi(x, y) = \langle Ax, x - y \rangle + \varphi(x) - \varphi(y)\ .$$

It is easy to see that conditions (1)–(4) of Theorem 2.5.6 are satisfied. Let us verify (5).

$\forall x, y \in E\ \forall t \in [0, 1]$, assume $x_j \rightharpoonup x$ and $\Phi(x_j, (1 - t)x + ty) \leqslant 0$. Setting $t = 0$ and 1 separately, we have

$$\overline{\lim}\, \langle A(x_j), x_j - x \rangle \leqslant \overline{\lim}\, [\varphi(x) - \varphi(x_j)] \leqslant 0 \tag{2.41}$$

and

$$\overline{\lim}\, \langle A(x_j), x_j - y \rangle \leqslant \overline{\lim}\, [\varphi(y) - \varphi(x_j)] \leqslant \varphi(y) - \varphi(x)\ . \tag{2.42}$$

From the pseudo monotonicity, it follows that

$$\underline{\lim}\, \langle A(x_j), x_j - y \rangle \geqslant \langle A(x), x - y \rangle\ . \tag{2.43}$$

Combining (2.42) and (2.43), we obtain $\Phi(x, y) \leqslant 0$.

(2) Now we apply Theorem 2.5.6 and conclude the existence of $x_0 \in E$ satisfying $\Phi(x_0, y) \leqslant 0$. This is the desired conclusion. □

Theorem 2.5.8 *(F. Browder) Suppose that X is a real reflexive Banach space, and that $A : X \to X^*$ is pseudo monotone and coercive, i.e., $\| x \|^{-1} \langle A(x), x \rangle \to +\infty$ as $\| x \| \to +\infty$. Then A is surjective.*

Proof. We shall prove that $\forall z \in X^*$, $\exists x_0 \in X$ satisfying $A(x_0) = z$. Define $T : x \mapsto A(x) - z$. Then T is pseudo monotone and satisfies

$$\langle Tx, x \rangle > 0 \quad \text{as } \| x \| > R_0\ ,$$

for some $R_0 > 0$, provided by the coerciveness of A. We apply the Hartman–Stampacchia theorem to conclude the existence of $x_0 \in X$ satisfying $\langle Tx_0, x_0 - y \rangle \leqslant 0\ \forall y \in X$. Since y is arbitrary in the linear space X, it follows that $Ax_0 = z$. □

As a consequence, we return to Example 4, the $p-$ Laplacian $-\Delta_p : W_0^{1,p}(\Omega) \to W^{-1,p'}(\Omega), 1 < p < \infty$, is a homeomorphism. The surjection follows from Browder's theorem, and the injection as well as the continuity of the inverse mapping follow from the inequality in the verification of the monotonicity.

Corollary 2.5.9 *(Hartman–Stampacchia) Suppose that A is a hemi-continuous monotone operator defined on the unit ball B centered at θ of a Hilbert space H. If $Ax \neq \lambda x$, $\forall x \in \partial B$, $\forall \lambda < 0$ then $\exists x_0 \in \overline{B}$ such that $Ax_0 = \theta$.*

Proof. Since A is pseudo monotone, we apply Theorem 2.5.7 to conclude the existence of $x_0 \in \overline{B}$ satisfying

$$(Ax_0, x_0 - y) \leqslant 0 \quad \forall y \in \overline{B} \, .$$

If $x_0 \in \overset{\circ}{B}$, then we have $Ax_0 = \theta$. Otherwise $x_0 \in \partial B_R$ $\forall y \in B_R$, we decompose $y = (1-t)x_0 + y^\perp$, and $Ax_0 = \lambda x_0 + y_0^\perp$, where $t > 0$, $y^\perp \perp x_0$, $y_0^\perp \perp x_0$, and $\lambda \in \mathbb{R}^1$. Then

$$-\lambda t \parallel x_0 \parallel^2 + (y_0^\perp, y^\perp) \geqslant 0 \, . \tag{2.44}$$

Since $y^\perp$ is arbitrary, first letting $t \to +0$, we obtain $y_0^\perp = \theta$, i.e., $Ax_0 = \lambda x_0$. By the assumption, $\lambda \geqslant 0$. Again by (2.44)

$$\lambda t \parallel x_0 \parallel^2 \leqslant 0 \, ,$$

which implies that $\lambda = 0$. Therefore $Ax_0 = \theta$. □

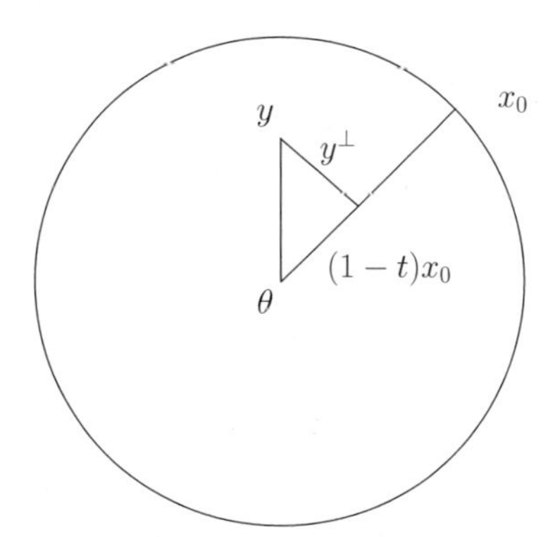

Fig. 2.6.

Corollary 2.5.10 *(Minty) Suppose that H is a real Hilbert space, and that A is a continuous strongly monotone operator, i.e., $\exists c > 0$ such that*

$$(Ax - Ay, x - y) \geqslant c \parallel x - y \parallel^2 \quad \forall x, y \in H \, . \tag{2.45}$$

Then A is a homeomorphism.

Proof. Obviously, A is pseudo monotone and coercive. As a consequence of the Browder theorem, A is surjective. The injectivity of A as well as the continuity of A^{-1} follows from the inequality (2.45). □

Applications

1. Elliptic variatiational inequality

Let A be the p-Laplacian defined in Example 4 and let $\Omega \subset \mathbb{R}^n$ be a bounded domain. We consider the closed convex subset of $X = \overset{\circ}{W}{}_p^1(\Omega)$ with $1 < p < \infty$:

$$E = \{u \in \overset{\circ}{W}{}_p^1(\Omega) |\ |\nabla u(x)| \leqslant 1 \text{ a.e.}\} .$$

Given $f \in W_{p'}^1(\Omega)$, where $p' = \frac{p}{p-1}$, find $u \in E$ such that

$$\langle Au, v - u\rangle \geqslant \langle f, v - u\rangle \quad \forall v \in E .$$

The existence of a solution follows from the Hartman–Stampacchia theorem directly. This is a free boundary problem, in fact the solution u, if it is regular, satisfies the equation:

$$\begin{cases} -\operatorname{div}(|\nabla u|^{p-2}\nabla u) = f & \text{in } \Omega_1, \\ |\nabla u| = 1 & \text{in } \Omega_2, \\ \text{u and } \nabla u \text{ coincide on the interface of } \Omega_1 \text{ and } \Omega_2 , \end{cases}$$

where

$$\Omega_1 = \{x \in \Omega |\ |\nabla u(x)| < 1\} ,$$

and

$$\Omega_2 = \Omega \backslash \Omega_1 .$$

The same method can be applied to the obstacle problem and the Stefan problem studied in Sect. 2.3. But in these problems, the working spaces should be taken to be the following Hilbert spaces: $H_0^1(\Omega)$ and $H^1([0,T] \times \Omega)$ respectively.

2. Weak solutions for quasi-linear elliptic equations

Let Ω be a bounded domain of $\mathbb{R}^n$. We study the weak solution of the quasi-linear elliptic equation:

$$\begin{cases} \sum_{i=1}^n \frac{\partial}{\partial x_i} A_i(x, u(x), \nabla u(x)) + B(x, u(x), \nabla u(x)) = f(x) & \text{in } \Omega , \\ u|_{\partial\Omega} = 0 , \end{cases}$$

where we make use of the following structural condition on $\vec{A} = \{A_i\}_1^n$ and B:

(1) $B \in C(\overline{\Omega} \times \mathbb{R}^1 \times \mathbb{R}^n, \mathbb{R}^1)$, $\vec{A} \in C(\overline{\Omega} \times \mathbb{R}^1 \times \mathbb{R}^n, \mathbb{R}^n)$,

(2) $\exists p \in (1,\infty)$, $\exists g \in L^{p'}(\Omega)$, $p' = \frac{p}{p-1}$, $\exists C > 0$ such that

$$\begin{aligned} \| \vec{A}(x,u,\xi) \| &\leqslant g(x) + C(|u| + |\xi|)^{p-1} , \\ |B(x,u,\xi)| &\leqslant g(x) + C(|u| + |\xi|)^{p-1} , \end{aligned}$$

(3)

$$[\vec{A}(x,u,\xi) - \vec{A}(x,u,\xi')] \cdot (\xi - \xi') > 0, \ \forall (x,u) \in \Omega \times \mathbb{R}^1, \ \forall \xi, \xi' \in \mathbb{R}^n, \ \xi \neq \xi' ,$$

and

(4)

$$\frac{1}{1 + |\xi| + |\xi|^{p-1}} \vec{A}(x,u,\xi) \cdot \xi \to +\infty \quad \text{as } |\xi| \to \infty, (x,u,\xi) \in \Omega \times \mathbb{R}^1 \times \mathbb{R}^n .$$

Remark 2.5.11 *Condition (3) (i.e., the monotone condition) implies the ellipticity of the differential operator:* $u(x) \mapsto \operatorname{div} \vec{A}(x,u(x),\nabla u(x))$.

Let $X = \overset{\circ}{W}{}_p^1(\Omega)$; we define a form on $X \times X$:

$$a(u,v) = \int_\Omega [\vec{A}(x,u(x),\nabla u(x)) \cdot \nabla v(x) + B(x,u(x),\nabla u(x)) v(x)] dx \\ \times \ \forall (u,v) \in X \times X .$$

From (1) and (2),

$$|a(u,v)| \leqslant C(\| u \|_{1,p}^{p-1} + \| g \|_{p'}) \| v \|_{1,p} .$$

We define $T : X \to X^*$ by $\langle Tu, v \rangle = a(u,v)$, where $\langle , \rangle$ is the duality between X^* and X.

In order to verify that T is pseudo monotone, we define $A : X \times X \to X^*$ as follow:

$$\langle A(u,w), v \rangle = \int_\Omega \vec{A}(x,u,\nabla w) \cdot \nabla v + \int_\Omega B(x,u,\nabla u) v \ \forall u,v,w \in X .$$

We conclude:

(a) $\langle A(u,u) - A(u,w), u - w \rangle \geqslant 0$, $\forall u, w \in X$.
(b) $\forall u \in X, w \mapsto A(u,w)$ is bounded and demi-continuous.
(c) If $u_j \rightharpoonup u$ in X and if $\langle A(u_j,u_j) - A(u_j,u), u_j - u \rangle \to 0$, then $A(u_j,w) \rightharpoonup A(u,w)$ in X^* $\forall w \in X$.

(a) and (b) are obviously true; we prove (c).

Proof. From $u_j \rightharpoonup u$ in X, we have $u_j \to u$(in $L^q(\Omega)$) and $u_j \to u$ a.e. after a subsequence, and then $\forall w \in X$,

$$\vec{A}(x,u_j(x),\nabla w(x)) \to \vec{A}(x,u(x),\nabla w(x)) \quad \text{a.e.}$$

Combining with the growth condition (2),

$$\vec{A}(\cdot, u_j, \nabla w) \to \vec{A}(\cdot, u, \nabla w) \text{ in } L^{p'}(\Omega, \mathbb{R}^n) .$$

Thus

$$\int_\Omega \vec{A}(x, u_j, \nabla w)\nabla u_j \to \int_\Omega \vec{A}(x, u, \nabla w)\nabla u , \tag{2.46}$$

and

$$\int_\Omega \vec{A}(x, u_j, \nabla w)\nabla v \to \int_\Omega \vec{A}(x, u, \nabla w)\nabla v . \tag{2.47}$$

It remains to verify that $g_j := B(\cdot, u_j, \nabla u_j) \rightharpoonup g := B(\cdot, u, \nabla u)$ $(L^{p'}(\Omega))$. The difficulty is that we only have $\nabla u_j \rightharpoonup \nabla u$ in $L^p(\Omega, \mathbb{R}^n)$, and that B is nonlinear with respect to ∇u. How can we pass the limit?

If one can show that $\nabla u_j \to \nabla u$ a.e., then $g_j \to g$ a.e., and then we claim that $g_j \rightharpoonup g$ in $L^{p'}(\Omega)$.

In fact, let $E_k = \{x \in \Omega \| g_j(x) - g(x)| \le 1, j \ge k\}$; we have $E_1 \subset E_2 \subset \ldots$, and $\text{mes}(E_k) \to \text{mes}(\Omega)$. Since $\forall \chi \in L^p(\Omega)$ with support in E_k, we have $\int_\Omega (g_j - g)\chi \to 0$, from the Lebesgue dominance theorem, and those χ span a dense set in $L^p(\Omega)$; in combination with the fact that $\|g_j\|_{p'}$ is bounded, the conclusion follows.

We turn to verifying that $\nabla u_j \to \nabla u$ a.e., and define the function:

$$F_j(x) = [\vec{A}(x, u_j(x), \nabla u_j(x)) - \vec{A}(x, u_j(x), \nabla u(x))] \cdot \nabla(u_j(x) - u(x)) .$$

By (3), $F_j(x) \geqslant 0$. As a consequence of our assumption, $\int_\Omega F_j(x)dx \to 0$. From Fatou's lemma, it follows that $F_j(x) \to 0$ a.e. Therefore there exists a null set Z such that $u_j(x) \to u(x)$, and $F_j(x) \to 0$ $\forall x \notin Z$.

Firstly, we claim that $\exists$ a measurable function $M(x) < +\infty$, such that

$$|\nabla u_j(x)| \leqslant M(x) \quad \forall x \notin Z . \tag{2.48}$$

From (2),

$$F_j(x) \ge \vec{A}(x, u_j(x), \nabla u_j(x))\nabla u_j(x) - C_1(x)(1 + |\nabla u_j(x)| + |\nabla u_j(x)|^{p-1})$$

where $C_1(x)$ depends on $u(x)$ and $\nabla u(x)$.

If (2.48) were false, then there would be a positive measure set E, on which $|\nabla u_j(x)| \to +\infty$. It follows that $F_j(x) \to +\infty$, from (4). But, this is a contradiction.

Secondly, we claim that $\forall x \notin Z$, $\nabla u_j(x) \to \nabla u(x)$.

In fact, let $\xi(x)$ be a limit point of $\nabla u_j(x)$. From $F_j(x) \to 0$ it follows that:

$$[\vec{A}(x, u(x), \xi(x)) - \vec{A}(x, u(x), \nabla u(x))] \cdot (\xi(x) - \nabla u(x))dx = 0 .$$

Applying the monotone relation (3), we have $\xi(x) = \nabla u(x)$. Since all limit points of $\nabla u_j(x)$ equal $\nabla u(x)$, we obtain $\nabla u_j(x) \to \nabla u(x)$ a.e.

Combining (2.47) with this fact, we obtain $A(u_j, w) \rightharpoonup A(u, w)$ in X^* $\forall w \in X$. □

(d) If $u_j \rightharpoonup u$ in X and $A(u_j, v) \rightharpoonup \psi$ in X^*, then $\langle A(u_j, v), u_j\rangle \to \langle \psi, u\rangle$.

In fact, we already have $u_j \to u$ $(L^p(\Omega, \mathbb{R}^1))$ and $\vec{A}(\cdot, u_j, \nabla v) \to \vec{A}(\cdot, u, \nabla v)$ $(L^{p'}(\Omega, \mathbb{R}^n))$. Thus

$$\langle A(u_j, v), u_j\rangle = \int_\Omega \vec{A}(x, u_j, \nabla v)\nabla(u_j - u) + B(x, u_j, \nabla u_j)(u_j - u) \\ + \langle A(u_j, v), u\rangle \to \langle \psi, u\rangle \,.$$

After these preparations, we turn to verifying that the operator T, defined by the form a, is pseudo monotone. Since $u \mapsto T(u)$ is obviously demicontinuous, it is sufficient to prove that $u_j \rightharpoonup u$ and $\overline{\lim}\, \langle T(u_j), u_j - u\rangle \leqslant 0$ implies that $\underline{\lim}\, \langle T(u_j), u_j - v\rangle \geqslant \langle T(u), u - v\rangle$ $\forall v \in X$.

First we show that $\langle T(u_j), u_j - u\rangle \to 0$, and define the sequence:

$$b_j = \langle A(u_j, u_j) - A(u_j, u), u_j - u\rangle \,.$$

By (a), $b_j \geqslant 0$. Since $A(u_j, u)$ is bounded in X^*, there exists a subsequence, for which we don't change the subscript, such that $A(u_j, u) \rightharpoonup \psi$ in X^*. By (d), $\langle A(u_j, u), u_j\rangle \to \langle \psi, u\rangle$. Thus, $\langle A(u_j, u), u_j - u\rangle \to 0$. But by assumption,

$$\overline{\lim}\, \langle A(u_j, u_j), u_j - u\rangle = \overline{\lim}\langle T(u_j), u_j - u\rangle \leqslant 0 \,,$$

therefore $b_j \to 0$. It follows $\langle T(u_j), u_j - u\rangle \to 0$.

It remains to show: $\underline{\lim}\langle T(u_j), u - v\rangle \geq \langle T(u), u - v\rangle$, from (a),

$$\langle T(u_j) - A(u_j, w), u_j - w\rangle \geqslant 0, \;\; \forall w \in X \,.$$

Letting $w = (1-\theta)u + \theta v$, $\theta \in (0,1)$, $\forall v \in X$, we have

$$\theta\langle T(u_j), u - v\rangle \geqslant \theta\langle A(u_j, w), u - v\rangle + \langle A(u_j, w), u_j - u\rangle - \langle T(u_j), u_j - u\rangle \,.$$

By taking limit,

$$\underline{\lim}\, \langle T(u_j), u - v\rangle \geqslant \langle A(u, w), u - v\rangle$$

or

$$\underline{\lim}\, \langle T(u_j), u - v\rangle \geqslant \langle A(u, (1-\theta)u + \theta v), u - v\rangle \quad \forall \theta \in (0,1)$$

letting $\theta \to 0$, by (b), the RHS $= \langle T(u), u - v\rangle$, this is the desired inequality.

Statement 2.5.12 *Under the above structural conditions, we assume further:*

(5) $\exists C_1 > 0$ $\exists h \in L^1(\Omega)$ *such that*

$$\vec{A}(x, u, \xi) \cdot \xi > C_1|\xi|^p - h(x) \quad \forall (x, u, \xi) \in \Omega \times \mathbb{R}^1 \times \mathbb{R}^n \,.$$

Then $\forall f \in L^{p'}(\Omega)$, *there exists a weak solution* $u \in \overset{\circ}{W}{}^1_p(\Omega)$ *of the quasilinear elliptic equation:*

$$\int_{\Omega} \vec{A}(x,u(x),\nabla u(x)) \cdot \nabla v(x) + B(x,u(x),\nabla u(x))v(x) = \int_{\Omega} f(x)v(x)$$
$$\times \ \forall v \in \overset{\circ}{W}{}^1_p(\Omega) \, .$$

Proof. It is reduced to find the fixed point of the pseudo monotone operator T. Since (5) implies the coerciveness:

$$\frac{1}{\| u \|_{1,p}} \langle Tu, u \rangle = \frac{1}{\| u \|_{1,p}} a(u,u) \to +\infty \, .$$

The existence of u follows from Browder's theorem. □

It is worth noting that in the above statement, no a priori estimate has been used. In the verification of the pseudo monotonicity, a crucial point is the ellipticity condition (or the monotone condition (3)). The statement can be extended to high-order quasi-elliptic equations. This is an approach where no compactness is concerned.

2.6 Maximal Monotone Mapping

To a set-valued mapping A, $\forall x \in D(A)$, one may take any nonempty subset $D(A_1)$ of $D(A)$ and a nonempty subset $A_1(x)$ of $A(x)$, and define a new set-valued mapping $A_1 : x \to A_1(x) \, \forall x \in D(A_1)$. If A is monotone, so is A_1. In order to avoid the indetermination of domain and the image of set-valued monotone mappings in the study of their surjectivity, one introduces the notion of maximal monotone mappings.

First, we define the graph $\Gamma_A := \{(x,y) \in X \times Y | \, x \in D(A), y \in A(x)\}$ of a set-valued mapping $A : X \to 2^Y$.

Definition 2.6.1 *Assume that A_1 and A_2 are two monotone set-valued mappings from X to 2^{X^*}. If $\Gamma_{A_1} \subset \Gamma_{A_2}$, then we say that A_2 is a monotone extension of A_1. A monotone mapping, which does not have any proper monotone extension, is called a maximal monotone mapping.*

In other words, $A : X \to 2^{X^*}$ is maximal monotone iff for $(x,x^*) \in X \times X^*$,

$$\langle y^* - x^*, y - x \rangle \geq 0 \; \forall \, (y, y^*) \in \Gamma_A \; \Rightarrow \; (x, x^*) \in \Gamma_A \, .$$

Example 1. Assume that $\phi : \mathbf{R}^1 \to \mathbf{R}^1$ is a nondecreasing function. The map $A : u \mapsto [\phi(u-0), \phi(u+0)]$ is maximal monotone, but the map $B : u \mapsto \phi(u-0)$ is not, if $\phi(u-0) \neq \phi(u+0)$.

Example 2. A hemi-continuous monotone operator A is maximal monotone. In fact, for $(x,x^*) \in X \times X^*$, if

$$\langle Ay - x^*, y - x \rangle \geq 0 \; \forall \, y \in X \, ,$$

then $\forall z \in X, \forall t \in [0,1]$, we have

$$\langle A((1-t)x + tz) - x^*, z - x\rangle \geq 0 .$$

Letting $t \to 0$, it follows that $\langle Ax - x^*, z - x\rangle \geq 0$. Since $z \in X$ is arbitrary, we obtain $Ax = x^*$.

Moreover, we shall prove later that the subdifferential of a proper l.s.c. convex function is maximal monotone.

Since maximal monotone mappings occur in convex analysis, free boundary boundary problems in mathematical physics, and nonlinear semigroups. we shall extend most of the results of monotone operators to their set-valued counterpart-maximal monotone mappings. The central problem is on the surjectivity.

Now we are going to extend the notions of the demi-continuity and the hemi-continuity for single-valued mappings to set-valued mappings.

Definition 2.6.2 *Suppose that E is a nonempty subset of a real Banach space X. A mapping $A : E \to 2^{X^*}$ is called weakly * upper hemi-continuous (w*u.h.c.,) if $\forall x, y \in E$ $\forall z \in X$, the function $t \mapsto \langle A((1-t)x + ty), z\rangle$ is a u.s.c. set-valued function at $t = 0$. It is called upper demi-continuous (u.d.c.) if $\forall z \in X$, $x \mapsto \langle A(x), z\rangle$ is u.s.c. from E to $2^{\mathbb{R}^1}$.*

Obviously, u.d.c. implies w*u.h.c., and for single-valued mappings, w*u.h.c. is hemi-continuous, and u.d.c. is demi-continuous.

Note that for any set-valued mapping $A : D(A) \to 2^{X^*}$, the inverse mapping $A^{-1} : X^* \to 2^X$, $x^* \mapsto \{x \in D(A) |\ x^* \in A(x)\}$ is well defined (in case $x^* \notin \text{rang}\,(A)$, one defines $A^{-1}(x^*) = \emptyset$). In particular, if X is reflexive then A is (maximal) monotone $\Longleftrightarrow$ A^{-1} is (maximal) monotone.

Lemma 2.6.3 *Assume that $A : X \to 2^{X^*}$ is monotone. Then $\forall x_0 \in X$, $\exists \varepsilon > 0$ such that A is bounded on $B_\varepsilon(x_0)$.*

Proof. We may assume $x_0 = \theta$. We prove it by contradiction. Assume that A is not bounded in any neighborhood of θ, i.e., $\exists x_n \to \theta$ $\exists x_n^* \in A(x_n)$ such that $\| x_n^* \| \to \infty$. We claim that $\forall \varepsilon > 0$ $\exists z \in B_\varepsilon(\theta)$ and a subsequence $\{n'\}$ such that

$$\langle x_{n'}^*, x_{n'} - z\rangle \to -\infty . \tag{2.49}$$

Then from the monotonicity, $\forall z^* \in A(z)$, we have

$$\langle x_{n'}^* - z^*, x_{n'} - z\rangle \geqslant 0 .$$

(2.49) implies that $\langle z^*, z\rangle = +\infty$. Obviously, this is impossible.

Let us return to proving the existence of z satisfying (2.49). If it is not true, then $\exists \varepsilon > 0$ $\forall z \in \overline{B}_\varepsilon(\theta)$ $\exists C_z \in \mathbb{R}^1$ such that

$$\langle x_n^*, x_n - z\rangle \geqslant C_z \quad \forall n .$$

Now, $\forall k \in \mathbb{N}$, let

$$E_k = \{u \in \overline{B}_\varepsilon(\theta) |\ \langle x_n^*, x_n - u \rangle \geqslant -k,\ \forall n\}\ .$$

Obviously, E_k are closed, and $\bigcup\limits_{k=1}^{\infty} E_k = \overline{B}_\varepsilon(\theta)$. From the Baire category argument, $\exists r > 0, \exists y_0 \in \overline{B}_\epsilon(\theta)$ and $\exists k_0 \in \mathbb{N}$ such that $B_r(y_0) \subset E_{k_0}$. Setting $C = C_{-y_0}$, we have

$$\langle x_n^*, x_n + y_0 \rangle \geqslant C \quad \forall n\ ,$$

and

$$\langle x_n^*, x_n - u \rangle \geq -k_0,\ \forall u \in \overline{B}_r(y_0)\ .$$

By addition,

$$\langle x_n^*, 2x_n + y_0 - u \rangle \geqslant C - k_0 \quad \forall u \in \overline{B}_r(y_0)\ .$$

Setting $v = 2x_n + y_0 - u$, for sufficiently large n, such that $\| x_n \| < \frac{r}{4}$, we obtain

$$\langle x_n^*, v \rangle \geqslant C - k_0 \quad \forall v \in B_{\frac{r}{2}}(\theta)\ .$$

This means $\| x_n^* \| \leqslant \frac{2|C-k_0|}{r}$, which contradicts $\|x_n^*\| \to \infty$; (2.49) is proved. □

Proposition 2.6.4 *Assume that $A : X \to 2^{X^*}$ is a maximal monotone mapping with $D(A) = X$ on a reflexive real Banach space X. Then:*

(1) $\forall x \in X$ $A(x)$ is a nonempty closed convex set.
(2) A is u.d.c.
(3) If there is a sequence $\{x_j\}$, and a sequence $x_j^* \in A(x_j)$ satisfying $x_j \rightharpoonup x$ and $\overline{\lim}\, \langle x_j^*, x_j - x \rangle \leqslant 0$, then $\exists x^* \in A(x)$ such that

$$\underline{\lim}\, \langle x_j^*, x_j - y \rangle \geqslant \langle x^*, x - y \rangle \quad \forall y \in X\ .$$

Proof. (1) If $x_i^* \in A(x)$ $i = 1, 2$, then $\forall y \in X$, $\forall y^* \in A(y)$ we have

$$\langle y^* - x_i^*, y - x \rangle \geqslant 0, \quad i = 1, 2\ .$$

Therefore

$$\langle y^* - (\lambda x_1^* + (1-\lambda) x_2^*), y - x \rangle \geqslant 0 \quad \forall \lambda \in [0, 1]\ .$$

By the maximality, $\lambda x_1^* + (1-\lambda)x_2^* \in A(x)$. Then $A(x)$ is convex. Similarly, $A(x)$ is closed.

(2) In order to show that A is u.d.c., i.e., $\forall z \in X$, $x \mapsto \langle A(x), z \rangle$ is a u.s.c. set-valued mapping: $X \to 2^{\mathbb{R}^1}$. We prove it by contradiction. If $\exists z_0 \in X$ $\exists \varepsilon_0 > 0$ $\exists x_n \to x_0$ and $\exists x_n^* \in A(x_n)$ such that $\langle x_n^*, z_0 \rangle \notin \langle A(x_0), z_0 \rangle + (-\varepsilon_0, \varepsilon_0)$. From Lemma 2.6.3, $\{x_n^*\}$ is bounded, and then it possesses a weakly convergent subsequence: $x_n^* \rightharpoonup x_0^*$ in X^*. Since A is maximal, $x_0^* \in A(x_0)$. This is a contradiction.

(3) The proof is the same as that for single-valued monotone operators; cf. Lemma 2.5.3. □

Conversely, we have:

Proposition 2.6.5 *Assume that X is a reflexive real Banach space and that $A : X \to 2^{X^*}$ is a w^*.u.h.c. monotone mapping. If $\forall x \in X$, $A(x)$ is a nonempty closed convex set, then A is maximal monotone.*

Proof. We shall verify that for $(x, x^*) \in X \times X^*$, if

$$\langle y^* - x^*, y - x \rangle \geqslant 0 \quad \forall (y, y^*) \in \Gamma_A \,, \tag{2.50}$$

then $(x, x^*) \in \Gamma_A$. For otherwise, according to the Ascoli separation theorem, $\exists z \in X$ such that

$$\langle x^*, z \rangle > \sup \{ \langle w^*, z \rangle | \; w^* \in A(x) \} \,.$$

Since A is w*.u.h.c. $\exists \delta > 0$ such that

$$\langle x^*, z \rangle > \langle y_t^*, z \rangle \quad \forall y_t^* \in A(y_t) \; \forall t \in (0, \delta) \,,$$

where $y_t = x + tz$. But from the monotonicity, we have

$$t \langle y_t^* - x^*, z \rangle = \langle y_t^* - x^*, y_t - x \rangle \geqslant 0 \,.$$

This is a contradiction. □

The following theorem is a characterization of maximal monotone mappings on Hilbert spaces.

Theorem 2.6.6 *(Minty) Suppose that H is a Hilbert space, and that $A : H \to 2^H$ is a set-valued mapping. The following statements are equivalent:*

(1) A is maximal monotone.
(2) A is monotone and $I + A$ is surjective.
(3) $\forall \lambda > 0$ $(I + \lambda A)^{-1}$ is nonexpansive.

Before going on to prove Theorem 2.6.6, we now introduce the following notion.

Definition 2.6.7 *Suppose that E is a nonempty subset of a real Hilbert space H. A set-valued mapping $T : E \to 2^H$ is called expansive if $\forall x, y \in E$, $\forall x^* \in T(x)$, $\forall y^* \in T(y)$, $\| x^* - y^* \| \geqslant \| x - y \|$.*

Obviously, $T : H \to 2^H$ is surjective and expansive iff T^{-1} is nonexpansive. In this case, T^{-1} is single valued and injective.

Lemma 2.6.8 *$A : E \to 2^H$ is monotone if and only if $\forall \lambda > 0$, $T_\lambda = I + \lambda A$ is expansive.*

Proof. Since $\forall x, y \in E$, $\forall x^* \in A(x)$, $\forall y^* \in A(y)$, we have

$$\| y-x+\lambda(y^*-x^*) \|^2 = \| y-x \|^2 + 2\lambda(y^*-x^*, y-x) + \lambda^2 \| y^*-x^* \|^2 \ . \tag{2.51}$$

"$\Longrightarrow$" If A is monotone, then the RHS of (2.51) $\geqslant \| y-x \|^2$, i.e., T_λ is expansive.

"$\Longleftarrow$" If T_λ is expansive, then the LHS of (2.51) $\geqslant \| y-x \|^2$. Letting $\lambda \to 0$, A is monotone. □

Lemma 2.6.9 *(Deburnner–Flor) Suppose that E is a nonempty closed convex subset of a real Hilbert space H. If $A : E \to 2^H$ is monotone, then $\forall y \in H$, $\exists x \in E$ such that*

$$(\eta + x, \xi - x) \geqslant (y, \xi - x) \quad \forall (\xi, \eta) \in \Gamma_A \ .$$

Proof. $\forall (\xi, \eta) \in \Gamma_A$, let us define

$$C(\xi, \eta) = \{x \in E |\ (\eta + x - y, \xi - x) \geqslant 0\} \ .$$

Our conclusion is equivalent to $\bigcap\{C(\xi,\eta)|\, (\xi,\eta) \in \Gamma_A\} \neq \emptyset$. Since $C(\xi,\eta)$ is a bounded closed convex set, it is weakly closed and E is weakly compact. It is sufficient to verify that $\{C(\xi,\eta)|\ (\xi,\eta) \in \Gamma_A\}$ has the finite intersection property. Indeed, $\forall \xi_1, \ldots, \xi_n \in E\ \forall \eta_i \in A(\xi_i)$, $i = 1, \ldots, n$, we consider the simplex

$$\triangle_n = \left\{\lambda = (\lambda_1, \ldots, \lambda_n) |\ \lambda_i \geqslant 0\ \forall i,\ \sum \lambda_i = 1\right\} \ ,$$

and the function

$$\Phi(\lambda, \mu) = \sum_{i=1}^{n} \mu_i (x(\lambda) + \eta_i - y, x(\lambda) - \xi_i) \quad \forall (\lambda, \mu) \in \triangle_n \times \triangle_n \ ,$$

where $x(\lambda) = \sum_{i=1}^n \lambda_i \xi_i$. It is easy to see that

$$\forall \mu \in \triangle_n,\ \lambda \mapsto \Phi(\lambda, \mu) \text{ is continuous,}$$
$$\forall \lambda \in \triangle_n,\ \mu \mapsto \Phi(\lambda, \mu) \text{ is linear ,}$$

and

$$\begin{aligned}
\Phi(\lambda, \lambda) &= \sum_{i=1}^{n} [\lambda_i (x(\lambda) - y, x(\lambda) - \xi_i) + \lambda_i (\eta_i, x(\lambda) - \xi_i)] \\
&= \sum_{i,j=1}^{n} \lambda_i \lambda_j (\eta_i, \xi_j - \xi_i) \\
&= \frac{1}{2} \sum_{i,j=1}^{n} \lambda_i \lambda_j (\eta_i - \eta_j, \xi_j - \xi_i) \leq 0 \ ,
\end{aligned}$$

because A is monotone. According to Ky Fan's inequality, $\exists \lambda_0 \in \triangle_n$ such that

$$\Phi(\lambda_0, \mu) \leqslant 0 \quad \forall \mu \in \triangle_n \ ,$$

i.e.,

$$(x(\lambda_0) + \eta_i - y, x(\lambda_0) - \xi_i) \leqslant 0 \quad \forall i \ .$$

Therefore $x(\lambda_0) \in \bigcap_{i=1}^{n} C(\xi_i, \eta_i)$. This proves the finite intersection property for $\{C(\xi, \eta) \mid (\xi, \eta) \in \Gamma_A\}$, and thus the lemma. □

Proof of Minty's theorem.

Proof. (1)$\Longrightarrow$(2) We only want to show that $\forall y \in H$, $\exists x \in H$ such that $y - x \in A(x)$. In fact, according to the Deburnner–Flor lemma, $\exists x \in H$ satisfying $(\eta - (y - x), \xi - x) \geqslant 0$, $\forall (\xi, \eta) \in \Gamma_A$. By maximality, $y - x \in A(x)$.

(1)$\Longleftarrow$(2) Assume that B is a monotone mapping with $\Gamma_A \subset \Gamma_B$; we want to show that $B = A$. Since $I + A$ is surjective, $\forall (x, y) \in \Gamma_B$, $\exists x' \in D(A)$ such that $x + y \in (I + A)(x')$, so is $x + y \in x' + B(x')$, i.e, $y' = y + x - x' \in B(x')$. Also from $(x, y) \in \Gamma_B$, $x + y \in x + B(x)$. Provided by the monotonicity of B, we have

$$\| x - x' \|^2 = (y' - y, x - x') \leqslant 0 \ ,$$

i.e., $x = x'$. Therefore $B = A$ and then A is maximal.

Since A is maximal monotone if and only if λA is $\forall \lambda > 0$, we may assume $\lambda = 1$ in proving the equivalence of (2) and (3). Also, by Lemma 2.6.8, $(I + A)$ is expansive $\Longleftrightarrow$ A is monotone.

(2) $\Longleftrightarrow$ (3) follows from the fact that $(I + A)$ is expansive and surjective $\Longleftrightarrow$ $(I + A)^{-1}$ is nonexpansive. □

In the literature, for a maximal monotone mapping A, $\forall \lambda > 0$, setting $J_\lambda = (I + \lambda A)^{-1}$ we call the operator $A_\lambda(x) = \frac{1}{\lambda}(I - J_\lambda)(x)$ the Yosida regularization of A. By definition, $A_\lambda(x) \in A(J_\lambda x)$, but in general, $A_\lambda(x) \neq A(J_\lambda x)$.

Theorem 2.6.10 *If $f : H \to \mathbb{R}^1 \cup \{+\infty\}$ is a proper, l.s.c., convex function, then ∂f is maximal monotone.*

Proof. The monotonicity has been known since the beginning of the last section. Due to Minty's theorem, it remains to verify the surjection of $I + \partial f$, i.e., $\forall y_0 \in H$, $\exists x_0 \in H$ such that $y_0 - x_0 \in \partial f(x_0)$.

To this end, we define a function:

$$\varphi(x) = \frac{1}{2} \| x \|^2 + f(x) - \langle y_0, x \rangle \ .$$

It is proper, l.s.c., and convex, and $\varphi(x) \to +\infty$ as $\| x \| \to \infty$. There exists $R_1 > 0$ s.t. $\varphi(x) > \varphi(\theta)$ as $\| x \| \geqslant R_1$. Applying the Hartman–Stampacchia theorem, $\exists x_0 \in D(\varphi) = D(f)$ such that

$$\varphi(x_0) \leqslant \varphi(x) \quad \forall x \in H .$$

Therefore, $\theta \in \partial\varphi(x_0)$, i.e., $y_0 - x_0 \in \partial f(x_0)$. □

In fact, the above conclusion can be extended to general real Banach spaces. In this extension, the normalized duality map $J : X \to 2^{X^*}$ plays an important role, $J(x) = \{x^* \in X^* \mid < x^*, x >= \|x\|^2 = \|x^*\|^2\}$, (cf. Example 1 in Sect. 2.2). It is easy to verify that J is an odd, positively homogeneous, bounded set-valued mapping, and it is known that $J(x) = \partial\varphi(x)$, where $\varphi(x) = \frac{1}{2}\|x\|^2$, so it is monotone.

The Browder theorem (Theorem 2.5.8) is also extended to maximal monotone mappings: Let X be a uniformly convex Banach space, if $D(A) \subset X$, and $A : D(A) \to X^*$ is maximal monotone and coercive: $\frac{\langle y,x\rangle}{\|x\|} \to +\infty$ $\forall (x, y) \in \Gamma_A$, as $\|x\| \to \infty$, then A is surjective.

For a proper l.s.c. convex function on a real Banach space, $f : X \to \mathbb{R}^1 \cup \{+\infty\}$, the conjugate function of $f, f^* : X^* \to \mathbb{R}^1 \cup \{+\infty\}$ will be defined in Sect. 4.1.3 to be

$$f^*(x^*) = \sup_{x \in X} \{\langle x^*, x\rangle - f(x)\} .$$

It will be proved therein that f^* is again a proper l.s.c. convex function, satisfying

$$f(x) + f^*(x^*) = \langle x^*, x\rangle \Leftrightarrow x^* \in \partial f(x) \Leftrightarrow x \in \partial f^*(x^*) .$$

Thus we have the following:

Corollary 2.6.11 *If $f : X \to \mathbb{R}^1$ is proper, l.s.c., and convex, then $\partial f^* = (\partial f)^{-1}$ is maximal monotone.*

Proof. According to the above discussion, $\partial f^* = (\partial f)^{-1}$. The maximality of ∂f^* follows from Theorem 2.6.10. □

3

Degree Theory and Applications

The Leray–Schauder degree is an important topological tool introduced by Leray and Schauder in the study of nonlinear partial differential equations in the early 1930s. The nontriviality of the degree ensures the existence of a fixed point of the compact mapping in the domain. It enjoys the properties of homotopy invariance and additivity, which make the topological tool more convenient in application, and provides more information on fixed points. In Sects. 3.5 and 3.6 we shall see how better results could be obtained by using this tool than by using any of the other methods that we discussed previously.

The Leray–Schauder degree is an extension of the Brouwer degree from finite–dimensional spaces to infinite–dimensional Banach spaces, while the Brouwer degree is a powerful tool in algebraic topology. We introduce the notion of the Brouwer degree in Sect. 3.1, and investigate its fundamental properties in Sect. 3.2. On the one hand, we introduce the definition of the Brouwer degree from the point of view of differential topology so that it closely ties with the counting of zeroes of mappings; on the other hand, we build up the relationship between this definition and that used in algebraic topology. The applications to the Brouwer fixed-point theorem and Borsuk–Ulam theorem, as well as the intersection number etc. are studied in Sect. 3.3.

The notion of the Leray–Schauder degree is defined by approximation. All its fundamental properties are transferred from those of the Brouwer degree directly. These are the contents of Sect. 3.4. We emphasize the computation of the degree, because the more precisely we know the degree the sharper we can estimate the number of fixed points. This opens a door to the study of multiple solutions in nonlinear analysis.

Rabinowitz's global bifurcation theorem, based on the computation of the Leray–Schauder degree, is an important part of nonlinear functional analysis and probably the only global result in bifurcation theory. It provides global information on the branch of solutions emanating from a bifurcation point. The applications to nonlinear Sturm–Liouville problems as well as nonlinear elliptic problems are presented. They are studied in Sect. 3.5.

In Sect. 3.6, we introduce various applications of the Leray–Schauder degree. Schaefer's fixed-point theorem is more convenient in application to partial differential equations. Relying on the degree arguments, multiple solutions for semilinear elliptic equations are studied. In particular, some interesting multiple results are obtained in combination with the super- and sub-solutions method. When we are concerned with positive solutions of differential equations, one modifies the degree theory to that for compact mappings defined on closed convex sets. Moreover, we present some other applications: The Krein–Rutmann theory for positive linear operators, the existence of a positive solution for the superlinear elliptic problem, and bridging solutions of disjoint domains etc. in order to expose the extensive applicability of the degree theory.

In Sect. 3.7, various extensions of the Leray–Schauder degree are discussed, including the degrees for set contraction mappings, condensing mappings, set-valued mappings, Fredholm mappings, etc.

3.1 The Notion of Topological Degree

The topological degree was originally introduced by H. Poincaré in the qualitative study of ODEs. Firstly, he defined the index of a singularity P_0 as follows: Let $P_0 = (x_0, y_0)$ be a singularity of the plane vector field (F(x,y), G(x,y)), i.e., $F(x_0, y_0) = G(x_0, y_0) = 0$. Let C be a closed curve surrounding P_0 in the phase plane. We count the variation of the angles of the tangents of the flow:

$$\begin{cases} \dot{x} = F(x, y) \\ \dot{y} = G(x, y) \end{cases}$$

generated by (F, G) along C counterclockwise. This amount should be a multiple of 2π, and is denoted by $\operatorname{ind}(P_0)$. One may measure this amount by observing the compass moving along the curve C in a magnetic field.

Afterwards, he defined the winding number of a closed curve C, on which there are no singularities of the vector field. Different from the index, at this time, C is not necessarily just enclosing one single singularity. He noticed:

(1) The winding number is invariant under deformation of curves C, if there is no singularity on these curves C.
(2) The winding number for a given curve C is invariant under deformation of vector fields, if there is no singularity on C.
A direct consequence follows:
(3) Let D be the domain enclosed by C. If the winding number of C is nonzero, then there must be a singularity of the vector field in D.

For a given vector field $f = (F, G)$, a closed curve C (or equivalently, the enclosed domain D), we denote the winding number by $\deg(f, D)$, More generally, instead of the zeroes of f, sometimes we study the solutions of

$$f(x,y) = P_0$$

for a given $P_0 \in \mathbb{R}^2$; we denote the winding number of $f - P_0$ by $\deg(f, D, P_0)$.

In order to extend the concept of winding numbers to higher-dimensional vector fields, we would rather make some stronger restrictions on f at first. It makes the geometric characterization of winding numbers easy to understand.

Let us first assume that f is an analytic function: $f(z) = F(z) + iG(z)$, where $z = x + iy$ and $w_0 = p_0 + iq_0$ are complex numbers. Let Ω be a bounded open domain in $\mathbb{C}$ with boundary $\partial\Omega$.

Suppose $f(z) \neq w_0\ \forall z \in \partial\Omega$, then we have

$$\begin{aligned}
\deg(f, \Omega, w_0) &= \frac{1}{2\pi}\int_{\partial\Omega} d\arg(f(z) - w_0) \\
&= \frac{1}{2\pi i}\int_{\partial\Omega} d\log(f(z) - w_0) \\
&= \frac{1}{2\pi i}\int_{\partial\Omega} \frac{f'(z)}{f(z) - w_0} dz \\
&= \sum_{z_j \in f^{-1}(w_0)\cap\Omega} \sigma_j ,
\end{aligned} \tag{3.1}$$

where σ_j is the multiplicity of z_j, i.e.,

$$f(z) = w_0 + c_j(z - z_j)^{\sigma_j} + \circ(|z - z_j|^{\sigma_j}), \tag{3.2}$$

as $|z - z_j| \to 0$, and $c_j \neq 0$.

If f is not analytic but differentiable, let $f = (F, G), z = (x, y)$ and $w_0 = (p_0, q_0)$. Assume $f(z) \neq w_0,\ \forall z \in \partial\Omega$ and

$$\det f'(z) = \det \frac{\partial(F,G)}{\partial(x,y)}(z) \neq 0 \quad \forall z \in f^{-1}(w_0)\,,$$

then the point set $\Omega \cap f^{-1}(w_0)$ is finite, from the IFT; and

$$\begin{aligned}
\deg(f, \Omega, w_0) &= \frac{1}{2\pi}\int_{\partial\Omega} d\arg(f(z) - w_0) \\
&= \frac{1}{2\pi}\int_{\partial\Omega} d\arctan\frac{G - q_0}{F - p_0} \\
&= \frac{1}{2\pi}\int_{\partial\Omega} \frac{(F - p_0)dG - (G - q_0)dF}{(F - p_0)^2 + (G - q_0)^2}\,.
\end{aligned}$$

Applying Green's formula, it equals

$$\sum_{z_j \in f^{-1}(w_0)\cap\Omega} \int_{\partial B_\varepsilon(z_j)} \frac{(F - p_0)dG - (G - q_0)dF}{(F - p_0)^2 + (G - q_0)^2} = \sum_{z_j \in f^{-1}(w_0)\cap\Omega} \operatorname{sgn}\det f'(z_j), \tag{3.3}$$

where $\varepsilon > 0$ is small enough that $B_\varepsilon(z_i) \cap B_\varepsilon(z_j) = \emptyset \; \forall z_i \neq z_j$ in $f^{-1}(w_0) \cap \Omega$, provided $\det f'(z) \neq 0 \; \forall z \in \bigcup_{z_i \in f^{-1}(w_0) \cap \Omega} B_\varepsilon(z_i)$.

The last formula provides a clue to defining the topological degree for higher-dimensional mappings. For each root of $f(z) = w_0$, we assign a signed number, which is the signature of the determinant at this point, and then the topological degree is the summation of all these signed numbers.

Before going to the definition, we need a special case of the Sard theorem. Since we have not proved the Sard theorem in Chap. 1, now we shall present a proof for the special case, due to its simplicity. However, we refer the reader to Milnor [Mi 2] for a proof of the general theorem, which can also be obtained by combining our proof for the special case with an inductive argument.

Theorem 3.1.1 *(Sard) Let $\Omega \subset \mathbb{R}^n$ be an open set, and let $f : \Omega \to \mathbb{R}^n$ be a C^1 map. Write*

$$\mathbf{Z} = \{x \in \Omega | \; \det f'(x) = 0\} \,.$$

Then $f(\mathbf{Z})$ is a zero measure set.

Proof. Consider a closed cube $C \subset \Omega$, each side with length a. We divide C into N^n configuration cubes $C_i^{(1)}, i = 1, \ldots, N^N$, with $\mathrm{int}(C_i^{(1)}) \cap \mathrm{int}(C_j^{(1)}) = \emptyset, i \neq j$, each side with length $\frac{a}{N}$. Thus on one hand,

$$\| \, f(x) - f(x_0) \, \| \leqslant K_C \, \| \, x - x_0 \, \| \leqslant K_C \frac{\sqrt{n}a}{N} \,,$$

$\forall x, x_0 \in C_i^{(1)}$ for some i, where $K_C = \max\{\| \, f'(x) \, \| \mid x \in C\}$.

On the other hand, $\forall \varepsilon > 0 \; \exists$ an integer N such that

$$\| \, f(x) - f(x_0) - f'(x_0)(x - x_0) \, \| < \varepsilon \, \| \, x - x_0 \, \| \leqslant \varepsilon \frac{\sqrt{n}a}{N} \,.$$

Let $x_0 \in \mathbf{Z}$, $f'(x_0)$ is not invertible. Thus $f'(x_0)\mathbb{R}^n$ cannot span the whole space $\mathbb{R}^n$, but is included in a $(n-1)$-dimensional linear subspace H. It follows that

$$\mathrm{dist}(f(x), f(x_0) + H) \leqslant \frac{\varepsilon\sqrt{n}a}{N} \quad \forall x \in C_i^{(1)} \,,$$

in which $C_i^{(1)}$ is the small cube including x_0. We conclude that $f(C_i^{(1)})$ is included in a cube, whose sides have length $\frac{2\varepsilon\sqrt{n}a}{N}$, $\frac{2K_C\sqrt{n}a}{N}$, $\ldots$ $\frac{2K_C\sqrt{n}a}{N}$, respectively. Therefore

$$\begin{aligned} m^*(f(C \cap \mathbf{Z})) &\leqslant \sum_{C_i^{(1)} \cap \mathbf{Z} \neq \emptyset} m^*(f(C_i^{(1)} \cap \mathbf{Z})) \\ &\leqslant \sum_{C_i^{(1)} \cap \mathbf{Z} \neq \emptyset} m^*(f(C_i^{(1)})) \\ &\leqslant 2^n K_C^{n-1} n^{\frac{n}{2}} a^n \varepsilon \,, \end{aligned}$$

where m^* is the Lebesgue outer measure. Since $\varepsilon > 0$ and the cube C are arbitrary, we obtain $m^*(f(\mathbf{Z})) = 0$. □

Suppose that Y is an n-manifold, a subset $W \subset Y$ is called a null set, if for each chart (φ, U) of Y, the set $\varphi(U \cap W) \subset \mathbb{R}^n$ is a null set.

Let us recall the definitions of regular/critical values/points of a map between two Banach manifolds. Sard's theorem asserts that if X and Y are n-manifolds, then for any $f \in C^1(X, Y)$, the set of critical values of f is a null set.

Let X_0, Y be two oriented smooth n-dimensional manifolds, and $X \subset X_0$ be an open subset satisfying the condition that $\overline{X} = X \cup \partial X$ is compact. If $f \in C(\overline{X}, Y) \cap C^1(X, Y)$, and if y_0 is a regular value of f then

$$f^{-1}(y_0) = \{x \in X |\ f(x) = y_0\}$$

must be a finite set, from the IFT. Define

$$f^{-1}(y_0) = \{x_1, \ldots, x_k\} .$$

Assuming that $y_0 \notin f(\partial X)$ is a regular value, we define

$$\deg(f, X, y_0) = \sum_{j=1}^{k} \operatorname{sgn} \det f'(x_j) . \tag{3.4}$$

Again we let $\mathbf{Z}$ denote the set of critical points of f.

We shall extend the definition to continuous mappings in three steps:

I. The special case $f \in C(\overline{X}, Y) \cap C^2(X, Y), y_0 \notin f(\mathbf{Z}) \cup f(\partial X)$.

In order to remove the assumptions on the regularity and the condition $f \in C^1(X, Y)$, we would rather express (3.4) in an integration form.

Let U_j be a neighborhood of x_j, such that

$$f : U_j \to f(U_j)$$

is a diffeomorphism, and that $U_i \cap U_j = \emptyset,\ \forall i \neq j$, where $i, j = 1, 2, \ldots, k$. Set

$$V = \bigcap_{j=1}^{k} f(U_j) .$$

This is a neighborhood of y_0. Let us choose a C^∞ n-form on Y:

$$\mu = \psi(y) dy_1 \wedge \ldots \wedge dy_n ,$$

(in the following, we write simply $dy = dy_1 \wedge \ldots \wedge dy_n$) such that

$$\operatorname{supp} \psi \subset V \cap (Y \backslash f(\partial X)) ,$$

and

$$\int_Y \mu = 1 .$$

Since $\det f'(x)$ has the same sign in each U_j, we have

$$\begin{aligned}
\int_X \mu \circ f &= \sum_{j=1}^k \int_{U_j} \mu \circ f \\
&= \sum_{j=1}^k \int_{U_j} \psi(f(x)) \det f'(x) dx \\
&= \sum_{j=1}^k \operatorname{sgn} \det f'(x_j) \int_{U_j} \psi \circ f(x) |\det f'(x)| dx \\
&= \sum_{j=1}^k \operatorname{sgn} \det f'(x_j) \int_{f(U_j)} \psi(y) dy \\
&= \sum_{j=1}^k \operatorname{sgn} \det(f'(x_j)) \\
&= \deg(x, X, y_0) . \qquad (3.5)
\end{aligned}$$

In the integration form, μ is arbitrarily chosen. We shall prove that the integral $\int_X \mu \circ f$ does not depend on the special choice of μ.

We call the chart (g, Ω) a good chart at $y_0 \in Y$, if

(1) $g(y_0) = \theta$,
(2) $g(\Omega) = \mathrm{I}_n = (-1, +1)^n$ in $\mathbb{R}^n$.

A $C^\infty - n$-form μ on Y is called admissible if $\mu = \psi(y)dy$ satisfies

(1) supp $\psi \subset \Omega \cap (Y \backslash f(\partial X))$ for a good chart (g, Ω) at y_0.
(2) $\int_Y \mu = 1$.

Lemma 3.1.2 *Suppose that $\mu = \psi(y)dy$ is a $C^\infty - n$-form on Y satisfying*

(1) supp $\psi \subset \Omega$, and (g, Ω) is a good chart at y_0,
(2) $\int_Y \mu = 0$.

Then there exists an $(n-1)$ form ω such that supp $\omega \subset \Omega$ and $d\omega = \mu$.

Proof. In the coordinates,

$$\begin{array}{ccc}
\Omega & \xrightarrow{g} & \mathrm{I}_n \\
\cap & & \cap \\
Y & & \mathbb{R}^n
\end{array}$$

one may assume supp$\psi \subset \mathrm{I}^n$ and $\int \psi(y)dy = 0$. What we want to prove is the existence of a function $\nu \in C^1(\mathbb{R}^n, \mathbb{R}^n)$ such that

$$\text{supp}\ \nu \subset \mathrm{I}_n \ \text{ and } \ \text{div}\ \nu = \psi\ .$$

Indeed, let $\nu = (\nu_1, \nu_2, \ldots, \nu_n)$, then

$$\omega = \nu_1 dy_2 \wedge \ldots \wedge dy_n + \ldots + (-1)^n \nu_n dy_1 \wedge \ldots \wedge dy_{n-1} \ \text{ satisfies } \ d\omega = \mu\ .$$

We prove the conclusion by mathematical induction. For $n = 1$, one takes

$$\nu(y) = \int_{-\infty}^{y} \psi(t)dt\ ,$$

then ν is as required.

Assume the lemma is true for $n = k$. Let

$$\phi(\hat{y}) = \int_{-\infty}^{\infty} \psi(\hat{y}, y_{k+1})dy_{k+1}\ ,$$

where $y = (\hat{y}, y_{k+1}),\ \hat{y} = (y_1, \ldots, y_k)$. Then

$$\text{supp}\ \phi \subset \mathrm{I}_k \triangleq (-1, 1)^k\ ,$$

and

$$\int \phi(\hat{y})d\hat{y} = \int \psi(y)dy = 0\ .$$

By the hypothesis of induction, there exists $\omega \in C^1(\mathbb{R}^k, \mathbb{R}^k)$ satisfying

$$\text{supp}\ \omega \subset \mathrm{I}_k,\ \text{div}\ \omega = \phi\ .$$

Choose

$$\tau \in C^1(\mathbb{R}^1),\ \text{supp}\ \tau \subset \mathrm{I}_1, \int_{-\infty}^{\infty} \tau(t)dt = 1\ ,$$

then

$$\int_{-\infty}^{\infty} [\psi(\hat{y}, y_{k+1}) - \phi(\hat{y})\tau(y_{k+1})]dy_{k+1} = 0\ .$$

Let

$$\nu_{k+1}(\hat{y}, y_{k+1}) = \int_{-\infty}^{y_{k+1}} [\psi(\hat{y}, t) - \phi(\hat{y})\tau(t)]dt\ ,$$

then

$$\frac{\partial}{\partial y_{k+1}} \nu_{k+1}(\hat{y}, y_{k+1}) = \psi(\hat{y}, y_{k+1}) - \phi(\hat{y})\tau(y_{k+1})\ .$$

This means that $\psi = \text{div}\ \nu$, where

$$\nu = (\omega_1(\hat{y})\tau(y_{k+1}), \ldots, \omega_k(\hat{y})\tau(y_{k+1}), \nu_{k+1}(\hat{y}, y_{k+1}))\ ,$$

and

$$(\omega_1, \ldots, \omega_k) = \omega\ .$$

□

In order to prove that the integral $\int_X \mu \circ f$ does not depend on μ, we assume that ν, μ are two admissible n forms with respect to a good chart (g, Ω) at y_0. According to Lemma 3.1.2, there exists an $(n-1)$ form ω such that $d\omega = \nu - \mu$, and then

$$\int_X \nu \circ f - \int_X \mu \circ f = \int_X d\omega \circ f .$$

If we can prove

$$\int_X (d\omega) \circ f = \int_X d(\omega \circ f), \tag{3.6}$$

then by the Stokes theorem, we conclude:

$$\int_X \nu \circ f = \int_X \mu \circ f .$$

We are going to prove (3.6). First, we need

Lemma 3.1.3 *If $f \in C(\overline{X}, Y) \cap C^2(X, Y)$, then for any $(n-1)$ form ω, we have*

$$d(\omega \circ f) = d\omega \circ f .$$

Proof. In local coordinates,

$$\omega = \sum_{j=1}^{n} \nu_j(y) dy_1 \wedge \ldots \wedge d\hat{y}_j \wedge \ldots \wedge dy_n,$$

where

$$dy_1 \wedge \ldots \wedge d\hat{y}_j \wedge \ldots \wedge dy_n = dy_1 \wedge \ldots \wedge dy_{j-1} \wedge dy_{j+1} \wedge \ldots \wedge dy_n .$$

Thus

$$\omega \circ f = \sum_{j=1}^{n} \left(\sum_{k=1}^{n} \nu_k \circ f(x) J_f^{kj}(x) \right) dx_1 \wedge \ldots \wedge d\hat{x}_j \wedge \ldots \wedge dx_n ,$$

where J_f^{ki} is the (k, i) cofactor of the Jacobian determinant $\det f'(x) = J_f(x)$. Hence

$$\begin{aligned} &d(\omega \circ f) \\ &= \sum_{i,k=1}^{n} \left[\sum_{l=1}^{n} \frac{\partial \nu_k \circ f(x)}{\partial y_l} J_f^{ki}(x) \cdot \frac{\partial f_l}{\partial x_i} + \nu_k \circ f(x) \frac{\partial J_f^{ki}(x)}{\partial x_i} \right] dx_1 \wedge \ldots \wedge dx_n \\ &= \left[\sum_{k,l=1}^{n} \frac{\partial \nu_k \circ f(x)}{\partial y_l} \sum_{i=1}^{n} \frac{\partial f_l}{\partial x_i} J_f^{ki}(x) + \sum_{k=1}^{n} \nu_k \circ f(x) \sum_{i} \frac{\partial J_f^{ki}(x)}{\partial x_i} \right] \\ &\quad \times dx_1 \wedge \ldots \wedge dx_n \end{aligned}$$

in which

$$f = (f_1, \ldots, f_n) .$$

Let

$$g = (-1)^{k-1}(f_1, \ldots, \hat{f}_k, \ldots, f_n) ,$$

then

$$J_f^{ki} = (-1)^{i-1} \det \left(\frac{\partial g}{\partial x_1}, \ldots, \widehat{\frac{\partial g}{\partial x_i}}, \ldots, \frac{\partial g}{\partial x_n} \right) .$$

Hence

$$\begin{aligned} \sum_{i=1}^n \frac{\partial}{\partial x_i} J_f^{ki} &= \sum_{i=1}^n (-1)^{i-1} \sum_{l \neq i} \det \left(\frac{\partial g}{\partial x_1}, \ldots, \frac{\partial^2 g}{\partial x_i \partial x_l}, \ldots, \widehat{\frac{\partial g}{\partial x_i}}, \ldots, \frac{\partial g}{\partial x_n} \right) \\ &= \sum_{i=1}^n \left(\sum_{l>i} (-1)^{i-1+l-2} \det \left\{ \frac{\partial^2 g}{\partial x_i \partial x_l}, \frac{\partial g}{\partial x_1}, \ldots, \widehat{\frac{\partial g}{\partial x_i}}, \ldots, \frac{\partial g}{\partial x_n} \right\} \right. \\ &\quad \left. + \sum_{i>l} (-1)^{i-1+l-1} \det \left\{ \frac{\partial^2 g}{\partial x_l \partial x_i}, \frac{\partial g}{\partial x_1}, \ldots, \widehat{\frac{\partial g}{\partial x_i}}, \ldots, \frac{\partial g}{\partial x_n} \right\} \right) = 0 . \end{aligned}$$

From this, together with

$$\sum_{i=1}^n \frac{\partial f_l}{\partial x_i} J_f^{ki}(x) = \delta_{kl} J_f(x)$$

follows

$$d(\omega \circ f) = \sum_{k=1}^n \frac{\partial \nu_k \circ f(x)}{\partial y_k} J_f(x) dx_1 \wedge \ldots \wedge dx_n = (d\omega) \circ f .$$

□

Now we arrive at:

Theorem 3.1.4 *Assume $f \in C(\overline{X}, Y) \cap C^2(X, Y)$. If $y_0 \notin f(\mathbf{Z}) \cap f(\partial X)$ and μ is an admissible C^∞ n form with respect to a good chart (g, Ω) at y_0, with supp(μ) $\subset \Omega \cap (Y \backslash f(\partial X))$, then $\int_X \mu \circ f$ is a constant independent of μ.*

II. Removal of the restriction $y_0 \notin f(\mathbf{Z})$

Combining Theorem 3.1.4 and (3.5), we know that

$$\deg(f, X, y_0) = \int_X \mu \circ f ,$$

if $f \in C(\overline{X}, Y) \cap C^2(X, Y)$ and $y_0 \notin f(\mathbf{Z}) \cup f(\partial X)$, where μ is an arbitrary admissible C^∞ n form with respect to a good chart at y_0. Note that the integral on the right-hand side of the above formula does not contain y_0 explicitly.

Further recall that the set of regular values is dense in $f(X)$ according to the Sard theorem, thus if we take the integral $\int_X \mu \circ f$ to be the definition of $\deg(f, X, y_0)$, the assumption $y_0 \notin f(\mathbf{Z})$ is not necessary. In fact, this integral is defined for all $y_0 \notin f(\partial X)$, and is integer valued (the left-hand side of the formula (3.5)) if y_0 belongs to the dense (regular value) set. Hence $\int_X \mu \circ f$ is integer valued. Therefore the restriction $y_0 \notin f(\mathbf{Z})$ can be removed when we use $\int_X \mu \circ f$ as the definition of the $\deg(f, X, y_0)$.

III. Extension to continuous mappings

We first prove:

Lemma 3.1.5 *Let* $\phi : \overline{X} \times [0,1] \xrightarrow{C} Y$ *satisfy* $\phi(\cdot, t) \in C^2(X, Y)$ $\forall t \in [0,1]$. *If* $y_0 \notin \phi(\partial X \times [0,1])$, *then* $\deg(\phi(\cdot, t), X, y_0)$ *is a constant independent of* t.

Proof. Since $\phi(\partial X \times [0,1])$ is closed, there is a good neighborhood Ω at y_0 such that $\Omega \cap \phi(\partial X \times [0,1]) = \emptyset$. If we choose $\psi \in C^\infty(Y, \mathbb{R}^1)$ such that $\mathrm{supp}\psi \subset \Omega$, $\int_Y \psi dy = 1$, then $\mu = \psi(y)dy$ is admissible, and

$$\deg(\phi(\cdot, t), X, y_0) = \int_X \mu \circ \phi(\cdot, t) . \tag{3.7}$$

The right-hand side is integer valued and is continuous in t, hence it must be a constant. □

Now we may extend the definition of degree to mappings in $C(\overline{X}, Y)$ by approximation.

By the use of the Whitney embedding theorem, we embed Y into $\mathbb{R}^N$ as a regular submanifold for large N. Thus there exist a tubular neighborhood W of Y and a C^∞ retract $r : W \to Y$. Since we assumed $y_0 \notin f(\partial X)$, one may choose $0 < \varepsilon < d_{R^N}(r^{-1}(y_0), f(\partial X))$ such that the ε neighborhood of Y is in W, where $d_{R^N}(\cdot, \cdot)$ is the distance in R^N. According to the Weierstrass approximation theorem and the partition of unity, there exists $g \in C^2(\overline{X}, \mathbb{R}^N)$ satisfying

$$\| f - g \|_{C(\overline{X}, \mathbb{R}^N)} < \varepsilon . \tag{3.8}$$

Thus for such g, $\deg(r \circ g, X, y_0)$ is well defined.

Moreover, we shall prove that

$$\deg(r \circ g_1, X, y_0) = \deg(r \circ g_2, X, y_0)$$

for $g_1, g_2 \in C^2(\overline{X}, \mathbb{R}^N)$ both satisfying (3.8).

Indeed, define $h : \overline{X} \times [0,1] \to \mathbb{R}^N$ by

$$h(x,t) = (1-t)g_1(x) + tg_2(x) ,$$

and $\phi : \overline{X} \times [0,1] \to Y$ by $\phi = r \circ h$. Then

$$\mathrm{dist}_{\mathbb{R}^N}(\phi(x,t), y_0) = 0 \Leftrightarrow \mathrm{dist}_{\mathbb{R}^N}(h(x,t), r^{-1}(y_0)) = 0 .$$

But,

$$\text{dist}_{\mathbb{R}^N}(h(\partial X, t), r^{-1}(y_0)) \geqslant \text{dist}_{\mathbb{R}^N}(f(\partial X), r^{-1}(y_0)) - \varepsilon > 0 ,$$

this proves that $y_0 \notin \phi(\partial X \times [0,1])$. Applying Lemma 3.1.5, we obtain

$$\deg(r \circ g_1, X, y_0) = \deg(r \circ g_2, X, y_0) .$$

We are now in a position to define the Brouwer degree for continuous mappings.

Definition 3.1.6 ***(Brouwer degree)*** *Let X_0, Y be two oriented smooth $n-$ manifolds, and let $X \subset X_0$ be an open subset with compact closure $\overline{X}$. For $f \in C(\overline{X}, Y), y_0 \notin f(\partial X)$, to the triple (f, X, y_0) we define*

$$\deg(f, X, y_0) = \deg(r \circ g, X, y_0) ,$$

where r is a C^∞ retract of a tubular neighborhood W of Y to Y, and g is a map defined in (3.8). This is called the Brouwer degree of f.

The Brouwer degree is integer valued.

It is not difficult to verify that Brouwer degree does not depend on the special choice of W and r.

By definition, if $X = \emptyset$, then $\deg(f, X, y_0) = 0$.

In dealing with mappings between complex manifolds, we identify C^n with R^{2n} by the canonical isomorphism $z \mapsto (x, y)$, where $z = x + iy$ for $z \in C^n$, $x, y \in R^n$. Similarly for the map $f \mapsto (u, v)$, where $f : \overline{\Omega} \to C^n$, and $f = u + iv$.

As an exercise, readers can verify that if f is analytic then $\det(f'(z)) > 0$.

3.2 Fundamental Properties and Calculations of Brouwer Degrees

The Brouwer degree has the following fundamental properties.

(1) (**Homotopy invariance**). If $\phi : \overline{X} \times [0,1] \to Y$ is continuous and $y_0 \notin \phi(\partial X \times [0,1])$, then

$$\deg(\phi(\cdot, t), X, y_0) = \text{constant} .$$

Proof. Using the above notations, we choose

$$0 < \epsilon < d_{R^N}(r^{-1}(y_0), \phi(\partial X \times [0,1])) ,$$

where $d_{R^N}(\cdot, \cdot)$ is the distance in R^N. By the above approximation, there is $\hat{\phi} \in C^2(\overline{X} \times [0,1], Y)$ such that

$$d_{C(\overline{X} \times [0,1], Y)}(\phi, \hat{\phi}) < \varepsilon .$$

From Definition 3.1.6 and Lemma 3.1.5 it follows that

$$\deg(\phi(\cdot,t),X,y_0)=\deg(\hat{\phi}(\cdot,t),X,y_0)=\text{constant}\,.$$

□

More generally, if $\phi:\overline{X}\times[0,1]\to Y$ is continuous, and if $t\mapsto X_t\subset\overset{\circ}{X}\ \forall t\in[0,1]$ are continuously deformed open domains, assume $y_0\notin\bigcup_{t\in[0,1]}\phi(\partial X_t,t)$, then

$$\deg(\phi(\cdot,t),X_t,y_0)=\text{constant}\,.$$

(2) If $y_1,y_2\in Y\backslash f(\partial X)$ are in the same component, then

$$\deg(f,X,y_1)=\deg(f,X,y_2)\,.$$

In particular, if $\partial X=\emptyset$, and Y is connected, then $\deg(f,X,y_0)$ is independent of y_0. (In this case, it is written as $\deg(f,X)$).

Proof. By definition, $\deg(f,X,y_0)$ is continuous in $y_0\in Y\backslash f(\partial X)$, and then is a constant, if $f\in C(\overline{X},Y)\cap C^2(X,Y)$. By approximation, we obtain the conclusion. □

Let Ω be a connected component of $Y\backslash f(\partial X)$. According to (2), $\forall y_0\in\Omega$, $\deg(f,X,y_0)$ is constant. One may write $\deg(f,X,\Omega)$ instead, if there is no confusion.

Corollary 3.2.1 *If $Y=\mathbb{R}^n$, $f\in C(\overline{X},Y)$ and $y_0\notin f(\partial X)$, then the degree* $\deg(f,X,y_0)$ *only depends on the restriction of f on ∂X, i.e., $\hat{f}=f|_{\partial X}$.*

Proof. Let $g\in C(\overline{X},Y)$ satisfy $\hat{g}=g|_{\partial X}=\hat{f}$. Define

$$F(x,t)=(1-t)f(x)+tg(x)\quad t\in[0,1]\,.$$

Then $y_0\notin F(\partial X\times[0,1])$. By homotopy invariance, we have

$$\deg(f,X,y_0)=\deg(g,X,y_0)\,.$$

□

Therefore, sometimes we write $\deg(\hat{f},\partial X,y_0)$ in place of $\deg(f,X,y_0)$. Returning to Lemma 3.1.5), we may rewrite (3.7) in a global version:

Corollary 3.2.2 *Suppose that $f\in C(\overline{X},Y)\cap C^2(X,Y)$ and that Ω is a connected component of $Y\backslash f(\partial X)$. If μ is a smooth n form with support in Ω and $\int_Y\mu\neq0$, then*

$$\deg(f,X,\Omega)=\frac{\int_X\mu\circ f}{\int_Y\mu}\,.$$

Proof. Let $\{\psi_\alpha | \alpha \in \wedge\}$ be a partition of unity of Y, such that $\text{supp}\psi_\alpha$ is in a "good" neighborhood Ω_α. Set $\mu_\alpha = \psi_\alpha \cdot \mu$, and choose $y_\alpha \in \text{supp}\ \mu_\alpha$. In the case where $\int_{\Omega_\alpha} \mu_\alpha \neq 0$, from (3.7) one has

$$\deg(f, X, \Omega) = \deg(f, X, y_\alpha) = \frac{\int_X \mu_\alpha \circ f}{\int_{\Omega_\alpha} \mu_\alpha} ,$$

i.e.,

$$\deg(f, X, \Omega) \int_{\Omega_\alpha} \mu_\alpha = \int_X \mu_\alpha \circ f . \tag{3.9}$$

In the case where $\int_{\Omega_\alpha} \mu_\alpha = 0$, by Lemma 3.1.2, there exists an $n-1$ form ω_α on Y such that

$$\text{supp}\ \omega_\alpha \subset \text{supp}\ \mu_\alpha \text{ and } d\omega_\alpha = \mu_\alpha .$$

Following Lemma 3.1.3,

$$\mu_\alpha \circ f = (d\omega_\alpha) \circ f = d(\omega_\alpha \circ f) ,$$

we obtain $\int_X \mu_\alpha \circ f = 0$ provided by the Stokes theorem. Again (3.9) holds. By addition, we have

$$\deg(f, X, \Omega) \int_Y \mu = \int_X \mu \circ f .$$

This is the conclusion. □

(3) (**Translation invariance**). If $Y = \mathbb{R}^n$, and $y_0 \notin f(\partial\Omega)$, then

$$\deg(f, X, y_0) = \deg(f - y_0, X, \theta) .$$

Proof. Let $f \in C^1(\overline{X}, \mathbb{R}^n)$ and $y_0 \notin f(\mathbf{Z})$, then

$$\begin{aligned}\deg(f, X, y_0) &= \sum_{x_i \in f^{-1}(y_0)} \text{sgn}\det f'(x_i) \\ &= \sum_{x_i \in (f-y_0)^{-1}(\theta)} \text{sgn}\det f'(x_i) \\ &= \deg(f - y_0, X, \theta) .\end{aligned}$$

The assertion can be obtained by approximations to y_0 and f. □

(4) (**Additivity**). Let $X_1, X_2 \subset X$ be open subsets in X such that $X_1 \cap X_2 = \emptyset$ and $y_0 \notin f(\overline{X} \backslash (X_1 \cup X_2))$, then

$$\deg(f, X, y_0) = \deg(f, X_1, y_0) + \deg(f, X_2, y_0) .$$

Proof. By the Whitney embedding theorem mentioned above, let

$$\varepsilon = d_{R^N}(r^{-1}(y_0), f(\overline{X}\backslash(X_1 \cup X_2)))(> 0) .$$

One may choose $g_0 \in C^2(\overline{X}, R^N)$ such that $d_{C(\overline{X},R^N)}(g, f) < \frac{\varepsilon}{2}$ for $g = r \circ g_0$, and choose $y_1 \notin g(\mathbf{Z})$ satisfying $d_{R^N}(y_0, y_1) < \frac{\varepsilon}{2}$, then

$$\text{dist}(y_1, g(\overline{X}\backslash(X_1 \cup X_2))) > 0 .$$

On the one hand, by (3.5) we have

$$\begin{aligned}\deg(g, X, y_1) &= \sum_{x_i \in g^{-1}(y_1)} \text{sgn} \det g'(x_i) \\ &= \sum_{x_i \in g^{-1}(y_1)\cap X_1} + \sum_{x_i \in g^{-1}(y_1)\cap X_2} \text{sgn} \det g'(x_i) \\ &= \deg(g, X_1, y_1) + \deg(g, X_2, y_1) .\end{aligned}$$

On the other hand, by property (2),

$$\begin{aligned}\deg(f, X, y_0) &= \deg(g, X, y_1), \\ \deg(f, X_i, y_0) &= \deg(g, X_i, y_1), \; i = 1, 2 .\end{aligned}$$

This is due to the fact that

$$\theta \neq tf(x) + (1 - t)g(x) - ty_0 - (1 - t)y_1 ,$$

as

$$x \in \partial X \cup \partial X_1 \cup \partial X_2 \subset \overline{X}\backslash(X_1 \cup X_2), \; t \in [0, 1] .$$

□

Corollary 3.2.3 ***(Excision)*** *If $K \subset \overline{X}$ is compact and $y_0 \notin f(K) \cup f(\partial X)$, then*

$$\deg(f, X, y_0) = \deg(f, X\backslash K, y_0) .$$

Proof. Let $X_1 = X\backslash K, X_2 = \emptyset$, then

$$\overline{X}\backslash(X_1 \cup X_2) = \partial X \cup K .$$

□

Corollary 3.2.4 ***(Kronecker existence)*** *If $y_0 \notin f(\partial X)$ and $\deg(f, X, y_0) \neq 0$, then $f^{-1}(y_0) \neq \emptyset$.*

Proof. Assume the contrary: $f^{-1}(y_0) = \emptyset$, i.e., $y_0 \notin f(\overline{X})$. By Corollary 3.2.3,

$$\deg(f, X, y_0) = \deg(f, X\backslash\overline{X}, y_0) = 0 .$$

□

Kronecker existence theorem is often used in the study of the solvability of the equation $f(x) = y_0$.

(5) (**Normality**). If $X \subset Y$, and $y_0 \notin \partial X$, then

$$\deg(id, X, y_0) = \begin{cases} 1, & y_0 \in \overset{\circ}{X} \\ 0, & y_0 \notin \overline{X} \,. \end{cases}$$

More generally, we have:

(6) If L is a $n \times n$ nondegenerate real matrix, and if X is a bounded open subset and $\theta \in X \subset \mathbb{R}^n$, then

$$\deg(L, X, \theta) = (-1)^{\beta} \,,$$

where $\beta = \sum_{\lambda_j < 0} \beta_j$ and β_j are the algebraic multiplicities of the negative eigenvalues λ_j of L, i.e.,

$$\beta_j = \dim \cup_{k=1}^{\infty} \ker (\lambda_j I - L)^k, \;\; j = 1, 2, \ldots \,.$$

Proof. Let $\lambda_1, \ldots, \lambda_n$ be all the eigenvalues of L. By definition,

$$\deg(L, X, \theta) = \operatorname{sgn} \det L = \operatorname{sgn} \prod_{j=1}^{n} \lambda_j \,.$$

Since complex roots appear in pairs and positive eigenvalues have no contribution to $\operatorname{sgn} \det L$, hence

$$\operatorname{sgn} \prod_{j=1}^{n} \lambda_j = \operatorname{sgn} \prod_{\lambda_j < 0} \lambda_j = \prod_{\lambda_i < 0} (-1)^{\beta_j} = (-1)^{\beta} \,.$$

□

Remark 3.2.5 *In the case where $X = Y = \mathbb{R}^n$, the fundamental properties (1),(3),(4),(5) uniquely determine the Brouwer degree. All the other properties are their consequences.*

(7) Let $f_i \in C(\overline{\Omega}_i, \mathbb{R}^{n_i})$ be open and bounded, $p_i \notin f_i(\partial \Omega_i)$, $i = 1, 2$. Then

$$\deg(f_1 \times f_2, \Omega_1 \times \Omega_2, (p_1, p_2)) = \deg(f_1, \Omega_1, p_1) \deg(f_2, \Omega_2, p_2) \,.$$

Proof. By approximation, we may assume that $f_i \in C^1(\overline{\Omega}_i, \mathbb{R}^{n_i})$ and that p_i is not a critical value of f_i, $i = 1, 2$.
Let μ_i be an admissible n_i form with support in a good chart (g_i, Ω_i) at p_i, $i = 1, 2$, respectively. Then $\mu_1 \times \mu_2$ is an admissible $n_1 + n_2$ form with support in a good chart $(g_1 \times g_2, \Omega_1 \times \Omega_2)$ at (p_1, p_2). We have

$$\int_{\Omega_1 \times \Omega_2} (\mu_1 \times \mu_2) \circ (f_1 \times f_2) = \int_{\Omega_1} \mu_1 \circ f_1 \int_{\Omega_2} \mu_2 \circ f_2 \,.$$

□

(8) (**Leray product formula**). Assume that $\Omega, M \subset \mathbb{R}^n$ are open subsets with compact closures. Let $f : \overline{\Omega} \to \mathbb{R}^n, g : \overline{M} \to \mathbb{R}^n$ be continuous, with $f(\overline{\Omega}) \subset M$. Let

$$\triangle = M \backslash f(\partial\Omega) = \cup_{j=1}^{\infty} \triangle_j ,$$

where each $\triangle_j$ is a connected component of $\triangle$. Suppose $p \notin g \circ f(\partial\Omega) \cup g(\partial M)$. Then

$$\deg(g \circ f, \Omega, p) = \sum_j \deg(g, \triangle_j, p) \deg(f, \Omega, \triangle_j) .$$

Proof. Before going to the proof, we should explain:
(a) The meaning of $\deg(f, \Omega, \triangle_j)$ is understood according to (2).
(b) There are at most finitely many terms on the right-hand side of the formula different from 0. In fact, $g^{-1}(p)$ is compact and

$$g^{-1}(p) \cap (\partial M \cup f(\partial\Omega)) = \emptyset .$$

This means $\triangle = \cup_{j=1}^{\infty} \triangle_j$ covers $g^{-1}(p)$. Since $\triangle_i \cap \triangle_j = \emptyset,\ i \neq j$, provided by finite covering, there are at most finitely many that $\triangle_j$ cover $g^{-1}(p)$.

Now we begin the proof.
1. First suppose $f \in C^1(\overline{\Omega}),\ g \in C^1(\overline{M})$ and that p is a regular value of $g \circ f$. We have

$$J_{g \circ f}(x) = J_g(f(x)) \cdot J_f(x) .$$

If $x \in (g \circ f)^{-1}(p)$, then $y = f(x)$ is a regular value of f, and p is a regular value of $g : \overline{M} \cap f(\overline{\Omega}) \to \mathbb{R}^n$, hence

$$\begin{aligned}
\deg(g \circ f, \Omega, p) &= \sum_{x_j \in (g \circ f)(p)} \operatorname{sgn} J_{g \circ f}(x_j) \\
&= \sum_{x_j \in (g \circ f)(p)} \operatorname{sgn} J_g(f(x_j)) \cdot \operatorname{sgn} J_f(x_j) \\
&= \sum_{y_k \in g^{-1}(p) \cap f(\Omega)} \operatorname{sgn} J_y(y_k) \left(\sum_{x_j \in f^{-1}(y_k)} \operatorname{sgn} J_f(x_j) \right) \\
&= \sum_{y_k \in g^{-1}(p) \cap f(\Omega)} \operatorname{sgn} J_y(y_k) \cdot \deg(f, \Omega, y_k) .
\end{aligned}$$

Putting together all these terms, in which y_k belongs to the same $\triangle_j$, from

$$\deg(f, \Omega, y_k) = \deg(f, \Omega, \triangle_j) ,$$

we have

$$\deg\,(g \circ f, \Omega, p) = \sum_{y_k \in g^{-1}(p) \cap \triangle_j} \operatorname{sgn} J_g(y_k) \deg(f, \Omega, \triangle_j)$$

$$= \sum_j \deg(g, \triangle_j, p) \deg(f, \Omega, \triangle_j) .$$

2. Passing to continuous mappings $f \in C(\overline{\Omega})$, $g \in C(\overline{M})$, the trouble lies in the fact that both $\triangle$ and $\{\triangle_j\}$ change as f varies. We have to be careful in making the $C^1(\overline{\Omega})-$ approximations. $\forall y \in f(\partial\Omega)$ let

$$O_k = \{y \in M | \ \deg(f, \Omega, y) = k\} .$$

Then

$$O_k = \cup\{\triangle_j | \ \deg(f, \Omega, \triangle_j) = k\}, \ k = 0, \pm 1, \pm 2, \dots .$$

Let

$$\varepsilon = \mathrm{dist}(g^{-1}(p), f(\partial\Omega)) .$$

We choose $\hat{f} \in C^1(\overline{\Omega})$ such that

$$\| \hat{f} - f \|_{C(\overline{\Omega})} < \frac{\varepsilon}{2} ,$$

then

$$p \notin g(\partial M) \cup g \circ \hat{f}(\partial\Omega) . \tag{3.10}$$

Let

$$\hat{O}_k = \{y \in M | \ \deg(\hat{f}, \Omega, y) = k\} .$$

Again we choose $\hat{g} \in C^1(\overline{M})$ such that

$$\| \hat{g} - g \|_{C(\overline{M})} < \mathrm{dist}(p, g(\partial M) \cup g \circ \hat{f}(\partial\Omega)) .$$

Since $\partial\hat{O}_k \subset \hat{f}(\partial\Omega) \cup \partial M$, it follows that

$$\| \hat{g} - g \|_{C(\overline{M})} < \mathrm{dist}(p, g(\partial\hat{O}_k)) .$$

Therefore $p \notin \hat{g}(\partial\hat{O}_k)$ and

$$\deg(g, \hat{O}_k, p) = \deg(\hat{g}, \hat{O}_k, p) . \tag{3.11}$$

It is easy to verify that

$$g^{-1}(p) \cap O_k = g^{-1}(p) \cap \hat{O}_k .$$

By excision,

$$\deg(g, O_k, p) = \deg(g, O_k \cap \hat{O}_k, p) = \deg(g, \hat{O}_k, p) \tag{3.12}$$

Combining (3.11) and (3.12) it follows that

$$\deg(g, O_k, p) = \deg(\hat{g}, \hat{O}_k, p) . \tag{3.13}$$

Finally, we obtain

$$\begin{aligned}\deg(g\circ f,\Omega,p) &= \deg(\hat{g}\circ \hat{f},\Omega,p)\\ &= \sum_k \deg(\hat{g},\hat{O}_k,p)\deg(\hat{f},\Omega,\hat{O}_k)\\ &= \sum_k k\deg(g,O_k,p)\\ &= \sum_k \deg(g,O_k,p)\deg(f,\Omega,O_k)\\ &= \sum_j \deg(g,\triangle_j,p)\deg(f,\Omega,\triangle_j)\,.\end{aligned}$$

□

Fixing the map f and the point y_0, the Brouwer degree is a function on the domain X. It provides the information on the behavior of the mapping f by all solutions included in X of the equation:

$$f(x) = y_0\,. \tag{3.14}$$

However, if we want to localize the notion to studying the local behavior of f at an isolated solution of equation (3.14), we are led to:

Definition 3.2.6 *(Brouwer index) Let x_0 be an isolated solution of (3.14), i.e., $\exists \varepsilon > 0$ such that there is no solution of (3.14) in $B_\varepsilon(x_0)\backslash\{x_0\}$. We call*

$$i(f,x_0,y_0) = \deg(f,B_\varepsilon(x_0),y_0)$$

the index of f at x_0 with respect to y_0.

The excision property guarantees that the definition is well defined, i.e., it does not depend on the special choice of $\varepsilon > 0$.
We have the following properties:

(9) Suppose that $f \in C(\overline{X},Y)\cap C^1(X,Y)$, and that x_0 is an isolated solution of (3.14). If $\det f'(x_0) \neq 0$, then

$$i(f,x_0,y_0) = (-1)^\beta\,,$$

where β is the sum of the algebraic multiplicities of the negative eigenvalues of $f'(x_0)$.

Proof. Since $f'(x_0)$ is invertible, there exists $\varepsilon > 0$ such that

$$\deg(f,B_\varepsilon(x_0),y_0) = \deg(f'(x_0),B_\varepsilon(\theta),\theta)$$

from the homotopy invariance and translation invariance. The conclusion then follows from (6). □

(10) If $f \in C(\overline{X}, Y)$ and $y_0 \notin f(\partial X)$ with $f^{-1}(y_0) = \{x_1, \ldots, x_p\}$, then

$$\deg(f, X, y_0) = \sum_{i=1}^{p} i(f, x_j, y_0) \, .$$

Proof. It follows from the excision property plus the additivity property. □

At the end of this section we establish the connection between the definition of the Brouwer degree given above with that in algebraic topology. Following Corollary 3.2.2, for $f \in C^2(X, Y)$, we have

$$\deg(f, X, y_0) = \frac{\int_X \mu \circ f}{\int_Y \mu} \, .$$

Let $f^* : \Omega^n(Y) \to \Omega^n(X), \mu \mapsto \mu \circ f$ be the pullback of f, where $\Omega^n(X)$ is the space of $n-$ forms over X. This means that $\deg(f, X, y_0)$ is the ratio of the integrations of $f^*\mu$ and μ, $\forall \mu \in \Omega^n(Y)$.

In algebraic topology, the Brouwer degree for maps f from the sphere into itself is defined to be the multiplier of the homomorphism $f^* : H^n(S^n) \to H^n(S^n)$, in which $H^n(S^n)$ stands for the cohomology group of the $n-$ sphere, and the map f maps the generator $[\omega]$ into $\lambda[\omega]$ for some integer λ.

According to de Rham theory, for a compact, oriented, connected n manifold without boundary X, the cohomology group $H^k(X)$ is defined to be

$$\ker d^k / \mathrm{Im}\ d^{k-1} \, ,$$

where $d^k : \Omega^k(X) \to \Omega^{k+1}(X)$ is the exterior differentiation, $k = 0, 1, \ldots, n$.

We shall prove that $\Omega^n(S^n)$ and then $H^n(S^n)$ is one dimensional. Namely:

Lemma 3.2.7 *For a compact, oriented, connected n-manifold X without boundary,* $\dim H^n(X) = 1$.

Proof. We want to show that $H^n(X)$ is generated by one generator. Let us choose an atlas $\{(U_i, \psi_i) | \ i = 1, \ldots, p\}$ such that

$$U_i \cap U_j \neq \emptyset \Rightarrow U_i \cup U_j \text{ is contained in a good chart} \, .$$

It is sufficient to prove that for $\forall \mu, \mu_0 \in \Omega^n(X)$ with $\mathrm{supp}\mu_0 \subset U_1$, and $\int_X \mu_0 = 1$, there exist $\lambda \in \mathbb{R}^1$ and $\omega \in \Omega^{n-1}(X)$ such that

$$\mu - \lambda \mu_0 = d\omega \, .$$

Since X is connected, for $\forall i, \exists$ a curve C starting from U_1 ending at U_i. Let $\{U_{i_0}, U_{i_1}, \ldots, U_{i_l}\}$ be a chain of neighborhoods in $\{U_i\}_1^p$, such that

$$U_{i_0} = U_1, \ U_{i_l} = U_i \text{ and } U_{i_k} \cap U_{i_{k-1}} \neq \emptyset \ k = 1, 2, \ldots, l \, .$$

We choose $\alpha_k \in \Omega^n(X)$ satisfying

$$\text{supp } \alpha_k \subset U_{i_k}, \ \int_X \alpha_k = 1, \ k = 1, 2, \ldots, l ,$$

and $\alpha_0 = \mu_0$. Since

$$\text{supp}(\alpha_k - \alpha_{k-1}) \subset U_{i_k} \cup U_{i_{k-1}}$$

is contained in a good neighborhood, and

$$\int_X (\alpha_k - \alpha_{k-1}) = 0 ,$$

we have $\omega_k^i \in \Omega^{n-1}(X)$ satisfying

$$\alpha_k - \alpha_{k-1} = d\omega_k^i, \ k = 1, 2, \ldots, l ,$$

provided by Lemma 3.1.2. Hence

$$\alpha_l = \mu_0 + d \sum_{k=1}^{l} \omega_k^i . \tag{3.15}$$

Since $U_{i_l} = U_i$, α_l depends on the index i, we write it as $\beta_i, i = 1, \ldots, p$.

Now, let $\{\chi_i\}_1^p$ be a partition of unity with respect to $\{U_i\}_1^p$, i.e.,

$$\text{supp } \chi_i \subset U_i \ \chi_i \geqslant 0, \ \sum_{i=1}^{p} \chi_i \equiv 1 .$$

Set

$$\mu_i = \chi_i \mu ,$$

then

$$\text{supp } \mu_i \subset U_i .$$

Denote $c_i = \int_X \mu_i$. There exists $\widetilde{\omega}_i \in \Omega^{n-1}(X)$ such that

$$\mu_i - c_i \beta_i = d\widetilde{\omega}_i, \tag{3.16}$$

from Lemma 3.1.2, $i = 1, 2, \ldots, p$.

Combining (3.15) and (3.16), we have

$$\begin{aligned} \mu = \sum_{i=1}^{p} \chi_i \mu &= \sum_{i=1}^{p} \mu_i \\ &= \sum_{i=1}^{p} c_i \beta_i + d \sum_{i=1}^{p} \widetilde{\omega}_i \end{aligned}$$

$$= \sum_{i=1}^{p} c_i \mu_0 + d \sum_{i=1}^{p} \left(\widetilde{\omega}_i + c_i \sum_{k=1}^{l} \omega_k^i \right)$$
$$= \lambda \mu_0 + d\omega \ ,$$

where $\lambda = \sum_{i=1}^{p} c_i = \int_X \mu_i$, and $\omega = \sum_{i=1}^{p} (\widetilde{\omega}_i + c_i \sum_{k=1}^{l} \omega_k^i)$. Thus we have proved that all elements in $H^n(X)$ are multipliers of $[\mu_0]$, i.e., $\dim H^n(X) = 1$. □

Combining Corollary 3.2.2 and Lemma 3.2.7, we obtain

Theorem 3.2.8 *Let X and Y be compact, oriented, connected n-manifolds without boundaries. Then for any $f \in C^2(X, Y)$, the following diagram commutes:*

$$\begin{array}{ccc} H^n(Y) & \xrightarrow{f^*} & H^n(X) \\ {\scriptstyle \int_Y} \downarrow & & {\scriptstyle \int_X} \downarrow \\ \mathbb{R}^1 & \xrightarrow[\deg f]{} & \mathbb{R}^1 \end{array}$$

Finally, we turn to studying the relationship between degrees on balls and those on spheres. Let $f : B^n \to \mathbb{R}^n$ be a continuous map satisfying $\theta \notin f(\partial B^n)$, so $\deg(f, B^n, \theta)$ is well defined. Let us define

$$\Phi(x) = \frac{f(x)}{\| f(x) \|} \quad \forall x \in \partial B^n \ .$$

Then $\Phi : S^{n-1} \to S^{n-1}$ defines a Brouwer degree: $\deg(\Phi, S^{n-1})$.

What is the relationship between $\deg(f, B^n, \theta)$ and $\deg(\Phi, S^{n-1})$?

Define a homotopy

$$\phi(x,t) = \begin{cases} \theta & \text{if } x = \theta \\ \frac{\|x\|^2}{\|f(\frac{x}{\|x\|})\|^t} f(\frac{x}{\|x\|}) & \text{if } x \neq \theta \ . \end{cases}$$

Since $f|_{\partial B^n} = \phi(\cdot, 0)|_{\partial B^n}$, from Corollary 3.2.1 and the homotopy invariance,

$$\begin{aligned} \deg(f, B^n, \theta) &= \deg(\phi(\cdot, 0), B^n, \theta) \\ &= \deg(\phi(\cdot, 1), B^n, \theta) \ . \end{aligned}$$

In the case where $f \in C^1$, and $y_0 \in S^{n-1}$ is a regular value of Φ, for $\forall \varepsilon > 0$ small, $\varepsilon^2 y_0$ is a regular value of $\phi(\cdot, 1)$, and $\Phi^{-1}(y_0) = \{x_1, \ldots, x_k\}$ if and only if $\phi(\cdot, 1)^{-1}(\varepsilon^2 y_0) = \{\varepsilon x_1, \ldots, \varepsilon x_k\}$.

Since

$$\operatorname{sgn} J_\Phi(x_j) = \operatorname{sgn} J_{\phi(\cdot,1)}(\varepsilon x_j), \ \ j = 1, 2, \ldots, k \ ,$$

we obtain $\deg(\phi(\cdot, 1), B^n, \theta) = \deg(\Phi, S^{n-1})$. The assumptions on f and on y_0 can easily be dropped. Namely, we have proved:

Corollary 3.2.9 $\deg(f, B^n, \theta) = \deg(\Phi, S^{n-1}) = \deg(\Phi)$.

Thus the notation $\deg(\Phi, S^{n-1})$ is in coincidence with the notation $\deg(\hat{f}, \partial B^n, \theta)$ introduced after Corollary 3.2.1.

3.3 Applications of Brouwer Degree

3.3.1 Brouwer Fixed-Point Theorem

The Brouwer degree is a fundamental tool in algebraic topology. It is widely used in topological arguments. We are satisfied having a glimpse of these applications.

Theorem 3.3.1 ***(Brouwer fixed-point theorem)*** *A continuous map f from $\overline{B}^n$ into itself has a fixed point $x_0 \in \overline{B}^n$, i.e., $f(x_0) = x_0$.*

Proof. Let $g = \mathrm{id} - f$. With no loss of generality, we may assume $\theta \notin g(\partial B^n)$. Since $\forall x \in \partial B^n$, $\forall \lambda \geqslant 1$, $f(x) \notin \lambda x$, we have $\theta \notin \phi(\partial B^n) \times [0,1]$, where

$$\phi(x,t) = (1-t)g(x) + tx \, .$$

Thus by the homotopy invariance and the normality,

$$\deg(g, \overset{\circ}{B^n}, \theta) = \deg(\mathrm{id}, \overset{\circ}{B^n}, \theta) = 1 \, .$$

Our conclusion follows from the Kronecker existence theorem. □

The following topological fact follows from a simple degree argument:

Theorem 3.3.2 *There is no continuous map $f : \overline{B}^n \to \partial B^n$ such that $f|_{\partial B^n} = id|_{\partial B^n}$, i.e., ∂B^n is not a retraction of $\overline{B}^n$.*

Proof. If not, there is a continuous map $f : \overline{B}^n \to \partial B^n$. $f|_{\partial B^n} = \mathrm{id}|_{\partial B^n}$. On one hand, $\deg(f, B^n, \theta) = 0$ due to the Kronecker existence theorem. On the other hand, by Corollary 3.2.1 and the normality

$$\deg(f, B^n, \theta) = \deg(\mathrm{id}, B^n, \theta) = 1 \, .$$

This is a contradiction. □

3.3.2 The Borsuk-Ulam Theorem and Its Consequences

Let $\Omega \subset \mathbb{R}^n$ be a bounded open set, which is symmetric with respect to the origin θ, i.e., $-\Omega = \Omega$. A map $f : \Omega \to \mathbb{R}^n$ is called odd, if

$$f(-x) = -f(x) \quad \forall x \in \Omega \, .$$

We are going to study the Brouwer degree of odd mappings.

Theorem 3.3.3 ***(Borsuk)*** *Suppose that $\Omega \subset \mathbb{R}^n$ is a bounded open set containing θ and is symmetric with respect to θ. If $f : \overline{\Omega} \to \mathbb{R}^n$ is an odd continuous map, then*

$$\deg(f, \Omega, \theta) = \mathit{odd} \, ,$$

whenever $\theta \notin f(\partial\Omega)$.

Proof. One chooses $\varepsilon > 0$ such that $B_\varepsilon = B_\varepsilon(\theta) \subset \Omega$. According to Tietze's theorem, there exists a continuous map $f_\varepsilon : \overline{\Omega} \to \mathbb{R}^n$ satisfying

$$f_\varepsilon(x) = \begin{cases} x & x \in \overline{B}_\varepsilon \\ f(x) & x \in \partial\Omega \,. \end{cases}$$

If we use $\frac{1}{2}(f_\varepsilon(x) - f_\varepsilon(-x))$ to replace $f_\varepsilon(x)$, then one may assume that f_ε is odd. Since $f_\varepsilon|_{\partial\Omega} = f|_{\partial\Omega}$, according to Corollary 3.2.1, the additivity and the normality,

$$\begin{aligned} \deg(f, \Omega, \theta) &= \deg(f_\varepsilon, \Omega, \theta) \\ &= \deg(f_\varepsilon, B_\varepsilon, \theta) + \deg(f_\varepsilon, \Omega\backslash\overline{B}_\varepsilon, \theta) \\ &= 1 + \deg(f_\varepsilon, \Omega\backslash\overline{B}_\varepsilon, \theta)\,. \end{aligned}$$

In the following, we want to prove that $\deg(f_\varepsilon, \Omega\backslash\overline{B}_\varepsilon, \theta)$ is even. Noticing that for a C^1 odd map g with regular value θ, elements in $g^{-1}(\theta) \cap (\Omega\backslash\overline{B}_\varepsilon)$ occur in pairs. Our strategy is to approximate f_ε by a C^1 odd map g with regular value θ. Once it is constructed, we have

$$\deg(f_\epsilon, \Omega\backslash\overline{B}_\epsilon, \theta) = \deg(g, \Omega\backslash\overline{B}_\epsilon, \theta) \equiv 0 \,(\mathrm{mod}\, 2)\,.$$

By the Weierstrass approximation theorem, we may assume that $f_\varepsilon \in C^\infty$. It remains to approximate it such that θ is a regular value. To this end we consider the map $F : (\Omega\backslash\bar{B}_\epsilon(\theta)) \times \mathbf{M}^{n\times n} \longrightarrow \mathrm{R}^n$ defined by $F(x, A) = f_\epsilon(x) + Ax$.

If we can show that $F \pitchfork \{\theta\}$, then according to the transversality theorem (Theorem 1.3.14), $g_A \pitchfork \{\theta\}$, for almost every $A \in \mathrm{M}^{n\times n}$, where $g_A(x) = F(x, A)$. In fact,

$$F'(x, A)(y, B) = f'_\epsilon(x)y + Ay + Bx, \ \forall (y, B) \in \mathrm{R}^n \times \mathrm{M}^{n\times n}\,.$$

Since $x \neq \theta, \forall z \in \mathrm{R}^n$, we take $B = \frac{<\cdot,x>}{\|x\|^2} z$, and $y = \theta$, we have $f'(x, A)(y, B) = z$, i.e., $F'(x, A)$ is surjective, or $F \pitchfork \{\theta\}$. Since $\|A\|$ can be chosen arbitrarily small, the proof is complete. □

Theorem 3.3.4 ***(Borsuk–Ulam)*** *Suppose that $\Omega \subset \mathbb{R}^n$ is a symmetric bounded open set including θ, and that $g : \partial\Omega \to \mathbb{R}^m$, $m < n$, is odd and continuous. Then there exists $x_0 \in \partial\Omega$ such that $g(x_0) = 0$.*

Proof. We prove by contradiction. Suppose $\theta \notin g(\partial\Omega)$. We define a continuous extension $\widetilde{g} : \Omega \to \mathbb{R}^m$ by the Tietze theorem. With no loss of generality, one may assume $\widetilde{g}$ is odd.

Now we apply Theorem 3.3.3, $\deg(\widetilde{g}, \Omega, \theta) \neq 0$. Choosing $y_0 \in \mathbb{R}^n\backslash\mathbb{R}^m$, with small norm, it follows that

$$\deg(\widetilde{g}, \Omega, y_0) \neq 0$$

Thus, $\widetilde{g}^{-1}(y_0) \neq \emptyset$, but this is impossible. □

Corollary 3.3.5 *Suppose that $\Omega \subset \mathbb{R}^n$ is a symmetric bounded open set including θ, and that $g : \partial\Omega \to \mathbb{R}^m$, with $m < n$, is continuous. Then there exists $x_0 \in \partial\Omega$ such that $g(x_0) = g(-x_0)$*

Proof. Set $g_1(x) = g(x) - g(-x)$, then apply Theorem 3.3.4 directly. □

Remark 3.3.6 *We consider a continuous vector field on S^n, i.e., a continuous map $g : S^n \to \mathbb{R}^n$. Corollary 3.3.5 means that there exists a pair of antipodal points $\pm x_0$, which have the same vector $g(x_0) = g(-x_0)$.*

Theorem 3.3.7 ***(Ljusternik–Schnirelmann–Borsuk)*** *Suppose that $\Omega \subset \mathbb{R}^n$ is a symmetric bounded open set including θ, and that $\{A_1, \ldots, A_p\}$ is a closed covering of $\partial\Omega$, satisfying $A_i \cap (-A_i) = \emptyset \ \forall i$. Then $p \geqslant n+1$.*

Proof. We may assume that $\cap_{i=1}^p A_i = \emptyset$. For otherwise, if $\exists x_0 \in \cap_{i=1}^p A_i$, then by covering $\partial\Omega = \cup_{i=1}^p (-A_i)$, there must be a j such that $x_0 \in A_j \cap (-A_j)$. But this is impossible. Now we set

$$d_i(x) = \text{dist}\ (x, A_i), \ \ i = 1, 2, \ldots, p\ ,$$

and $f : \partial\Omega \to \mathbb{R}^{p-1} \subset \mathbb{R}^{n-1}$ as follows:

$$f(x) = (d_1(x), d_2(x), \ldots, d_{p-1}(x))\ .$$

Supposing $p \leqslant n$, we apply Corollary 3.3.5, $\exists x_0 \in \partial\Omega$ satisfying $f(x_0) = f(-x_0)$. Since $\{A_1, A_2, \ldots, A_p\}$ is a covering, there exists $j \leqslant p$ such that $x_0 \in A_j$. There are two possibilities:

1. $j \leqslant p-1$. It follows that $d_j(-x_0) = d_j(x_0) = 0$, thus $-x_0 \in A_j$ or $x_0 \in A_j \cap (-A_j)$; this is a contradiction.
2. $x_0 \notin \bigcup_{j=1}^{p-1} A_j$, thus $j = p$. In this case, $d_j(-x_0) = d_j(x_0) > 0$, $\forall j = 1, 2, \ldots, p-1$, so $x_0 \notin \bigcup_{j=1}^{p-1} (-A_j)$. Again, by covering, $x_0 \in A_p \cap (-A_p)$, a contradiction.

□

The following theorem is an extension of the open mapping theorem to continuous mappings in finite-dimensional spaces.

Theorem 3.3.8 *(Invariance of domains) If $\Omega \subset \mathbb{R}^n$ is a nonempty open set, and if $f : \Omega \to \mathbb{R}^n$ is continuous and locally injective, then f is an open map.*

Proof. We want to show that $\forall x_0 \in \Omega$, $\forall \varepsilon > 0$ with $B_\varepsilon(x_0) \subset \Omega$, $\exists \delta > 0$ such that $f(B_\varepsilon(x_0)) \supset B_\delta(f(x_0))$.

One may assume $x_0 = \theta = f(\theta)$, and that $f : \overline{B}_\epsilon(\theta) \mapsto f(\overline{B}_\epsilon(\theta))$ is 1–1.

Since $\theta \notin f(\partial \overline{B}_\epsilon(\theta))$, $\exists \delta > 0$ such that $\overline{B}_\delta(\theta) \subset \mathbb{R}^n \backslash f(\partial B_\epsilon(\theta))$. Our problem is reduced to proving that $\forall y_0 \in B_\delta(\theta)$, $\exists x \in B_\epsilon(\theta)$ such that $f(x) = y_0$. Due to the Kronecker existence theorem, it is sufficient to prove that

$$\deg\,(f, B_\epsilon(\theta), y_0) \neq 0\ .$$

Since $\delta > 0$ can be chosen arbitrarily small, according to property (2) of the Brouwer degree,

$$\deg(f, B_\epsilon(\theta), y_0) = \deg(f, B_\epsilon(\theta), \theta) .$$

Let

$$H(x,t) = f\left(\frac{x}{1+t}\right) - f\left(\frac{-tx}{1+t}\right) \quad \forall (x,t) \in \overline{B}_\epsilon(\theta) \times [0,1], .$$

It is continuous on $\overline{B}_\epsilon(\theta) \times [0,1]$, and satisfies $H(x,0) = f(x)$, $H(x,1) = f(\frac{1}{2}x) - f(-\frac{1}{2}x)$. We claim that $\theta \notin H(\partial B_\varepsilon \times [0,1])$. For otherwise, there are $(x_0, t) \in \partial B_\epsilon(\theta) \times [0,1]$ satisfying

$$f\left(\frac{x_0}{1+t}\right) = f\left(\frac{-tx_0}{1+t}\right) .$$

Since f is $1-1$ on $B_\epsilon(\theta)$, we would have

$$\frac{x_0}{1+t} = \frac{-tx_0}{1+t} ,$$

i.e., $x_0 = \theta$. But, this is impossible.

Therefore by the homotopy invariance and Borsuk–Ulam theorem

$$\deg(f, B_\varepsilon(\theta), \theta) = \deg(H(\cdot,1), B_\varepsilon, \theta) \neq 0 .$$

and then $\deg(f, B_\varepsilon(\theta), y_0) \neq 0$. This proves our conclusion. □

Comparing with the global implicit function theorem, the following theorem is an extension to continuous mappings in finite-dimensional spaces:

Corollary 3.3.9 *If $f : \mathbb{R}^n \to \mathbb{R}^n$ is continuous and locally injective, and if $\| f(x) \| \to +\infty$ as $\| x \| \to +\infty$, then f is surjective.*

Proof. From Theorem 3.3.8, $f(\mathbb{R}^n)$ is open. We shall prove that $f(\mathbb{R}^n)$ is also closed. Suppose $y_n \in f(\mathbb{R}^n)$ with $y_n \to y_0$. We shall prove that $\exists x_0 \in \mathbb{R}^n$ satisfying $f(x_0) = y_0$. Indeed we have $x_n \in \mathbb{R}^n$ satisfying $f(x_n) = y_n \; \forall n$. According to the assumption, $\{x_n\}$ is bounded, and then there exists a convergent subsequence $x'_n \to x_0$. Thus $y_0 = f(x_0)$.

$f(\mathbb{R}^n)$ is both open and closed. Therefore $f(\mathbb{R}^n) = \mathbb{R}^n$. □

3.3.3 Degrees for S^1 Equivariant Mappings

We now turn to a computation of the degree of an S^1 group action equivariant map. Let us consider an S^1 representation on C^n: $T_\phi = e^{i\,\mathrm{diag}\,\{\lambda_1\phi,\ldots,\lambda_n\phi\}}$, where $\phi \in [0, 2\pi]$, and $\lambda_1, \ldots, \lambda_n \in \mathbf{Z}$.

Theorem 3.3.10 *Suppose that $\Omega \subset C^n$ is a $T(S^1)$ invariant open domain containing θ, and that $f \in C(\partial\Omega, C^n\backslash\{\theta\})$ satisfies*

$$f(T_\phi z) = e^{ik\phi} f(z), \ \ \forall z \in \partial\Omega, \ \ \forall \phi \in [0, 2\pi] .$$

Then

$$\deg(f, \Omega, \theta) = \frac{k^n}{\lambda_1 \dots \lambda_n} .$$

Proof. 1. We may assume that $\lambda_1 = \cdots = \lambda_n = 1$. In fact, if it is proved for the special case, then for general $\lambda_1, \dots, \lambda_n \in \mathbf{Z}_+$, we define the transform $g(z) = (z_1^{\lambda_1}, \dots, z_n^{\lambda_n})$, let $\tilde{\Omega} = g^{-1}(\Omega)$, then the map $f \circ g : \tilde{\Omega} \to C^n$ satisfies $T_\phi \circ g(z) = g \circ \tilde{T}_\phi z$, where $\tilde{T}_\phi = e^{i\phi} I_n$. Therefore

$$f \circ g(\tilde{T}_\phi z) = f(T_\phi \circ g(z)) = e^{ik\phi} f \circ g(z) .$$

Provided by the product formula, we have

$$\deg(f \circ g, \tilde{\Omega}, \theta) = \deg(f, \Omega, \theta) \cdot \deg(g, \tilde{\Omega}, \theta) .$$

We assume the conclusion holds for $\lambda_1 = \cdots = \lambda_n = 1$, therefore

$$\deg(f \circ g, \tilde{\Omega}, \theta) = k^n .$$

But,

$$\deg(g, \tilde{\Omega}, \theta) = \lambda_1 \cdots \lambda_n ,$$

we obtain our conclusion.

If $\lambda_1, \dots, \lambda_n$ are not all positive, then we introduce a transform $g : (z_1, \dots, z_j, \dots, z_n) \to (z_1, \dots, \bar{z}_j, \dots, z_n)$ for all negative λ_js. Again by the product formula, we have

$$\deg(f \circ g, g^{-1}(\Omega), \theta) = (-1)^{\mathrm{sgn}\, \lambda_1 \cdots \lambda_n} \deg(f, \Omega, \theta) .$$

Thus it is sufficient to prove the theorem for $\lambda_1 = \cdots = \lambda_n = 1$.

2. We may assume that f is smooth. This can be done by a mollifier; the argument is standard, so we omit it. The only thing that has to be verified is the equivariance of the mollified map.

3. Taking a smooth function on $R^1_+ : \eta(t) = 1, \text{ as } t \le \epsilon, \text{ and } 0, \text{ as } t > 2\epsilon$, where $\epsilon < \frac{1}{2}\mathrm{dist}\,(\partial\Omega, \theta)$, and define

$$\hat{f} = \eta(|z|)(z_1^k, \dots, z_n^k) + (1 - \eta(|z|)) f(z) .$$

Then $\hat{f}$ has the same degree as f on Ω.

In summary, we may assume that f is smooth and S^1-equivariant, with $\lambda_1 = \ldots = \lambda_n = 1$, and has the form $(z_1^k, \dots, z_n^k)$ in $B_\epsilon(\theta)$.

4. Now, we consider the degree of f on $\Omega\backslash\bar{B}_\varepsilon(\theta)$. Define a map $F : (\Omega\backslash\bar{B}_\varepsilon(\theta)) \times \mathrm{M}^{n\times n} \to C^n$ by

$$F(z, A) = f(z) + Az^k ,$$

where $z^k = (z_1^k, \ldots, z_n^k)$. We show that $F \pitchfork \{\theta\}$. In fact,

$$F'(z, A)(w, \bar{w}, B) = F_z \cdot w + F_{\bar{z}} \cdot \bar{w} + \left(\sum_1^n z_i^k B_{ij} \right)_1^n .$$

Since $z \notin \bar{B}_\epsilon(\theta), \forall \xi \in C^n$, the system

$$\sum_1^n B_{ij} z_j^k = \xi_i$$

has a solution B_{ij}. By taking $w = \bar{w} = 0$ the map $F'(z, A)$ is surjective.

Applying the transversality theorem, $f_A := F(\cdot, A) \pitchfork \{\theta\}$ for almost all $A \in \mathrm{M}^{n \times n}$.

Since $\|A\|$ can be chosen small, the critical set $S(f_A, \Omega \backslash B_\epsilon(\theta))$ of f_A on $\Omega \backslash B_\epsilon(\theta)$ consists of isolated points, but $\forall z^* \in S(f_A, \Omega \backslash B_\epsilon(\theta))$, $T_\phi z^* \in S(f_A, \Omega \backslash B_\epsilon(\theta))$. This is impossible if it is nonempty. This proves that $\deg(f_A, \Omega \backslash \bar{B}_\varepsilon(\theta), \theta) = 0$. Therefore

$$\deg(f_A, \Omega, \theta) = \deg(f_A, B_\epsilon(\theta), \theta) .$$

But on $B_\varepsilon(\theta), f_A(z) = z^k + Az^k$, for sufficiently small $\|A\|$, we obtain:

$$\deg(f_A, B_\epsilon(\theta), \theta) = \deg(f, B_\epsilon(\theta), \theta) = k^n .$$

Combining all together, we have proved the conclusion.

□

3.3.4 Intersection

The Brouwer degree is useful in the intersection theory. We briefly introduce it here. Let $\Omega_1, \Omega_2 \subset \mathbb{R}^n$ be two compact manifolds with $\partial\Omega_1 \neq \emptyset$ and $\partial\Omega_2 = \emptyset$, where $\dim \Omega_1 = k + 1, \dim \Omega_2 = n - k - 1$. Let $\phi_1 : \partial\Omega_1 \to \mathbb{R}^n, \phi_2 : \Omega_2 \to \mathbb{R}^n$ be two continuous mappings with $\phi_1(\partial\Omega_1) \cap \phi_2(\Omega_2) = \emptyset$. We say ϕ_1 and ϕ_2 link, if for any continuous extension of $\phi_1, \widetilde{\phi}_1 : \Omega_1 \to R^n$, we have

$$\widetilde{\phi}_1(\Omega_1) \cap \phi_2(\Omega_2) \neq \emptyset .$$

In other words,

$$\exists (x, y) \in \Omega_1 \times \Omega_2, \text{ such that } \widetilde{\phi_1}(x) = \phi_2(y) .$$

Define a mapping $F : \Omega_1 \times \Omega_2 \to \mathbb{R}^n$ by $F(x, y) = \widetilde{\phi}_1(x) - \phi_2(y)$; it is equivalent to saying that F has a zero in $\Omega_1 \times \Omega_2$. Since $\theta \notin F(\partial\Omega_1 \times \Omega_2) = F(\partial(\Omega_1 \times \Omega_2))$, $\deg(F, \Omega_1 \times \Omega_2, \theta)$ is well defined and depends on (ϕ_1, ϕ_2) only, on account of Corollary 3.2.1. It is sufficient to verify $\deg(F, \Omega_1 \times \Omega_2, \theta) \neq 0$. Thus the Brouwer degree is a tool in the study of linking.

Example 1. Let $r_1 > 1, r_2 > 0$. Let $\Omega_1, \Omega_2 \subset \mathbb{R}^n$ be defined as follows:

$$\begin{aligned} \Omega_1 &= B^k_{r_2}(\theta) \times [0, r_1] \\ &:= \left\{(x_1, x_2, \ldots, x_{k+1}, 0, \ldots, 0) \,\middle|\, x_1^2 + \ldots x_k^2 = r_2^2, 0 \le x_{k+1} \le r_1\right\}\,, \\ \Omega_2 &= S^{n-k-1} := \left\{(0, 0, \ldots, x_{k+1}, \ldots, x_n) \,\middle|\, x_{k+1}^2 + \ldots + x_n^2 = 1\right\}\,, \end{aligned}$$

and let $\phi_1 = \mathrm{id}|_{\partial\Omega_1}, \phi_2 = \mathrm{id}_{\Omega_2}$. Then ϕ_1 and ϕ_2 link.

In fact, let $F(x_1, x_2, \ldots, x_n) = (x_1, \ldots, x_k, x_{k+1} - \sqrt{(1 - x_{k+2}^2 - \ldots - x_n^2)}, x_{k+2}, \ldots, x_n)$, it has a unique zero in $\Omega_1 \times \Omega_2 : x_1 = \cdots = x_k = x_{k+2} = \cdots = x_n = 0, x_{k+1} = 1$. Therefore,

$$\deg(F, \Omega_1 \times \Omega_2, \theta) = 1\,.$$

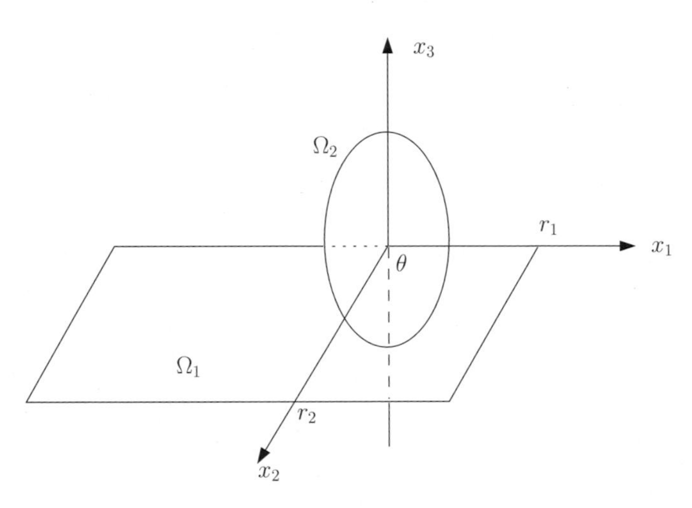

Fig. 3.1.

The Borsuk–Ulam theorem is applied to the study of the intersection of symmetric sets. Assume that X is a Banach space. To a closed symmetric set $A \subset X\backslash\{\theta\}$, we define the genus $\gamma(A)$ of A to be $\inf\{k \in \mathbb{N} \,|\, \exists$ an odd $\phi \in C(A, \mathbb{R}^k\backslash\{\theta\})\}$.

Example 2. Assume that $\Omega \subset \mathbb{R}^n$ is a symmetric bounded open set containing θ, and that there is an odd homeomorphism $h : A \to \partial\Omega$. Then $\gamma(A) = n$.

In fact, by definition, $\gamma(A) \le n$. Suppose $\gamma(A) < n$, then there exist $k < n$ and an odd ϕ such that $\phi \in C(A, \mathbb{R}^k\backslash\{\theta\})$. Let $g = \phi \circ h^{-1}$, then $g : \partial\Omega \to \mathbb{R}^k\backslash\{\theta\}$ is continuous and odd. According to the Borsuk–Ulam theorem, there exists $x_0 \in A$ such that $g(x_0) = \theta$. This is a contradiction.

3.4 Leray–Schauder Degrees

In analysis we study continuous mappings in infinite-dimensional spaces, so we should be concerned with extending the Brouwer degree from finite-dimensional spaces to infinite-dimensional Banach spaces. However, the generalization cannot suit arbitrary continuous mappings. This can be seen from the fact that if this sort of degree theory, which possesses the fundamental properties homotopy invariance, additivity and normality established in Sect. 3.2, had been built up, then the Brouwer fixed-point theorem would be extended directly to Banach spaces as follows.

Let B be the unit open ball at the origin in an infinite-dimensional real Banach space $\mathbb{X}$, $\phi \in C(\overline{B}, \overline{B})$. Let $f_t = \mathrm{id} - t\phi$, $t \in [0,1]$ and $\theta \notin f_t(\partial B)$. If $\deg(f_t, B, \theta)$ had been extended such that all the above properties of the Brouwer degree hold, then there would be a zero of $f_1 = \mathrm{id} - \phi$, i.e., a fixed point $x \in B$ of ϕ:

$$\phi(x) = x \, .$$

The "proof" is the same as in the previous section; we repeat it as follows: From the hypothesis $\theta \notin f_t(\partial B)$, according to the homotopy invariance and the normality,

$$\deg(f_1, B, \theta) = \deg(\mathrm{id}, B, \theta) = 1 \, .$$

Hence by the Kronecker existence, $f_1^{-1}(\theta) \cap B \neq \emptyset$, i.e., $\exists x \in B$ such that $\phi(x) = x$.

However, for an infinite-dimensional Banach space $\mathbb{X}$, the above conclusion cannot be true for any continuous mapping ϕ.

Example. Let

$$\mathbb{X} = l^2 = \left\{ x = (x_1, x_2, \ldots, x_n, \ldots) \mid \; \| x \|^2 = \sum_{n=1}^{\infty} x_n^2 < \infty \right\} .$$

and let ϕ be the mapping

$$x \mapsto (\sqrt{1 - \| x \|^2}, x_1, x_2, \ldots) \, .$$

Then $\phi : \overline{B} \to \overline{B}$ is continuous. But ϕ has no fixed point in $\overline{B}$. In fact, $\phi(\overline{B}) \subset \partial B$, if $\phi(\overline{x}) = \overline{x}$ for some $\overline{x} \in \overline{B}$, then $\| \overline{x} \| = 1$, but it is easily seen that $\overline{x_1} = 0, \overline{x_2} = \overline{x_1} = 0, \ldots$, therefore $\overline{x} = \theta$, a contradiction.

We are forced to restrict ourselves to a subfamily of continuous mappings. Let us recall:

Definition 3.4.1 *Let $\Omega \subset \mathbb{X}$ be a subset of a real Banach space. A continuous mapping $K : \Omega \to \mathbb{X}$ is said to be compact if it maps a bounded closed set into a compact set.*

As before, in the following we suppose that Ω is a bounded open set in the real Banach space $\mathbb{X}$.

Theorem 3.4.2 *If $K : \Omega \to \mathbb{X}$ is a compact map, and $\Omega \subset X$ is a bounded open set, then $\forall \varepsilon > 0$, there is a continuous operator K_ε taking values in a finite-dimensional linear subspace E_{m_ε} such that*

$$\| K(x) - K_\varepsilon(x) \| \leqslant \varepsilon \quad \forall x \in \overline{\Omega} .$$

Proof. Let $B_\varepsilon(y_j)$ $j = 1, \ldots, m_\varepsilon$, be finitely many $\varepsilon-$ balls covering $K(\overline{\Omega})$. Let

$$\psi_j(x) = (\varepsilon - \| x - y_j \|)_+$$

where

$$\lambda_+ = \begin{cases} \lambda, & \lambda \geqslant 0 , \\ 0, & \lambda < 0 . \end{cases}$$

Let

$$\varphi_i(x) = \frac{\psi_i(x)}{\sum_{j=1}^{m_\varepsilon} \psi_j(x)}, \quad i = 1, 2, \ldots, m_\varepsilon ,$$

and

$$K_\varepsilon(x) = \sum_{i=1}^{m_\varepsilon} \varphi_i(K(x)) y_i ,$$

then $K_\varepsilon(x) \in \text{span } \{y_1, \ldots, y_{m_\varepsilon}\}$ and

$$\begin{aligned} \| K(x) - K_\varepsilon(x) \| &\leqslant \sum_{i=1}^{m_\epsilon} \varphi_i(K(x)) \| K(x) - y_i \| \\ &\leqslant \varepsilon \sum_{i=1}^{m_\varepsilon} \varphi_i(K(x)) = \varepsilon . \end{aligned}$$

□

The idea in extending the degree to mapping $f = \text{id} - K$ on bounded open set $\Omega \subset X$, where K is a compact map, is by approximation: $\forall \varepsilon > 0$ arbitrarily small, we already have the Brouwer degree for $f_\varepsilon = \text{id} - K_\varepsilon$, then we want to define the degree for f by Brouwer degrees for these f_ε. In doing so, we should verify:

(1) If $y_0 \in \mathbb{X}$ satisfies $y_0 \notin f(\partial\Omega)$, then for small $\varepsilon > 0$, $y_0 \notin f_\varepsilon(\partial\Omega)$.
(2) For any two such f_{ε_1} and f_{ε_2}, the Brouwer degrees for $f_{\varepsilon_1}, f_{\varepsilon_2}$ are equal.

In proving (1), we need:

Lemma 3.4.3 *Let $K : \overline{\Omega} \to \mathbb{X}$ be compact, $f = id - K$ and $S \subset \overline{\Omega}$ be closed. Then $f(S)$ is closed.*

Proof. Let $\{x_n\} \subset S$ with $f(x_n) \to z^*$. We want to show that $z^* \in f(S)$. Indeed, there are a subsequence $\{n_i\}$ and $y^* \in \mathbb{X}$ such that $Kx_{n_i} \to y^*$, and then $x_{n_i} \to z^* - y^*$, which we write as x^*. Then $x^* \in S$. By the continuity of K, $Kx^* = y^*$, and hence $f(x^*) = z^*$, i.e., $z^* \in f(S)$. □

Thus, $y_0 \notin f(\partial\Omega)$ implies dist $(y_0, f(\partial\Omega)) > 0$. Choosing $0 < \varepsilon <$ dist $(y_0, f(\partial\Omega))$, we have

$$\text{dist }(y_0, f_\varepsilon(\partial\Omega)) > 0\,.$$

Now we turn to (2). Consider the mapping

$$f_\varepsilon : \overline{\Omega} \cap \mathbb{R}^{m_\varepsilon} \to \mathbb{R}^{m_\varepsilon}$$

where $m_\varepsilon = \dim \operatorname{span}\{\, K_\varepsilon(\Omega)\}$. The Brouwer degree $\deg(f_\epsilon, \Omega \cap \mathbb{R}^{m_\epsilon}, y_0)$ is well defined. Since m_ε depends on ε, we should compare all the degrees for these f_ε.

Namely, we have:

Lemma 3.4.4 *Let $\Omega \subset \mathbb{R}^n$ be a bounded open set, $\mathbb{R}^m \subset \mathbb{R}^n$, and $i : \mathbb{R}^m \to \mathbb{R}^n$ be the canonical immersion:*

$$x = (x_1, \ldots, x_m) \mapsto \hat{x} = (x_1, \ldots, x_m, 0, \ldots, 0)\,.$$

Let $K : \overline{\Omega} \to \mathbb{R}^m$ be continuous, $f = id - K$ and $p \in \mathbb{R}^m$ satisfying $\hat{p} \notin f(\partial\Omega)$. Then

$$\deg(f, \Omega, \hat{p}) = \deg(f|_{\mathbb{R}^m \cap \overline{\Omega}}, \mathbb{R}^m \cap \Omega, p)\,.$$

Proof. Let $\varepsilon =$ dist $(\hat{p}, f(\partial\Omega)) > 0$. Choose $\hat{K} \in C^1(\overline{\Omega}, \mathbb{R}^m)$ satisfying $\| K - \hat{K} \|_{C(\Omega)} < \frac{\varepsilon}{2}$. Let $y = \mathrm{id} - \hat{K}$. Then

$$J_g(y) = \det \begin{pmatrix} \mathrm{id}_m - \frac{\partial \hat{K}_i}{\partial x_j} & -\frac{\partial \hat{K}_i}{\partial x_k} \\ 0 & \mathrm{id}_{n-m} \end{pmatrix} \quad \begin{matrix} i,j = 1, \ldots, m, \\ k = m+1, \ldots, n, \end{matrix}, \forall y \in \bar{\Omega}\,.$$

By the Sard theorem the critical value set of $g|_{\mathbb{R}^m \cap \overline{\Omega}} : \mathbb{R}^m \cap \overline{\Omega} \to \mathbb{R}^m$ is an m-dimensional set of measure zero. It has a regular value $q \in \mathbb{R}^m$ such that $\| p - q \| < \varepsilon/2$, and $\hat{q} \notin g(\partial\Omega)$. Thus

$$\begin{aligned}
\deg(f, \Omega, \hat{p}) &= \deg(g, \Omega, \hat{q}) \text{ (homotopy invariance)} \\
&= \sum_{y_i \in g^{-1}(\hat{q})} \operatorname{sgn} J_g(y_i) \\
&= \sum_{y_i \in g|^{-1}_{\mathbb{R}^m \cap \overline{\Omega}}(\hat{q})} \operatorname{sgn} J_g|_{\mathbb{R}^m \cap \overline{\Omega}}(y_i) \; (g^{-1}(\hat{q}) \subset \mathbb{R}^m \cap \Omega) \\
&= \deg(g|_{\mathbb{R}^m \cap \overline{\Omega}}, \mathbb{R}^m \cap \Omega, q) \\
&= \deg(f|_{\mathbb{R}^m \cap \overline{\Omega}}, \mathbb{R}^m \cap \Omega, p) \text{ (homotopy invariance)}
\end{aligned}$$

□

Now we are ready to define the topological degree for $f = \mathrm{id} - K$ when K is compact.

Definition 3.4.5 *(Leray–Schauder degree) Let $\mathbb{X}$ be a real Banach space, and $\Omega \subset \overline{\mathbb{X}}$ be bounded and open. Let $K : \overline{\Omega} \to \mathbb{X}$ be compact, $f = id - K$ and $p \in \mathbb{X}\backslash f(\partial\Omega)$. Define*

$$\deg(f, \Omega, p) = \deg(f_\varepsilon, \Omega \cap E_\varepsilon, p) ,$$

where

$$\varepsilon \in (0, dist(p, f(\partial\Omega))), \;\; f_\varepsilon = id - K_\varepsilon ,$$

and K_ε is a continuous operator assuming values in a finite-dimensional space E_ε with $p \in E_\varepsilon$ and satisfies

$$\| Kx - K_\varepsilon x \| \leqslant \varepsilon \quad \forall x \in \overline{\Omega} .$$

Let us explain the legitimacy of the definition.

1. By Lemma 3.4.3, $\mathrm{dist}(p, f(\partial\Omega)) > 0$, according to Theorem 3.4.2, there exist such K_ε and E_ε.
2. We verify that $\deg(f_\varepsilon, \Omega \cap E_\varepsilon, p)$ is well defined for all such f_ε and E_ε and takes the same value. Since

$$\mathrm{dist}(p, f_\varepsilon(\partial\Omega \cap E_\varepsilon)) \geqslant \mathrm{dist}(p, f(\partial\Omega)) - \varepsilon > 0 ,$$

$\deg(f_\varepsilon, \Omega \cap E_\varepsilon, p)$ is well defined. Let $(f_{\varepsilon_0}, E_{\varepsilon_0})$ and $(f_{\varepsilon_1}, E_{\varepsilon_1})$ be two arbitrary pairs of mappings and finite-dimensional linear subspaces satisfying the hypotheses of the definition, and let

$$\hat{E} = \mathrm{span}\{E_{\varepsilon_0}, E_{\varepsilon_1}\}, \;\; \hat{f}_{\varepsilon_i} = \mathrm{id} - \hat{K}_{\varepsilon_i} ,$$

where $\hat{K}_{\varepsilon_i}(x) = (K_{\varepsilon_i}(x), 0)$ and 0 is the zero element in $\hat{E} \ominus E_{\varepsilon_i}$, $i = 0, 1$. By Lemma 3.4.4, we have

$$\deg(f_{\varepsilon_i}, E_{\varepsilon_i} \cap \Omega, p) = \deg(\hat{f}_{\varepsilon_i}, \hat{E} \cap \Omega, p), \;\; i = 0, 1 .$$

Let us define $\phi : (\Omega \cap \hat{E}) \times [0, 1] \to \hat{E}$ by

$$\phi(x, t) = t\hat{f}_{\varepsilon_0}(x) + (1 - t)\hat{f}_{\varepsilon_1}(x), \quad \forall t \in [0, 1] .$$

Applying the homotopy invariance of the Brouwer degree we obtain

$$\deg(\hat{f}_{\varepsilon_0}, \hat{E} \cap \Omega, p) = \deg(\hat{f}_{\varepsilon_1}, \hat{E} \cap \Omega, p) .$$

This proves

$$\deg(f_{\varepsilon_0}, E_{\varepsilon_0} \cap \Omega, p) = \deg(f_{\varepsilon_1}, E_{\varepsilon_1} \cap \Omega, p) .$$

Remark 3.4.6 *A set $\Omega \subset \mathbb{X}$ is called finitely bounded, if for all linear finite-dimensional subspaces $E \subset \mathbb{X}$, $E \cap \Omega$ is bounded. In Definition 3.4.5, one may assume that Ω is finitely bounded.*

More generally, for a given $\{e_1, \ldots, e_k\} \subset E$, if for all linear finite-dimensional subspaces $E \subset \mathbb{X}$ with $\{e_1, \ldots, e_k\} \subset E$, $E \cap \Omega$ is bounded, then Ω is called $\{e_1, \ldots, e_k\}$- finitely bounded. Again, the Leray–Schauder degree is well defined on open sets, which are $\{e_1, \ldots, e_k\}$-finitely bounded.

Similarly to the Brouwer degree, the Leray–Schauder degree enjoys the following fundamental properties:

(1) (Homotopy invariance) Let $K : \overline{\Omega} \times [0,1] \to \mathbb{X}$ be compact and $p \notin (\mathrm{id} - K)(\partial\Omega \times [0,1])$, then

$$\deg(\mathrm{id} - K(\cdot, t), \Omega, p) = \text{constant}\ .$$

(2) (Translation invariance)

$$\deg(\mathrm{id} - K, \Omega, p) = \deg(\mathrm{id} - K - p, \Omega, \theta)\ .$$

(3) (Additivity) Let $\Omega_1, \Omega_2 \subset \Omega, \Omega_1 \cap \Omega_2 = \emptyset$ and $p \notin (\mathrm{id} - K)(\overline{\Omega} \backslash (\Omega_1 \cup \Omega_2))$, then

$$\deg(\mathrm{id} - K, \Omega, p) = \deg(\mathrm{id} - K, \Omega_1, p) + \deg(\mathrm{id} - K, \Omega_2, p)\ .$$

(4) (Normality)

$$\deg(\mathrm{id}, \Omega, p) = \begin{cases} 1, & p \in \Omega\ , \\ 0, & p \notin \overline{\Omega}\ . \end{cases}$$

Similarly, all the other properties are consequences of these properties. Since the proofs are standard (reducing to finite-dimensional spaces by ε-approximations, and applying the corresponding properties of the Brouwer degree), we omit them.

In particular, we have:

(5) (Kronecker existence) Let $\Omega \subset \mathbb{X}$ be a bounded open set and let $K : \overline{\Omega} \to \mathbb{X}$ be compact. If $y_0 \notin (\mathrm{id} - K)(\partial\Omega)$ and $\deg(\mathrm{id} - K, \Omega, y_0) \neq 0$, then there exists $x_0 \in \Omega$ satisfying $x_0 = Kx_0 + y_0$.

(6) Let K be a compact linear operator, $1 \notin \sigma(K)$ (the spectrum of K) and $\theta \in \Omega$, then

$$\deg(\mathrm{id} - K, \Omega, \theta) = (-1)^\beta\ ,$$

where

$$\beta = \sum_{\lambda_j > 1, \lambda_j \in \sigma(K)} \beta_j,\ \beta_j = \dim \cup_{k=1}^{\infty} \ker(\lambda_j \mathrm{I} - K)^k\ .$$

Proof. According to the Riesz–Schauder theory, $\sigma(K)$ has only a point spectrum except 0. Let E_1 be the finite-dimensional subspace spanned by all the generalized eigenvectors corresponding to eigenvalues > 1, then we have the direct sum decomposition $\mathbb{X} = E_1 \oplus E_2$. Both E_1 and E_2 are invariant subspaces of K. Define $K_{E_i} = K_i$, then, for $\forall t \in [0,1]$, for $\forall x \in E_2 \backslash \{\theta\}$, $tK_2 x \neq x$. The homotopy invariance and the excision ensure that $\exists \varepsilon > 0$ such that

$$\begin{aligned}\deg(\mathrm{id} - K, \Omega, \theta) &= \deg(\mathrm{id} - K, B_\varepsilon, \theta) \\ &= \deg(\mathrm{id} - (K_1 \oplus tK_2), B_\varepsilon, \theta) \\ &= \deg(\mathrm{id} - K_1, B_\varepsilon, \theta) = (-1)^\beta .\end{aligned}$$

□

Now, let us return to the discussion in the beginning of this section. We have the special form of the Schauder fixed-point theorem: if $K : \overline{B} \to \overline{B}$ is compact, then K has a fixed point in $\overline{B}$.

However, the unit ball B may be replaced by any bounded closed convex set. Because it is known from Dugundji's theorem (see Sect. 3.6) that for any closed convex subset C of $\mathbb{X}$, there is a retract $r : \mathbb{X} \to C$, i.e., r is continuous, and satisfies $r \circ i = \mathrm{id}|_C$, and $i \circ r \sim \mathrm{id}|_{\mathbb{X}}$, where i is the injection $C \to \mathbb{X}$. The following Schauder fixed-point theorem, which we have met in Sect. 2.2 is a consequence of the Kronecker existence property of the Leray–Schauder degree.

Theorem 3.4.7 *(Schauder fixed point) Let C be a bounded closed convex set in $\mathbb{X}$. If $K : C \to C$ is compact. Then K has a fixed point.*

Proof. We choose $R > 0$ large enough such that $C \subset B_R(\theta)$. Let $r : \overline{B}_R(\theta) \to C$ be a retract, and $i : C \to \overline{B}_R(\theta)$ be the injection. We have a compact mapping $f = i \circ K \circ r : \overline{B}_R(\theta) \to \overline{B}_R(\theta)$:

$$\begin{array}{ccc} \overline{B}_R(\theta) & & \overline{B}_R(\theta) \\ r \big\downarrow & & \big\uparrow i \\ C & \xrightarrow{K} & C \end{array}$$

Following the steps in the proof of the Brouwer fixed-point theorem, there is a fixed point $x_0 \in \overline{B}_R$ of f, i.e., $x_0 = i \circ K \circ r(x_0)$. Thus $x_0 \in C$ and then $r(x_0) = x_0$. It follows that $x_0 = Kx_0$. □

Another fixed-point theorem, which is also frequently used in the theory of differential equations in studying the existence of solutions by a priori estimates, reads as:

Theorem 3.4.8 *(Schaefer) Suppose that $K : \mathbb{X} \times [0,1] \to \mathbb{X}$ is a compact mapping, satisfying $K(x,0) = \theta$. If the set $S = \{x \in \mathbb{X} | \ \exists t \in [0,1]$ such that $x = K(x,t)\}$ is bounded. Then $K = K(\cdot,1)$ has a fixed point.*

Proof. Taking $r > 0$ large such that $S \subset \overset{\circ}{B}_r$, we define

$$f(x,t) = x - K(x,t) \quad \forall (x,t) \in \overline{B}_r \times [0,1]$$

then $\theta \notin f(\partial B_r \times [0,1])$. According to the homotopy invariance

$$\deg(\mathrm{id} - K, B_r, \theta) = \deg(\mathrm{id}, B_r, \theta) = 1 .$$

□

Corollary 3.4.9 *Suppose that $K : \mathbb{X} \to \mathbb{X}$ is compact and that the set*

$$S = \{x \in \mathbb{X} |\ \exists t \in [0,1] \text{ such that } x = tK(x)\}$$

is bounded. Then K has a fixed point.

Proof. Set

$$K(x,t) = tK(x)$$

□

As an application of Schaefer's fixed-point theorem, we study the following semi-linear elliptic BVP, which we have met in Sect. 1.2.

Let $\Omega \subset \mathbb{R}^n$ be a bounded open domain with smooth boundary.

Theorem 3.4.10 *Suppose that $f \in C^\gamma(\overline{\Omega} \times \mathbb{R}^1 \times \mathbb{R}^n, \mathbb{R}^1)$, for some $\gamma \in (0,1)$, satisfies*

(1) $\exists$ an increasing function $c : \mathbb{R}^1_+ \to \mathbb{R}^1_+$ such that

$$|f(x,\eta,\xi)| \leqslant c(|\eta|)(1 + |\xi|^2) \quad \forall (x,\eta,\xi) \in \overline{\Omega} \times \mathbb{R}^1 \times \mathbb{R}^n .$$

(2) $\exists$ a constant $M > 0$ such that

$$f(x,\eta,\theta) \begin{cases} < 0 & \text{as } \eta > M \\ > 0 & \text{as } \eta < -M. \end{cases}$$

Assume $\phi \in C^{2,\gamma}(\partial\Omega)$. Then the equation

$$\begin{cases} -\triangle u = f(x, u(x), \nabla u(x)) \\ u|_{\partial\Omega} = \phi \end{cases} \tag{3.17}$$

has a solution $u \in C^{2,\gamma}(\overline{\Omega})$.

Proof. Fixing $\overline{\gamma} \in (0,\gamma)$, we define a map T from $C^{2,\overline{\gamma}}(\overline{\Omega}) = X$ into itself by $u \to v$, where v is the solution of the following BVP:

$$\begin{cases} -\triangle v = f(x, u(x), \nabla u(x)) \\ v|_{\partial\Omega} = \phi . \end{cases}$$

By the Schauder estimate, we obtain

$$\| v \|_{C^{2,\gamma}(\overline{\Omega})} \leqslant C(\| u \|_{C^{2,\overline{\gamma}}(\overline{\Omega})}, \| \phi \|_{C^{2,\gamma}(\partial\Omega)}) .$$

Thus, T is compact.

Let us introduce a parameter t, and define $v = T(u,t)$ to be the solution of the equation:

$$\begin{cases} -\triangle v = tf(x, u(x), \nabla u(x)) \\ v|_{\partial\Omega} = \phi . \end{cases}$$

We intend to apply Schaefer's fixed-point theorem; it is sufficient to verify the boundedness of the set

$$S = \{u \in X | \ \exists t \in [0,1] \text{ such that } u = T(u,t)\} ,$$

i.e., $\exists$ a constant $C > 0$ such that solutions u_t of the equations

$$\begin{cases} -\triangle u_t = tf(x, u_t(x), \nabla u_t(x)) \\ u_t|_{\partial\Omega} = \phi \end{cases} \tag{3.18}$$

satisfy the following a priori estimate of the solution of (3.17):

$$\| u_t \|_{C^{2,\gamma}} \leqslant C \tag{3.19}$$

However, this has been done in Sect. 1.2. □

Comparing the continuity method with the fixed-point method based on Schaefer's theorem, the latter does not require the invertible of the linearized equation. Moreover, less regularity on the nonlinear term is assumed. Of course, there is only the existence of a solution but not uniqueness.

The Borsuk theorem is also extended.

Theorem 3.4.11 *Let $\Omega \subset \mathbb{X}$ be a symmetric bounded open set including θ. If $K : \overline{\Omega} \to \mathbb{X}$ is an odd compact map with $\theta \notin f(\partial\Omega)$, where $f = id - K$, then $\deg(f, \Omega, \theta)$ is odd.*

Proof. We approximate f by finite-dimensional odd maps. Indeed, in Theorem 3.4.2, let K be approximated by K_ε, with Im $K_\varepsilon \subset \text{span}\{y_1, \ldots, y_{m_\varepsilon}\}$, and let

$$\hat{K}_\varepsilon(x) = \frac{1}{2}[K_\varepsilon(x) - K_\varepsilon(-x)] .$$

Then

$$\| K(x) - \hat{K}_\varepsilon(x) \| \leqslant \frac{1}{2} \| K(x) - K_\varepsilon(x) \| + \frac{1}{2} \| K(-x) - K_\varepsilon(-x) \| < \varepsilon ,$$

and

$$\text{Im } \hat{K}_\varepsilon \subset \text{span}\{y_1, \ldots, y_{m_\varepsilon}\} .$$

The conclusion follows from Theorem 3.3.3 directly. □

Corollary 3.4.12 *Let $\Omega \subset \mathbb{X}$ be a symmetric bounded open set including θ. If $K : \overline{\Omega} \to \mathbb{X}$ is compact, $f = id - K$, and if*

$$f(x) \neq tf(-x) \quad \forall (x,t) \in \partial\Omega \times [0,1] ,$$

then $\deg(f, \Omega, \theta)$ *is odd.*

Proof. Define

$$\begin{aligned}\phi(x,t) &= \mathrm{id} - \left[\frac{1}{1+t}K(x) - \frac{t}{1+t}K(-x)\right] \\ &= \frac{1}{1+t}f(x) - \frac{t}{1+t}f(-x).\end{aligned}$$

Thus $\theta \notin (\partial\Omega \times [0,1])$, $\phi(\cdot, 0) = f$ and $\phi(\cdot, 1) = \frac{1}{2}(f(x) - f(-x))$ is odd. We obtain

$$\deg(f, \Omega, \theta) = \deg(\phi(\cdot, 1), \Omega, \theta) = \text{odd number} .$$

□

By the same proof of the invariance of domains theorem for the Brouwer degree, we have its infinite-dimensional version:

Corollary 3.4.13 ***(Invariance of domains)*** *Assume that $\Omega \subset \mathbb{X}$ is a non-empty open set, and that $K : \overline{\Omega} \to \mathbb{X}$ is compact. If $f = id - K$ is locally injective, then f is an open map.*

The notion of the index of an isolated solution is also extended. For $f = \mathrm{id} - K$, where K is a compact map, we define the index of f at an isolated fixed point x_0 as follows:

$$i(f, x_0, \theta) = \deg(f, B_\varepsilon(x_0), \theta)$$

for sufficiently small $\varepsilon > 0$.

(7) In particular, if K is differentiable at x_0, where $f(x_0) = 0$, and if $f'(x_0)$ is invertible, then we have

$$i(f, x_0, \theta) = (-1)^\beta ,$$

where $\beta = \sum_{\{\lambda_j > 1 \,|\, \lambda_j \in \sigma(K'(x_0))\}} \beta_j$ and $\beta_j = \dim \cup_{k=1}^\infty \ker(\lambda_j \mathrm{I} - K'(x_0))^k$.

Proof. By homotopy invariance, $\exists \varepsilon > 0$ such that

$$\begin{aligned}i(f, x_0, \theta) &= \deg(f, B_\varepsilon(x_0), \theta) \\ &= \deg(\mathrm{id} - K'(x_0), B_\varepsilon(x_0), \theta) .\end{aligned}$$

It is sufficient to verify that $K'(x_0)$ is a compact linear operator, because if it is so, then the conclusion follows directly from (6). The verification is as follows:

If $T = K'(x_0)$ were not compact, then there would be $\{x_j\} \subset B_1(\theta)$ and $\varepsilon > 0$ such that

$$\| Tx_i - Tx_j \| \geqslant \varepsilon \quad \text{as } i \neq j \, .$$

Choose $\delta > 0$ small such that $\forall k = 1, 2, \ldots,$

$$\| K(x_0 + \delta x_k) - K(x_0) - \delta T x_k \| \leqslant \frac{\varepsilon\delta}{4} \, ,$$

thus

$$\begin{aligned}
\frac{\varepsilon\delta}{2} &\geqslant \| K(x_0 + \delta x_i) - K(x_0 + \delta x_j) - \delta T(x_i - x_j) \| \\
&\geqslant \delta \| T(x_i - x_j) \| - \| K(x_0 + \delta x_i) - K(x_0 + \delta x_j) \| \\
&\geqslant \delta\varepsilon - \| K(x_0 + \delta x_i) - K(x_0 + \delta x_j) \| \, ,
\end{aligned}$$

that is

$$\| K(x_0 + \delta x_i) - K(x_0 + \delta x_j) \| \geqslant \frac{1}{2}\varepsilon\delta, \ i \neq j \, .$$

It contradicts with the compactness of K. □

The Leray-Schauder degree theory is also applied to the study of the intersections of infinite-dimensional manifolds, in which the continuous mappings should be replaced by compact vector fields.

3.5 The Global Bifurcation

We have studied in Sect. 1.3, the local bifurcation phenomenon, which describes a branch of solutions splitting into several branches. Having the topological tools at hand, we are able to provide more precise local information and to investigate the global behavior of bifurcating branches. Consider $F : X \times \mathbb{R}^1 \to X$ in the following form:

$$F(x, \lambda) = Lx - \lambda x + N(x, \lambda) \, , \tag{3.20}$$

where X is a real Banach space, $L \in \mathcal{L}(X, X)$, $\lambda \in \mathbb{R}^1$, and $\|N(x, \lambda)\| = o(\|x\|)$ as $x \to \theta$ uniformly in any finite interval of λ. As we know, bifurcation points (θ, λ_0) only occur at $\lambda_0 \in \sigma(L)$, i.e., λ_0 is a spectrum of L. In particular, if L is compact, and $\lambda_0 \neq 0$, then λ_0 is an eigenvalue of L. Conversely, we have:

Theorem 3.5.1 ***(Knasnoselski)*** *Suppose that L is a linear compact operator on X, and that $\lambda_0 \neq 0$ is an eigenvalue of L with odd multiplicity, i.e., $\beta = \dim \bigcup_{k=1}^{\infty} \ker(L - \lambda_0 I)^k$ is odd. If $\forall\lambda, N(\cdot, \lambda)$ is compact, and N is continuous in x and λ, and satisfies*

$$\|N(x, \lambda)\| = o(\|x\|)$$

uniformly in any finite interval of λ, *then* (θ, λ_0) *is a bifurcation point of the equation* $F(x, \lambda) = \theta$.

Proof. Set

$$S = \{(x, \lambda) \in X \times \mathbb{R}^1 \mid F(x, \lambda) = \theta\} ,$$

and

$$S_+ = \overline{S \setminus (\{\theta\} \times \mathbb{R}^1)} .$$

If (θ, λ_0) is not a bifurcation point, then there is a closed interval $[\lambda_-, \lambda_+]$ not including 0 and satisfying:

1. $\sigma(L) \cap [\lambda_-, \lambda_+] = \{\lambda_0\}$,
2. $\exists r > 0$ such that $(\overline{B_r}(\theta) \times [\lambda_-, \lambda_+]) \cap S_+ = \varnothing$.

In the sequel we write $B_r = B_r(\theta)$ briefly. Define a deformation:

$$\Phi(x, t) = x - \frac{1}{\lambda(t)}(Lx + N(x, \lambda(t))) \qquad t \in [0, 1] ,$$

where $\lambda(t) = t\lambda_- + (1 - t)\lambda_+$. Hence $\theta \notin \Phi(\partial B_r \times [0, 1])$.
It follows that

$$\deg\left(\mathrm{id} - \frac{1}{\lambda_+}(L + N(\cdot, \lambda_+)), B_r, \theta\right) = \deg\left(\mathrm{id} - \frac{1}{\lambda_-}(L + N(\cdot, \lambda_-)), B_r, \theta\right) .$$

For sufficiently small $r > 0$, again by the homotopy invariance, we obtain

$$\deg\left(\mathrm{id} - \frac{1}{\lambda_+}L, B_r, \theta\right) = \deg\left(\mathrm{id} - \frac{1}{\lambda_-}L, B_r, \theta\right) ,$$

i.e.,

$$(-1)^{\sum\limits_{\lambda_j > \lambda_+} \beta_j} = (-1)^{\sum\limits_{\lambda_j > \lambda_-} \beta_j} ,$$

where $\lambda_j \in \sigma(L)$, and β_j is the multiplicity of λ_j. This is

$$(-1)^{\beta} = 1 .$$

By assumption β is odd. This is impossible.

The oddness of the algebraic multiplicity of λ_0 is a sufficient condition for bifurcation points, but not necessary (see Chap. 5, Theorem 5.1.37.)

We turn to studying some global results.

Lemma 3.5.2 *Let* K *be a compact metric space,* K_1, K_2 *be disjoint closed subsets. Then the following alternatives hold: either*

1. $\exists$ *a component of* K *intersecting* K_1 *and* K_2, *or*
2. $\exists$ *compact subsets* $\widehat{K}_1$, $\widehat{K}_2$ *such that*

$$K_i \subset \widehat{K}_i, i = 1, 2,$$
$$K = \widehat{K}_1 \cup \widehat{K}_2, \qquad \text{and} \qquad \widehat{K}_1 \cap \widehat{K}_2 = \varnothing .$$

Proof. If (1) is not true, then $\exists\,\varepsilon_0 > 0$ such that each ε_0- chain cannot intersect with K_1 and K_2 simultaneously. For otherwise, $\forall\,\varepsilon > 0\ \exists\, a_i^\varepsilon \in K_i$ and ε-chain C_ε connecting $a_i^\varepsilon, i = 1,2$. Since K_i, $i = 1,2$, are compact, $\{a_i^{\frac{1}{n}}\}$ has a limiting point $a_i \in K_i, i = 1,2$.

Set

$$C_{a_1} = \{x \in K|\ \forall\,\varepsilon > 0 \quad a_1 \text{ can be connected with } x \text{ by an } \varepsilon\text{-chain}\}\ .$$

Since K is compact, C_{a_1} is a closed connected set. Therefore, $a_2 \in C_{a_1}$. This means that (1) holds. We arrive at a contradiction.

To the ε_0, let $\widehat{K}_1 = \{y \in K|\ \exists x \in K_1 \text{ and } \exists \varepsilon_0 \text{ chain connecting } x \text{ and } y\}$. Obviously, $\widehat{K}_1 \cap K_2 = \varnothing$ and $K_1 \subset \widehat{K}_1$.

We shall prove that $\widehat{K}_1$ is both open and closed.

On the one hand, $\forall\, x \in \overline{\widehat{K}_1}$, $B_{\varepsilon_0}(x) \cap \widehat{K}_1 \neq \varnothing$, which implies $x \in \widehat{K}_1$, therefore $\widehat{K}_1$ is closed. On the other hand, $\forall x \in \widehat{K}_1, B_{\varepsilon_0}(x) \subset \widehat{K}_1$, so it is also open. We set $\widehat{K}_2 = K \backslash \widehat{K}_1$. Then $\widehat{K}_1$ and $\widehat{K}_2$ meet all conditions we need. □

Theorem 3.5.3 ***(Leray-Schauder)*** *Let X be a real Banach space, $T : X \times \mathbb{R}^1 \to X$ be a compact map satisfying $T(x,0) = \theta$, and $f(x,\lambda) = x - T(x,\lambda)$. Let*

$$S = \{(x,\lambda) \in X \times \mathbb{R}^1\,|\,f(x,\lambda) = \theta\}\ ,$$

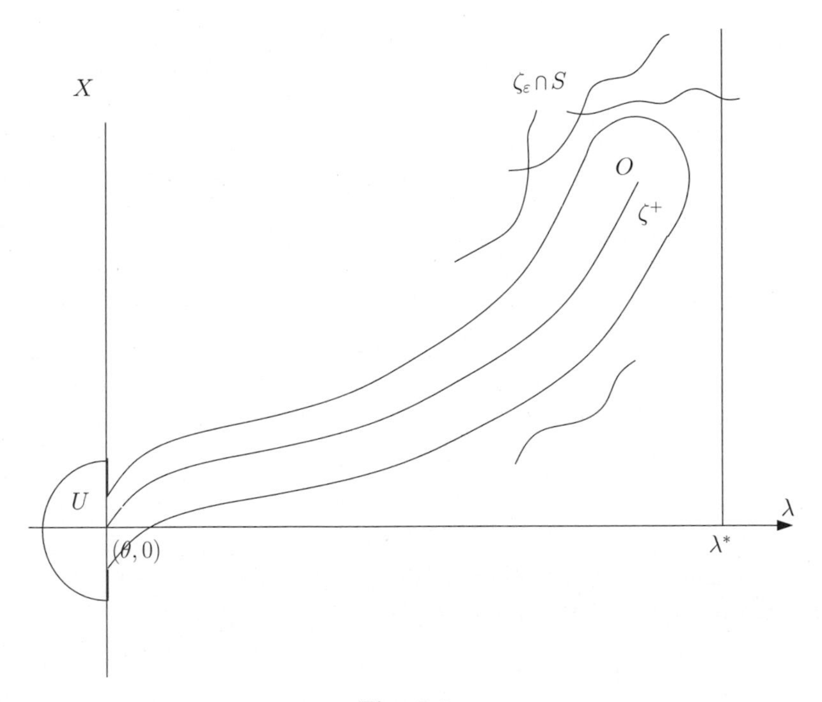

Fig. 3.2.

and let ζ be the component of S passing through $(\theta, 0)$. If

$$\zeta^{\pm} = \zeta \cap (X \times \mathbb{R}^1_{\pm}) ,$$

then both ζ^+ and ζ^- are unbounded.

Proof. Since T is compact, if ζ^+ (or ζ^-) is bounded, then it is compact. We consider a $\varepsilon-$ neighborhood ζ_ε of ζ^+. Let $K = \overline{\zeta_\varepsilon} \cap S$. This is a compact metric space.

Set $K_1 = \zeta^+$, $K_2 = \partial\zeta_\varepsilon \cap S$. According to Lemma 3.5.2, $\exists \widehat{K}_i, i = 1, 2$, such that

$$\begin{aligned} &K_i \subset \widehat{K}_i \qquad i = 1, 2, \\ &K = \widehat{K}_1 \cup \widehat{K}_2, \qquad \text{and} \qquad \widehat{K}_1 \cap \widehat{K}_2 = \varnothing . \end{aligned}$$

Choosing

$$0 < \delta < \min\{\text{dist}(\widehat{K}_1, \widehat{K}_2), \text{dist}(\widehat{K}_1, \partial\zeta_\varepsilon)\} ,$$

and defining a $\delta/2-$ neighborhood O of $\widehat{K}_1$, we have

$$\begin{aligned} &\partial O \cap S = \varnothing, \\ &\zeta^+ \subset O . \end{aligned}$$

Setting

$$U = O \cup \{(x, \lambda) \in X \times \mathbb{R}^1_- \mid \|x\|^2 + \lambda^2 < \delta^2\} ,$$

and λ^* larger than the projection of O onto $\mathbb{R}^1_+$, we consider the map $F : (X \times \mathbb{R}^1) \times [0, 1] \to X \times \mathbb{R}^1$ as follows:

$$(x, \lambda, t) \mapsto (f(x, \lambda), \lambda - t\lambda^*) .$$

Then

$$F(x, \lambda, t) = (\theta, 0) \quad \Longleftrightarrow \quad \begin{cases} f(x, \lambda) = \theta \\ \lambda = t\lambda^* \end{cases} .$$

Now,

$$\begin{aligned} &\forall (x, \lambda) \in \partial O, \text{ by construction, } f(x, \lambda) \neq \theta; \\ &\forall (x, 0) \in U, \text{ if } f(x, 0) = \theta, \text{ then } x = \theta; \\ &\forall (x, \lambda) \in X \times \mathbb{R}^1_- \text{ with } \|x\|^2 + \lambda^2 = \delta^2, \text{ we have } \lambda \neq t\lambda^*. \end{aligned}$$

Since

$$\partial U \subset \partial O \cup \left\{(x, 0) \,\middle|\, \frac{\delta}{2} \le \|x\| \le \delta\right\} \cup \{(x, \lambda) \in X \times \mathbb{R}^1_- \mid \|x\|^2 + \lambda^2 = \delta^2\} ,$$

we have $(\theta, 0) \notin F(\partial U \times [0, 1])$. Thus

$$\deg(F(\cdot, \cdot, 0), U, (\theta, 0)) = \deg(F(\cdot, \cdot, 1), U, (\theta, 0)) .$$

On the one hand, since $\lambda \neq \lambda^*$ on U, from Knonecker existence theorem, we have

$$\deg(\, F(\,\cdot\,,\,\cdot\,,1), U,\, (\theta, 0)\,) = 0\ .$$

On the other hand, $\forall t$,

$$\begin{cases} x - T(x, t\lambda) = \theta, \\ \lambda = 0, \end{cases} \qquad \Longleftrightarrow \qquad \begin{cases} x = \theta, \\ \lambda = 0. \end{cases}$$

This implies $(x - T(x, t\lambda), \lambda) \neq (\theta, 0)$ on ∂U. Then, by the homotopy invariance and the excision property,

$$\deg(\, F(\,\cdot\,,\,\cdot\,,0), U,\, (\theta, 0)\,) = \deg(\, id_{X\times\mathbb{R}^1}, B_{\frac{\delta}{2}}(\theta, 0), (\theta, 0)\,) = 1\ .$$

This is the contradiction. □

Although the Leray–Schauder theorem is not directly related to bifurcation problems (if $T(\theta, \lambda) = \theta$, then the conclusion is trivial: $\zeta = \{\theta\} \times \mathbb{R}^1$), but in applications, it can be used in the spirit of the following global bifurcation theorem due to P. Rabinowitz and improved by Ize [Iz].

Theorem 3.5.4 ***(Rabinowitz)*** *Let X be a real Banach space and $F(x, \lambda) = x - \lambda Lx - N(x, \lambda)$, where $L \in \mathcal{L}(X, X)$ and $N : X \times \mathbb{R}^1 \to X$ are compact. Let S be the solution set of $F(x, \lambda) = \theta$, $S_+ = \overline{S\backslash(\{\theta\} \times \mathbb{R}^1)}$, and let ζ be the component of S_+, containing (θ, λ_1). Assume that $N(x, \lambda) = o(\|\,x\|)$ uniformly on any finite interval in λ and that $\lambda_1^{-1} \in \sigma(L)$ is an eigenvalue of odd multiplicity. Then the following alternatives hold: Either*

1. *ζ is unbounded; or*
2. *there are only finite number of points $\{(\theta, \lambda_i)\,|\, i = 1, , \dots, l\}$ lying on ζ, where $\lambda_i^{-1} \in \sigma(L), i = 1, 2, \dots, l$. Furthermore, if β_i is the algebraic multiplicity of λ_i^{-1}, then $\sum_{i=1}^{l} \beta_i$ is even.*

Proof. If ζ is bounded, then ζ is compact, because both L and N are compact. Since L is a compact linear operator, there are at most finitely many points $\{(\theta, \lambda_j)\,|\,\lambda_j \in \sigma(L^{-1}), j = 1, \dots, l\}$ lying on ζ. Let $0 < \varepsilon < \mathrm{dist}(\zeta, \{\lambda\,|\,\lambda^{-1} \in \sigma(L), \lambda \neq \lambda_1, \lambda_2, \dots, \lambda_l\})$, and let ζ_ε be the $\varepsilon-$ neighborhood of ζ. Set $K = \zeta_\varepsilon \cap S_+$. Then K is a compact metric space, and $\zeta \cap \partial\zeta_\varepsilon = \varnothing$. Set $K_1 = \zeta$, $K_2 = S_+ \cap \partial\zeta_\varepsilon$. According to Lemma 3.5.2, $\exists$ compact sets $\widehat{K}_i, i = 1, 2$, such that $K_i \subset \widehat{K}_i, i = 1, 2,\ K = \widehat{K}_1 \cup \widehat{K}_2$ and $\widehat{K}_1 \cap \widehat{K}_2 = \varnothing$. Set $0 < \delta < \min\{\mathrm{dist}(\widehat{K}_1, \widehat{K}_2), \mathrm{dist}(\widehat{K}_1, \partial\zeta_\varepsilon)\}$, and let O be the $\frac{1}{2}\delta$ neighborhood of $\widehat{K}_1$. Then we have

$$\begin{cases} \zeta \subset O \subset \overline{O} \subset \zeta_\varepsilon, \\ S_+ \cap \partial O = \varnothing \end{cases}.$$

If $\lambda^{-1} \in \sigma(L)$, and $\lambda \neq \lambda_1, \lambda_2, \dots, \lambda_l$, then $(\theta, \lambda) \notin O$. We consider the following map:

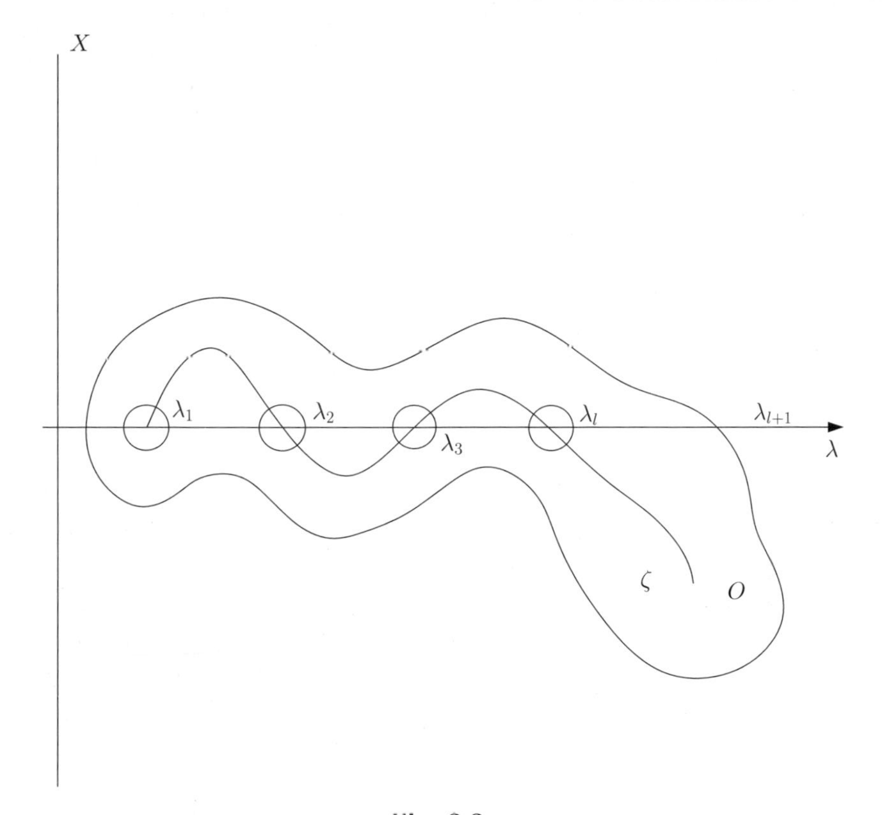

Fig. 3.3.

$$\Phi : O \times \mathbb{R}^1 \to X \times \mathbb{R}^1,$$
$$\Phi(x, \lambda; t) = (F(x, \lambda),\ \| x\|^2 - t^2)\ .$$

We claim that $(\theta, 0) \notin \Phi(\partial O \times \{t\})$ as $t \neq 0$. Indeed, if $\Phi(x, \lambda; t) = (\theta, 0)$, i.e.,

$$\begin{cases} F(x, \lambda) = \theta \\ \| x\|^2 = t^2 \neq 0 \end{cases};$$

then $(x, \lambda) \in S_+$, but $S_+ \cap \partial O = \varnothing$.

By the homotopy invariance, for $0 < r < R$, we have

$$\deg(\Phi(\,\cdot\,, R), O, (\theta, 0)) = \deg(\Phi(\,\cdot\,, r), O, (\theta, 0))\ .$$

We choose R large enough such that $O \subset B_R(\theta, 0)$, the ball in $X \times \mathbb{R}^1$. From the Knonecker existence theorem,

$$\deg(\Phi(\,\cdot\,, R), O, (\theta, 0)) = 0\ .$$

Then we choose $\varepsilon > 0$ such that the ε-balls $B_\varepsilon((\theta, \lambda_j)) \subset O, j = 1, 2, \ldots, l$. For small $r > 0$, one wishes to prove:

$$\deg(\Phi(\,\cdot\,,r),O,(\theta,0)) = \sum_{i=1}^{l} \deg(\Phi(\,\cdot\,,r),B_{\sqrt{r^2+\varepsilon^2}}((\theta,\lambda_j)),(\theta,0))\,. \tag{3.21}$$

Let $K > 0$ be an upper bound:

$$\|(\mathrm{id} - \lambda L)^{-1}\| \leqslant K\,,$$

for λ in the projection of O onto $\mathbb{R}^1$ subtracting the set $\bigcup_{j=1}^{l} (\lambda_j - \varepsilon, \lambda_j + \varepsilon)$. We choose $r > 0$ such that

$$\|N(x,\lambda)\| < \frac{1}{2K}\|\,x\| \qquad \forall (x,\lambda) \in O, \|\,x\| \leqslant r\,.$$

In order to prove (3.21), according to the excision property, it is sufficient to prove:

$$\Phi(x,\lambda,r) \neq \theta \qquad \forall (x,\lambda) \in O \backslash \bigcup_{j=1}^{l} B_{\sqrt{r^2+\varepsilon^2}}(\theta,\lambda_j)\,.$$

However, if $\Phi(x,\lambda,r) = \theta$ and $(x,\lambda) \notin \bigcup_{j=1}^{l} B_{\sqrt{r^2+\varepsilon^2}}(\theta,\lambda_j)$, then $0 = \|\,x\|^2 - r^2 \geqslant \varepsilon^2 - (\lambda - \lambda_j)^2$. This implies $|\lambda - \lambda_j| \geqslant \varepsilon$.

Since $(x,\lambda) \in O$, we have

$$\|\,x\| = \|(\mathrm{id} - \lambda L)^{-1} N(x,\lambda)\| \leqslant \frac{1}{2}\|\,x\|\,,$$

therefore $x = \theta$. This is impossible.

Finally, we compute $\deg(\,\Phi(\,\cdot\,,r),B_{\sqrt{r^2+\varepsilon^2}}((\theta,\lambda_j)),(\theta,0))$.

Let us define

$$\Psi(x,\lambda,s) = (x - \lambda Lx - sN(x,\lambda), \|\,x\|^2 - r^2), \qquad s \in [0,1]\,.$$

Provided by the homotopy invariance, we obtain

$$\begin{aligned}&\deg(\Phi(\,\cdot\,,r),B_{\sqrt{r^2+\varepsilon^2}}((\theta,\lambda_j)),(\theta,0))\\ &\quad = \deg((\mathrm{id} - \lambda L, \|\,\cdot\,\|^2 - r^2),B_{\sqrt{r^2+\varepsilon^2}}(\theta,\lambda_j),(\theta,0))\,.\end{aligned}$$

Again, we define

$$\Psi_1(x,\lambda,t) = (x - \lambda Lx\,,\, t(\|\,x\|^2 - r^2) + (1-t)(\varepsilon^2 - (\lambda - \lambda_j)^2)) \qquad \forall t \in [0,1]\,.$$

Now, $\forall (x,\lambda) \in \partial B_{\sqrt{r^2+\varepsilon^2}}(\theta,\lambda_j)$, if

$$0 = t(\|\,x\|^2 - r^2) + (1-t)(\varepsilon^2 - (\lambda - \lambda_j)^2) = \varepsilon^2 - (\lambda - \lambda_j)^2\,,$$

then $|\lambda - \lambda_j| = \varepsilon$ and $\|\,x\| = r$, we have $x - \lambda Lx \neq \theta$.

Therefore by the homotopy invariance,

$$\begin{aligned}
&\deg\Big((\mathrm{id} - \lambda L, \|\cdot\|^2 - r^2)\,,\, B_{\sqrt{r^2+\varepsilon^2}}(\theta, \lambda_j)\,,\, (\theta, 0) \Big) \\
&= \deg\big((\mathrm{id} - (\lambda + \lambda_j)L, (\varepsilon^2 - \lambda^2))\,,\, B_{\sqrt{r^2+\varepsilon^2}}(\theta, 0)\,,\, (\theta, 0) \big) \\
&= i(\mathrm{id} - (\lambda_j - \varepsilon)L, \theta, \theta) - i(\mathrm{id} - (\lambda_j + \varepsilon)L, \theta, \theta) \\
&= \begin{cases} 1 & \\ & \pmod 2 \\ 0 & \end{cases} \quad \begin{array}{l} \text{if } \beta_j \text{ is odd,} \\ \text{if } \beta_j \text{ is even.} \end{array}
\end{aligned}$$

If $\sum_{j=1}^{l} \beta_j = \text{odd}$, then

$$\sum_{j=1}^{l} \deg\big(\Phi(\,\cdot\,, r)\,,\, B_{\sqrt{r^2+\varepsilon^2}}(\theta, \lambda_j)\,,\, (\theta, 0) \big) = 1 \pmod 2\,.$$

This contradicts (3.21). □

In many PDE and ODE problems, which we shall see later, the second alternative is excluded. But here we shall present an example showing that the second alternative occurs. Let $X = R^2, L = \begin{pmatrix} 1 & 0 \\ 0 & \frac{1}{2} \end{pmatrix}$, and $N(x, \lambda) = L \begin{pmatrix} -x_2^3 \\ x_1^3 \end{pmatrix}$ for $x = \begin{pmatrix} x_1 \\ x_2 \end{pmatrix}$. Then $\lambda_1 = 1, \lambda_2 = 2$. For $\lambda \in [1, 2]$, we have a pair of solutions passing through the points $(\theta, 1)$ and $(\theta, 2)$:

$$(x_1, x_2) = \pm\big((\lambda - 1)^{\frac{1}{8}} (2 - \lambda)^{\frac{3}{8}}, (2 - \lambda)^{\frac{1}{8}} (\lambda - 1)^{\frac{3}{8}} \big)\,.$$

The above global bifurcation theorem has a powerful application in the nonlinear Sturm–Liouville problem.

Recall the equation:

$$\begin{cases} \ddot{y} + \lambda \sin y = 0 & \text{in } (0, \pi)\,, \\ y(0) = y(\pi) = 0\,, \end{cases}$$

which we studied in Sect. 1.3. We know that $\lambda = n^2, n = 1, 2, \ldots$ are bifurcation points, i.e., in the neighborhood of (θ, n^2) there are nontrivial solutions. However, we did not know how is the global behavior of those components of solutions passing through these bifurcation points. Now, we shall apply the Rabinowitz bifurcation theorem to understand it better.

We start with slightly general equations. Let

$$Au = -\frac{d}{dx}\left(p(x) \frac{du}{dx} \right) + q(x)u \qquad x \in (0, \pi)\,,$$

where $p \in C^1([0, \pi])$, $p(x) > 0$, $q \in C^0([0, \pi])$, with $q \geq 0$, and let $D(A) = H^2 \cap H_0^1([0, \pi])$. Then A is a self-adjoint operator with simple eigenvalues $0 < \lambda_1 < \lambda_2 < \ldots$, and $L = A^{-1}$ is compact from $L^2([0, \pi])$ to $C_0^1([0, \pi])$, provided by the Sobolev embedding theorem.

Suppose $h \in C\left([0, \pi] \times \mathbb{R}^2 \times \mathbb{R}^1, \mathbb{R}^1\right)$. We consider the nonlinear eigenvalue problem:

$$\begin{cases} Au = \lambda u + h(x, u(x), u'(x), \lambda) & x \in (0, \pi) \\ u \in D(A) \end{cases} \tag{3.22}$$

The equation is reduced to the following:

$$u = \lambda L u + N(u, \lambda) \qquad u \in C_0^1(\,[0, \pi], \mathbb{R}^1\,)\,. \tag{3.23}$$

where $N(u, \lambda) = Lh(x, u(x), u'(x), \lambda)$ is compact from $C_0^1(\,[0, \pi], \mathbb{R}^1\,) \times \mathbb{R}^1$ to $C_0^1(\,[0, \pi], \mathbb{R}^1\,)$. If further, we assume:

$$h(x, \xi, \eta; \lambda) = o(|\xi|^2 + |\eta|^2)^{\frac{1}{2}} \tag{3.24}$$

as $(\xi, \eta) \to (0, 0)$ uniformly in x and λ (in finite intervals), then in the Banach space $C_0^1([0, \pi], \mathbb{R}^1)$

$$\|N(u, \lambda)\| = o(\|u\|)\,.$$

According to the global bifurcation theorem, the component ζ_j of nontrivial solutions passing through (θ, λ_j), $\lambda_j^{-1} \in \sigma(L)$ (or equivalently $\lambda_j \in \sigma(A)$), $j = 1, 2, \ldots$ has two possibilities: either ζ_j is unbounded or ζ_j meets some $\zeta_i, i \neq j$.

We shall prove that the latter possibility actually does not occur in this problem.

Let us recall the linear ODE (i.e., $h = 0$)

$$\begin{cases} Au = -(pu')' + qu = \lambda u & \text{in } (0, \pi)\,, \\ u(0) = u(\pi) = 0\,. \end{cases}$$

The nontrivial solution set consists of $\bigcup\limits_{j=1}^{\infty} \phi_j \mathbb{R}^1$, where ϕ_j is the eigenfunction associate with $\lambda_j, j = 1, 2, \ldots$ according to the Sturm-Liouville Theory, ϕ_j has exactly $j - 1$ simple roots in the open interval $(0, \pi)$. We may assume ${\phi_j}'(0) > 0$. Let us extend this conclusion to nonlinear problems (3.22). Set

$$S_j = \{v \in C_0^1(\,[0, \pi]\,) \mid v \text{ has exactly } (j-1) \text{ simple roots in } (0, \pi), \text{ and } v'(0), \\ v'(\pi) \neq 0\}$$

$j = 1, 2, \ldots$.

Thus S_j is an open set in $C_0^1(\,[0, \pi]\,)$, and

$$S_i \cap S_j = \varnothing \qquad \text{as} \qquad i \neq j\,,$$

$i, j = 1, 2, \ldots$.

We shall prove that $\zeta_i \cap \zeta_j = \varnothing$ as $i \neq j$.

It is sufficient to prove that

$$\zeta_j \subset (S_j \times \mathbb{R}^1) \cup \{(\theta, \lambda_j)\}, \qquad j = 1, 2, \ldots$$

Lemma 3.5.5 *There exists a neighborhood O_j of (θ, λ_j) such that if $(u, \lambda) \in O_j \backslash (\{\theta\} \times \mathbb{R}^1)$ is a solution of (3.22) then $u \in S_j, \forall j$.*

Proof. We prove by contradiction. If $\exists$ solutions (u_n, α_n) of (3.23), with $u_n \neq \theta$, satisfying $(u_n, \alpha_n) \to (\theta, \lambda_j)$, but $u_n \notin S_j$. Let $v_n = \frac{u_n}{\|u_n\|}$, then

$$v_n = \alpha_n L v_n + \|u_n\|^{-1} N(u_n, \alpha_n) .$$

Since $\|v_n\| = 1, \alpha_n \to \lambda_j$, $\|u_n\|^{-1} \|N(u_n, \alpha_n)\| = o(1)$, and L is compact, we have $v_n \to v$, $\|v\| = 1$, satisfying

$$v = \lambda_j L v .$$

According to the Sturm-Liouville theorem, $v \in S_j$. This contradicts the openness of S_j. □

Now, we are going to prove:

Theorem 3.5.6 *Suppose that h satisfies (3.24). Then the branch of nontrivial solutions ζ_j passing through (θ, λ_j) is included in $(S_j \times \mathbb{R}^1) \cup \{(\theta, \lambda_j)\}$. In particular, ζ_j is unbounded.*

Proof. From Lemma 3.5.5, $\zeta_j \cap O_j \subset (S_j \times \mathbb{R}^1) \cup \{(\theta, \lambda_j)\}$. We shall prove that ζ_j is entirely contained in the above set. If not, $\exists (u^*, \lambda^*) \in \zeta_j \cap \partial(S_j \times \mathbb{R}^1) \backslash (\theta, \lambda_j)$. Since $\partial(S_j \times \mathbb{R}^1) = \partial S_j \times \mathbb{R}^1$, and $u^* \in \partial S_j$ implies that u^* has at least a double zero, i.e., $\exists \xi \in [0, \pi]$ such that $u^*(\xi) = u^{*\prime}(\xi) = 0$, it follows from the uniqueness of ODE, $u^* = \theta$. Thus $(u^*, \lambda^*) = (\theta, \lambda_i), \lambda_i^{-1} \in \sigma(L), i \neq j$, according to the necessary condition of bifurcation. Again, by Lemma 3.5.5, $i = j$. This is a contradiction. Our conclusion follows from Theorem 3.5.4 directly. □

Remark 3.5.7 *One may consider other boundary conditions than the Dirichlet condition.*

Let us return to the example in Sect. 1.3:

$$\begin{cases} -u'' = \lambda \sin u & x \in (0, \pi) , \\ u(0) = u(\pi) = 0 . \end{cases} \tag{3.25}$$

Applying Theorem 3.5.6, we conclude that $\forall n = 1, 2, \ldots$, there is an unbounded branch of solutions ζ_n passing through the bifurcation point (θ, n^2).

Again, one may prove that if $(u, \lambda) \in \zeta_n$, then

$$\lambda \geqslant n^2 . \tag{3.26}$$

Indeed, if $\lambda < n^2$, then

$$\lambda \frac{\sin u}{u} < n^2 .$$

Comparing (3.25) with the equation

$$\begin{cases} -v'' = n^2 v & x \in (0, \pi) \\ v(0) = v(\pi) = 0, \end{cases} \tag{3.27}$$

we would have $v \in S_{n+1}$, on account of Sturm's theorem. But $v \in S_n$. This is a contradiction.

Moreover, $\forall (u, \lambda) \in \zeta_n$,

$$u'(x) = \lambda \int_x^\xi \sin u(t) dt \,,$$

where $\xi \in (0, \pi)$ is a root of $u'(x) = 0$. Thus

$$\|u'\|_C \leqslant \lambda\pi \,,$$

and

$$\|u\|_{C^1} \leqslant \lambda\pi(1 + \pi) \,. \tag{3.28}$$

Combining (3.26) with (3.28), we have

$$\zeta_n \subset \{(u, \lambda) \in C_0^1([0, \pi]) \times \mathbb{R}^1 \,|\, \lambda \geqslant n^2, \|u\|_{C^1} \leqslant \lambda\pi(1 + \pi)\} \,.$$

Hence, the projection of ζ_n onto $\mathbb{R}^1$ is the interval (n^2, ∞). For any $\lambda_0 \in (n^2, \infty)$, there exist at least n branches $\zeta_1, \zeta_2, \ldots, \zeta_n$ that meet $\lambda = \lambda_0$. On each ζ_j there exists at least a pair of nontrivial solutions $\pm u_{\lambda_0}^{(j)}$, $j = 1, 2, \ldots, n$. Namely:

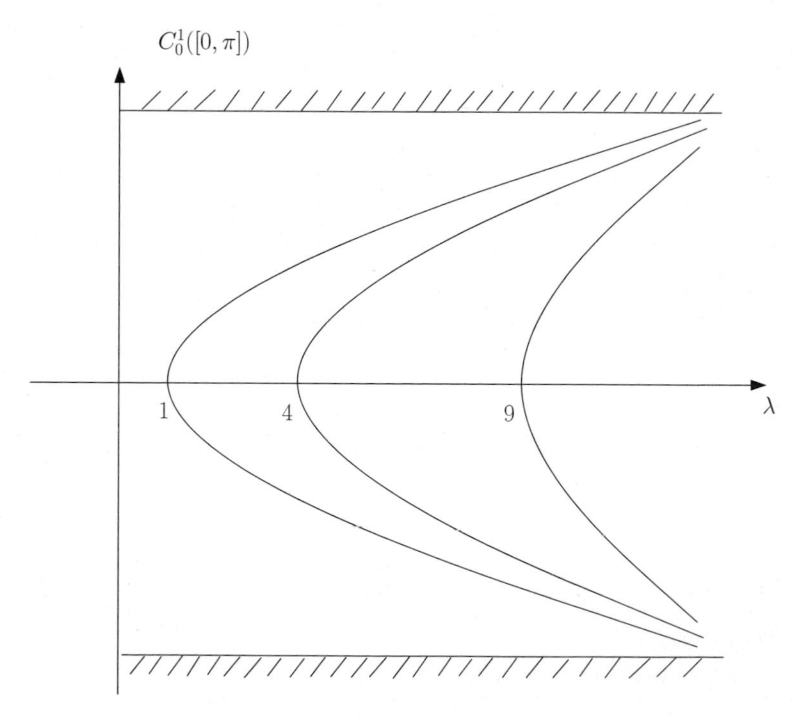

Fig. 3.4.

Theorem 3.5.8 *$\forall\lambda \in (n^2, (n+1)^2]$ equation (3.26) has at least n distinct pairs of nontrivial solutions.*

Remark 3.5.9 *It is natural to ask if we can extend the above method to study elliptic nonlinear eigenvalue problems as that for ODE. The main obstruction is that there is no counterpart of the Sturm–Liouville theory for PDEs. There are only a few partial results in this direction.*

Remark 3.5.10 *(Bifurcation at infinity)*

We study the bifurcation at infinity by transforming the problem into the bifurcation at zero. We consider the equation:

$$x = \lambda Lu + M(x, \lambda) , \tag{3.29}$$

where L is a linear compact operator on X, and $M : X \times R^1 \to X$ satisfies $\|M(x, \lambda)\| = o(\|x\|)$ uniformly in any finite interval as $\|x\| \to \infty$.

(λ_0, ∞) is called a bifurcation point at infinity, if there exists a sequence of solutions (λ_n, x_n) of (3.29), such that $\|x_n\| \to \infty$, and $\lambda_n \to \lambda_0$.

Now we define $y = \|x\|^{-2}x$, then $\|y\| = \|x\|^{-1}$. Equation (3.29) is transformed into:

$$y = \lambda Ly + N(y, \lambda) , \tag{3.30}$$

where $N(y, \lambda) = \|y\|^2 M(\|y\|^{-2}y, \lambda)$. Since

$$\|N(y, \lambda)\| = o(\|y\|) \text{ as } y \to \theta ,$$

(3.30) is exactly the equation we studied previously.

3.6 Applications

3.6.1 Degree Theory on Closed Convex Sets

In applications of the Leray–Schauder degree theory, boundary conditions should be carefully investigated, i.e., on the boundary there is no fixed point of the compact vector field. However, for mappings with images in a closed convex set, the notion of the boundary can be reduced. Let us present an extension version of the Tietze theorem, Dugundji's theorem, concerning the extension of continuous mappings on closed convex sets, which we have met in Sect. 3.4.

Theorem 3.6.1 *(Dugundji) Suppose that A is a closed subset of a metric space X, and that C is a convex subset of a Banach space Y. Then for any continuous map $f : A \to C$, there exists a continuous extension $\widetilde{f} : X \to C$.*

Proof. Since $X \setminus A$ is a metric subspace, it is paracompact. $\forall x \in X\backslash A$, let $0 < r(x) < \frac{1}{2}\text{dist}(x, A)$, and let $B_{r(x)}(x)$ be the ball with radius $r(x)$ and center x . All these balls form a covering of $X\backslash A$. The covering has a locally finite refinement $\{U_\alpha|\ \alpha \in \wedge\}$. On the set $X\backslash A$, we define

$$\lambda_\alpha(x) = \frac{\text{dist}(x, X\backslash U_\alpha)}{\sum_{(\beta\in\Lambda)} \text{dist}(x, X\backslash U_\beta)} \quad \forall\alpha \in \Lambda .$$

They satisfy:

$$\text{supp }\lambda_\alpha \subset U_\alpha,\ 0 \leqslant \lambda_\alpha(x) \leqslant 1, \text{and} \sum_{\alpha\in\Lambda} \lambda_\alpha(x) = 1 .$$

$\forall\alpha \in \Lambda$ we choose $a_\alpha \in A$, such that

$$\text{dist}(a_\alpha, U_\alpha) < 2\text{dist}(A, U_\alpha) ,$$

and let

$$\widetilde{f}(x) = \begin{cases} \sum \lambda_\alpha(x) f(a_\alpha) & x \in X\backslash A \\ f(x) & x \in A . \end{cases}$$

Since $\{U_\alpha \,|\, \alpha \in \Lambda\}$ is a locally finite covering, the sum is finite. We claim that $\widetilde{f}$ is continuous. It is sufficient to prove that $\widetilde{f}$ is continuous on the boundary ∂A, i.e., for $\forall x_0 \in \partial A$. $\forall\varepsilon > 0$, $\exists\delta > 0$ such that

$$\| \widetilde{f}(x) - f(x_0) \| < \varepsilon ,$$

whenever $\text{dist}(x, x_0) < \delta$.

In fact, $\exists\delta > 0$, such that if $\text{dist}(x, x_0) < 6\delta$ and $x \in A$, then

$$\| f(x) - f(x_0) \| < \varepsilon .$$

Now, $\forall x \in X$, if $\text{dist}(x, a_\alpha) < 3\delta$, from

$$\text{dist}(x_0, a_\alpha) \leq \text{dist}(x_0, x) + \text{dist}(x, a_\alpha) ,$$

it follows that $\text{dist}(x_0, a_\alpha) < 4\delta$, then

$$\|f(a_\alpha) - f(x_0)\| < \epsilon . \tag{3.31}$$

Otherwise, $\text{dist}(x, a_\alpha) \geq 3\delta$, then

$$\text{dist}(x, a_\alpha) \geq 3\text{dist}(x, x_0) \geq 3\text{dist}(x, A) .$$

In this case we claim that $x \notin U_\alpha$. If not, i.e., $x \in U_\alpha$, then we would have

$$\begin{aligned} \text{dist}(x, a_\alpha) &\leqslant \text{diam}(U_\alpha) + \text{dist}(U_\alpha, a_\alpha) \\ &\leqslant 2r(x) + 2\text{dist}(A, U_\alpha) \\ &\leqslant \text{dist}(A, x) + 2\text{dist}(A, x) \\ &= 3\text{dist}(A, x) \end{aligned}$$

This is a contradiction, therefore, $x \notin U_\alpha$ and then

$$\lambda_\alpha(x) = 0 . \tag{3.32}$$

Combining equations (3.31) and (3.32), we have

$$\| \widetilde{f}(x) - f(x_0) \| \leqslant \sum_{\mathrm{dist}(x,a_\alpha)<3\delta} \lambda_\alpha(x) \| f(a_\alpha) - f(x_0) \| < \varepsilon .$$

The proof is finished. □

Suppose that C is a closed convex subset of a real Banach space $\mathbb{X}$, and that $U \subset C$ is a bounded relatively open set. Assume that $K : \overline{U} \to C$ is compact, and that $\theta \notin (\mathrm{id} - K)(\partial U)$, where ∂U is the boundary of U relative to C. Then we can define

$$\deg_C (\mathrm{id} - K, U, \theta) = \deg (\mathrm{id} - K \circ r, B_R \cap r^{-1}(U), \theta) ,$$

where r is the retraction of C, and $R > 0$ is a large number such that $U \subset B_R(\theta)$.

Firstly, we point out that the definition makes sense, i.e, $\theta \notin (\mathrm{id} - K \circ r)(\partial(B_R(\theta) \cap r^{-1}(U)))$. In fact, if it is not true, then $\exists p \in \partial(B_R(\theta) \cap r^{-1}(U))$ such that $p = K \circ r(p)$. Thus, $p \in C$ and then $p = Kp$, which implies $p \in C \cap \partial(B_R(\theta) \cap r^{-1}(U))$, i.e., $p \in \partial U$. This is a contradiction.

Secondly, one verifies that the degree so defined does not depend on the retraction r, and the radius R. In fact, let $R_1 < R_2$ such that $U \subset \mathrm{int}(B_{R_1}(\theta))$. All fixed points of $\mathrm{id} - K$ in U must be in $B_{R_1}(\theta)$. By excision,

$$\deg (\mathrm{id} - K \circ r, r^{-1}(U) \cap B_{R_2}(\theta), \theta) = \deg (\mathrm{id} - K \circ r, r^{-1}(U) \cap B_{R_1}(\theta), \theta) .$$

Again let r_1, r_2 be two retractions: $\mathbb{X} \to C$, by excision,

$$\begin{aligned} &\deg (\mathrm{id} - K \circ r_i, r_i^{-1}(U) \cap B_R(\theta), \theta) \\ &= \deg (\mathrm{id} - K \circ r_i, r_1^{-1}(U) \cap r_2^{-1}(U) \cap B_R(\theta), \theta) . \end{aligned}$$

for $i = 1, 2$. We define

$$F(x, t) = \mathrm{id} - (tK \circ r_1 + (1 - t)K \circ r_2) \ \forall t \in [0, 1] .$$

From the homotopy invariance, $\deg_C (\mathrm{id} - K, U, \theta)$ is independent of r_i, $i = 1, 2$.

In particular, if $\mathbb{X}$ is a real OBS, i.e., $\mathbb{X}$ is a real Banach space with a closed positive cone P (see Sect. 2.1), then P is a closed convex subset, and then $\deg_P (\mathrm{id} - K, U, \theta)$ makes sense.

Now, we present a few degree computations for special boundary conditions in an OBS.

Lemma 3.6.2 *Suppose that $(\mathbb{X}, P)$ is an OBS, and that $U \subset P$ is bounded and open. Assume that $K : \overline{U} \to P$ is compact, satisfying*

$$\exists y \in P\backslash\{\theta\}, \text{ such that } x - Kx \neq ty \; \forall t \geq 0, \; \forall x \in \partial U \; ;$$

then

$$\deg_P(\mathrm{id} - K, U, \theta) = 0 \; .$$

Proof. From the homotopy invariance,

$$\deg(\mathrm{id} - K - ty, U, \theta) = constant, \; \forall t \geq 0 \; .$$

Since K is compact, $\exists C > 0$ such that $\|x - Kx\| \leq C \, \forall x \in U$. One chooses $t > \frac{C}{\|y\|}$, then $x - Kx \neq ty \, \forall x \in U$. It follows that $\deg_P(\mathrm{id} - K, U, \theta) = 0$. □

Theorem 3.6.3 *Suppose that $(\mathbb{X}, P)$ is an OBS, $U \subset P$ is bounded open and contains θ. Assume that $\exists \rho > 0$ such that $B_\rho(\theta) \cap P \subset U$ and that $K : \overline{U} \to P$ is compact and satisfies:*

(1) $\forall x \in P$, with $\|x\| = \rho, \forall \lambda \in [0, 1), x \neq \lambda K(x)$,
(2) $\exists y \in P\backslash\{\theta\}$, such that $x - K(x) \neq ty, \forall x \in \partial U, \forall t \geq 0$.

Then K possesses a fixed point on $\overline{U_\rho}$, where $U_\rho = U \backslash B_\rho(\theta)$.

Proof. We may assume that K has no fixed point on $\partial B_\rho(\theta)$ then

$$\deg_P(\mathrm{id} - K, U, \theta) = \deg_P(\mathrm{id} - K, U_\rho, \theta) + \deg_P(\mathrm{id} - K, B_\rho(\theta) \cap P, \theta) \; .$$

According to Lemma 3.6.2 and assumption (2), LHS $= 0$, and from assumption (1) and the homotopy invariance, the second term on the RHS $= 1$. Therefore, $\deg_P(\mathrm{id} - K, U_\rho, \theta) = -1$. □

Theorem 3.6.4 *Suppose that $(\mathbb{X}, P)$ is an OBS, $U \subset P$ is bounded and open, and contains θ. Assume that $\exists \rho > 0$ such that $U \subset B_\rho(\theta) \cap P$, and that $U_\rho := (B_\rho(\theta) \cap P) \backslash U$ possesses an interior point. If $K : \overline{B}_\rho(\theta) \cap P \to P$ is compact and satisfies:*

(1) $\forall x \in P, \|x\| = \rho, \forall \lambda \in [0, 1), x \neq \lambda K(x)$.
(2) $\exists y \in P\backslash\{\theta\}$ such that $x - K(x) \neq ty, \; \forall t \geq 0, \; \forall x \in \partial U$.

Then K has a fixed point in $\overline{U}_\rho$.

Proof. The proof is similar to the previous theorem: $\deg(\mathrm{id} - K, U_\rho, \theta) = 1$. We omit it. □

Remark 3.6.5 *Let (X, P) be an OBS, and let $0 < \rho_1 < \rho_2$, and $P_{\rho_i} = P \cap B_{\rho_i}(\theta)$. We write $\partial P_{\rho_i} = P \cap \partial B_{\rho_i}(\theta)$, $i = 1, 2$. A map $f : P_{\rho_2} \backslash P_{\rho_1} \to P$ is called a cone compression, if $x \nleq f(x), \; \forall x \in \partial P_{\rho_2}$, and $f(x) \nleq x, \; \forall x \in \partial P_{\rho_1}$. It is called a cone expansion if $f(x) \nleq x, \; \forall x \in \partial P_{\rho_2}$, and $x \nleq f(x) \; \forall x \in \partial P_{\rho_1}$. Noticing that*

$$f(x) \nleq x \text{ implies } \forall y \in P\backslash\{\theta\}, \forall t \geq 0, x - f(x) \neq ty ,$$

and

$$x \nleq f(x) \text{ implies } \forall \lambda \in [0,1], x \neq \lambda f(x) ,$$

Theorems 3.6.3 and 3.6.4 extend the earlier results of Krasnoselski on cone compression and cone expansion mappings.

Let (X, P) be an OBS satisfying $X = \overline{P - P}$. A map $T : P \to P$ is called positive. We shall study the bifurcation problem for positive mappings on OBS. Let $R_+ = [0, +\infty)$ and $f : P \times R_+ \to P$ be a continuous mapping satisfying $f(\theta, \lambda) = \theta \ \forall \lambda \in R_+$. We are looking for $\lambda_0 \in R_+$, such that $\forall \epsilon > 0, \exists (x, \lambda) \in (B_\epsilon^+ \backslash \{\theta\}) \times (\lambda_0 - \epsilon, \lambda_0 + \epsilon)$ satisfying

$$x = f(x, \lambda) ,$$

where $B_\epsilon^+ = B_\epsilon(\theta) \cap P$. At this time, (θ, λ_0) is a bifurcation point of positive solutions.

Lemma 3.6.6 *Let $(\mathbb{X}, P)$ be defined above. Let $f : P \times R_+ \to P$ be of the form:*

$$f(x, \lambda) = \lambda Tx + g(x, \lambda) ,$$

where $T \in \mathcal{L}(\mathbb{X}, \mathbb{X})$ is a positive and compact operator, and $g : P \times R_+ \to P$ is compact, satisfying $\|g(x, \lambda)\| = o(\|x\|)$ as $\|x\| \to 0$ and $g(x, 0) = \theta$. If (θ, λ_0) is a bifurcation point with $\lambda_0 \geq 0$, then $\lambda_0 > 0$, and λ_0^{-1} is an eigenvalue of T with positive eigenvector.

Proof. By definition, $\exists (x_n, \lambda_n) \in (P\backslash\{\theta\}) \times R_+$, such that $(x_n, \lambda_n) \to (\theta, \lambda_0)$, with $x_n = f(x_n, \lambda_n)$. Setting $y_n = \frac{x_n}{\|x_n\|}$, we have

$$y_n - \lambda_0 T y_n = (\lambda_n - \lambda_0) T y_n + g(x_n, \lambda_n) \|x_n\|^{-1} .$$

Define $S_+ = \partial B_1(\theta) \cap P$, then $y_n \in S_+$, and then, after a subsequence we have $Ty_n \to z_0$. Let $y_0 = \lambda_0 z_0$, then $y_0 \in S_+$. Obviously, $y_0 = \lambda_0 T y_0$, and $\lambda_0 > 0$. □

One has the following variant of the global bifurcation theorem.

Theorem 3.6.7 *(Dancer) Let (X, B) and f, T, g be as in Lemma 3.6.6. Assume that T has finitely many positive eigenvalues with positive eigenvectors. Let $\Sigma = \{(x, \lambda) \in P \times R_+ \mid x = f(x, \lambda)\}$ and $\Sigma^+ = \overline{\Sigma \cap ((P\backslash\{\theta\}) \times (R^+\backslash\{0\}))}$. Then Σ^+ contains an unbounded connected branch emanating from the reciprocal of one of these eigenvalues.*

Proof. Let λ_0^{-1} be the smallest positive eigenvalue of T, and let ζ be the connected component of $\Sigma^+ \cup (\{\theta\} \times [0, \lambda_0])$, containing $\{\theta\} \times [0, \lambda_0]$. If ζ is bounded, then $\exists \mu > \lambda_0$ such that $\zeta \subset Q_\mu := B_\mu^+ \times [0, \mu]$, and $\zeta \cap \partial Q_\mu = \emptyset$, where $B_\mu^+ = B_\mu(\theta) \cap P$.

Since μ^{-1} is not an eigenvalue with positive eigenvector of T, $\exists\alpha > 0$ such that

$$\|x - \mu Tx\| \geq \alpha\mu\|x\|, \quad \forall x \in P .$$

One chooses $\rho \in (0, \min\{\frac{\alpha\mu}{2}, \mu\})$, such that $\|g(x,\mu)\| \leq \frac{\alpha\mu}{2}\|x\|$ as $x \in \overline{B_\rho^+}$. Choose $\epsilon \in (0, \min\{\mu, \rho\})$. Let $\zeta_1 = \zeta \cup (\{\theta\} \times [0,\mu])$, $D = ((\overline{B_\mu^+} \backslash B_\epsilon^+ \times \{0\}) \cup ((\partial B_\mu(\theta) \cap P) \times [0,\mu]) \cup ((\overline{B_\mu^+} \backslash B_\epsilon^+) \times \{\mu\})$. Then $D \cap \zeta_1 = \emptyset$. Similar to the proof in the global bifurcation theorem, $\exists U \subset P \times [0,\mu]$ open, such that $\Sigma \cap \partial U = \emptyset, \zeta_1 \subset U$, and $\overline{U} \cap D = \emptyset$. Thus,

$$\begin{aligned}
1 &= \deg_P(\mathrm{id}, B_\epsilon^+, \theta) \\
&= \deg_P(\mathrm{id} - f(\cdot, 0), U_0, \theta) \\
&= \deg_P(\mathrm{id} - f(\cdot, \mu), U_\mu, \theta) \\
&= \deg_P(\mathrm{id} - f(\cdot, \mu), B_\epsilon^+, \theta) ,
\end{aligned}$$

where $U_\lambda := \{x \in P | (x, \lambda) \in U\}$.

Let $\phi \in S^+$ be a positive eigenvector associated with λ_0^{-1} with norm 1. We claim that the equation

$$x - \mu Tx = \beta\phi \quad \forall\beta > 0$$

has no positive solution. For otherwise, if $\exists\beta_0 > 0, x_0 \in P$, satisfying

$$x_0 - \mu Tx_0 = \beta_0\phi .$$

Then

$$x_0 \geq (\mu\tau_0\lambda_0^{-1} + \beta_0)\phi > (\tau_0 + \beta_0)\phi ,$$

where $\tau_0 = \sup\{\tau \in R^+ | x_0 \geq \tau\phi\}$. This contradicts the definition of τ_0. Setting $\beta \in (0, \frac{\epsilon\rho}{2})$, we have

$$\|x - \mu Tx + (1-\lambda)\beta\phi + \lambda g(x,\mu)\| \geq \alpha\mu\|x\| - \frac{\epsilon\rho}{2} - \frac{\alpha\mu}{2}\|x\| > 0, \quad \text{as } \|x\| = \varepsilon .$$

Then, by the homotopy invariance,

$$\deg_P(\mathrm{id} - f(\cdot,\mu), B_\epsilon^+, \theta) = \deg_P(\mathrm{id} - \mu T - \beta\phi, B_\epsilon^+, \theta) = 0 .$$

The contradiction shows that ζ is unbounded. □

3.6.2 Positive Solutions and the Scaling Method

The above degree computation is applied to the study of the positive solution of certain superlinear elliptic equations. Let $\Omega \subset \mathbb{R}^n$ be an open bounded domain with smooth boundary. Assume that $a_{ij}, b_j \in C(\overline{\Omega}), i,j = 1, \ldots, n,$ $\phi \in C^1(\partial\Omega)$ is nonnegative, and $f \in C(\overline{\Omega} \times \mathbb{R}^1)$. Let

$$Lu(x) = \sum_{i,j=1}^{n} \frac{\partial}{\partial x_i} \left(a_{i,j}(x) \frac{\partial u}{\partial x_j} \right) + \sum_{j=1}^{n} b_j(x) \frac{\partial u}{\partial x_j} \,. \tag{3.33}$$

We study positive solutions of the equation:

$$\begin{cases} Lu + f(x,u) = 0, & \text{in } \Omega \,, \\ u = \phi, & \text{on } \partial\Omega \,, \end{cases} \tag{3.34}$$

in which f is superlinear in u. In order to apply degree theory, we should estimate a priori bounds for positive solutions u. We introduce here a useful method – the scaling method – based on the following global results of Liouville-type theorems, cf. Gidas and Spruck [GS], Chen and Li [ChL 1] and Y. Li [Li 1].

Theorem. Assume $p \in (1, \frac{n+2}{n-2})$ for $n \geq 3$, and $p > 1$ for $n = 1, 2$. If $u \in C^2(\mathbb{R}^n)$ is a nonnegative solution of the equation:

$$\Delta u + u^p = 0 \ \text{ in } \ \mathbb{R}^n, \tag{3.35}$$

then $u \equiv 0$.

Proof. (We only prove the theorem under a stronger condition: $1 < p < \frac{n}{n-2}$, for general case see the above-mentioned references.) Let $\varphi_1 > 0$ be the first eigenfunction of the Laplacian on $B_1(\theta) \subset R^n$, i.e.,

$$\begin{cases} -\Delta\varphi_1 = \lambda_1\varphi_1, & \text{in } B_1(0) \,, \\ \varphi_1 = 0 & \text{on } \ \partial B_1(\theta) \,. \end{cases} \tag{3.36}$$

For $\forall R > 0$, let $\varphi_R(x) = \varphi_1(\frac{x}{R})$, we have

$$\begin{cases} -\Delta\varphi_R = \frac{\lambda_1}{R^2}\varphi_R, & \text{in } B_R(\theta) \,, \\ \varphi_R = 0 & \text{on } \ \partial B_R(\theta) \,. \end{cases} \tag{3.37}$$

Integrating by parts,

$$\int_{B_R} u^p \varphi_R^p = -\int_{B_R} \Delta u \varphi_R^p \tag{3.38}$$

$$= \int_{B_R} \nabla u \nabla \varphi_R^p \tag{3.39}$$

$$= -\int_{B_R} u \Delta \varphi_R^p + p \int_{\partial B_R} u \frac{\partial \varphi_R}{\partial n} \varphi_R^{p-1} \tag{3.40}$$

$$= -\int_{B_R} u \Delta \varphi_R^p \,. \tag{3.41}$$

$$\tag{3.42}$$

Since

$$\int_{B_R} u\Delta\varphi_R^p = p(p-1)\int_{B_R} u\varphi_R^{p-2}|\nabla\varphi_R|^2 - \frac{p\lambda_1}{R^2}\int_{B_R} u\varphi_R^p\,, \tag{3.43}$$

it follows that

$$\int_{B_R}(u\varphi_R)^p \le \frac{\lambda_1 p}{R^2}\int_{B_R} u\varphi_R^p \tag{3.44}$$

$$\le \frac{\lambda_1 p}{R^2}\left(\int_{B_R}(u\varphi_R)^p\right)^{\frac{1}{p}}\left(\int_{B_R}\varphi_R^p\right)^{\frac{p-1}{p}}. \tag{3.45}$$

Therefore,

$$\int_{B_R}(u\varphi_R)^p \le \left(\frac{\lambda_1 p}{R^2}\right)^{\frac{p}{p-1}}\int_{B_R}\varphi_R^p = (\lambda_1 p)^{\frac{p}{p-1}} R^{n-\frac{2p}{p-1}}\int_{B_1}\varphi_1^p\,. \tag{3.46}$$

We obtain $\lim_{R\to 0}\int_{B_R}(u\varphi_R)^p = 0$. Thus $u \equiv 0$. □

Similarly, one has the Liouville theorem for half space:

Theorem. Assume $p \in (1, \frac{n+2}{n-2})$ for $n \ge 3$, and $p > 1$ for $n = 1, 2$. If $u \in C^2(\mathbb{R}^n) \cap C(\overline{\mathbb{R}}_+^n)$ is a nonnegative solution of the equation:

$$\begin{cases} \Delta u + u^p = 0, & \text{in } \overline{\mathbb{R}}_+^n\,, \\ u = 0 & \text{on } x_n = 0\,, \end{cases} \tag{3.47}$$

then $u \equiv 0$.

Theorem 3.6.8 *Assume $p \in (1, \frac{n+2}{n-2})$ for $n \ge 3$, and $p > 1$ for $n = 1, 2$. Suppose that $f \in C(\bar{\Omega} \times R^1)$ satisfies*

$$\lim_{t\to+\infty}\frac{f(x,t)}{t^p} = h(x) \;\; \textit{uniformly in } x \in \overline{\Omega}, \tag{3.48}$$

where h is a positive function. If $u \ge 0$ is a solution of equation (3.34), then there exists a constant C depending on p, Ω, ϕ and f only such that $u \le C$.

Proof. One proves the theorem by contradiction. If the conclusion is not true, then there exist a sequence of solutions u_k of equation (3.34) and points $P_k \in \overline{\Omega}$ satisfying

$$M_k := \sup_{x\in\overline{\Omega}} u_k(x) = u_k(P_k) \to +\infty \text{ as } k \to \infty\,.$$

We may assume $P_k \to P \in \overline{\Omega}$ as $k \to \infty$. There are two possibilities: either (1) $P \in \Omega^\circ$, or (2) $P \in \partial\Omega$.

In case (1): Let $d = \frac{1}{2}\mathrm{dist}(P, \partial\Omega)$, $\lambda_k^{\frac{2}{p-1}} M_k = 1$, $y = \frac{x-P_k}{\lambda_k}$, and set

$$v_k(y) = \lambda_k^{\frac{2}{p-1}} u_k(x) .$$

It is defined on $B_{\frac{d}{\lambda_k}}(\theta)$ and satisfies

$$\sup_{y \in B_{\frac{d}{\lambda_k}}(\theta)} v_k(y) = v_k(\theta) = 1 .$$

and

$$\sum_{i,j=1}^{n} \frac{\partial}{\partial y_j} \left(a_{ij}^k(y) \frac{\partial v_k(y)}{\partial y_i} \right)$$
$$+ \lambda_k \sum_{j=1}^{n} b_j^k(y) \frac{\partial v_k(y)}{\partial y_j} + \lambda_k^{\frac{2p}{p-1}} f\left(\lambda_k y + P_k, \lambda_k^{\frac{-2}{p-1}} v_k(y) \right) = 0 ,$$

where $a_{ij}^k(y) = a_{ij}(\lambda_k y + P_k) \to a_{ij}(P)$, $\lambda_k b_j^k(y) = \lambda_k b_j(\lambda_k y + P_k) \to 0$, and the last term is asymptotic to $h(\lambda_k y + P_k) v_k^p$.

$\forall R > 0$, $\exists k$ such that $B_R(\theta) \subset B_{\frac{d}{\lambda_k}}(\theta)$. Since v_k is uniformly bounded on $B_R(\theta)$, from the L^q estimates, $1 < q < \infty$, we have $\|v_k\|_{W^{2,q}(B_R(\theta))} \leq C$, a uniform constant. According to the Sobolev embedding theorem and the diagonal principle, one has a subsequence $v_{k_j} \to v$ in $C^{1,\beta} \cap W^{2,q}(B_R(\theta))$ for some $\beta \in (0,1), \forall R > 0$, and $v(\theta) = 1$. Thus as a weak solution v satisfies

$$\begin{cases} \sum_{i,j=1}^{n} a_{ij}(P) \frac{\partial^2 v(y)}{\partial y_i \partial y_j} + h(P) v^p(y) = 0, \text{ in } \mathbb{R}^n . \\ v(\theta) = 1. \end{cases}$$

After scaling and rotation, it is reduced to equation (3.35). Again by bootstrap iteration, $v \in C^2(\mathbb{R}^n)$, it follows from the Liouville-type theorem that $v \equiv 0$. This contradicts $v(\theta) = 1$.

In case (2), $P \in \partial\Omega$. Without loss of generality, we may assume $\exists \delta > 0$ such that $B_\delta(P) \cap \partial\Omega$ is contained in the hyperplane $x_n = 0$. One defines v_k, λ_k as before. Let $d_k = d(P_k, \partial\Omega)$, then $d_k = P_k \cdot e_n$, where e_n is the unit normal vector of $\partial\Omega$ near P. Therefore v_k is well defined in $B_{\frac{\delta}{\lambda_k}}(\theta) \cap \{y_n > -\frac{d_k}{\lambda_k}\}$. Again we have $\sup v_k(y) = v_k(\theta) = 1$.

We claim that there is a constant $C_1 > 0$ such that $\frac{d_k}{\lambda_k} \geq C_1$. In fact, according to elliptic regularity up to the boundary, $|\nabla v_k|$ is uniformly bounded, so is

$$\left| v_k(\theta) - v_k\left(0, \ldots, 0, -\frac{d_k}{\lambda_k}\right) \right| \leq C_1^{-1} \frac{d_k}{\lambda_k} ,$$

i.e.,

$$1 - \lambda_k^{\frac{2}{p-1}} \sup_{x \in \partial\Omega} \phi(x) \leq C_1^{-1} \frac{d_k}{\lambda_k} .$$

Since $\lambda_k \to 0$, our claim is proved.

If there is a subsequence $\frac{d_{k_j}}{\lambda_{k_j}} \to \infty$, then the discussion is reduced to case (1). Therefore we may assume that there is another constant C_2 such that $C_1 \leq \frac{d_k}{\lambda_k} \leq C_2$, after a subsequence, $\frac{d_k}{\lambda_k} \to s$. By taking the limit, we obtain:

$$\begin{cases} \sum_{i,j=1}^{n} a_{ij}(P) \frac{\partial^2 v}{\partial y_i \partial y_j} + h(P) v^p(y) = 0 & \text{in } y_n > -s \,, \\ v(y) = 0 & \text{on } y_n = -s \,. \end{cases}$$

After scaling and rotation, it is reduced to equation (3.47). By the same reasoning, we have $v \equiv 0$, which contradicts $v(\theta) = 1$.

Combining these two cases, we have proved the theorem. □

Lemma 3.6.9 *Under the assumptions of Theorem 3.6.8, if further $f(x,t) \geq 0$, and*

$$\liminf_{t \to +\infty} \frac{f(x,t)}{t} > \lambda_1, \ \textit{uniformly in } x \in \overline{\Omega} \,, \tag{3.49}$$

where λ_1 is the first eigenvalue of the linear elliptic operator $-L$ (see equation (3.33)) with eigenfunction φ_1. Then there exists a constant $C > 0$ such that for all solutions u of the equation:

$$\begin{cases} Lu + f(x,u) + t\varphi_1 = 0, \ \textit{in } \Omega, \ \forall t \geq 0 \,, \\ u = 0 \qquad \textit{on } \partial\Omega \,, \end{cases} \tag{3.50}$$

we have

$$\int_\Omega f(x,u(x)) \varphi_1(x) dx \leq C \,,$$

and

$$t \leq C \,.$$

Proof. We may assume $\|\varphi_1\|_{L^2} = 1$. By the condition (3.49), there are constants $M > 0, k \in (\lambda_1, \liminf_{t \to +\infty} \frac{f(x,t)}{t})$ such that

$$f(x,t) \geq kt - M, \quad \forall t > 0 \,.$$

For any solution u of equation (3.50), by integration, we have

$$-\int_\Omega \varphi_1(x) Lu(x) dx = \int_\Omega \varphi_1(x) f(x,u(x)) dx + t \,,$$

thus

$$\lambda_1 \int_\Omega \varphi_1(x) u(x) dx \geq k \int_\Omega \varphi_1(x) u(x) dx - M \int_\Omega \varphi_1(x) + t \,,$$

i.e.,

$$t + (k - \lambda_1) \int_\Omega \varphi_1(x) u(x) dx \leq M \int_\Omega \varphi_1(x) dx \,.$$

Since $f(x,t) \geq 0$, from the maximum principle, $u \geq 0$. The conclusions follow. □

Theorem 3.6.10 *Suppose that $f \in C(\overline{\Omega} \times \mathbb{R}^1)$ is positive, which satisfies the conditions (3.48) (3.49) and*

$$\limsup_{t \to 0} \frac{f(x,t)}{t} < \lambda_1 \text{ uniformly in } \quad x \in \overline{\Omega}\,. \tag{3.51}$$

Then equation (3.34) has a positive solution in which $\phi = 0$.

Proof. Define $F : u(x) \mapsto f(x, u(x))$ from $C(\overline{\Omega})$ into itself, and $K = (-L)^{-1}$ being the linear compact operator on $C(\overline{\Omega})$.

Then $T = K \circ F$ is a compact mapping from $C(\overline{\Omega})$ into its positive cone. Applying Lemma 3.6.9 and Theorem 3.6.8 we take $R > 0$ large enough such that equation (3.50) has no positive solution u with $C-$ norm $\|u\| = R$, and it follows from Lemma 3.6.2 that $\deg_P(id - T, B_R(\theta) \cap P, \theta) = 0$. According to condition (3.51) and the homotopy invariance, we have $\rho > 0$ small such that $\deg_P(id - T, B_\rho(\theta) \cap P, \theta) = 1$. Therefore there is a nontrivial positive solution. □

3.6.3 Krein–Rutman Theory for Positive Linear Operators

Let X be a real Banach space, and $P \subset X$ be a positive closed cone with nonempty interior $\overset{\circ}{P}$. A linear continuous operator $L \in \mathcal{L}(X, X)$ is called positive if $L(P) \subset P$; it is called strictly positive if $L(P\backslash\{\theta\}) \subset \overset{\circ}{P}$.

Example. Let M be a compact topological space with a Radon measure μ. Let $X = C(M), P = \{u \in C(M) | u \geq 0\}$, then (X, P) is an OBS. Given a nonnegative continuous function $K : M \times M \to R^1$, we define:

$$(Lu)(x) = \int_M K(x,y)u(y)d\mu\,.$$

Then L is a positive operator.

Example. We consider the second-order elliptic operator:

$$Lu = -\Sigma_{i,j=1}^n a_{ij}\partial_{ij}^2 u + \Sigma_{i=1}^n b_i \partial_i u + cu\,, \tag{3.52}$$

where $a_{ij}, b_i, c \in C(\overline{\Omega})$, $i, j = 1, 2, \ldots, n$, $\Omega \subset R^n$ is an open domain with smooth boundary (cf. Protter and Weinberger [PW]). Assume the ellipticity condition, i.e., $\exists \alpha > 0$ such that

$$\Sigma_{i,j=1}^n a_{ij}(x)\xi_i\xi_j \geq \alpha|\xi|^2,\ \forall \xi = (\xi_1, \ldots, \xi_n) \in R^n, \forall x \in \overline{\Omega}\,,$$

and

$$c \geq 0\,.$$

It is known that $\forall f \in L^p, 1 < p < \infty$, the equation:

$$\begin{cases} Lu = f , \\ u|_{\partial\Omega} = 0 , \end{cases}$$

has a unique solution $u \in W^{2,p}(\Omega)$; the inverse operator $K = L^{-1} : f \to u$ is positive and bounded on $L^p(\Omega)$(and also on $C_0(\overline{\Omega})$ the subspace of $C(\overline{\Omega})$ with boundary value zero), but not strictly positive. However, we have:

Lemma 3.6.11 *Let $X = C_0^1(\overline{\Omega})$, and $P = \{u \in X | u \geq 0\}$. Then $K = L^{-1}$ is a strictly positive compact operator.*

Proof. We only want to prove the strict positivity, i.e., $\forall f \in C_0^1(\overline{\Omega})$ if $f \geq 0$, but $f \neq \theta$, then $u = Kf \in \overset{\circ}{P}$.

In fact, by the strong maximum principle, $u(x) > 0\ \forall x \in \Omega$. Again by the Hopf maximum principle, $\partial_n u|_{\partial\Omega} < 0$, where ∂_n is the outer normal derivative. Since $\partial\Omega$ is compact, $\exists \delta > 0$ such that $\sup_{x\in\partial\Omega} \partial_n u(x) < -\delta < 0$. Now, $u \in C^1(\overline{\Omega})$, there is a neighborhood N of $\partial\Omega$ on which $\partial_\nu u \leq -\delta/2$, where ν is the direction connecting $x \in N$ to the closest point on $\partial\Omega$. Setting $\alpha = \inf\{u(x)|x \in \Omega\backslash N\}$ and $\beta = \mathrm{Min}\{\alpha, \delta/2\}$, then the open ball, centered at u with radius β, is contained in $\overset{\circ}{P}$. □

The first eigenvalue λ_1 and the first eigenfunction ϕ play important role in the study of second-order elliptic equations. It is known that $\lambda_1 > 0$ is algebraically simple, and that the normalized $\phi > 0$ is unique. In fact, the conclusion is a special case of the following Krein–Rutman theorem:

Theorem 3.6.12 *(Krein–Rutman) Suppose that (X, P) is an OBS with non-empty interior $\overset{\circ}{P}$, and that T is a compact strictly positive operator. Then T possesses a unique positive eigenvector $\phi \in \overset{\circ}{P}$ with $\|\phi\| = 1$, associated with a algebraically simple eigenvalue $\frac{1}{\lambda_1} = \lim_{n\to\infty} \|T^n\|^{\frac{1}{n}} > 0$, satisfying:*

$$\lambda_1 \leq |\lambda| \qquad \forall \lambda^{-1} \in \sigma(T) . \tag{3.53}$$

Proof. (1). (The existence of a positive eigenvector.) $\forall x \in P\backslash\{\theta\}, \exists M > 0$, such that $MTx \geq x$. For otherwise, it must be $Tx - \frac{1}{n}x \notin P, \forall n \in \mathcal{N}$ large, and then $Tx \notin \overset{\circ}{P}$; this contradicts the strict positiveness of T. According to Dugundji's theorem, there exists a retraction $r : X \to P$. We define $\mathcal{T} : X \times R^1 \to P$ as follows:

$$\mathcal{T}(y, \lambda) = \lambda r \circ T(y + x) .$$

Thus, $\forall \lambda, \mathcal{T}(\cdot, \lambda)$ is compact, and satisfies $\mathcal{T}(y, 0) = \theta$, $\forall y \in X$. We notice that for $\lambda \geq 0, y = \mathcal{T}(y, \lambda)$ is equivalent to $y \in P$, and $y = \lambda T(y + x)$. From the Leray–Schauder Theorem, there is an unbounded connected branch of solutions $\zeta^+ \subset P \times \mathbb{R}^1_+$ passing through $(\theta, 0)$.

We claim that if $(y,\lambda) \in \zeta^+$, then $\lambda \in [0,M]$. In fact, since $MTx \geq x$, $y = \lambda T(y+x) \geq \frac{\lambda}{M}x$, we obtain

$$y \geq \lambda Ty \geq \left(\frac{\lambda}{M}\right)^n x, \quad \forall n \in \mathcal{N} .$$

If $\lambda > M$, then $x = \theta$. This is a contradiction. Therefore, $\exists (y_n, \lambda_n) \in \zeta_+$ such that $\lambda_n \to \lambda^* \in [0,M], \|y_n\| > n$. Setting $z_n = \frac{y_n}{\|y_n\|}$, we have

$$z_n = \lambda_n T z_n + \frac{\lambda_n}{\|y_n\|} Tx .$$

Since T is compact, we have $z_n \to z^* \neq \theta$ satisfying $z^* = \lambda^* T z^*$. This implies that $z^* \in \overset{\circ}{P}, \lambda^* > 0, \|z^*\| = 1$. Let $\phi = z^*$, and $\lambda_1 = \lambda^*$; these are what we need.

(2). (The uniqueness of the positive eigenvector.) $\forall x \notin P, \forall y \in \overset{\circ}{P}$, let us define $\delta_y(x) = \sup\{\lambda \geq 0 | y + \lambda x \in P\}$. It is easily seen that $\delta_y(x) > 0$ is a continuous function of $x \in X \backslash P$. By definition,

$$\begin{cases} \lambda \in [0, \delta_y(x)] \text{ implies } y + \lambda x \in P , \\ \lambda > \delta_y(x) \text{ implies } y + \lambda x \notin P . \end{cases}$$

Now, suppose that $\phi_i \in \overset{\circ}{P}$ satisfy $\|\phi_i\| = 1,\ \phi_i = \lambda_i T \phi_i,\ i = 1,2$. Set

$$\gamma_1 = \delta_{\phi_1}(-\phi_2), \quad \gamma_2 = \delta_{\phi_2}(-\phi_1) .$$

Then

$$\begin{cases} T(\phi_1 - \gamma_1 \phi_2) = \frac{1}{\lambda_1}(\phi_1 - \gamma_1 \frac{\lambda_1}{\lambda_2}\phi_2) , \\ T(\phi_2 - \gamma_2 \phi_1) = \frac{1}{\lambda_2}(\phi_2 - \gamma_2 \frac{\lambda_2}{\lambda_1}\phi_1) . \end{cases}$$

If $\phi_2 - \gamma_2\phi_1 \neq \theta$, then $T(\phi_2 - \gamma_2\phi_1) \in \overset{\circ}{P}$. This implies $\lambda_2 < \lambda_1$. But $T(\phi_1 - \gamma_1\phi_2) \in P$ implies $\lambda_1 \leq \lambda_2$. This is a contradiction. Therefore, $\phi_1 = \phi_2$. Again, by the normality, $\lambda_1 = \lambda_2$.

(3). (The inequality (3.52)) Assume that $\psi \in X \backslash (P \cup (-P))$ satisfying $\lambda T\psi = \psi$, and $\|\psi\| = 1$.

In the case where λ is real: From $\phi \pm \delta_\phi(\pm\psi)\psi \neq \theta$, it follows that

$$\frac{1}{\lambda_1}\left(\phi \pm \frac{\lambda_1}{\lambda}\delta_\phi(\pm\psi)\psi\right) = T(\phi \pm \delta_\phi(\pm\psi)\psi) \in \overset{\circ}{P} .$$

Therefore $\lambda_1 < |\lambda|$.

In the case where λ is not real: We write $\lambda = |\lambda| e^{i\theta}$, then $\exists x_1, x_2 \in X$, such that

$$x_1 + x_2 = \lambda(Tx_1 + iTx_2) .$$

Let $H = \text{span}\{x_1, x_2\}$, then $T|_H = \frac{1}{|\lambda|}R_\theta$, where R_θ is the following matrix:

$$\begin{pmatrix} \cos\theta & -\sin\theta \\ \sin\theta & \cos\theta \end{pmatrix}$$

We claim that $P \cap H = \{\theta\}$. For otherwise, since both P and H are invariant with respect to T, $T|_{P\cap H}$ is positive, according to the conclusion of the first paragraph, it should have a positive eigenvalue. But this contradicts the representation of R_θ.

Now, $\forall z \in H\backslash\{\theta\}$,

$$\frac{1}{\lambda_1}\left(\phi + \frac{\lambda_1}{|\lambda|}\delta_\phi(z)R_\theta z\right) = T(\phi + \delta_\phi(z)z) \in \overset{\circ}{P} .$$

Thus $\frac{\lambda_1}{|\lambda|}\delta_\phi(z) < \delta_\phi(R_\theta z)$. We choose $z_0 \in H\backslash\{\theta\}$ such that $\delta_\phi(z_0) = \sup\{\delta_\phi(z)|z \in C\}$, where $C = \{\cos\theta x_1 + \sin\theta x_2 | \theta \in [0, 2\pi]\}$, it follows that $\lambda_1 < |\lambda|$.

(4). (The algebraic simplicity of λ_1). From (1), we have shown that $\ker(\text{id} - \lambda_1 T) = \text{span}\{\phi\}$. According to the Riesz–Schauder theory, it is sufficient to show that $\ker(\text{id} - \lambda_1 T)^2 = \ker(\text{id} - \lambda_1 T)$, i.e., if $x \in X$ satisfies $x - \lambda_1 Tx = -\phi$, then $x = \theta$. To this end, it is sufficient to show that $x \in P \cap (-P)$.

In fact, if $x \notin P$, let $\gamma = \delta_\phi(-x)$, then

$$T(\phi - \gamma x) = \frac{1}{\lambda_1}\phi - \frac{\gamma}{\lambda_1}(x + \phi) .$$

It follows that

$$\frac{1}{\lambda_1}(\phi - \gamma x) = T(\phi - \gamma x) + \frac{\gamma}{\lambda_1}\phi \in \overset{\circ}{P} .$$

It contradicts the definition of γ, therefore $x \in P$. Similarly, $x \in -P$. Therefore $x \in P \cap (-P) = \theta$. □

Corollary 3.6.13 *Let L be the linear second-order elliptic operator (3.52) with the Dirichlet boundary condition, defined on a bounded open domain Ω in R^n, the boundary $\partial\Omega$ being smooth. Then the eigenvalue problem:*

$$\begin{cases} Lu = \lambda u & \text{in } \Omega \\ u = 0 & \text{on } \partial\Omega . \end{cases}$$

has a positive eigenfuction ϕ with positive, algebraically and geometrically simple eigenvalue λ_1, satisfying

$$\lambda_1 \le |\lambda| \quad \forall \lambda \in \sigma(L) .$$

Corollary 3.6.14 *([Tu]). Let (X, P) be an OBS, and let f, T, g, Σ^+ be as in Theorem 3.6.7, in which T is a strictly positive compact operator. Then (θ, λ_1) is a bifurcation point and Σ^+ contains an unbounded connected component in P emanating from (θ, λ_1).*

Also, we have the following result on the bifurcation at infinity:

Corollary 3.6.15 *Let (X, P) be an OBS. Let $T \in \mathcal{L}(X, X)$ be strictly positive and compact, and $g : P \times R^+ \to P$ be compact, satisfying $g(x, 0) = \theta, g(x, \lambda) = o(\|x\|)$, as $\|x\| \to \infty$ uniformly on each finite interval. Define*

$$f(x, \lambda) = \lambda Tx + g(x, \lambda) \ \forall (x, \lambda) \in P \times R^+ .$$

Let $\Sigma = \{(x, \lambda) \in P \times R^+ | f(x, \lambda) = x\}$ and $\Sigma^+ = \overline{\Sigma \cap (P \backslash \{\theta\} \times R^+ \backslash \{0\})}$. Then there are only two possibilities for the connected component ζ of Σ^+, passing through the point (∞, λ_1), where λ_1 is the eigenvalue of T with positive eigenvector:

(1) The projection of ζ onto R^1_+ is unbounded.
(2) $\zeta \cap (\{\theta\} \times R^1_+) \neq \emptyset$.

3.6.4 Multiple Solutions

In this subsection, we present a few examples showing that degree theory is applicable to the study of multiple solutions for certain nonlinear elliptic equations.

Example 1. Let L be the second-order elliptic operator defined in (3.52), and let $g \in C(R^1)$ satisfy $g(0) = 0, |g(t)| \leq C$, and $g(t) - \lambda t = o(t)$, as $|t| \to 0$, where C is a constant, and $\lambda \in \mathbb{R}^1$ is between the first and the second eigenvalues of L. We consider the equation:

$$\begin{cases} Lu = g(u) & \text{in } \Omega , \\ u = 0. & \text{on } \partial\Omega . \end{cases} \tag{3.54}$$

Obviously, $u = 0$ is a trivial solution. We shall show that there exists a nontrivial solution. This can be seen by the following degree computation.

Let $X = C^1_0(\overline{\Omega})$, since λ is between λ_1 and λ_2, by homotopy invariance of the index,

$$i(\text{id} - L^{-1}g, \theta) = i(\text{id} - \lambda L^{-1}, \theta) = -1 .$$

Since all solutions are bounded in X, again by the homotopy invariance of the degree,

$$\deg (\text{id} - L^{-1}g, B_R(\theta), \theta) = \deg (\text{id}, B_R(\theta), \theta) = 1 .$$

If there were no nontrivial solution, the above identities would contradict the additivity of the degree.

If one slightly changes the above assumptions then there exists a positive solution.

Example 2. Let L be as in (3.52), and let $g \in C^1(R^1)$ satisfy $g(0) = 0, 0 \leq g \leq C$, and $g(t) - \lambda t = o(t)$, as $t \to 0$ with $\lambda > \lambda_1$. Then there exists a positive solution of (3.54).

This can be shown by the sub- and super-solutions method. Obviously, $\overline{u} = L^{-1}C$ is an super-solution. Let $\underline{u} = \epsilon\phi_1$, where ϕ_1 is the positive eigenfunction of L, and $\epsilon > 0$ small. By the condition of g near zero, it is a sub-solution of L. From the Hopf maximum principle, we have $\underline{u} \leq \overline{u}$. Thus we obtain a positive solution.

We are inspired by Example 2 to compute the degree of an order-preserving compact map on an order interval.

Theorem 3.6.16 *Let (X, P) be an OBS with $\overset{\circ}{P} \neq \emptyset$, and let $[\underline{u}, \overline{u}] \subset X$ be a $\{e_1, \ldots, e_k\}$ finitely bounded order interval for some $\{e_1, \ldots, e_k\} \subset X$. If $f : [\underline{u}, \overline{u}] \to X$ is order preserving, compact and satisfying:*

$$f(\underline{u}) \in \underline{u} + \overset{\circ}{P}, f(\overline{u}) \in \overline{u} - \overset{\circ}{P} ;$$

Then $U := [\underline{u}, \overline{u}]^\circ \neq \emptyset$ and

$$\deg(id - f, U, \theta) = 1 .$$

Proof. Since f is order preserving, $\forall u \in [\underline{u}, \overline{u}],\ f(\underline{u}) \leq f(u) \leq f(\overline{u})$.

Moreover, we have $f([\underline{u}, \overline{u}]) \subset U$. In fact, $\exists \epsilon > 0$, such that

$$f(\underline{u}) + B_\epsilon(\theta) \subset \underline{u} + P, f(\overline{u}) + B_\epsilon(\theta) \subset \overline{u} - P .$$

Therefore

$$f(u) + B_\epsilon(\theta) \in (\underline{u} + P) \cap (\overline{u} - P) ,$$

i.e., $f(u) \in U$. This implies that U is a nonempty open convex set.

Choosing arbitrarily $x \in U, \forall t \in [0, 1]$, define

$$\phi(u, t) = t(u - f(u)) + (1 - t)(u - x) .$$

Since $f(\partial U) \subset U$, $tf(u) + (1 - t)x \in U\ \forall (u, t) \in \partial U \times [0, 1]$. Thus $\theta \notin \phi(\partial U \times [0, 1])$. From the homotopy invariance of the degree,

$$\deg(\text{id} - f, U, \theta) = \deg(\phi(\cdot, 1), U, \theta) = \deg(\phi(\cdot, 0), U, \theta) = \deg(\text{id}, U, x) = 1 .$$

□

The above theorem is due to Amann [Am 3].

Corollary 3.6.17 *(Amann) Let (X, P) be an OBS with $\overset{\circ}{P} \neq \emptyset$, and let $\underline{u_i}, \overline{u_i}$, $i = 1, 2$, satisfy $\underline{u_1} \leq \overline{u_1} \leq \underline{u_2}, \underline{u_1} \leq \underline{u_2} \leq \overline{u_2}$; and $[\underline{u_1}, \overline{u_1}] \cap [\underline{u_2}, \overline{u_2}] = \emptyset$. If $[\underline{u_1}, \overline{u_2}]$ is $\{e_1, \ldots, e_k\}$- finitely bounded for some $\{e_1, \ldots, e_k\} \subset X$, and if $f : [\underline{u_1}, \overline{u_2}] \to X$ is an order-preserving compact map satisfying:*

$$f(\underline{u_i}) \in \underline{u_i} + \overset{\circ}{P}, f(\overline{u_i}) \in \overline{u_i} - \overset{\circ}{P},\ i = 1, 2 .$$

Then there are at least three distinct fixed points of f on the order interval $[\underline{u_1}, \overline{u_2}]$.

Proof. Set $U_i = [\underline{u_i}, \overline{u_i}])^\circ$, $i = 1, 2$, and $U = [\underline{u_1}, \overline{u_2}]^\circ$. From the additivity,

$$\begin{aligned}\deg(\mathrm{id} - f, U, \theta) &= \deg(\mathrm{id} - f, U_1, \theta) + \deg(\mathrm{id} - f, U_2, \theta) \\ &+ \deg(\mathrm{id} - f, U \backslash \overline{(U_1 \cup U_2)}, \theta) .\end{aligned}$$

Applying the above theorem, we have

$$\deg(\mathrm{id} - f, U \backslash \overline{(U_1 \cup U_2)}, \theta) = -1 .$$

According to the Kronecker existence property, $\exists$ fixed points u_i, $i = 1, 2, 3$, such that $u_i \in U_i$, $i = 1, 2$. and $u_3 \in U \backslash \overline{(U_1 \cup U_2)}$. □

In the applications to PDEs, let $\Omega \subset R^n$ be a bounded domain with smooth boundary. We take $X = C_0^1(\overline{\Omega})$, and $e \in X$ is a positive function in Ω and satisfies $\partial_n e(x) < 0$, $\forall x \in \partial\Omega$, where ∂_n is the outer normal derivative. Then it is easy to verify that any order interval $[u, v]$ is $e-$ finitely bounded.

Example 3. Let L be the operator defined in (3.52), and let $g \in C^1(R, \mathbb{R}^1_+)$ be nondecreasing and satisfying $g(0) = 0, g'(0) = 1, t < g(t) \forall t \in (0, t_0]$ for some small $t_0 > 0$, and $g(t) \leq C$ as $t > 0$, where $C > 0$ is a constant. Then $\exists \lambda^* \in (0, \lambda_1)$ such that the equation:

$$\begin{cases} Lu = \lambda g(u), & \text{in } \Omega \\ u = 0, & \text{on } \partial\Omega , \end{cases} \tag{3.55}$$

possesses at least one positive solution $\forall \lambda \geq \lambda^*$ and at least two distinct positive solutions $\forall \lambda \in (\lambda^*, \lambda_1)$.

Indeed, we extend g to R^1 oddly; the mapping $f(u) = L^{-1} g(u)$ is order preserving. $\forall \lambda > 0$, it has a pair of strict sub- and super-solutions $\{-\lambda C e, +\lambda C e\}$, where $e = L^{-1} 1$. Thus, $v_\lambda := \lim_{n \to \infty} (\lambda f)^n (\lambda C)$ exists and is the maximal solution of (3.55). Obviously, v_λ is upper semi-continuous and monotone nondecreasing in λ. Define $\lambda^* = \inf\{\lambda \in R^1_+ | v_\lambda > 0\}$. We claim that $\lambda^* < \lambda_1$, where λ_1 is the first eigenvalue of L. In fact, near the bifurcation point (θ, λ_1) we find $u_\lambda > 0$, a solution of (3.55) with small $\|u_\lambda\|_C < t_0$. One has

$$\int_\Omega L u_\lambda \phi_1 dx = \lambda \int_\Omega g(u_\lambda) \phi_1 dx ,$$

where $\phi_1 > 0$ is the first normalized eigenfunction of L^*. Then

$$\lambda_1 \int_\Omega u_\lambda \phi_1 dx > \lambda \int u_\lambda \phi_1 dx .$$

It follows that $\lambda < \lambda_1$, Since $v_\lambda \geq u_\lambda > 0$, then $\lambda^* < \lambda_1$.

For $\forall \lambda \in (\lambda^*, \lambda_1)$ $\exists \delta > 0$, such that $\lambda g(t) < \lambda_1 t$, as $|t| < \delta$. Let ϕ_1 be the normalized first eigenfunction; as $\epsilon \in (0, \delta)$, $\pm\epsilon\phi_1$ is a pair of sub- and super-solutions of (3.55). However, as $\lambda' = \frac{\lambda^* + \lambda}{2}$, $v_{\lambda'}$ is a sub-solution of (3.55). Set $\epsilon > 0$ small enough such that $\epsilon\phi_1 < v_{\lambda'}$. Then we find two pairs of sub- and super solutions: $[-\epsilon\phi_1, +\epsilon\phi_1], [v_{\lambda'}, \lambda C e]$. According to the Amann theorem, there must be two distinct nontrivial solutions of (3.55): $u_\lambda^1 \in [-\epsilon\phi_1, \lambda C e] \backslash ([-\epsilon\phi_1, \epsilon\phi_1] \cup [v_{\lambda'}, \lambda C e])$, and $u_\lambda^2 \in [v_{\lambda'}, \lambda C e]$. Since $\epsilon > 0$ can be arbitrarily small, both of u_λ^1, u_λ^2 are positive. This is our conclusion.

3.6.5 A Free Boundary Problem

We turn to an application of the bifurcation at infinity to a free boundary problem for the flux equation in the confined plasma. Let $\Omega \subset R^2$ be the section of the container, a bounded domain with smooth boundary rotated along and off the z axis in R^3, which contains plasma confined by an external magnetic field. Let u be the flux function of the magnetic field. The current I is a given positive constant. The following equation is derived from the Maxwell equation:

$$\begin{cases} Lu := -[\frac{\partial}{\partial r}(\frac{1}{r}\frac{\partial}{\partial r}) + \frac{1}{r}\frac{\partial^2}{\partial z}]u = \lambda u_+ \text{ in } \Omega, \\ \qquad u = C, \text{ on } \partial\Omega, \\ \qquad -\int_{\partial\Omega} \frac{1}{r}\frac{\partial u}{\partial n} = I, \end{cases} \tag{3.56}$$

where C is a constant to be determined, and $u_+ = \max\{u, 0\}$.

The domain $\Omega_p := \{x \in \Omega \,|\, u(x) > 0\}$, which is the space occupied by the plasma, is unknown. It will be solved simultaneously with the flux function u. Physically, we are interested in the case where $\Omega_p \subset \overline{\Omega_p} \subset \Omega$, and $\Omega_p \neq \emptyset$.

From Green's formula,

$$\lambda \int_\Omega u_+ = \int_\Omega Lu = -\int_{\partial\Omega} \frac{1}{r}\frac{\partial u}{\partial n} = I\,.$$

We see that the problem is solvable only if $\lambda > 0$.

Again, from Green's formula,

$$\lambda \int_\Omega u_+\phi_1 = \lambda_1 \int_\Omega u\phi_1 + C\int_{\partial\Omega} \frac{1}{r}\frac{\partial\phi_1}{\partial n}\,,$$

where ϕ_1 is the first eigenfunction of L with respect to Dirichlet boundary condition.

Combining with the maximum principle it is seen that $C > 0 \Rightarrow \lambda < \lambda_1$, and $C = 0 \Rightarrow \lambda = \lambda_1$. In all these cases, the solution $u(x) > 0\,\forall\, x \in \Omega$, from the maximum principle. i.e., $\Omega_p = \Omega$. They are not of interest in physics.

We reduce the problem (3.56) to the following: Given a constant $C_1 < 0$, find (v, λ) satisfying

$$\begin{cases} Lv = \lambda v_+ \text{ in } \Omega\,, \\ \;v = C_1 \text{ on } \partial\Omega\,, \\ \quad\lambda > 0\,. \end{cases} \tag{3.57}$$

Indeed, if (v, λ) is a solution of (3.57), let $v = \alpha u, C_1 = \alpha C$, where $\alpha > 0$ is an adjustment constant satisfying

$$\int_{\partial\Omega} \frac{1}{r}\frac{\partial u}{\partial n} = -I\,.$$

Then (u, λ) is the solution of equation (3.56). We have the following conclusion:

Statement 3.6.18 $\forall I > 0, \exists u \in C^{2,\gamma}(\overline{\Omega})$ $\gamma \in (0,1)$ with $C < 0$, which solves equation (3.56) if and only if $\lambda > \lambda_1$.

Proof. After the above discussion, we just take $C_1 = -1$. Let $w = v + 1$, then equation (3.57) is reduced to

$$\begin{cases} Lw = \lambda(w-1)_+ \text{ in } \Omega\,, \\ w = 0 \text{ on } \partial\Omega\,, \\ \lambda > 0\,. \end{cases} \tag{3.58}$$

Let us take $X = C(\overline{\Omega})$, $F : w \mapsto (w-1)_+$, $K = L^{-1}$, our problem is reduced to finding the solution of $w = \lambda K \circ F(w)$. It is easily seen that there is no bifurcation point located at $\{\theta\} \times R^1_+$. According to Corollary 3.6.15, the projection of the connected component ζ of the positive solution set passing through the bifurcation point at infinity (∞, λ_1) onto R^1_+ is unbounded. It must include the interval $(\lambda_1, +\infty)$. Therefore, $\forall \lambda \in (\lambda_1, +\infty)$, there is a solution w_λ on ζ.

On the other hand, if there is a nontrivial solution w_λ of equation (3.58), then $w_\lambda \geq 0$ and $F(w_\lambda) < w_\lambda$. Thus

$$\begin{aligned} \int_\Omega w_\lambda \phi_1 dx &= \frac{1}{\lambda_1} \int_\Omega w_\lambda L\phi_1 dx \\ &= \frac{1}{\lambda_1} \int_\Omega L w_\lambda \phi_1 dx \\ &= \frac{\lambda}{\lambda_1} \int_\Omega F(w_\lambda)\phi_1 dx \\ &< \frac{\lambda}{\lambda_1} \int_\Omega w_\lambda \phi_1 dx\,. \end{aligned}$$

This proves $\lambda > \lambda_1$. The proof of the regularity is standard. □

3.6.6 Bridging

We present an example to express the fact that solutions of a differential equation on disjoint domains can be glued up as a new solution in a larger domain. A typical example we consider here is a dumbbell $D \subset R^2$, i.e., two disks $B_\pm = B_{1/2}((\pm 1, 0))$ connected by a segment $E = \{(x,y) \in R^2 \,|\, y = 0, x \in [-1/2, 1/2]\}$. Let $B = B_+ \cup B_-$, and $D = B \cup E$. Assuming that $f \in C^1$ is of power growth with $f(0) = 0$, and that $u_\pm \in H^1_0(B_\pm)$ are solutions of the equation:

$$-\triangle u = f(u)\,. \tag{3.59}$$

We consider a sequence of domains D_n squeezing to D, and want to find a solution $u_n \in H^1_0(D_n)$ of (3.59) such that u_n is closed to $u_\pm$ in certain sense.

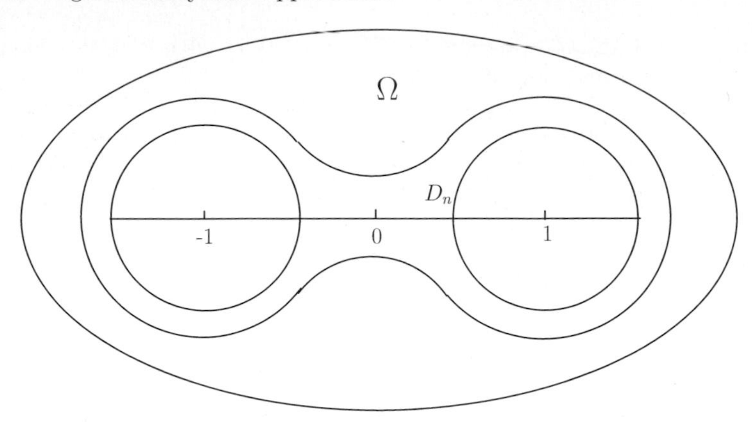

Fig. 3.5.

Let Ω be a bounded domain containing D_n, and $D \subset D_n$. Assuming that $U \subseteq \Omega$ is a bounded domain, the composition operator $u \mapsto f(u)$ is C^1 from $H^1(U)$ to $L^q(U)$, $\forall q \in (1, \infty)$. Let $r : H^1(\Omega) \to H^1(U)$ be the restriction $u \to u|_U$, and $e : H_0^1(U) \to H_0^1(\Omega)$ be the extension $(eu)(x) = u(x)$ if $x \in \overline{U}$ and 0 elsewhere. Let $K = (-\Delta)^{-1} : L^q(U) \to H_0^1(U), 1 \leq q < \infty$, then K is linear and compact. We define $T = e \cdot K \cdot f \cdot r : H_0^1(\Omega) \to H_0^1(\Omega)$. The following statements are easy to verify:

(1) $\forall u \in H_0^1(U)$, we have $(r \cdot e)u = u$.
(2) $\forall v \in H_0^1(\Omega)$, if $rv \in H_0^1(U)$, then $(e \cdot r)v = v$.
(3) For $u_0 \in H_0^1(U), u_0$ is a solution of (3.59) if and only if $e \cdot u_0$ is a fixed point of T.
(4) If u_0 is nondegenerate (i.e., $\ker -\Delta - f'(u_0)I) = \{\theta\}$), then $I - T'(u_0)$ is invertible on $H_0^1(\Omega)$.

Let us choose $X = H_0^1(\Omega)$. Let T_n, T_B and e_n, e_B be the operators and the extensions on $D_n\, \forall n \in \mathcal{N}$ and $B = B_+ \cup B_-$, respectively.

Statement 3.6.19 For any nondegenerate fixed point $e_B u_0$ of T_B, $u_0 \in H_0^1(B)$, there exists a fixed point $e_n u_n$ of T_n for n large, such that $\|e_n u_n - e_0 u_0\|_X \to 0$ as $n \to \infty$.

Proof. By the nondegeneracy assumption, $e_B u_0$ is isolated and then $i(I - T, e_B u_0) = \pm 1$. In order to show the existence of a fixed point $e_n u_n$ of $T_n, u_n \in H_0^1(D_n)$, it is sufficient to verify that for all small $\delta > 0, I - T_n$ is homotopic to $I - T_B$ on ∂V, where $V = B_\delta(e_B u_0)$ is a ball in $H_0^1(\Omega)$. If it is so, then

$$\deg(I - T_n, V, \theta) = \deg(I - T_B, V, \theta) \neq 0 \,.$$

Our conclusion follows.

Now, we verify it by contradiction. Suppose that $\exists t_n \in [0,1], \exists u_n \in H_0^1(D_n), e_n u_n \in \partial V$, such that $e_n u_n = t_n T_B e_n u_n + (1 - t_n) T_n e_n u_n$. Since

$e_n u_n$ is bounded, and T_n, T_B are compact, there exists a subsequence u_{n_j} such that $e_{n_j} u_{n_j}$ converges to some v and then $v \in \partial V$.

We verify that v is a solution of (3.59), and that $v(x) = 0$, a.e., on $\Omega \backslash B$. Since V can be chosen arbitrarily close to $e_B u_0$, it contradicts the nondegeneracy of u_0.

(1) $\forall \phi \in H_0^1(B)$, from

$$\int_B \nabla u_n \nabla \phi = \int_B f(u_n)\phi\,,$$

it follows that

$$\int_B \nabla v \nabla \phi = \int_B f(v)\phi\,;$$

i.e., v is a solution (3.59) on B.

(2) From $e_n u_n = 0$ a.e., on $\Omega \backslash D_n$, we have $v = 0$ a.e., on $\Omega \backslash D$. Since $|E| = 0, v = 0$, a.e., on $\Omega \backslash B$. Therefore, $v \in H_0^1(\Omega)$ is in fact a solution of (3.59). □

In other words, u_+ and u_- are nondegenerate solutions of (3.59) on $H_0^1(B_+)$ and $H_0^1(B_-)$, respectively. We define $u_0(x) = u_\pm(x)$ as $x \in B_\pm$, then $e_B u_0$ is a degenerate fixed point of T_B. By Statement 3.6.19 there are a sequence of domains D_n squeezing to D and a sequence of solutions $u_n \in H_0^1(D_n)$ of the equation (3.59) such that u_n is close to u_0.

3.7 Extensions

We briefly introduce a few directions in generalizing the Leray-Schauder degree.

3.7.1 Set-Valued Mappings

We have used the fixed-point theorem for set-valued mappings in Chap. 2. It is natural to ask if the degree theory for set-valued mappings can be set up.

Let X be a Banach space, let $\Gamma(X)$ be the set of all nonempty closed convex subsets of X.

Definition 3.7.1 *(Compact convex set-valued mapping) Let $\Omega \subset X$ be a bounded subset, and let $\phi : \Omega \to \Gamma(X)$. It is called a compact convex set-valued mapping, if (1) it is upper semi-continuous, and (2) $\overline{\phi(\overline{\Omega})}$ is compact.*

Obviously, any compact single-valued mapping is a compact convex set-valued mapping.

The main idea in extending the Leray–Schauder degree to compact convex set-valued mappings is to approximate the set-valued mapping by single-valued mappings. The following notion is useful:

Definition 3.7.2 *(Single-valued approximation) Let (M, d) be a metric space, and let Y be a Banach space. Let $T : M \to 2^Y$ and $\tilde{T} : M \to Y$. We say that $\tilde{T}$ is a $\epsilon-$ single-valued approximation of T for $\epsilon > 0$, if*

1. *$\tilde{T}(M) \subset \overline{\text{conv}}T(M)$,*
2. *$\forall x \in M$, $\exists y \in M$ and $z \in T(y)$ satisfying $d(x, y) < \epsilon$, and $\|\tilde{T}(x) - z\| < \epsilon$.*

The following picture shows the $\epsilon-$ approximation of a set-valued mapping.

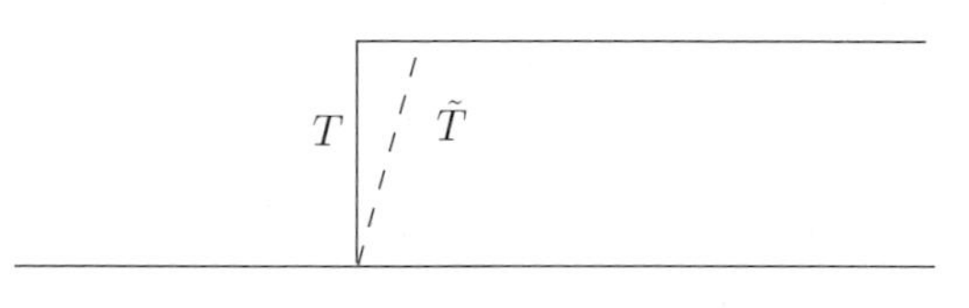

Fig. 3.6.

Theorem 3.7.3 *Let (M, d) be a metric space, X be a Banach space, and let $\phi : M \to \Gamma(X)$ be upper semi-continuous. Then $\forall \epsilon > 0$ there exists an $\epsilon-$ approximation ϕ_ϵ of ϕ.*

Proof. By the upper semi-continuity, $\forall x \in M, \exists \delta_x > 0$, such that $\phi(B_{\delta_x}(x)) \subset \phi(x) + B_\epsilon(\theta)$. One may choose $\delta_x < \frac{\epsilon}{2}$. Then $\mathrm{A} = \{B_{\delta_x}(x) \mid x \in M\}$ is an open covering of M. It has an open locally finite star-refinement $\mathrm{B} = \{V_\beta \mid \beta \in \Lambda\}$ (see Dugundji [Du 2]). The so-called star-refinement means that $\forall \beta \in \Lambda, \exists x_\beta \in M$ such that $St(V_\beta, \mathrm{B}) \subset B_{\delta_{x_\beta}}(x_\beta)$, where $St(V_\beta, \mathrm{B}) = \bigcup\{V_\alpha \in \mathrm{B} \mid V_\alpha \bigcap V_\beta \neq \varnothing\}$. Define a partition of unity with respect to $\mathrm{B} : \{\lambda_\beta \mid \beta \in \Lambda\}$, and

$$\phi_\epsilon(x) = \Sigma_{\beta\in\Lambda}\lambda_\beta(x)z_\beta ,$$

where $z_\beta \in \phi(V_\beta)$. We are going to verify that ϕ_ϵ is an $\epsilon-$ approximation of ϕ. Indeed, $\forall x_0 \in M$, if $x_0 \in V_\beta$, then there must be $y \in M$ and $\delta_y < \frac{\epsilon}{2}$ such that $V_\beta \subset B_{\delta_y}(y)$. Thus, $d(x_0, y) < 2\delta_y < \epsilon$. However, $z_\beta \in \phi(V_\beta) \subset \phi(B_{\delta_y}(y)) \subset \phi(y) + B_\epsilon(\theta)$, we have $\phi_\epsilon(x_0) = \Sigma_{i=1}^n \lambda_{\beta_i}(x_0)z_{\beta_i} \in \phi(y) + B_\epsilon(\theta)$, provided by the convexity of $\phi(y)$, and the fact $\{V_{\beta_i}\}_{i=1}^n = \{V_\beta \mid x_0 \in V_\beta, \beta \in \Lambda\}$. We may choose $z \in \phi(y)$ such that $\|\phi_\epsilon(x_0) - z\| < \epsilon$. Obviously, $\phi_\epsilon(M) \subset \overline{\text{conv}}\phi(M)$. □

What is the homotopy equivalence for compact convex set-valued mappings?

Definition 3.7.4 *Let Ω be a bounded open set in a Banach space X, and let $\phi_1, \phi_2 : \bar{\Omega} \to \Gamma(X)$ be two compact convex set-valued mappings. We say that ϕ_1 is homotopically equivalent to ϕ_2, denoted by $\phi_1 \simeq \phi_2$, if there exists a family of compact convex set-valued mappings $\Phi : [0, 1] \times \bar{\Omega} \to \Gamma(X)$, such that $\theta \notin F([0, 1] \times \partial\Omega)$, and $\Phi(i, \cdot) = \phi_i(\cdot)$, where $F(t, x) = x - \Phi(t, x)$.*

Lemma 3.7.5 *Suppose that $\phi : \bar{\Omega} \to \Gamma(X)$ is a compact convex set-valued mapping and $f = id - \phi$. If $\theta \notin f(\partial\Omega)$, then $\exists \epsilon_0 > 0$ such that $\forall \epsilon \in (0, \epsilon_0)$, for any $\epsilon-$ single-valued approximation $\tilde{\phi}$ of ϕ, we have $\theta \notin F([0,1] \times \partial\Omega)$, where $F(t,x) = x - (1-t)\tilde{\phi}(x) - t\phi(x)$.*

Proof. We prove the lemma by contradiction. If there exist single-valued continuous mappings $\tilde{\phi}_n$, and $t_n \in [0,1], x_n \in \partial\Omega, y_n \in \Omega, z_n \in \phi(y_n)$ such that:

$$x_n \in (1-t_n)\tilde{\phi}_n(x_n) + t_n\phi(x_n) ,$$

$$\|x_n - y_n\| < \frac{1}{n} ,$$

$$\|\tilde{\phi}_n(x_n) - z_n\| < \frac{1}{2n} .$$

Since $\overline{\phi(\overline{\Omega})}$ is compact, and $\tilde{\phi}_n(\Omega) \subset \overline{\text{conv}}\phi(\overline{\Omega})$, after a subsequence, we have $x_n \to x^* \in \partial\Omega$, $t_n \to t^*$, and then $y_n \to x^*$. Since ϕ is u.s.c., for n large, we have $\phi(x_n) \subset \phi(x^*) + B_{\frac{1}{n}}(\theta), \phi(y_n) \subset \phi(x^*) + B_{\frac{1}{2n}}(\theta)$. It follows, $\tilde{\phi}_n(x_n) \in \phi(x^*) + B_{\frac{1}{n}}(\theta)$, then

$$(1-t_n)\tilde{\phi}_n(x_n) + t_n\phi(x_n) \subset \phi(x^*) + B_{\frac{1}{n}}(\theta) ,$$

and then $x_n \in \phi(x^*) + B_{\frac{1}{n}}(\theta)$ as n large. Thus $x^* \in \phi(x^*)$. This is a contradiction.

□

Lemma 3.7.6 *If ϕ_1, ϕ_2 are two homotopically equivalent compact convex set-valued mappings, then there exists $\epsilon > 0$ such that for any $\epsilon-$ approximation $\tilde{\phi}_1, \tilde{\phi}_2$ of ϕ_1 and ϕ_2, respectively, we have $\tilde{\phi}_1 \simeq \tilde{\phi}_2$.*

Proof. According to Lemma 3.7.5, we have $\tilde{\phi}_0 \simeq \phi_0 \simeq \phi_1 \simeq \tilde{\phi}_1$ in the sense of set-valued mappings. We shall prove that $\tilde{\phi}_0 \simeq \tilde{\phi}_1$ is also true in single-valued sense.

In fact there exists a compact convex set-valued mapping $h : [0,1] \times \bar{\Omega} \to \Gamma(X)$ satisfying $h(i,x) = \tilde{\phi}_i(x), i = 0,1$ and $\epsilon := \text{dist}(H([0,1] \times \partial\Omega), \theta) > 0$, where $H = \text{id} - h$. By Lemma 3.7.3, and Lemma 3.7.5, there exists a $\frac{\epsilon}{2}-$ approximation $\tilde{h} : [0,1] \times \bar{\Omega} \to X$ of h such that $h(\mu, \cdot) \simeq \tilde{h}(\mu, \cdot)$. More precisely, let

$$p(\lambda, \mu; x) = (1-\lambda)\tilde{h}(\mu, x) + \lambda h(\mu, x) ,$$

we have $\theta \notin P([0,1] \times [0,1] \times \partial\Omega)$, where $P(\lambda, \mu; x) = x - p(\lambda, \mu; x)$.

Define a family of single-valued mappings:

$$g(t,x) = \begin{cases} p(1-3t, 0; x) & \text{if } 0 \le t \le \frac{1}{3} \\ p(0, 3t-1; x) & \text{if } \frac{1}{3} \le t \le \frac{2}{3} \\ p(3t-2, 1; x) & \text{if } \frac{2}{3} \le t \le 1 . \end{cases}$$

It is easy to verify that this is the required homotopy. □

Now, we are ready to define the degree of id $-\ \phi$, where ϕ is a compact convex set-valued mapping.

Definition 3.7.7 *Let Ω be an open bounded set in a Banach space X, let $\phi : \bar{\Omega} \to \Gamma(X)$ be a compact convex-valued mapping, and $f = id - \phi$. If $\theta \notin f(\partial\Omega)$, define*

$$\deg(f, \Omega, \theta) = \lim_{\epsilon \to 0} \deg(\tilde{f}_\epsilon, \Omega, \theta) ,$$

where $\tilde{f}_\epsilon = id - \tilde{\phi}_\epsilon$ and $\tilde{\phi}_\epsilon$ is an $\epsilon-$ approximation of ϕ.

According to Lemmas 3.7.5 and 3.7.6, the degree is well defined.

Readers can easily verify that the degree enjoys the same basic properties as the Leray–Schauder degree: Homotopy invariance, additivity, translation invariance and normality. So do the Kronecker existence and the excision.

The results of Examples 1–3 in Sect. 3.6 can be extended to case where g has jump discontinuity by the use of the degree of set-valued mappings, see K. C. Chang [Ch 1], [Ch 2].

3.7.2 Strict Set Contraction Mappings and Condensing Mappings

The motivation of introducing the strict set contraction mapping is to extend the notion of compactness and then to extend the L-S degree.

Let X be a Banach space, and let $A \subset X$ be a bounded subset. $\forall \epsilon > 0$, we consider a $\epsilon-$ net N_ϵ of A. Let

$$\alpha(A) = \inf\{\epsilon > 0 \mid \exists \text{a finite}\, \epsilon - \text{net}\, N_\epsilon \,\text{of}\, A\} .$$

By definition, for $\alpha(A) = 0$ if and only if $\bar{A}$ is compact. It is easy to verify the following simple properties of α:

1. If $A \subset B$, then $\alpha(A) \leq \alpha(B)$.
2. $\alpha(A \cup B) \leq \alpha(A) + \alpha(B)$.
3. $\alpha(\bar{A}) = \alpha(A)$.
4. $\alpha(A + B) \leq \alpha(A) + \alpha(B),\ \ \alpha(\lambda A) = |\lambda|\alpha(A),\ \forall \lambda \in C$.
5. For a sequence of bounded closed nonempty subsets $\cdots \subset A_n \subset \cdots \subset A_2 \subset A_1$ satisfying $\alpha(A_n) \to 0$, we have $A = \cap_{n=1}^{\infty} A_n \neq \varnothing$, and $\alpha(A) = 0$.
6. **(Mazur)** $\alpha(\overline{\text{conv}}(A)) = \alpha(A)$.

Definition 3.7.8 *(k-set contraction map) Let X be a Banach space, $\Omega \subset X$ be a bounded subset. A continuous map $\phi : \Omega \to X$ is called a $k-$ set contraction mapping, $k \geq 0$, if*

$$\alpha(\phi(A)) \leq k\alpha(A) \ \ \forall\, \textit{bounded}\ A \subset \Omega .$$

Remark 3.7.9 We assign $\infty \cdot 0 = 0$, then continuous mappings can be considered as ∞ – set contraction mappings.

Example 1. If $\phi : \Omega \to X$ is compact, the ϕ is a $0-$ set contraction.

Example 2. If $\phi : \Omega \to X$ is a Lipschitz mapping with Lipschitz constant L, then ϕ is an $L-$ set contraction.

Definition 3.7.10 *(Strict set contraction mapping and condensing mapping) A $k-$ set contraction mapping ϕ is called strict if $k \in [0,1)$. It is called a condensing mapping if*

$$\alpha(\phi(A)) < \alpha(A) \ \ \forall \text{ bounded } A\,.$$

According to (4): If $\phi, \psi : \Omega \to X$ are k_1- and k_2- set contraction mappings respectively, then $\phi + \psi$ is a $k_1 + k_2-$ set contraction. In particular, if ϕ is a contraction mapping (i.e., a Lipschitz mapping with Lipschitz constant $L < 1$), and if ψ is compact, then $\phi + \psi$ is a strict set contraction mapping.

By the definition, if $\phi : \Omega \to X$ is a k_1- set contraction, and $\psi : \phi(\Omega) \to X$ is a k_2- set contraction. (In case, $k_1 = \infty$, we assume further ϕ maps bounded set to bounded set); then $\psi \circ \phi$ is a $k_1 k_2-$ set contraction.

First we are going to define the degree for strict set contraction mappings. We intend to reduce it to compact mappings. Let $f = \mathrm{id} - \phi$, where ϕ is a $k-$ contraction mapping, where $0 \leq k < 1$.

If $\theta \notin f(\bar{\Omega})$, then we define $\deg(f, \Omega, \theta) = 0$.

Otherwise the fixed-point set of ϕ is not empty. Define $A_1 = \overline{\mathrm{conv}}\phi(\bar{\Omega})$, and $A_{k+1} = \overline{\mathrm{conv}}\phi(\bar{\Omega} \cap A_k), k = 1, 2, \ldots$. Since the fixed point set of ϕ is not empty, $A := \bigcap_{n=1}^{\infty} A_n \neq \varnothing$. However, we have

$$\begin{aligned} \alpha(A_{n+1}) &= \alpha(\overline{\mathrm{conv}}\phi(A_n \cap \bar{\Omega})) \\ &\leq k\alpha(A_n) \\ &\leq \cdots \\ &\leq k^n \alpha(A_1) \to 0\,. \end{aligned}$$

Therefore $\alpha(A) = 0$, A is a compact convex set, and then $\phi : \bar{\Omega} \cap A \to A$.

According to the Dugundji extension theorem, there exists a continuous $\tilde{\phi} : \bar{\Omega} \to A$ such that $\tilde{\phi}|_{\bar{\Omega} \cap A} = \phi|_{\bar{\Omega} \cap A}$. Then $\tilde{\phi}$ is compact and shares the same fixed point set with ϕ.

We claim that $\theta \notin (\mathrm{id} - \phi)(\partial\Omega) \Rightarrow \theta \notin (\mathrm{id} - \tilde{\phi})(\partial\Omega)$. In fact, if $x \in \partial\Omega$ satisfies $x = \tilde{\phi}(x)$, then $x \in \partial\Omega \cap A$, so is $\tilde{\phi}(x) = \phi(x)$ and then $\theta \in (\mathrm{id} - \phi)(\partial\Omega)$.

With the above notations, we define the degree for strict set contraction mappings.

Definition 3.7.11 *Let ϕ be a strict set contraction mapping; we define*

$$\deg(\mathrm{id} - \phi, \Omega, \theta) = \deg(\mathrm{id} - \tilde{\phi}, \Omega, \theta)\,.$$

To verify that the degree is well defined, we shall prove that the definition does not depend on the special choice of $\tilde{\phi}$, i.e., if $\tilde{\phi}_1, \tilde{\phi}_2$ are two such extensions: $\tilde{\phi}_i : \bar{\Omega} \to A, \tilde{\phi}_i|_{\bar{\Omega} \cap A} = \phi|_{\bar{\Omega} \cap A}, i = 1, 2$, then we define $F(t, x) = x - [(1-t)\tilde{\phi}_1(x) + t\tilde{\phi}_2(x)]$. From $\theta \notin (\mathrm{id} - \phi)(\partial\Omega)$, it follows that

$\theta \notin (\mathrm{id}-F)([0,1]\times\partial\Omega)$, i.e., $\tilde{\phi}_1 \simeq \tilde{\phi}_2$. According to the homotopy invariance of the Leray–Schauder degree, we have $\deg(\mathrm{id}-\tilde{\phi}_1,\Omega,\theta) = \deg(\mathrm{id}-\tilde{\phi}_2,\Omega,\theta)$.

It is easy to verify that the degree enjoys the homotopy invariance, additivity, translation invariance and normality. They are left to readers as exercises.

Accordingly, this enables us to apply the degree theory to a map, which can be decomposed into a summation of a contraction mapping and a compact map.

Finally, we extend the degree to condensing mappings. If ϕ is a condensing mapping, $\phi(\bar{\Omega})$ is bounded, say it is included in the ball centered at θ with radius $R > 0$. For any $\epsilon > 0$, setting $\lambda \in (1-\frac{\epsilon}{R},1)$, and $\phi_\lambda = \lambda\phi$, we have $\|\phi(x)-\phi_\lambda(x)\| \le \epsilon$, and

$$\alpha(\phi_\lambda(A)) \le \lambda\alpha(\phi(A)) \le \lambda\alpha(A)\ \forall\,\text{bounded } A\,,$$

i.e., ϕ_λ is a strict set contraction mapping which is closed to ϕ. We define

$$\deg(\mathrm{id}-\phi,\Omega,\theta) = \deg(\mathrm{id}-\phi_\lambda,\Omega,\theta)\,.$$

Again, it is easy to verify that the degree is also well defined and enjoys all basic properties of the Leray–Schauder degree. Again, this are left to readers.

3.7.3 Fredholm Mappings

We know that the Leray–Schauder degree can be applied to quasilinear elliptic equations: Find $u \in C^{2,\gamma}\cap C_0(\bar{\Omega}), \gamma \in (0,1)$ satisfying

$$\sum_{i,j=1}^{n} a_{ij}(x,u,\nabla u)\frac{\partial^2 u}{\partial x_i \partial x_j} + f(x,u,\nabla u) = 0, \text{ in } \Omega\,. \tag{3.60}$$

It is executed as follows. Define a mapping $K: C_0^1(\bar{\Omega}) \to C_0^1(\bar{\Omega})$ such that $\forall u \in C_0^1(\bar{\Omega})$, $v = Ku \in C^{2,\gamma}\cap C_0(\bar{\Omega})$ satisfies the following linear equation:

$$\sum_{i,j}^{n} a_{ij}(x,u,\nabla u)\frac{\partial^2 v}{\partial x_i \partial x_j} + f(x,u,\nabla u) = 0 \text{ in } \Omega\,.$$

Thus the fixed points of K are the solutions of equation (3.60).

However, if we consider a general elliptic equation:

$$A(u) := A(x,u,\nabla u,\nabla^2 u) = g(x) \text{ in } \Omega\,, \tag{3.61}$$

where g is a given function and the quadratic form is positive definite:

$$\sum_{i,j=1}^{n} \frac{\partial A}{\partial u_{x_i x_j}}\xi_i\xi_j \ge \alpha|\xi|^2,\ \ \alpha > 0\,,$$

then it seems that the Leray–Schauder degree argument is not applicable, because we do not know how to recast it as a fixed-point problem for compact mappings. However, the linearization of $A(u)$ is a linear second-order elliptic operator, so is a Fredholm operator with index zero. People have made great efforts in defining an integer-valued degree theory for C^1 Fredholm mappings of index 0 between Banach manifolds. Among them we should mention Caccioppoli [Cac 2], Smale [Sm 3](for $\mathbf{Z}_2$ valued), Elworthy and Tromba [ElT 1],[ElT 2], Borisovich, Zvyagin, and Sapronov [BZS] (for C^2 mappings and $\mathbf{Z}$ valued), and Fitzpatrick, Pejsachowicz, and Rabier [FP],[FPR] and [PR 1] [PR 2] etc. We are satisfied to introduce the idea of the definition; details are to be found in the above-mentioned references.

The main difficulty in extending the Leray–Schauder degree theory to Fredholm mappings lies in the fact discovered by Kuiper [Kui] that the general linear group of infinite-dimensional separable Hilbert space is connected and even contractible. A new ingredient has to be introduced to define the orientation of Fredholm mappings. Following [FPR], one defines the parity of curves of Fredholm operators. Let us denote by $\mathcal{K}(X)$ the space of all compact linear operators, and by $\mathbf{\Phi}_0(X,Y),(GL(X,Y))$ the space of all Fredholm operators of index 0 (and isomorphisms resp.) between Banach spaces X and Y.

For a curve of linear compact operators $K \in C([0,1],\mathcal{K}(X))$, if $I-K(i), i = 0,1$ are invertible, then we define the parity by

$$\sigma(I-K) = i_{LS}(I-K(0),\theta)i_{LS}(I-K(1),\theta) ,$$

where $i_{LS}(I-K(i),\theta), i=0,1$, are the Leray–Schauder indices.

The notion of parity is extended to curves of Fredholm mappings with index 0: $\forall A \in C([0,1],\mathbf{\Phi}_0(X,Y))$, if $A(i) \in GL(X,Y)$, then there exists $N \in C([0,1],GL(Y,X))$ such that $N(t)A(t) = I-K(t), \forall t \in [0,1]$, where $K \in C([0,1],\mathcal{K}(X))$. If $A(i) \in GL(X,Y), i=0,1$, then we define

$$\sigma(A) = \sigma(I-K) .$$

One can show that $\sigma(A)$ does not depend on the special choice of N.

In particular, if $A \in C([0,1],GL(X,Y))$, then we take $N(t) = A^{-1}(t)$, and then $\sigma(A) = 1$.

If $X = Y$ is a finite-dimensional Banach space, by definition, $\sigma(A) = \pm 1$ if and only if $A(0)$ and $A(1)$ lie in the same/different connected component(s) in $GL(X)$. But this is not true for infinite-dimensional space due to the above-mentioned Kuiper's theorem. A geometric interpretation is given in [FP]: Let $S_j = \{L \in \mathbf{\Phi}(X,Y) \,|\, \mathrm{dimker} L = j\}, j = 1,2,\ldots$, and $S = \mathbf{\Phi}_0(X,Y)\backslash GL(X,Y)$, then $S = \bigcup_{j=1}^\infty S_j$, and $\sigma(A)$ is the number of points of transversal intersection of a generic path A with S_1.

The parity of the curve $A \in C([0,1],\mathbf{\Phi}_0)$ enjoys the following homotopy invariance and the invariance under reparametrizations:

(Homotopy invariance) Let $H \in C([0,1]^2,\mathbf{\Phi}_0(X,Y))$ and suppose that $H(t,i) \in GL(X,Y), i=0,1, \forall t \in [0,1]$. Then $\sigma(H(t,\cdot))$ is a constant.

(Invariance under reparametrizations) Let $A \in C([0,1], \Phi_0(X,Y))$ with $A(i) \in GL(X,Y), i = 0,1$, and let $\gamma \in C([0,1],[0,1])$ satisfy $\gamma(i) = i, i = 0,1$. Then $\sigma(A \circ \gamma) = \sigma(A)$.

Let $\Omega \subset U \subset X$ be open subsets, and U be connected and simply connected. In the following the closure $\bar{\Omega}$ and the boundary $\partial\Omega$ are understood relative to U.

Definition 3.7.12 *A Fredholm mapping $F \in C^1(U,Y)$ of index 0 is called Ω-admissible if $F|_{\bar{\Omega}}$ is proper. A Fredholm homotopy $H \in C^1([0,1] \times U, Y)$ of index 1 is called Ω-admissible if $H|_{[0,1]\times\bar{\Omega}}$ is proper.*

Let $p \in U$ be a regular point of a C^1 Ω-admissible Fredholm mapping F; we take it as a base point. Let $y \in Y \backslash F(\partial\Omega)$ be a regular value of $F|_\Omega$, then $F^{-1}(y) \cap \Omega$ is a finite set (may be empty): $\{x_1, x_2, \ldots, x_k\}$. If it is not empty, let $\gamma_i \in C([0,1], U)$, with $\gamma_i(0) = p, \gamma_i(1) = x_i$, be curves connecting p and $x_i, i = 1,2,\ldots,k$. Then the parities $\sigma_i = \sigma(F' \circ \gamma_i), i = 1,2,\ldots,k$, are all well defined, and are independent of γ_i. Thus for a regular value y, we define the degree of F with the base point p by

$$\deg_p(F, \Omega, y) = \sum_{i=1}^{k} \sigma_i \,.$$

If $F^{-1}(y) \cap \Omega = \emptyset$, then we define $\deg_p(F, \Omega, y) = 0$.

Quinn and Sard [QS] avoided the requirement of the separability of the spaces X and Y, and obtained an improved version of Sard–Smale Theorem, by which one shows that for any $y \in Y \backslash F(\partial\Omega)$ there exists $\epsilon > 0$ such that $B_\epsilon(y) \subset Y \backslash F(\partial\Omega)$ contains a regular value z of F. In combining with an approximation theorem for C^1 Fredholm mappings of any index in [PR 2], we can define

Definition 3.7.13 *Assume that $F \in C^1(U,Y)$ is Ω-admissible, $p \in U$ is a base point of F, and $y \in Y \backslash F(\partial\Omega)$. We define the degree $\deg_p(F, \Omega, y) = \deg_p(F, \Omega, z)$, where $z \in B_\epsilon(y) \subset Y \backslash F(\partial\Omega)$ is a regular value of $F|_\Omega$.*

Again, the degree is independent of the choice of z.

For different base points p_0, p_1, one has

$$\deg_{p_0}(F, \Omega, y) = \sigma(F' \circ \gamma) \deg_{p_1}(F, \Omega, y) \,,$$

where $\gamma \in C([0,1], U), \gamma(i) = i, i = 0,1$.

The following fundamental properties hold:

(Homotopy invariance) Let H be an Ω-admissible homotopy. Suppose that $y \in Y \backslash ([0,1] \times \partial\Omega)$ and that p is a base point of $H(t,\cdot) \forall t \in [0,1]$. Then

$$\deg_p(H(1,\cdot), \Omega, y) = \deg_p(H(0,\cdot), \Omega, y) \,.$$

(Additivity) Suppose the $\Omega = \Omega_1 \cup \Omega_2$, where $\Omega_i, i = 1,2$ are disjoint open subsets of U, and that $F \in C^1(U,Y)$ is Ω-admissible, $p \in U$ is a base point of F. Then F is Ω_i-admissible for $i = 1, 2$. Moreover, if $y \in Y \backslash F(\partial\Omega)$, then $y \in Y \backslash F(\partial\Omega_i), i = 1, 2$, and

$$\deg_p(F, \Omega, y) = \deg_p(F, \Omega_1, y) + \deg_p(F, \Omega_2, y) .$$

(Normality) Let Ω be an open subset of X, $\forall p \in X$, and $y \notin \partial\Omega$. Then $\deg_p(\mathrm{id}, \Omega, y) = 1$ if $y \in \Omega$, $= 0$ if $y \notin \Omega$.

Obviously, the excision property and the Kronecker existence theorem hold as well.

The generalized degree has been used to extend Rabinowitz global bifurcation theorem [PR1], and is applied to the study of bifurcation problems for semi-linear elliptic equations on R^n (see [JLS]).

4 Minimization Methods

The calculus of variations studies the optimal shape, time, velocity, energy, volume or gain etc. under certain conditions. Laws in astronomy, mechanics, physics, all natural sciences and engineering technologies, as well as in economic behavior obey variational principles. The main object of the calculus of variations is to find out the solutions governed by these principles. Tracing back to Fermat, who postulated that light follows a path of least possible time, this is a subject in finding the minimizers of a given functional. Starting from the brothers Johann and Jakob Bernoulli and L. Euler, the calculus of variations has a long history, and renews itself according to the developments of mathematics and other sciences.

The problem is formulated as follows: Assume that $f : \mathbb{R}^n \times \mathbb{R}^N \times \mathbb{R}^{nN} \to \mathbb{R}^1$ is a continuous function, and that E is a set of N-vector functions. Let J be a functional defined on E:

$$J(u) = \int f(x, u(x), \nabla u(x)) dx \, .$$

Find $u_0 \in E$, such that

$$J(u_0) = \text{Min}\{J(u) \mid u \in E\} \, .$$

The central problems in the calculus of variations are the existence and the regularity of the minimizers. These are the 19th and the 20th problems among the 23 problems posed by Hilbert in his famous lecture delivered at International Congress of Mathematicians in 1900.

This chapter is devoted to an introduction of the minimization method. We pay attention only to the existence of minimizers, but not to the regularity, although the latter is a very important and rich part of the theory of the calculus of variations. The direct method, studied in Sect. 4.2, is the core of the minimization method, in which w*-compactness and w* lower semi-continuity (w*l.s.c.) play crucial roles.

A necessary and sufficient condition on the integrand f for the w*l.s.c. of the functional J on the Sobolev space $W^{1,p}$, $p \in (1, \infty]$ is studied in Sect. 4.3.

In the case when w*l.s.c. fails, either the minimizing sequence does not converge or it does not converge to a minimizer. The Young measure and the relaxation functional are introduced in Sect. 4.4.

In the spaces $W^{1,1}$ and L^1 the closed balls are no longer w* compact. Instead, we consider the BV space and the Hardy space, respectively. They are studied in Sect. 4.5.

Two interesting applications are given in Sect. 4.6. One is on the phase transitions and the other is the segmentation in the image processing.

The concentration phenomenon, which happens in many problems from geometry to physics, concerns the lack of compactness. We give a brief introduction to the method of managing this phenomenon in Sect. 4.7.

The minimax method dealing with saddle points is briefly introduced in Sect. 4.8. With the aid of the Ekeland variational principle and the Palais–Smale condition, it is studied in the spirit of the minimization method.

Section 4.1 is an introduction, where various variational principles and their reductions are introduced.

4.1 Variational Principles

Let X be a real Banach space, and $U \subset X$ be an open set. A point $x_0 \in U$ is called a local maximum (or minimum) point of $f : U \to \mathbb{R}^1$, if

$$f(x) \leqslant f(x_0) \quad (\text{or } f(x) \geqslant f(x_0)) \qquad \forall x \in B_\varepsilon(x_0) \subset U ,$$

for some $\varepsilon > 0$.

If further, f is G-differentiable at x_0, then

$$df(x_0, h) = \frac{d}{dt} f(x_0 + th)\big|_{t=0} = \theta \ \forall h \in X ,$$

or simply

$$df(x_0) = \theta . \tag{4.1}$$

Moreover, if f has second-order G-derivatives at x_0, then

$$d^2 f(x_0)(h, h) \leqslant 0 \quad (\text{or } \geqslant 0) \qquad \forall h \in X .$$

In particular, if X is a Hilbert space, and $f \in C^2$ then $d^2 f(x_0)$ is a self-adjoint operator. We conclude that $d^2 f(x_0)$ is nonnegative (nonpositive) if x_0 is a local minimum (or maximum) point. Conversely, x_0 is a local minimum (or maximum) point if $d^2 f(x_0)$ is positive (or negative) definite.

4.1.1 Constraint Problems

Let X, Y be real Banach spaces, $U \subset X$ be an open set. Suppose that $f : U \to \mathbb{R}^1, g : U \to Y$ are C^1 mappings. Let

$$M = \{x \in U \,|\, g(x) = \theta\}\,.$$

Find the necessary condition for

$$\min_{x \in M} f(x)\,. \tag{4.2}$$

Theorem 4.1.1 *(Ljusternik) Suppose that $x_0 \in M$ solves (4.2), and that* $\operatorname{Im} g'(x_0)$ *is closed. Then $\exists (\lambda, y^*) \in \mathbb{R}^1 \times Y^*$ such that $(\lambda, y^*) \neq (0, \theta)$, and*

$$\lambda f'(x_0) + g'(x_0)^* y^* = \theta\,. \tag{4.3}$$

Furthermore, if $\operatorname{Im}\, g'(x_0) = Y$*, then* $\lambda \neq 0$.

Proof. In the case where $Y_1 = \operatorname{Im} g'(x_0) \subsetneqq Y$, the conclusion (4.3) is trivial; one may choose $\lambda = 0$ and $y^* \in Y_1^\perp := \{z^* \in Y^* \,|\, \langle z^*, z\rangle = 0 \;\forall z \in Y_1\}$.

We assume $Y_1 = Y$. The tangent space $T_{x_0}(M)$ of M at x_0 is as follows:

$$\begin{aligned} &T_{x_0}(M) \\ &\quad = \{h \in X \,|\, \exists \varepsilon > 0, \exists v \in C^1((-\varepsilon, \varepsilon), X), x_0 + v(t) \in M, v(0) = \theta, \dot{v}(0) = h\}\,. \end{aligned}$$

We want to prove that $T_{x_0}(M) = \ker g'(x_0)$. In order to avoid technical complication, we make an additional assumption: Either $g'(x_0)$ is a Fredholm operator, or X is a Hilbert space. The assumption is superfluous, because a modified IFT has been studied in [De] (pp. 334) to improve the proof. Indeed, from

$$g(x_0 + v(t)) = \theta\,,$$

it follows that

$$g'(x_0)h = \frac{d}{dt} g(x_0 + v(t))\big|_{t=0} = 0 \;\;\forall\, h \in T_{x_0}(M)\,,$$

i.e., $T_{x_0}(M) \subset \ker g'(x_0)$. On the other hand, if $h \in \ker g'(x_0)$, one solves the equation:

$$g(x_0 + th + w(t)) = \theta\,,$$

for $w \in C^1((-\varepsilon, \varepsilon), X_1), w(0) = \theta$, where X_1 is the complement of $\ker g'(x_0)$ and $\varepsilon > 0$ is small. Since $g(x_0) = \theta$, $g'(x_0) : X_1 \to Y$ is an isomorphism, one may apply the IFT to obtain such a solution w. Setting $v(t) = th + w(t)$, we have $v(0) = \theta$ and $\dot{v}(0) = h + \dot{w}(0)$.

From

$$g'(x_0)(h + \dot{w}(0)) = \theta\,,$$

it follows that $\dot{w}(0) \in \ker g'(x_0)$, but $\dot{w}(0) \in X_1$ which implies that $\dot{w}(0) = \theta$, i.e., $\dot{v}(0) = h$.

Now, $\forall\, h \in T_{x_0}(M)$, $\exists v(t)$ satisfying $v(0) = \theta, \dot{v} = h$ and $x_0 + v(t) \in M$, so that

$$f(x_0 + v(t)) \geqslant f(x_0)\,,$$

which implies that

$$\langle f'(x_0), h \rangle = 0 ,$$

i.e. $f'(x_0) \in \ker g'(x_0)^{\perp}$. By the closed range theorem, $\exists y^* \in Y^*$ such that

$$-f'(x_0) = g'(x_0)^* y^* .$$

This completes the proof.

Corollary 4.1.2 *Suppose that $g_1, \ldots, g_m : U \to \mathbb{R}^1$ are C^1 functions, and that x_0 solves (4.2) with*

$$M = \{x \in U \,|\, g_i(x) = 0, i = 1, 2, \ldots, m\} .$$

If $\{g_i'(x_0)\}_1^m$ is linearly independent, then $\exists\, \lambda_1, \ldots, \lambda_m$ such that

$$f'(x_0) + \sum_{i=1}^{m} \lambda_i g_i'(x_0) = 0 .$$

Moreover, we may also consider inequality constraints: Given $f, g_1, \ldots, g_m, h_1, \ldots, h_l \in C^1(X, R^1)$ find

$$\min \{f(x) | g_i(x) = 0,\ i = 1, \ldots, m;\ h_j(x) \le 0,\ j = 1, \ldots, l\} .$$

In the same manner, we find the necessary condition of an extremum point x_0:

$$\exists\, \lambda_1, \ldots, \lambda_m \in \mathbb{R}^1 ,$$

$$\exists\, \mu_0, \mu_1, \ldots, \mu_l \ge 0, \text{ but not all zero} ,$$

such that

$$\mu_0 f'(x_0) + \sum_{i=1}^{m} \lambda_i g_i'(x_0) + \sum_{j=1}^{l} \mu_j h_j'(x_0) = \theta, \text{ and } \mu_j h_j(x_0) = 0,\ \forall j = 1, \ldots, l .$$

This is called the Kuhn–Tucker condition in the mathematical programming.

In particular, if all the functions $h_j, j = 1, \ldots, l$ are convex, which implies that the set $C = \{x \in X | h_j(x) \le 0, j = 1, \ldots, l\}$ is convex, then a necessary condition for an extremum $x_0 \in C$ is the following variational inequality:

$$\left\langle f'(x_0) + \sum_{1}^{m} \lambda_i g_i'(x_0), y - x_0 \right\rangle \ge 0,\ \forall y \in C .$$

Remark 4.1.3 $\lambda_1, \ldots, \lambda_m; \mu_0, \mu_1, \ldots, \mu_l$, *are called the Lagrangian multipliers.*

4.1.2 Euler–Lagrange Equation

In case f is a functional of the following form:

$$f(u) = \int_\Omega \varphi(x, u(x), \nabla u(x))dx \qquad \Omega \subset \mathbb{R}^n, \tag{4.4}$$

where $\varphi : \overline{\Omega} \times \mathbb{R}^N \times \mathbb{R}^{nN} \to \mathbb{R}^1$ is a C^2 function, let us turn to the 1st and 2nd variations of the functional (4.4). $\forall u \in C^1(\overline{\Omega})$

$$\begin{aligned} &\langle f'(u), v\rangle \\ &= \int_\Omega \left\{ \sum_{i=1}^N \sum_{\alpha=1}^n \frac{\partial \varphi(x, u(x), \nabla u(x))}{\partial \xi_\alpha^i} \frac{\partial v_i}{\partial x_\alpha} + \sum_{i=1}^N \frac{\partial \varphi(x, u(x), \nabla u(x))}{\partial u_i} v_i \right\}, \end{aligned}$$

and

$$\begin{aligned} \langle f''(u), v \otimes v\rangle = \int_\Omega \Bigg\{ &\sum_{i,k=1}^N \sum_{\alpha,\beta=1}^n \frac{\partial^2 \varphi(x, u(x), \nabla u(x))}{\partial \xi_\alpha^i \partial \xi_\beta^k} \frac{\partial v_i}{\partial x_\alpha} \frac{\partial v_k}{\partial x_\beta} \\ &+ 2 \sum_{i,k=1}^N \sum_{\alpha=1}^n \frac{\partial^2 \varphi}{\partial \xi_\alpha^i \partial u_k} \frac{\partial v_i}{\partial x_\alpha} \cdot v_k + \sum_{i,k=1}^N \frac{\partial^2 \varphi}{\partial u_i \partial u_k} v_i v_k \Bigg\}, \end{aligned}$$

$\forall v \in C^1(\overline{\Omega}, \mathbb{R}^N)$.

Thus the Euler–Lagrange equation under certain boundary conditions reads as:

$$-\sum_{\alpha=1}^n \frac{\partial}{\partial x_\alpha} \left(\frac{\partial \varphi(x, u(x), \nabla u(x))}{\partial \xi_\alpha^i} \right) + \frac{\partial \varphi(x, u(x), \nabla u(x))}{\partial u_i} = 0, \;\; i = 1, 2, \ldots, N\,.$$

It is a second-order differential system.

If the functional f is only defined on a closed convex subset C of the function space $C^1(\overline{\Omega})$, and u is a minimizer in C, then we only have a variational inequality:

$$\begin{aligned} &\int_\Omega \left\{ \sum_{i=1}^N \sum_{\alpha=1}^n \frac{\partial \varphi(x, u(x), \nabla u(x))}{\partial \xi_\alpha^i} \frac{\partial (v_i - u_i)}{\partial x_\alpha} + \frac{\partial \varphi(x, u(x), \nabla u(x))}{\partial u_i}(v_i - u_i) \right\} \\ &\geq 0\,, \end{aligned}$$

for all $v \in C$.

In case $N = 1$, the Euler–Lagrange equation reads as

$$-\mathrm{div}\varphi_\xi(x, u(x), \nabla u(x)) + \varphi_u(x, u(x), \nabla u(x)) = 0\,.$$

This is a differential equation. We give here a few examples.

1. (Geodesics) Let U be an open set of R^n, and let $g_{ij} : U \to R^1, \forall i,j = 1,2,\ldots,n$ be symmetric and positive definite. For $u = (u^1,\ldots,u^n) \in C^1([0,1],U)$, one defines

$$\varphi(u,\xi) = \sum_{i,j=1}^{n} g_{ij}(u)\xi_i\xi_j,\ \forall\, \xi = (\xi_1,\ldots,\xi_n) \in R^n\ .$$

Then the Euler–Lagrange equation is the following differential system:

$$\frac{d^2u^i}{dt^2} + \Gamma^i_{j,k}\frac{du^j}{dt}\frac{du^k}{dt} = 0,\quad \forall i = 1,2,\ldots,n,\ t \in [0,1]\ ,$$

where

$$\Gamma^i_{j,k} = \frac{1}{2}[\partial_j g_{lk} + \partial_k g_{lj} - \partial_l g_{jk}]g^{li},$$

and (g^{li}) is the inverse matrix of (g_{il}), i.e., $g_{ij}g^{ik} = \delta^k_j$.
2. (Poisson equation) Let $\varphi(x,u,\xi) = c(x)u + \frac{1}{2}|\xi|^2$. The Euler–Lagrange equation is the Poisson equation:

$$\Delta u = c(x) \qquad \text{in } \Omega\ .$$

3. (Hamiltonian systems)
For $H \in C^1(\mathbb{R}^1 \times \mathbb{R}^n \times \mathbb{R}^n, \mathbb{R}^1)$, the following ODE system

$$\begin{cases} \dot{x} = -H_p(t,x,p), \\ \dot{p} = H_x(t,x,p), \end{cases}$$

with $(t,x,p) \in \mathbb{R}^1 \times \mathbb{R}^n \times \mathbb{R}^n$, is called a Hamiltonian system, and H is called a Hamiltonian function. Sometimes we use the notations:

$$z = (x,p)\ ,$$

and

$$J = \begin{pmatrix} 0 & -I \\ I & 0 \end{pmatrix}\ ,$$

where I is the $n \times n$ unit matrix. The above system then has a simple form:

$$\dot{z} = J\,\mathrm{grad}\, H(t,z)\ .$$

$J\,\mathrm{grad}$ sometimes is also called a symplectic gradient. The 2π-periodic solution can be seen as a critical point of the following functional:

$$f(z) = \int_0^{2\pi} \frac{1}{2}\langle z, J\dot{z}\rangle_{\mathbb{R}^{2n}} + H(t,z)$$

on the space $C^1([0,2\pi],\mathbb{R}^{2n})$. In other words, the Euler–Lagrange equation of this functional is the Hamiltonian system.

4. (Minimal surface) Let $u \in C^1(\overline{\Omega}, \mathbb{R}^1)$, and $\varphi(\xi) = [1+|\xi|^2]^{\frac{1}{2}}$. Then the Euler–Lagrange equation reads as

$$\operatorname{div}\left\{\frac{\nabla u}{[1+|\nabla u|^2]^{\frac{1}{2}}}\right\} = 0 \,.$$

5. (Obstacle problem revisit) Let $\Omega \subset \mathbb{R}^n$ be a bounded domain, $g \in L^2(\Omega), \psi \in H^1(\Omega)$. Find a minimizer of the problem: $\text{Min}\{f(u) \,|\, u \in E\}$, where

$$f(u) = \int_\Omega \left(\frac{1}{2}|\nabla u|^2 - g \cdot u\right) ,$$

and

$$E = \{u \in H_0^1(\Omega) \,|\, u(x) \le \psi(x) \text{ a.e. } x \in \Omega\} \,.$$

The variational inequality reads as

$$\int_\Omega \nabla u \cdot \nabla(v-u) \ge \int_\Omega g(v-u), \ \forall v \in E \,.$$

Now we derive the necessary condition on φ for a minimizer of f under Dirichlet boundary conditions. If u is a minimizer, then by Taylor expansion,

$$\langle f''(u), v \otimes v\rangle \geqslant 0, \quad \forall v \in W_0^{1,\infty}(\Omega, \mathbb{R}^N) \,. \tag{4.5}$$

Let $\rho(t) = 1+t$ for $t \in [-1,0]$, and $\rho(t) = 1-t$ for $t \in [0,1]$. $\forall x_0 = (x_{0,1} \dots, x_{0,n}) \in \Omega$, $\exists \varepsilon > 0$ such that the cube centered at $x_0 : \prod_{\alpha=1}^n [x_{0,\alpha} - \epsilon, x_{0,\alpha} + \epsilon] \subset \Omega$. $\forall \lambda = (\lambda^1, \dots, \lambda^n) \in \mathbf{S}^{n-1}$ and $\forall \xi = (\xi_1, \dots, \xi_N) \in \mathbb{R}^N$, we define

$$v_\varepsilon(x) = \Pi_{\alpha=1}^n \rho\left(\frac{\lambda^\alpha(x_\alpha - x_{0,\alpha})}{\varepsilon}\right) \cdot \xi \,,$$

where ρ is understood to be 0 outside $(-1,1)$. Substituting v_ε into (4.5), and letting $\varepsilon \longrightarrow 0$, we obtain

$$\sum_{i,k=1}^N \sum_{\alpha,\beta=1}^n \frac{\partial^2 \varphi(x_0, u(x_0), \nabla u(x_0))}{\partial \xi_\alpha^i \partial \xi_\beta^k} \lambda^\alpha \lambda^\beta \xi_i \xi_k \ge 0 \ \forall (\lambda, \xi) \in (R^n \backslash \{\theta\}) \times R^N \,.$$

This is called the Legendre–Hardamard condition, which can be rewritten in a compact form:

$$(\lambda \otimes \xi)^T \partial^2 \varphi(x, u(x), \nabla u(x))(\lambda \otimes \xi) \geqslant 0 \,, \tag{4.6}$$

where $\lambda \otimes \xi$ denotes the rank-one $n \times N$ matrix. In the next section, this condition on φ is called rank-one convexity. Thus, rank-one convexity is the necessary condition for φ at a minimizer.

In particular, if $N = 1$, condition (4.6) is reduced to

$$\sum_{j,l=1}^n \frac{\partial^2 \varphi(x, u(x), \nabla u(x))}{\partial \xi_j \partial \xi_l} \lambda^j \lambda^l \geqslant 0, \quad \forall \lambda \in \mathbb{R}^n \backslash \{\theta\}, \ \forall x \in \Omega \,. \tag{4.7}$$

However, if the strict inequality in (4.7) holds for all $\lambda \in \mathbb{R}^n \backslash \{\theta\}$, i.e., φ is strictly convex with respect to ξ. This is the ellipticity condition. To the vectorial case, the condition:

$$\sum_{i,k=1}^{N} \sum_{\alpha,\beta=1}^{n} \frac{\partial^2 \varphi}{\partial \xi_\alpha^i \partial \xi_\beta^k}(u, u(x), \nabla u(x)) \xi_\alpha^i \xi_\beta^k > 0, \forall \xi = \{\xi_\alpha^i\} \in \mathbb{M}^{n \times N}, \text{with } |\xi| \neq 0 ,$$

is called the strong ellipticity.

4.1.3 Dual Variational Principle

The following notion was introduced by Fenchel, and is very important in convex analysis.

Definition 4.1.4 *Suppose that X is a real Banach space and that $f : X \to \mathbb{R}^1 \cup \{+\infty\}$ is proper. The conjugate function of $f, f^* : X^* \to \mathbb{R}^1 \cup \{+\infty\}$ is defined by*

$$f^*(x^*) = \sup_{x \in X} \{\langle x^*, x\rangle - f(x)\} .$$

The notion was initiated by Young's inequality: if $f(x) = \frac{1}{p}|x|^p, 1 < p < \infty$, then $f^*(x) = \frac{1}{p'}|x|^{p'}$, where $\frac{1}{p} + \frac{1}{p'} = 1$.

In particular, if $f(x) = \frac{1}{2}\langle Ax, x\rangle_{R^n}$, where A is positive definite, then $f^*(p) = \frac{1}{2}\langle p, A^{-1}p\rangle_{R^n}$.

The following propositions hold:

1. f^* is convex and l.s.c. Moreover, it is proper, if f is proper l.s.c. and convex.

   Proof. Only the properness of f^* needs to be proved, i.e., one should find $x_0^* \in X^*$ such that $f^*(x_0^*) < +\infty$.

   We consider the closed convex set $\text{epi}(f) = \{(x,t) \in X \times \mathbb{R}^1 |\ f(x) \leqslant t\}$. Since f is proper, $\exists x_0 \in X$ such that $f(x_0) < +\infty$. One chooses $t_0 < f(x_0)$, then $(x_0, t_0) \notin \text{epi}(f)$. By the Ascoli separation theorem, $\exists (x_0^*, \lambda) \in X^* \times \mathbb{R}^1$, $\exists \alpha \in \mathbb{R}^1$, satisfying

   $$\langle x_0^*, x\rangle + \lambda t > \alpha > \langle x_0^*, x_0\rangle + \lambda t_0, \quad \forall (x,t) \in \text{epi}(f) . \tag{4.8}$$

   In particular,

   $$\langle x_0^*, x_0\rangle + \lambda f(x_0) > \langle x_0^*, x_0\rangle + \lambda t_0 .$$

   It follows that $\lambda > 0$, and then

   $$\left\langle -\frac{1}{\lambda} x_0^*, x \right\rangle - f(x) < -\frac{\alpha}{\lambda} \quad \forall x \in D(f) ,$$

i.e.,

$$f^*\left(-\frac{x_0^*}{\lambda}\right) < -\frac{\alpha}{\lambda} < +\infty\,.$$

Since f^* is proper, the conjugate function f^{**} for f^* is well defined. $\square$

2. If $f \leqslant g$, then $g^* \leqslant f^*$.
3. (Young's inequality) $\langle x^*, x\rangle \leqslant f(x) + f^*(x^*)$.
4. $f(x) + f^*(x^*) = \langle x^*, x\rangle \Longleftrightarrow x^* \in \partial f(x)$.

Proof. By definition, we have

$$\begin{aligned} x^* \in \partial f(x) &\Longleftrightarrow \langle x^*, y - x\rangle \leqslant f(y) - f(x), \quad \forall y \in X, \\ &\Longleftrightarrow \langle x^*, y\rangle - f(y) \leqslant \langle x^*, x\rangle - f(x), \quad \forall y \in X, \\ &\Longleftrightarrow f^*(x^*) \leqslant \langle x^*, x\rangle - f(x)\,. \end{aligned}$$

Combining with Young's inequality, the last inequality is equivalent to

$$f(x) + f^*(x^*) = \langle x^*, x\rangle\,.$$

$\square$

5. If $g(x) = f(x - x_0) + \langle x_0^*, x\rangle + a$, then $g^*(x^*) = f^*(x^* - x_0^*) + \langle x^*, x_0\rangle - (a + \langle x_0^*, x_0\rangle)$.
6. If $g(x) = f(\lambda x)$, then $g^*(x^*) = f^*(\frac{x^*}{\lambda})$ for $\lambda \neq 0$.

Theorem 4.1.5 *(Fenchel–Moreau) If f is a proper, l.s.c., convex function, then $f^{**} = f$.*

Proof. By Young's inequality, $f^{**} \leqslant f$. It remains to show the reversed inequality. We prove it by contradiction, i.e., assume that $\exists x_0 \in X$ such that $f^{**}(x_0) < f(x_0)$.

Similar to the proof of proposition 1, we separate epi(f) with the point $(x_0, f^{**}(x_0))$, and obtain (4.8), in which $t_0 = f^{**}(x_0)$ and $\lambda \geq 0$.

If $\lambda > 0$, then

$$\langle x_0^*, x\rangle + \lambda f(x) > \alpha \quad \forall x \in \mathrm{dom}(f)\,.$$

It follows that

$$f^*\left(-\frac{x_0^*}{\lambda}\right) \leqslant -\frac{\alpha}{\lambda}\,.$$

However, by the definition of f^{**},

$$f^{**}(x_0) \geqslant \left\langle -\frac{x_0^*}{\lambda}, x_0\right\rangle - f^*\left(-\frac{x_0^*}{\lambda}\right)\,.$$

It follows that

$$\langle x_0^*, x_0\rangle + \lambda f^{**}(x_0) \geqslant \alpha\,.$$

This contradicts (4.8), in which $t_0 = f^{**}(x_0)$. In the case where $f(x_0) < +\infty$, from (4.8), $\lambda > 0$. It remains to verify that $f^{**}(x_0) = +\infty$ if $f(x_0) = +\infty$ and $\lambda = 0$. Again from (4.8), $\exists \epsilon > 0$ such that

$$\langle x_0^*, x - x_0 \rangle \geq \epsilon \ \forall x \in \text{dom}(f) .$$

Since f^* is proper, there is an $x_1^* \in X^*$ such that $f^*(x_1^*) < +\infty$, and

$$\langle x_1^*, x \rangle - f(x) - f^*(x_1^*) \leq 0 \ \forall x \in \text{dom}(f) .$$

Putting them together, $\forall n \in \mathbf{N}$, we have

$$\langle x_1^* - n x_0^*, x \rangle + n \langle x_0^*, x_0 \rangle + n\epsilon - f(x) - f^*(x_1^*) \leq 0, \ \forall x \in \text{dom}(f) ,$$

it follows that

$$f^*(x_1^* - n x_0^*) + n \langle x_0^*, x_0 \rangle + n\epsilon - f^*(x_1^*) \leq 0 ,$$

or

$$n\epsilon + \langle x_1^*, x_0 \rangle - f^*(x_1^*) \leq \langle x_1^* - n x_0^*, x_0 \rangle - f^*(x_1^* - n x_0^*) \leq f^{**}(x_0) .$$

Letting $n \to \infty$, we obtain $f^{**}(x_0) = +\infty$. Therefore, $f = f^{**}$. □

Corollary 4.1.6 *For a proper, l.s.c., convex function f, $x^* \in \partial f(x) \Longleftrightarrow x \in \partial f^*(x^*)$.*

Proof. It is a direct consequence of proposition 4 and Theorem 4.1.5. □

In the case where both ∂f and ∂f^* are single valued, Corollary 4.1.6 means that they are mutually inverse. In this sense, the conjugate function f^* of f is called the Legendre transform of f.

Recall the variational formulation of the Hamiltonian system, which describes the motion of particle systems. The functional

$$f(z) = \int_0^{2\pi} \left[\frac{1}{2} \langle z, J\dot{z} \rangle_{\mathbb{R}^{2n}} + H(t, z) \right]$$

is very indefinite. One cannot pose the minimization problem. However, in some cases, the Legendre transform may help.

(1) Assume that $H(t, x, p)$ is proper, l.s.c., and convex in the variables p, we define the Legendre transform of H with respect to p, i.e.,

$$L(t, x, q) = \sup_p \{ \langle p, q \rangle_{\mathbb{R}^n} - H(t, x, p) \} .$$

It is called the Lagrangian. In other words, $\forall (x, t)$ as a function of q, L is the conjugate function of the function $H(t, x, p)$. From proposition 4 and the Hamiltonian system, one has

$$\langle -\dot{x}, p\rangle_{\mathbb{R}^n} = L(t,x,-\dot{x}) + H(t,x,p) \,.$$

If both H and L are differentiable, then,

$$L_x(t,x,-\dot{x}) = -H_x(t,x,p) \,.$$

But from Corollary 4.1.6,

$$\dot{x} = -H_p(t,x,p) \;\Leftrightarrow\; p = L_q(t,x,-\dot{x}) \,.$$

This shows that $(x, \dot{x})$ satisfies the system:

$$\frac{d}{dt}L_q(t,x,-\dot{x}) + L_x(t,x,-\dot{x}) = 0 \,. \tag{4.9}$$

However, (4.9) is the Euler–Lagrange equation of the functional:

$$I(x) = \int_0^{2\pi} L(t,x,-\dot{x})dt \,.$$

The system (4.9) is called the associate Lagrange system.

For example, in a system of particles with generalized coordinates $q = (q_1,\ldots,q_n)$, the kinetic energy of the system is a positive definite quadratic form: $T = \frac{1}{2}\sum a_{ij}q_iq_j$, where $a_{ij} = a_{ij}(t,q)$, and the potential energy is a continuous function bounded from below: $U = U(q)$. The total energy $H = T + U$ is called the Hamiltonian, and $L = T - U$ is called the Lagrangian. In this case, the functional I is bounded from below.

(2) Assume that $H(t,x,p)$ is strictly convex in $z = (x,p)$, then the Legendre transform of H with respect to z reads as

$$G(t,w) = \sup_z\{\langle z,w\rangle_{\mathbb{R}^{2n}} - H(t,z)\} \,.$$

Assume $H_z(t,\theta) = \theta$. We study nontrivial 2π-periodic solutions of the Hamiltonian system. Let us consider a constraint variational problem:

$$I(w) = \int_0^{2\pi} G(t,-J\dot{w})dt \,,$$

with

$$g(w) = \int_0^{2\pi} \langle w, J\dot{w}\rangle_{\mathbf{R}^{2n}} dt = c \neq 0 \,,$$

where c is a parameter. The Euler–Lagrange equation reads as

$$G_w(t,-J\dot{w}) = \lambda w \,.$$

Claim: $\lambda \neq 0$. If not, provided by

$$w = H_z(t,z) \;\Leftrightarrow\; z = G_w(t,w) \,,$$

we have $-J\dot{w} = H_z(t, \theta) = \theta$. This contradicts $c \neq 0$.

Let $z = \lambda w$. Again by duality, we have

$$\dot{z} = \lambda J H_z(t, z) .$$

One can adjust c such that $\lambda = 1$.

For example, if $H(t, z)$ is bounded by two quadratic functions: $0 < m|z|^2 \leq H(t, z) \leq M|z|^2$, then $G(t, w)$ is also. Thus the functional I is bounded from below.

4.2 Direct Method

4.2.1 Fundamental Principle

Given a topological space X and a function $f : X \to \mathbb{R}^1 \cup \{+\infty\}$, which is bounded from below, we seek the minimizers of f on X, if they exist. It is natural to require the l.s.c. of f (i.e., $\forall t \in \mathbb{R}^1$, the level set $f_t := \{x \in X |\ f(x) \leqslant t\}$ is closed) and certain compactness on X. However, if X is not compact, the coercive condition on f should be assumed as a replacement: f is proper and $\forall t \in \mathbb{R}^1$ $\exists$ a compact subset $K_t \subset X$ such that the level set $f_t \subset K_t$.

Indeed, we have:

Theorem 4.2.1 *If $f : X \to \mathbb{R}^1 \cup \{+\infty\}$ is l.s.c. and coercive, then it attains a minimizer on X. In particular, f is bounded from below.*

Proof. Since f is coercive, it is proper. Then $-\infty \leq m := \inf f < +\infty$. By the assumption, $\forall t > m$, the set f_t is compact. Therefore by the finite intersection property, $\bigcap_{t>m} f_t \neq \emptyset$. A point x_0 in the intersection achieves $f(x_0) = m$. Since f does not assume the value $-\infty, m > -\infty$, and then f is bounded from below. □

However, in analysis, people prefer to use a minimizing sequence (i.e., a sequence $\{x_j\} \subset X$ such that $f(x_j) \to m := \inf f$) to approach the minimizer. In this case, it is required that f is sequentially lower semi-continuous (s.l.s.c., for short) and we assume that K_t is sequentially compact, $\forall t \in \mathbb{R}^1$. These two notions are defined as follows:

$$f \text{ is s.l.s.c.} \Longleftrightarrow \underline{\lim} f(z_j) \geqslant f(z) \text{ if } z_j \to z .$$

and

$$\begin{aligned} K \text{ is sequentially compact} &\Longleftrightarrow \text{any sequence } \{z_j\} \subset K \\ &\text{contains a subsequence } z_j \to z \in K . \end{aligned}$$

Namely,

Theorem 4.2.2 *If $f : X \to \mathbb{R}^1 \cup \{+\infty\}$ is s.l.s.c., and that $\forall t \in \mathbb{R}^1$ $\exists$ a sequentially compact set $K_t \subset X$ such that $f_t \subset K_t$, then f attains a minimizer in X.*

Recall the Eberlein–Schmulian theorem, for weakly closed subsets of a Banach space,

weakly compact = sequentially weakly compact.

According to the Banach–Alaoglu theorem, every w*-closed norm bounded set in the dual space of a Banach space is w* compact. Thus, for every weakly closed subset of a reflexive Banach space,

weakly closed plus norm bounded = weakly compact = sequentially weakly compact.

In this case, the coerciveness of f is equivalent to that $\forall t \in R^1$ the level set f_t of f is bounded in norm, or equivalently, that f is proper and $f(x) \to +\infty$ as $\|x\| \to \infty$.

However, how do we verify the weakly* lower semi-continuity (w*l.s.c., in short) or sequentially weakly* lower semi-continuity (s.w*l.s.c., in short) of f?

We notice that in a Banach space, closed convex set = weakly closed convex set (Hahn–Banach theorem) = sequentially weakly closed convex set (Mazur theorem).

Thus, if f is convex, then l.s.c. = w.l.s.c. = s.w.l.s.c.

Combining the above discussions, both Theorem 4.2.1 and Theorem 4.2.2 imply Corollary 2.3.7 as a special case. Because of the importance of this statement, we rewrite it as follow:

Theorem 4.2.3 *Let X be a reflexive Banach space and let $E \subset X$ be a weakly (or weakly sequentially) closed nonempty subset. If $f : E \to \mathbb{R}^1 \cup \{+\infty\}$ is a l.s.c., convex and coercive function, then f has a minimizer on E.*

In fact, Theorem 4.2.3 is a general principle in the proof of the existence of a minimizer. We shall present several classical examples to show how it works.

4.2.2 Examples

Example 1. (Dirichlet problem for Poisson equation)

Given $f \in L^2(\Omega)$, where $\Omega \subset \mathbb{R}^n$ is a bounded domain, find a minimizer of the following functional on $H_0^1(\Omega)$:

$$J(u) = \int_\Omega \left(\frac{1}{2}|\nabla u|^2 - f \cdot u\right) dx \, .$$

Obviously, J is convex and l.s.c.. We verify the coerciveness: From the Poincaré inequality, we have a constant $C > 0$ such that

$$\int_\Omega |u|^2 dx \leqslant C \int_\Omega |\nabla u|^2 .$$

Thus, we may assume $\|u\|_{H_0^1} = (\int_\Omega |\nabla u|^2)^{\frac{1}{2}}$.

Then $\exists C_1 > 0$, such that

$$\begin{aligned} J(u) &\geqslant \frac{1}{2}\int_\Omega |\nabla u|^2 - \| f \|_2 \cdot \| u \|_2 \\ &\geqslant \frac{1}{2}\int_\Omega |\nabla u|^2 - \frac{1}{4}\int_\Omega |\nabla u|^2 - C_1 \| f \|_2^2 \\ &= \frac{1}{4} \| u \|_{H_0^1}^2 - C_1 \| f \|_2^2 \to +\infty \end{aligned}$$

as $\| u \|_{H_0^1} \to +\infty$.

It is well known that the Euler–Lagrange equation of J is the Poisson equation with Dirichlet boundary conditions:

$$\begin{cases} -\triangle u = f & \text{in } \Omega , \\ u = 0 & \text{on } \partial\Omega . \end{cases}$$

Example 2. (Harmonic map) Let $\Omega \subset \mathbb{R}^m$ be a bounded open domain with smooth boundary $\partial\Omega$. A map $u = (u^1, \ldots, u^{n+1}) : \Omega \to S^n \subset \mathbb{R}^{n+1}$, where S^n is the unit sphere, is called harmonic if

$$-\triangle u^k = u^k |\nabla u|^2 \quad \text{in } \Omega,\ k = 1, 2, \ldots, n+1 ,$$

where $|\nabla u|^2 = \sum_{k=1}^{n+1} |\nabla u^k|^2$.

Given $\varphi = (\varphi^1, \ldots, \varphi^{n+1}) : \partial\Omega \to S^n$, find a harmonic map u with prescribed boundary condition $u|_{\partial\Omega} = \varphi$.

Let us consider a subset of the Banach space $X = H^1(\Omega, \mathbb{R}^{n+1})$:

$$M = \{u \in X |\ u|_{\partial\Omega} = \varphi,\ u(x) \in S^n \text{ a.e.}, x \in \Omega\}$$

and define the functional

$$E(u) = \frac{1}{2}\int_\Omega |\nabla u|^2 dx .$$

Firstly, M is a weakly sequentially closed set. Indeed, if $u_j \rightharpoonup u$ in $H^1(\Omega, \mathbb{R}^{n+1})$, then modulo a subsequence $\{u_j\}$ we have $u_j \to u$ in $L^2(\Omega, \mathbb{R}^{n+1})$ and then $u_j(x) \to u(x)$ a.e. From $u_j \in M\ \forall j$, we have $u \in M$. Obviously, E is l.s.c., convex and coercive.

We are then able to apply Theorem 4.2.3 to obtain the existence of a minimizer $u^* \in M$. What is the Euler–Lagrange equation for E? From

$$dE(u,v) = \int_\Omega \nabla u \cdot \nabla v = 0 ,$$

$\forall v \in H_0^1(\Omega, \mathbb{R}^{n+1})$ satisfying

$$v(x) \in T_{u^*(x)} S^n \quad \text{a.e. in } \Omega ,$$

we obtain

$$(\triangle u^*)^T(x) = 0 \text{ a.e. in } \Omega ,$$

where $(\triangle u^*)^T(x)$ is the tangential projection of $\triangle u^*(x)$ at $u(x)$.

Noticing that the normal projection of $\triangle u(x)$ at $u(x)$ reads as

$$(\triangle u)^N(x) = \triangle u(x) \cdot u(x) ,$$

and then by differentiation twice of the constraint $|u(x)|^2 = 1$, we obtain

$$\triangle u(x) \cdot u(x) = -|\nabla u(x)|^2 ,$$

from which follows

$$-\triangle u^*(x) = u^*(x)|\nabla u^*(x)|^2 .$$

However, we have only proved that $u^* \in H^1(\Omega, \mathbb{R}^{n+1})$. The rest of the problem is about the regularity of u^*. For $m = 2$, the harmonic map u^* with minimal energy is smooth if φ is, according to a result due to Morrey [Mo 2], but for $m > 2$, generally speaking, there is no such regularity (see Schoen and Uhlenbeck [ScU 1, ScU 2], Lin [Lin]. The definition and an existence result for harmonic maps between two Riemannian manifolds can be found in Eells and Sampson [ES]. The regularity problem for harmonic maps has attracted many authors. As a special case for elliptic systems, this regularity problem is related to the Hilbert 20th problem. It is proved in Helein [Hel 1] that for $m = 2$ any harmonic map is regular, but for $m = 3$ Riviere [Ri] showed that there exists a non-minimal energy harmonic map discontinuous everywhere.

Example 3. Nonlinear eigenvalue problem
Let $\Omega \subset \mathbb{R}^n$ be a bounded open domain with smooth boundary, and let $\phi \in C(\overline{\Omega} \times \mathbb{R}^1, \mathbb{R}^1)$ satisfy

(1) $|\phi(x,t)| \leqslant C(1 + |t|^{q-1}), \quad q < 2^* = \frac{2n}{n-2}$ if $n \geqslant 3$,
(2) $t\phi(x,t) > 0 \quad$ for $t \neq 0$;

then $\exists c_0 > 0$ such that $\forall c \in (0, c_0]$ the equation

$$\begin{cases} -\triangle u(x) = \lambda \phi(x, u(x)) & x \in \Omega , \\ u|_{\partial\Omega} = 0 , \end{cases}$$

has a solution (λ_c, u_c) satisfying

$$\int_\Omega \int_0^{u_c(x)} \phi(x,t) dt dx = c .$$

Proof. Set $X = H_0^1(\Omega)$, $g(u) = \int_\Omega \Phi(x, u(x))dx$, where $\Phi(x,t) = \int_0^t \phi(x, s)dx = O(|t|^q)$, $M_c = g^{-1}(c)$, and $f(u) = \frac{1}{2}\int_\Omega |\nabla u|^2$.

Since $\Phi(x,t) \geqslant 0$ but not identical to 0, we have $g(u) \geqslant 0$ but not identical to 0. We have $c_0 > 0$ such that $g^{-1}(c_0) \neq \emptyset$; let $u_0 \in g^{-1}(c_0)$, $t \mapsto g(tu_0)$ is continuous on $[0,1]$, and the function ranges over $[0, c_0]$. Therefore $g^{-1}(c) \neq \emptyset$ $\forall c \in [0, c_0]$. Since the embedding $H_0^1(\Omega) \to L^q(\Omega)$ is compact, g is completely continuous, i.e., $u_j \rightharpoonup u^* \Longrightarrow g(u_j) \to g(u^*)$. Thus M_c is a sequential weakly closed subset. Obviously, f is l.s.c., convex and coercive, so we have proved the existence of a minimum point u_c.

It remains to verify that $g'(u^*) = \phi(x, u^*(x)) \neq \theta$. In fact, $\theta \notin M_c$ for $c \neq 0$, then $u^* \neq \theta$, therefore, $\phi(x, u^*(x)) \neq \theta$.

As a corollary, we consider the nontrivial solution of the equation:

$$\begin{cases} -\Delta u = |u|^{q-2}u & \text{in } \Omega,\ 2 < q < 2^* \\ u|_{\partial\Omega} = 0\ . \end{cases}$$

Let $(\lambda_0, u_0) \in \mathbb{R}^1 \times H_0^1(\Omega)$ be a solution of the nonlinear eigenvalue problem:

$$\begin{cases} -\Delta u = \lambda |u|^{q-2}u \\ u|_{\partial\Omega} = 0\ , \end{cases}$$

with

$$\int_\Omega |u_0|^q > 0\ .$$

Let

$$\lambda_0 = \int_\Omega |\nabla u_0|^2 \Big/ \int_\Omega |u_0|^q > 0\ .$$

Setting

$$u^* = \lambda_0^{\frac{1}{q-2}} u_0\ ,$$

we obtain

$$\begin{cases} -\Delta u^* = |u^*|^{q-2}u^* & \text{in } \Omega \\ u^*|_{\partial\Omega} = 0\ . \end{cases}$$

□

Remark 4.2.4 *If $-\Delta$ is replaced by $-\Delta + I$, then the same conclusion holds.*

Example 4. (Prescribing constant mean curvature problem)

Given a constant $H > 0$ and a curve $\gamma : S^1 \to \mathbb{R}^3$. Suppose diam$(\gamma) = R$ with $HR < 1$. Find a disc-type surface $u : D^2 \xrightarrow{C^2} \mathbb{R}^3$ satisfying

$$\begin{cases} \Delta u = 2H u_x \wedge u_y & \text{in } D \\ u|_{\partial D} = \gamma\ , \end{cases} \tag{4.10}$$

where

$$u = (u^1, u^2, u^3)$$

and

$$u_x \wedge u_y = \left(\begin{vmatrix} u_x^2 & u_y^2 \\ u_x^3 & u_y^3 \end{vmatrix}, \begin{vmatrix} u_x^3 & u_y^3 \\ u_x^1 & u_y^1 \end{vmatrix}, \begin{vmatrix} u_x^1 & u_y^1 \\ u_x^2 & u_y^2 \end{vmatrix} \right) .$$

This is a geometric problem, in which H is a prescribed constant mean curvature, and γ is a prescribed boundary value.

Write the problem in its variational form; define

$$J(u) = \int_D |\nabla u|^2 + \frac{4H}{3} \int_D u \cdot (u_x \wedge u_y) \text{ on } H^1 \cap L^\infty(D, R^3) .$$

Since

$$\langle J'(u), v \rangle = 2\langle -\Delta u + 2H(u_x \wedge u_y), v \rangle ,$$

$\forall v \in C_0^\infty(D, \mathbb{R}^3)$. The Euler–Lagrange equation is exactly (4.10). In appearance, $\inf J(u) = -\infty$, there is no minimum. However, if we add a constraint condition:

$$\| u \|_{L^\infty(D,\mathbb{R}^3)} \leq R' ,$$

where $R < R'$, and $HR' < 1$, then by

$$|u_x \wedge u_y| \leqslant |u_x||u_y| \leqslant \frac{1}{2}(|u_x|^2 + |u_y|^2) ,$$

we have

$$J(u) \geqslant \int_D \left(|\nabla u|^2 - \frac{2}{3}|\nabla u|^2 \right) = \frac{1}{3} \int_D |\nabla u|^2 .$$

Moreover, on the convex set

$$C_{R'} = \{u \in H^1(D, \mathbb{R}^3) | \ u|_{\partial D} = \gamma, \ \| u \|_{L^\infty} \leq R'\} ,$$

which is sequentially weakly closed in $H^1(D, R^3)$, J is nonnegative and coercive. We verify the w.s.l.s.c. of J on $C_{R'}$ under H^1 norm. Suppose $u_j \rightharpoonup u^*$ $(H^1(D, \mathbb{R}^3))$, with $u_j \in C_{R'}$; we have a subsequence, for which we do not change the subscripts:

$$\begin{cases} u_j \to u^*, \text{ strongly in } L^2(D, \mathbb{R}^3) , \\ u_{j,x} \rightharpoonup u_x^*, \text{ weakly in } L^2(D, \mathbb{R}^3) , \\ u_{j,y} \rightharpoonup u_y^*, \text{ weakly in } L^2(D, \mathbb{R}^3) . \end{cases}$$

and

$$u_j(x) \to u^*(x) \quad \text{a.e. in } D .$$

Rewrite

$$u_j \cdot (u_{jx} \wedge u_{jy}) - u^* \cdot (u_x^* \wedge u_y^*) = v_j \cdot (u_x^* \wedge u_y^*) + u_j \cdot (v_{jx} \wedge u_y^* + u_x^* \wedge v_{jy}) + u_j \cdot (v_{jx} \wedge v_{jy}) ,$$

where $v_j = u_j - u^*$. Noticing

$$|u_j \wedge u_y^*| \leqslant R'|u_y^*| \,,$$

we obtain

$$\int_D |u_j \wedge u_y^* - u^* \wedge u_y^*|^2 \to 0 \,,$$

provided by Lebesgue's dominance theorem. Therefore

$$\int_D u_j \cdot (v_{jx} \wedge u_y^*) = -\int_D v_{jx} \cdot (u_j \wedge u_y^*) \to 0 \,.$$

Similarly

$$\int_D u_j \cdot (u_x^* \wedge v_{jy}) \to 0 \,.$$

Since

$$\int_D |u_x^* \wedge v_y^*| \leqslant \frac{1}{2}\int_D |\nabla u^*|^2 \,,$$

and $|v_j| \leqslant 2R'$, again by Lebesgue's dominance theorem, we have

$$\int_D v_j \cdot (u_x^* \wedge u_y^*) \to 0 \,.$$

Hence,

$$\begin{aligned} J(u_j) - J(u^*) &= \int_D |\nabla v_j|^2 + 2\nabla v_j \nabla u^* + \frac{4H}{3} u_j \cdot (v_{jx} \wedge v_{jy}) + \circ(1) \\ &\geqslant \frac{1}{3}\int_D |\nabla v_j|^2 + \circ(1) \,; \end{aligned}$$

it follows that

$$\underline{\lim}_{j\to\infty} J(u_j) \geqslant J(u^*) \,.$$

How do we drop the constraint? By definition, if $u^* \in C_{R'}$ achieves the minimum point of J on $C_{R'}$, then u^* satisfies the variational inequality:

$$\int_D \nabla u^* \nabla v + 2Hv \cdot (u_x^* \wedge u_y^*) \geqslant 0, \ \forall v = w - u^*, \tag{4.11}$$

where $w \in C_{R'}$. In particular, for any $\eta \in C_0^\infty(D, \mathbb{R}^1), \eta \geq 0$, one chooses $\varepsilon > 0$ such that $(1 - \varepsilon\eta)u^* \in C_{R'}$. Substituting $v = -\varepsilon\eta u^*$ into (4.11), we have

$$-\frac{1}{2}\triangle|u^*|^2 + |\nabla u^*|^2 + 2Hu^* \cdot (u_x^* \wedge u_y^*) \leqslant 0 \ \text{ in } D \,.$$

It follows that

$$\begin{cases} -\triangle|u^*|^2 \leqslant 0, \ \text{in } D \,, \\ |u^*| = |\gamma| \ \text{ on } \partial D \,. \end{cases}$$

Thus, $|u^*|^2$ is subharmonic. Applying the maximum principle, we have $\| u^* \|_{L^\infty} < R < R'$, i.e., u^* is strictly in the interior of $C_{R'}$ under the $L^\infty(D, \mathbb{R}^3)$ norm. Therefore we may choose arbitrarily $v \in C_0^\infty(D, R^3)$ with $\| v \|_{L^\infty} < R' - R$ as variations. Thus u^* is a weak solution of equation (4.10).

4.2.3 The Prescribing Gaussian Curvature Problem and the Schwarz Symmetric Rearrangement

We turn to studying a problem arising in differential geometry. Let (M, g_0) be a smooth compact two-dimensional Riemannian manifold with metric g_0; let $k(x)$ be its Gaussian curvature.

Given a function $K(x)$ on M, does there exist a metric g, which is pointwisely conformal to g_0, such that K is the Gaussian curvature with respect to g?

Setting a function u on M satisfying $g = e^{2u} g_0$, the problem is reduced to solving the following PDE:

$$\triangle u = k - K e^{2u} \text{ on } M \, , \tag{4.12}$$

where $\triangle$ is the Laplace–Beltrami operator with respect to g_0.

According to the Gauss–Bonnet formula:

$$\int_M K dV_g = 2\pi \chi(M) \, , \tag{4.13}$$

where $\chi(M)$ is the Euler characteristic of M, and V_g is the volume form with respect to g.

The curvature K is restricted by the topology $\chi(M)$; e.g., if $\chi(M) > 0$ (or < 0) then $\max K > 0$ (or $\min K < 0$ resp.), and if $\chi(M) = 0$, then either $K \equiv 0$ or K changes sign.

A result in the case where $\chi(M) < 0$ has been studied in Sect. 2.1, Example 5. Namely:

Statement. Assume $k(x) < 0$ and

$$K(x) < 0 \quad \forall x \in M \, . \tag{4.14}$$

Then (4.12) possesses a solution.

However, (4.14) is not a necessary condition; the problem remains open if one merely assumes $\min K < 0$.

In contrast, the case $\chi(M) = 0$ was completely solved by a variational argument. A necessary condition for the solvability of (4.12) with $\chi(M) = 0$ reads as

$$\begin{cases} \text{either } K \equiv 0 \, , \\ \text{or } K \text{ changes sign and } \int_M K e^{2v} dV_{g_0} < 0 \, , \end{cases} \tag{4.15}$$

where v is the solution of

$$\triangle v = k, \ \overline{v} = \frac{1}{\text{Vol}(M)} \int_M v = 0 \, . \tag{4.16}$$

In fact, if u is a solution of (4.12), and let $w = u - v$, then

$$-\triangle w = K e^{2(w+v)} \, . \tag{4.17}$$

(4.15) follows from

$$\int_M Ke^{2v} dV_{g_0} = -\int_M \triangle w e^{-2w} dV_{g_0}$$
$$= -2\int_M |\nabla w|^2 e^{-2w} dV_{g_0} < 0 .$$

Conversely, the condition (4.15) is also sufficient. Without loss of generality, we assume $K \not\equiv 0$ (otherwise, this is a linear problem). Let us consider a variational problem associated with (4.17):

$$J(w) = \frac{1}{2}\int_M |\nabla w|^2 dV_{g_0}$$

defined on

$$\mathcal{M} = \{w \in H^1(M) |\ g_1(w) = g_2(w) = 0\} ,$$

where

$$g_1(w) = \int_M w dV_{g_0} ,$$

and

$$g_2(w) = \int_M Ke^{2(w+v)} dV_{g_0} .$$

If $\mathcal{M}$ is a nonempty weakly closed Banach manifold, and if w is a local minimum point, then there exist Lagrange multipliers $\lambda, \mu \in R^1$ such that

$$\int_M \nabla w \cdot \nabla\varphi + \lambda\varphi + 2\mu K e^{2(w+v)}\varphi = 0 ,$$

$\forall \varphi \in H^1(M)$, i.e.,

$$-\triangle w + \lambda + 2\mu K e^{2(w+v)} = 0 .$$

From $g_2(w) = 0$, it follows that $\lambda = 0$; and according to (4.15) and integration by parts, we have $\mu < 0$. Let us choose $r = \frac{1}{2}\log(-2\mu)$, then $u = v+w+r$ solves (4.12).

We introduce the notion of the Schwarz symmetric rearrangement of a function and discuss its main properties, which will be often used in analysis.

For a nonnegative measurable function u defined on a n-dimensional measurable set Ω with $m(\Omega) < \infty$, the Schwarz symmetric rearrangement of u is defined to be the following:

$$u^*(y) = \sup\{t \geq 0 \,|\, |y|^n \leq C_n^{-1} m\{x \in \Omega \,|\, u(x) \geq t\}\} ,$$

where C_n is the volume of the unit ball in R^n. It possesses the following properties:

1. Let $u^*(y) = g(|y|)\, \forall y \in \Omega^* := B_R(\theta)$, where $R^n = \frac{1}{C_n} m(\Omega)$. Then g is nonincreasing and $g(R) = 0$.

2. Let $u^t = \{x \in \Omega \,|\, u(x) \geq t\} \, \forall t \geq 0$. Then $m(u^t) = m(u^{*t}) \, \forall t \geq 0$.
3. $\int_\Omega u^p = \int_{\Omega^*} u^{*p} \, \forall p \in [1, \infty)$.
4. (Faber–Krahn inequality, [Fa], [Krh]) If $u \in H^1(\Omega)$, where $\Omega \subset R^n$ is a bounded domain, then

$$\int_{\Omega^*} |\nabla u^*|^2 \leq \int_\Omega |\nabla u|^2 \,.$$

Proof. Let $H^{n-1}(E)$ denote the $(n-1)$-dimensional Hausdorff measure. From 2, the isoperimetric inequality, and the rotational invariance of u^*, we have

$$H^{n-1}(u^{-1}(t))^2 = \left(\int_{u^{-1}(t)} 1\right)^2 \leq \left(\int_{u^{-1}(t)} |\nabla u|\right)\left(\int_{u^{-1}} \frac{1}{|\nabla u|}\right) ,$$

$$H^{n-1}(u^{*-1}(t))^2 = \left(\int_{u^{*-1}(t)} 1\right)^2 = \left(\int_{u^{*-1}} |\nabla u^*|^2\right)\left(\int_{u^{*-1}(t)} \frac{1}{|\nabla u^*|}\right) ,$$

and

$$H^{n-1}(u^{*-1}(t)) \leq H^{n-1}(u^{-1}(t)) \,.$$

By the co-area formula and 2, it follows that

$$-\int_{u^{-1}(t)} \frac{1}{|\nabla u|} = \frac{dm(u^t)}{dt} = \frac{dm(u^{*t})}{dt} = -\int_{u^{*-1}(t)} \frac{1}{|\nabla u^*|} \,.$$

Thus

$$\int_{u^{*-1}(t)} |\nabla u^*| \leq \int_{u^{-1}(t)} |\nabla u| \,.$$

Again by the co-area formula

$$\int_{\Omega^*} |\nabla u^*|^2 = \int_0^\infty \int_{u^{*-1}(t)} |\nabla u^*| dt \leq \int_0^\infty \int_{u^{-1}(t)} |\nabla u| dt = \int_\Omega |\nabla u|^2 \,.$$

□

We now turn to the variational problem.

Lemma 4.2.5 *(Trudinger) Assume that Ω is a bounded planar domain with smooth boundary. Then $\forall u \in H_0^1(\Omega)$ with $\| u \|^2 := \int_\Omega |\nabla u|^2 \leq 1$, $\forall \beta < 4\pi$, $\exists \gamma = \gamma(\beta)$ such that $\int_\Omega e^{\beta u^2} \leq \gamma$.*

Proof. Since $\| \, |u| \, \| \leq \| u \|$, we may assume $u \geq 0$. Let $u^*(y)$ be the Schwarz symmetric rearrangement of u.

Thus $u^*(y)$ is radially symmetric, nonincreasing and satisfies

$$|\{y \in R^2 |\ u^*(y) \geq t\}| = |\{x \in \Omega |\ u(x) \geq t\}| \ \forall t \in R^1 \,.$$

Let $u^*(y) = g(|y|)$, and $R = (\pi^{-1}|\Omega|)^{\frac{1}{2}}$. We have $g(R) = 0$. By changing variables:

$$e^{-s} = \left(\frac{r}{R}\right)^2 \text{ and } f(s) = g(r) ,$$

we have $f(0) = 0,\ f'(s) = -\frac{R}{2}e^{-\frac{s}{2}}g'(r)$ and

$$\begin{aligned}
|f(s)| &\leq \int_0^s |f'(\sigma)|d\sigma \\
&\leq \sqrt{s}\left(\int_0^s |f'(\sigma)|^2 d\sigma\right)^{\frac{1}{2}} \\
&\leq \sqrt{\frac{s}{2}}\left(\int_0^R |g'(r)|^2\, r dr\right)^{\frac{1}{2}} \\
&= \sqrt{\frac{s}{4\pi}}\left(\int_{B_R} |\nabla u^*|^2\right) \\
&\leq \sqrt{\frac{s}{4\pi}}\left(\int_\Omega |\nabla u|^2\right)^{\frac{1}{2}} ,
\end{aligned}$$

provided by the Faber–Krahn inequality. Thus

$$\begin{aligned}
\int_\Omega \exp[\beta u^2] &= \int_{B_R} \exp[\beta (u^*)^2] \\
&= 2\pi \int_0^R \exp[\beta g(r)^2] r dr \\
&= |\Omega| \int_0^\infty \exp[\beta f(s)^2 - s] ds \\
&\leq \left(1 - \frac{\beta}{4\pi}\right)^{-1} |\Omega| .
\end{aligned}$$

□

Corollary 4.2.6 *There are constants $\beta > 0$, and $\gamma > 0$ such that*

$$\int_M \exp[\beta u^2] dV_{g_0} \leq \gamma ,$$

for all $u \in H^1(M)$ with $\overline{u} := \frac{1}{\mathrm{Vol}(M)}\int_M u = 0$ and $\| u \|^2 = \int_M |\nabla u|^2 dV_{g_0} = 1$.

Proof. By the use of the partition of unity, we have $u = \sum_{j=1}^n \chi_j u$, where the support of the function $\chi_j \geq 0\, \forall j$ is contained in a chart, and $\sum \chi_j \equiv 1$. Since $\overline{u} = 0$, from Poincaré's inequality, there is a constant $c > 0$ such that $|u|_2 := (\int_M |u|^2)^{\frac{1}{2}} \leq c \| u \|$, and then $\| \chi_j u \| \leq \| \nabla \chi_j \|_\infty |u|_2 + |\chi_j|_\infty \| u \| \leq c_1 \| u \|$ for some constant c_1. We obtain from Lemma 4.2.5:

$$\int_M \exp[\beta u^2] \leq \sum_{j=1}^{n} \frac{1}{n} \int_M \exp[n^2 \beta (\chi_j u)^2] \leq \gamma$$

for sufficiently small β. □

Corollary 4.2.7 *There exist constants $\beta > 0$, and $\gamma > 0$ such that*

$$\int_M e^{2u} dV_{g_0} \leq \gamma \exp[\beta^{-1} \parallel u \parallel^2 + 2\overline{u}] \; \forall u \in H^1(M) \, . \tag{4.18}$$

Proof. Let $a = \|u\|$, $v = a^{-1}(u - \overline{u})$, then $\overline{v} = 0$ and $\|v\| = 1$. From $2av \leq \beta v^2 + \frac{u^2}{\beta}$ and Corollary 4.2.6, we obtain

$$\int_M e^{2(u-\overline{u})} dV_{g_0} \leq \gamma \exp(\beta^{-1} \parallel u \parallel^2) \, .$$

The proof is complete. □

Lemma 4.2.8 *If $u_n \rightharpoonup u_0$ in $H^1(M)$, then after a subsequence, $e^{u_n} \to e^{u_0}$ in $L^2(M)$.*

Proof. We have

$$\begin{aligned} |e^{u_n} - e^{u_0}|_2^2 &= \int e^{2u_0} |e^{(u_n - u_0)} - 1|^2 \\ &\leq \int e^{2u_0} |u_n - u_0|^2 (1 + e^{2(u_n - u_0)}) \\ &\leq 2|e^{2u_0}|_4 (1 + |e^{2(u_n - u_0)}|_4) |u_n - u_0|_4^2 \, . \end{aligned}$$

On account of Corollary 4.2.7 the first two factors are bounded. Then by the compactness of Sobolev embedding, one has a subsequence such that $|u_n - u_0|_4 \to 0$. □

Corollary 4.2.9 *The functional g_2 is C^1 and is weakly continuous on $H^1(M)$ with $\langle g_2'(w), \varphi \rangle = 2 \int K e^{2(w+v)} \varphi \; \forall \varphi \in H^1(M)$.*

It remains to verify the following:

(1) $\mathcal{M}$ is a nonempty weakly closed Banach manifold,
(2) $J|\mathcal{M}$ is weakly lower semi-continuous,
(3) $J|\mathcal{M}$ is coercive.

Indeed, (2), (3) are trivial and the nonemptyness of $\mathcal{M}$ is easy to verify. The weak closedness of $\mathcal{M}$ follows from Corollary 4.2.9. Since the functions 1 and $Ke^{2(w+v)}$ are linear independent $\forall w \in \mathcal{M}$, $\mathcal{M}$ is a submanifold of $H^1(M)$. Namely, we have:

Statement. (M. S. Berger) Assume $\chi(M) = 0$. (4.12) is solvable if and only if K satisfies (4.15).

For $\chi(M) > 0$, topologically $M = S^2$ or RP^2. As we have seen in (4.13), a necessary condition of the solvability of the problem (4.12) is $\max K > 0$. If we follow the above procedure, let v be the solution of

$$-\triangle v = \overline{k} - k, \ \overline{v} = 0 .$$

We turn to the equation as before:

$$-\triangle w = K e^{2(w+v)} - \overline{k} , \tag{4.19}$$

and introduce a functional

$$J(w) = \frac{1}{2} \int_M (|\nabla w|^2 + 2\overline{k} w) dV_{g_0}$$

defined on

$$\mathcal{M} = \{ w \in H^1(M) | \ g_2(w) = \overline{k} \} .$$

If w is a local minimum point of J, then there exists $\mu \in R^1$ such that

$$-\triangle w = 2\mu K e^{2(w+v)} - \overline{k} .$$

From $w \in \mathcal{M}$, it follows that $\mu = 1/2$, i.e., (4.19) is the Euler–Lagrange equation for J.

From the Gauss–Bonnet formula, $\overline{k} > 0$, therefore if $\max K > 0$, $\mathcal{M} \neq \emptyset$. Similarly, one can show that $\mathcal{M}$ is a weakly closed Banach manifold, and that J is weakly lower semi-continuous. If J were coercive, then (4.12) would be solvable. But this is not true in general. In fact, set $\widetilde{w} = w - \overline{w}$, we have

$$2\overline{w} = \log \overline{k} \mathrm{Vol}(M) - \log \int_M K e^{2(v+\widetilde{w})} dV_{g_0} ,$$

and

$$\begin{aligned} J(w) &\geq \frac{1}{2} \left[\int_M |\nabla w|^2 + \overline{k} \, \mathrm{Vol}(M) \log \overline{k} \mathrm{Vol}(M) - \overline{k} \, \mathrm{Vol}(M) \log \int_M K e^{2(v+\widetilde{w})} \right] \\ &\geq \frac{1}{2} \left(1 - \frac{4\pi}{\beta} \right) \int_M |\nabla w|^2 + \mathrm{const.} \end{aligned}$$

due to (4.18) and the Gauss–Bonnet formula

$$\overline{k} \, \mathrm{Vol}(M) = \int_M k dV_{g_0} = 4\pi .$$

If $\beta > 4\pi$, then the coerciveness follows. But Moser [Mos 4] proved that the best constant $\beta = 4\pi$ for $M = S^2$ with the canonical metric g_0. This is the point where our argument breaks down.

However, for some special K, e.g., if K is even, we shall improve the estimate in (4.18). In this case, the sphere S^2 is reduced to the real projective space RP^2 geometrically.

Lemma 4.2.10 *(Aubin) For a function $w \in H^1(S^2)$, if there exist $f_j \in C^1(S^2), j = 1, 2 \ldots, k$ and $\alpha > 0$, satisfying*

$$\int_{S^2} f_j e^{2w} dV_{g_0} = 0, \ j = 1, 2, \ldots, k \ ,$$

and

$$\sum_{j=1}^{k} |f_j(x)| \geq \alpha \ \forall x \in S^2 \ ,$$

then $\forall \epsilon > 0 \ \exists C_\epsilon > 0$ such that

$$\int_{S^2} e^{2w} \leq C_\epsilon \exp \left[\frac{1}{8\pi - \epsilon} \int_{S^2} |\nabla w|^2 + 2\overline{w} \right] \quad \forall w \in H^1(S^2) \ .$$

Proof. Set

$$\Omega_j^{\pm} = \left\{ x \in S^2 | \ \pm f_j(x) \geq \frac{\alpha}{k} \right\} \ ,$$

and $g_j^{\pm}, \ h_j^{\pm} \in C^1(S^2)$ satisfying

$$\begin{aligned} &\text{supp} g_j^+ \cap \text{supp} h_j^- = \text{supp} g_j^- \cap \text{supp} h_j^+ = \emptyset, \ 0 \leq g_j^{\pm}, \ h^{\pm} \leq 1 \ , \\ &h_j^{\pm}(x) = 1 \ \forall x \in \Omega_j^{\pm}, \ \text{and} \ g_j^{\pm}(x) = 1 \ \text{if} \ \pm f_j(x) \geq 0 \ \forall j \ . \end{aligned}$$

Then

$$S^2 = \cup_{j=1}^k (\Omega_j^+ \cup \Omega_j^-) \ ,$$

and then

$$\int_{S^2} e^{2w} \leq \sum_{j=1}^{k} \left(\int_{\Omega_j^+} + \int_{\Omega_j^-} \right) e^{2w} \ .$$

Without loss of generality, one many assume that $\int_{\Omega_1^+} e^{2w}$ is the largest among $\int_{\Omega_j^{\pm}} e^{2w}, \ j = 1, 2, \ldots$.

First, $\forall w \in H^1(S^2)$, if $\| h_1^+ w \| \leq \| g_1^- w \|$, then $\forall \epsilon_1 > 0, \exists C_1 > 0$, such that

$$\begin{aligned} 2 \| h_1^+ w \|^2 &\leq \| h_1^+ w \|^2 + \| g_1^- w \|^2 \\ &\leq \| w \|^2 + C_1 (\| w \| \, |w|_2 + |w|_2^2) \\ &\leq (1 + \epsilon_1) \| w \|^2 + C_1 |w|_2^2 \ . \end{aligned}$$

Following Lemma 4.2.5 and Corollary 4.2.7,

$$\begin{aligned} \int_{S^2} e^{2h_1^+ w} &\leq C_{\epsilon_1} \exp \left[\frac{1}{4\pi - \epsilon_1} \| h_1^+ w \|^2 + 2C_2 |w|_2^2 \right] \\ &\leq C_{\epsilon_1} \exp \left[\frac{1 + \epsilon_1}{2(4\pi - \epsilon_1)} \|w\|^2 + C_3(\epsilon_1) |w|_2^2 \right] \ . \end{aligned}$$

Otherwise, we have

$$\begin{aligned} 2 \parallel g_1^- w \parallel^2 &\leq \parallel g_1^- w \parallel^2 + \parallel h_1^+ w \parallel^2 \\ &\leq (1+\epsilon_1) \parallel w \parallel^2 + C_1 |w|_2^2 , \end{aligned}$$

and

$$\begin{aligned} \int_{S^2} e^{2g_1^- w} &\leq C_{\epsilon_1} \exp\left[\frac{1}{4\pi - \epsilon_1} \parallel g_1^- w \parallel^2 + 2C_2 |w|_2^2\right] \\ &\leq C_{\epsilon_1} \exp\left[\frac{1+\epsilon_1}{2(4\pi - \epsilon_1)} \|w\|^2 + C_3(\epsilon_1)|w|_2^2\right]. \end{aligned}$$

Now let $u = w - \overline{w}$. $\forall \epsilon_0 > 0$, $\exists C_4 = C_4(\epsilon_0)$ such that

$$|u|_1 \leq \epsilon_0 \parallel u \parallel^2 + C_4 . \tag{4.20}$$

$\forall \epsilon_2 > 0$ small, we choose $a = a(\epsilon_2) > 0$ such that

$$|\{x \in S^2 | \, u(x) \geq a\}| = \epsilon_2 .$$

Therefore, we obtain

$$a \leq \frac{1}{\epsilon_2}|u|_1 \leq \frac{\epsilon_0}{\epsilon_2} \parallel u \parallel^2 + \frac{1}{\epsilon_2} C_4 , \tag{4.21}$$

from (4.20), and

$$|(u-a)_+|_2^2 \leq \epsilon_2^{\frac{1}{2}} |(u-a)_+|_4^2 \leq C_5 \epsilon_2^{\frac{1}{2}} \parallel u \parallel^2 . \tag{4.22}$$

Setting $\epsilon_0 = \frac{1}{2}\epsilon_2^2$, and choosing $\epsilon_1, \epsilon_2 > 0$ so small that

$$\frac{1+\epsilon_1}{2(4\pi - \epsilon_1)} + \epsilon_2 + C_3(\epsilon_1) C_5 \epsilon_2^{\frac{1}{2}} < \frac{1}{8\pi - \epsilon} ,$$

Substituting $w = (u-a)_+$ in the estimates of the integrals: $\int_{S^2} e^{2h_1^+ w}$ and $\int_{S^2} e^{2g_1^- w}$, since $\int_{S^2} e^{2h_1^+ u} \leq e^{2a} \int_{S^2} e^{2(h_1^+ u - a)_+}$ and $\int_{S^2} e^{2g_1^- u} \leq e^{2a} \int_{S^2} e^{2(g_1^- u - a)_+}$, the right-hand sides become $B_\epsilon \exp[\frac{1}{8\pi-\epsilon}\|u\|^2]$.

Next, for the specified w in the assumption, we have

$$\int_{S^2} e^{2u} \leq 2k \int_{\Omega_1^+} e^{2u} \leq 2k \int_{S^2} e^{2h_1^+ u} .$$

In particular, according to $\int f_1 e^{2u} = 0$, we also have

$$\int_{\Omega_1^+} e^{2u} \leq \frac{k}{\alpha} \int_{S^2} (f_1)_+ e^{2u} = \frac{k}{\alpha} \int_{S^2} (f_1)_- e^{2u}$$

$$\leq \text{Max}\{|f_i| \mid i = 1, \ldots, k\} \frac{k}{\alpha} \int_{S^2} e^{2g_1^- u} .$$

In both cases,

$$\int_{S^2} e^{2u} \leq C_\epsilon \exp \left[\frac{1}{8\pi - \epsilon} \parallel u \parallel^2 \right] . \tag{4.23}$$

i.e.,

$$\int_{S^2} e^{2w} \leq C_\epsilon \exp \left[\frac{1}{8\pi - \epsilon} \parallel w \parallel^2 + 2\overline{w} \right] .$$

□

Now, we return to the case $(M, g_0) = (S^2, g_0)$ where g_0 is the canonical metric, but assuming $K(x) = K(-x)$, $\forall x \in S^2$.

Let us define

$$\mathcal{M}^e = \{w \in H^1(S^2) | \ g_2(w) = \overline{k} \text{ and } w(x) = w(-x)\} .$$

Then, again, $\mathcal{M}^e$ is a nonempty weakly closed Banach manifold. The functional

$$J(w) = \frac{1}{2} \int_M (|\nabla w|^2 + 2\overline{k} w) dV_{g_0}$$

is again weakly lower semi-continuous.

Since $\forall w \in \mathcal{M}^e$

$$\int_{S^2} x_i e^{2w(x)} dV_{g_0} = 0 \quad i = 1, 2, 3 ,$$

where x_i are the Euclidean coordinates, when we embed S^2 into R^3 canonically, Lemma 4.2.10 is applicable, and then the coerciveness of J on $\mathcal{M}^e$ follows. Thus, we have:

Statement (J. Moser) For $M = RP^2$ with the canonical metric g_0, (4.12) is solvable if and only if $\max K > 0$.

Remark 4.2.11 *Under certain symmetric conditions on K other than the evenness, Hong [Hon 2] obtained the existence of solutions for equation (4.12).*

Remark 4.2.12 *The best constant $\beta = 4\pi$ is due to Moser; there are several different proofs and extensions, see Adams [Ad], Carleson and Chang [CC] and Ding and Tian [DiT].*

4.3 Quasi-Convexity

In the calculus of variations, we pay attention to the following functional:

$$J(u) = \int_\Omega f(x, u(x), \nabla u(x))dx \ , \tag{4.24}$$

where $f : \Omega \times \mathbb{R}^N \times \mathbb{R}^{nN} \to \mathbb{R}^1$ is a Caratheodory funciton, and $\Omega \subset \mathbb{R}^n$ is an open domain.

As we have seen, the w.l.s.c. (or w*l.s.c.) condition plays an important role in the proof of the existence of a minimizer.

4.3.1 Weak Continuity and Quasi-Convexity

From the abstract theory, we know that an l.s.c. function is w.l.s.c. and s.w.l.s.c. if it is convex. But to the above integral functional, what is the relationship between the s.w.l.s.c. (or s.w*.l.s.c.) with the convexity of the function f with respect to its variables (x, u, ξ)?

To this end we would like to understand better the weak convergence in the spaces $W^{1,p}(\Omega, \mathbb{R}^n)$, $1 \leqslant p \leqslant \infty$. Perhaps the simplest example of a weakly convergent but not strongly convergent sequence in our mind is $\{\sin(mt)\}_1^\infty$ in $L^2([0, 2\pi])$. The phenomenon is caused by the large oscillations of the sequence of functions. It can be extended as follows:

Lemma 4.3.1 *Let* $D = \prod_{j=1}^n (a_j, b_j)$ *be a rectangle in* $\mathbb{R}^n$*, and let* $\varphi \in L^p(D)$*,* $1 \leqslant p \leqslant \infty$*, which is extended periodically to* $\mathbb{R}^n$*. Let* $\varphi_m(x) = \varphi(mx)$*,* $\forall m \in \mathbb{N}$*, and*

$$\overline{\varphi} = \frac{1}{m(D)} \int_D \varphi(x)dx \ .$$

Then $\varphi_m \rightharpoonup \overline{\varphi}$ *in* $L^p(D)$ $1 < p < \infty$ *and* $\varphi_m \overset{*}{\rightharpoonup} \overline{\varphi}$ *in* $L^\infty(D)$*.*

Proof. First, we may assume $\overline{\varphi} = 0$. Otherwise, we consider $\widetilde{\varphi} = \varphi - \overline{\varphi}$ instead.
Second,

$$\| \varphi_m \|_p^p = \int_D |\varphi(mx)|^p dx = \frac{1}{m^n} \int_{m \cdot D} |\varphi(y)|^p dy = \| \varphi \|_p^p, \ \forall 1 \leqslant p \leqslant \infty \ .$$

Third, we define a signed set function $\Phi(E) = \int_E \varphi$ for any measurable set E. It is σ-additive, and satisfies $\Phi(x + D) = 0, \quad \forall x \in \mathbb{R}^n$.

Now, $\forall$ rectangles $Q = \prod_{i=1}^n (c_i, d_i)$, by cancelling the nonoverlapping translations of D in kQ, we have the estimate:

$$\left| \int_D \varphi_k \chi_Q \right| = \left| \int_Q \varphi_k \right| = \frac{1}{k^n} \left| \Phi(kQ) \right| \leqslant \frac{n}{k} \int_D |\varphi| \ .$$

Thus, for simple functions of the form $\xi = \sum \alpha_i \chi_{Q_i}$, $Q_i \cap Q_j = \emptyset \ i \neq j$, we have

$$\int_D \varphi_k \cdot \xi \to 0 \quad \text{as } k \to \infty \ .$$

As $1 < p \leqslant \infty$, the simple functions of the above form consist of a dense subset of $L^{p'}(D)$, $p' = \frac{p}{p-1}$. $\forall f \in L^{p'}(D)$, $\exists \xi$ a simple function as above such that $\| f - \xi \|_{p'} < \varepsilon/(2 \| \varphi \|_p)$. Therefore

$$\left| \int_D \varphi_k \cdot f \right| \leqslant \| \varphi_k \|_p \| f - \xi \|_{p'} + \left| \int_D \varphi_k \cdot \xi \right| < \varepsilon$$

for k large enough. □

Remark 4.3.2 *Lemma 4.3.1 also holds for $p = 1$.* In fact, the only problem in the above proof for $p = 1$ is that the simple functions ξ cannot approximate any $f \in L^\infty$ in L^∞ norm. But, one can choose ξ such that $\| f - \xi \|_1 < \varepsilon$, with $|\alpha_i| \leqslant 2 \| f \|_\infty$ $\forall i$. Now, $\forall \lambda > 0$, let $E_{k,\lambda} = \{x \in D | \, |\varphi_k(x)| \geqslant \lambda\}$, $\forall k$. Since $\varphi_1 = \varphi \in L^1(D)$, $\forall \varepsilon > 0$, $\exists \lambda = \lambda(\varepsilon)$ such that $\int_{E_{1,\lambda}} |\varphi| < \varepsilon$. From the definition of φ_k, so is $\int_{E_{k,\lambda}} |\varphi_k| < \varepsilon$. Again,

$$\left| \int_D \varphi_k \cdot f \right| \leqslant \int_D |\varphi_k (f - \xi)| + \left| \int_D \varphi_k \cdot \xi \right| = I + II$$

where $II \to 0$ as $k \to \infty$, and

$$\begin{aligned} I &= \int_D |\varphi_k (f - \xi)| \\ &= \int_{E_{k,\lambda}} |\varphi_k| |f - \xi| + \int_{D \setminus E_{k,\lambda}} |\varphi_k (f - \xi)| \\ &\leqslant 3 \| f \|_\infty \cdot \int_{E_{k,\lambda}} |\varphi_k| + \lambda \| f - \xi \|_1 . \end{aligned}$$

Again, we have proved that $\varphi_k \rightharpoonup \overline{\varphi}$ in $L^1(D)$.

Corollary 4.3.3 *Let $\Omega = (0,1)$, $0 < \lambda < 1$, α, $\beta \in \mathbb{R}^1$, and*

$$\varphi(x) = \begin{cases} \alpha & \text{if } x \in (0, \lambda) \\ \beta & \text{if } x \in (\lambda, 1) \, , \end{cases}$$

then $\varphi_k \rightharpoonup (* \rightharpoonup) \lambda\alpha + (1 - \lambda)\beta$ in L^p, $p \in [1, \infty), (p = \infty$, resp.).

Now, we turn to studying the s.w.l.s.c. (or s.w*.l.s.c.) of J in $W^{1,p}(\Omega), 1 \leq p \leq \infty$. For simplicity, we assume $D = [0,1]^n$, and $f = f(\xi)$. $\forall k \in \mathcal{N}$, let D_l^k be a sub-cube of D with length 2^{-k} on each side and centered at $c_l^k = 2^{-k}(y_1^l + \frac{1}{2}, \ldots, y_n^l + \frac{1}{2})$, where $(y_1^l, \ldots, y_n^l)$ runs over the lattice points $(0, 1, \ldots, 2^k - 1)^n$, $l = 1, \ldots, 2^{kn}$. Then $D = \bigcup_{l=1}^{2^{kn}} D_l^k, \forall k \in \mathcal{N}$. $\forall v \in C_0^\infty(D, \mathbb{R}^N)$. Let us define

$$w_k(x) = \frac{1}{2^k} v(2^k(x - c_l^k)), \ \forall x \in D_l^k, \ \forall l = 1, \ldots, 2^{kn} . \tag{4.25}$$

Then

$$\nabla w_k(x) = \nabla v(2^k(x - c_l^k)) \quad \forall x \in D_l^k,\ \forall l = 1, \dots, 2^{kn}$$

and

$$\begin{cases} w_k \to 0 \quad L^\infty(D)\ , \\ \nabla w_k \rightharpoonup (^*\rightharpoonup) 0 \text{ in } L^p(D)\ p \in [1, \infty) (p = \infty \,\text{resp.})\ . \end{cases}$$

provided by Lemma 4.3.1. Now we arrive at:

Lemma 4.3.4 *Let $\Omega \subset \mathbb{R}^n$ be a domain, and let $J(u) = \int_\Omega f(\nabla u)$. If J is s.w.l.s.c. on $W^{1,p}(\Omega, \mathbb{R}^N)$ $1 \leqslant p < \infty$ (s.w*l.s.c. on $W^{1,\infty}(\Omega, \mathbb{R}^N)$), then for $\forall A \in M^{n \times N}$, $\forall$ cube $D \subset \overline{D} \subset \Omega$, we have*

$$f(A) \leqslant \frac{1}{m(D)} \int_D f(A + \nabla v) \quad \forall v \in W_0^{1,\infty}(D, \mathbb{R}^N)\ ,$$

where $M^{n \times N}$ denotes the $n \times N$ matrix space.

Proof. We choose a sequence in the Banach space $X = W^{1,p}(\Omega, \mathbb{R}^N)$ as follows:

$$u_k(x) = Ax + w_k(x) \quad k = 1, 2, \dots\ ,$$

where the sequence $\{w_k\}$ is defined in (4.25) and equals to zero outside D. Then it weakly converges to $u(x) = Ax$ (w^*-converges as $p = \infty$,). It follows that

$$m(\Omega) f(A) = J(u) \leqslant \liminf_{k \to \infty} J(u_k)\ .$$

But

$$\begin{aligned} J(u_k) &= \int_D f(A + \nabla w_k(x)) dx + \int_{\Omega \backslash D} f(A) dx \\ &= \sum_{l=1}^{2^{kn}} \int_{D_l^k} f(A + \nabla v(2^k(x - c_l^k))) dx + f(A) m(\Omega \backslash D) \\ &= \int_D f(A + \nabla v(x)) dx + f(A) m(\Omega \backslash D)\ . \end{aligned}$$

This is our conclusion. □

According to Lemma 4.3.4, we introduce the following:

Definition 4.3.5 *A Borel measurable and locally integrable function $f: \mathbb{R}^{nN} \to \mathbb{R}^1$ is called quasi-convex (in the Morrey sense), if for $\forall A \in M^{n \times N}$, $\forall v \in W_0^{1,\infty}(D, \mathbb{R}^N)$, and for any cube $D \subset \mathbb{R}^\kappa$, the inequality*

$$m(D) f(A) \leqslant \int_D f(A + \nabla v(x)) dx \tag{4.26}$$

holds.

In the sequel of this chapter, we shall specify the terminology "quasi-convexity" to be quasi-convexity in the Morrey sense.

Remark 4.3.6 *The cube D in the above definition can be replaced by any bounded domains, from a simple scaling and covering argument.*

Let us study the relationship between the convexity and the quasi-convexity.

On the one hand, according to Jessen's inequality, it is easy to see that if f is convex, then

$$\frac{1}{m(D)}\int_D f\left(A+\nabla v(x)\right)dx \geqslant f\left(\frac{1}{m(D)}\int_D \left(A+\nabla v(x)\right)dx\right) = f(A)\,;$$

i.e., convexity $\Longrightarrow$ quasi-convexity.

On the other hand, $\forall B, C \in M^{n\times N}$, with rank $(B-C)=1$, $\forall\lambda\in(0,1)$, let $A=\lambda B+(1-\lambda)C$. After suitable translation and rotation, we may assume $B=(1-\lambda)a\otimes e_1$ and $C=-\lambda a\otimes e_1$ for some $a\in\mathbb{R}^N$ and $e_1\in\mathbb{R}^n$. Let us introduce a 1-periodic sawtooth function:

$$\varphi(t)=\begin{cases}(1-\lambda)t, & t\in[0,\lambda]\,,\\ -\lambda(t-1), & t\in[\lambda,1]\,.\end{cases}$$

For $\forall x\in D=[0,1]^n$, let

$$u_k(x)=ak^{-1}\varphi(kx^1)\,,$$

then

$$\nabla u_k(x)=a\otimes e_1\begin{cases}(1-\lambda) & \{kx^1\}\in(0,\lambda)\\ -\lambda & \{kx^1\}\in(\lambda,1)\,,\end{cases}$$

where $\{y\}$ denotes the fractional part of $y\in\mathbb{R}^1$; and let

$$v_k(x)=a\min\{k^{-1}\varphi(kx^1),\ \mathrm{dist}\,(x,\partial D)\}\,,$$

where $\mathrm{dist}\,(x,\partial D)=\inf\{\sup_{1\leqslant i\leqslant n}\|\,x_i-y_i\,\|\mid y=(y_1,\dots,y_n)\in\partial D\}$. Then we have $v_k|_{\partial D}=0$ and there is a constant $K>0$ such that $|v_k(x)-v_k(y)|\le\|x-y\|$. Therefore $v_k\in W_0^{1,p}(D,R^N)$. Furthermore, we have

$$m\{x\in D|\ \nabla u_k(x)\neq\nabla v_k(x)\}\to 0\,.$$

as $k\to\infty$.

Let us divide D into two disjoint parts D_1 and D_2, where $D_1=\{x\in D|\nabla u_k(x)=(1-\lambda)a\otimes e_1\}$ and $D_2=\{x\in D|\nabla u_k=-\lambda a\otimes e_1\}$. Thus $D=D_1\cup D_2$.

If f is quasi-convex, then

$$m(D)f(\lambda B+(1-\lambda)C)\le\int_D f(\lambda B+(1-\lambda)C+\nabla v_k)\,.$$

On the account of the local integrability of f and the absolute continuity of the integral, we have

$$\begin{aligned}\lim \int_D f(\lambda B + (1-\lambda)C + \nabla v_k) &= \lim \int_D f(\lambda B + (1-\lambda)C + \nabla u_k) \\ &= \lim \int_{D_1} f(B) + \int_{D_2} f(C) \\ &= \lambda m(D) f(B) + (1-\lambda) m(D) f(C)\,.\end{aligned}$$

Finally, we obtain:

$$f(\lambda B + (1-\lambda)C) \leqslant \lambda f(B) + (1-\lambda) f(C)\,,$$

i.e., f is convex along segments connecting two matrices with rank-1 difference.

Definition 4.3.7 *A function $f : \mathbb{M}^{n\times N} \to \mathbb{R}^1 \cup \{+\infty\}$ is said to be rank one convex, if*

$$f(\lambda B + (1-\lambda)C) \leqslant \lambda f(B) + (1-\lambda) f(C)$$

$\forall \lambda \in [0,1]$, $\forall B, C \in \mathbb{M}^{n\times N}$ *with* $\mathrm{rank}\{B - C\} \leqslant 1$.

As a direct consequence, we see that either $n = 1$ or $N = 1$, rank one convex = convex.

In summary, we have proved:

convex $\Longrightarrow$ quasi-convex $\Longrightarrow$ rank one convex.

In particular, either $n = 1$ or $N = 1$, quasi-convex = convex.

A natural question: are there any counterexamples of these reverse implications?

1. Quasi-convex in Morrey sense $\not\Longleftarrow$ convex. For $A \in \mathbb{M}^{n\times n}$, $\det A$ is quasi-convex in Morrey sense, but not convex.

It is easily seen that $\det A$ is not convex, even for $n = 2$.

Let us show that it is quasi-convex in Morrey sense. More precisely, we have

$$\frac{1}{m(D)} \int_D \det(A + \nabla v) = \det(A) \quad \forall D \subset \Omega,\ \forall v \in C_0^\infty(D, \mathbb{R}^n)\,.$$

For simplifying the computation, we only verify the cases $n \leqslant 3$. Higher-dimension cases can be shown by induction.

For $n = 2$, let

$$A = \begin{pmatrix} a_{11} & a_{12} \\ a_{21} & a_{22} \end{pmatrix}, \quad v = \begin{pmatrix} v_1 \\ v_2 \end{pmatrix} \in C_0^\infty(D, \mathbb{R}^2)\,.$$

From

$$\det(\nabla v) = \partial_{x_1}(v_1 \partial_{x_2} v_2) - \partial_{x_2}(v_1 \partial_{x_1} v_2)\,,$$

we have

$$\int_D \det(\nabla v) = 0\,,$$

and then

$$\begin{aligned}&\frac{1}{m(D)}\int_D \det(A+\nabla v)\\&=\frac{1}{m(D)}\int_D[\det(A)+a_{11}\partial_{x_2}v_2+a_{22}\partial_{x_1}v_1-a_{12}\partial_{x_1}v_2-a_{21}\partial_{x_2}v_1+\det(\nabla v)]\\&=\det(A)\,.\end{aligned}$$

For $n=3$, let $A=(a_{ij})_{1\leqslant i,j\leqslant 3}$, $\mathrm{adj}_2(A)=(A_{ij})_{1\leqslant i,j\leqslant 3}$ where A_{ij} is the (ij) minor of A,

$$\begin{aligned}\det(A+\nabla v)&=\langle (A+\nabla v)^1,(\mathrm{adj}_2(A+\nabla v))^1\rangle\\&=\langle A^1,(\mathrm{adj}_2(A+\nabla v))^1\rangle+\langle(\nabla v)^1,(\mathrm{adj}_2(A+\nabla v))^1\rangle\\&=\sum_{i=1}^3\left[A_i^1(\mathrm{adj}_2(A+\nabla v))_i^1+\frac{\partial v_1}{\partial x_i}(\mathrm{adj}_2(A+\nabla v))_i^1\right],\end{aligned}$$

where $B^1=(b_{11},\dots,b_{1n})$ is the first row vector of the matrix $B=(b_{ij})_{1\leqslant i,j\leqslant n}$, and $\langle,\rangle$ is the scalar product in $\mathbb{R}^n$. By the conclusion for case $n=2$, we have

$$\begin{aligned}&\frac{1}{m(D)}\sum_{i=1}^3\int_D A_i^1(\mathrm{adj}_2(A+\nabla v))_i^1\\&=\frac{1}{m(D)}\sum_{i=1}^3 A_i^1(\mathrm{adj}_2(A))_i^1\\&=\det(A)\,.\end{aligned}$$

Since

$$\begin{aligned}&\sum_{i=1}^3\frac{\partial}{\partial x_i}(\mathrm{adj}_2(A+\nabla v))_i^1\\&=\frac{\partial}{\partial x_1}\begin{vmatrix}\frac{\partial v_2}{\partial x_2}&\frac{\partial v_2}{\partial x_3}\\\frac{\partial v_3}{\partial x_2}&\frac{\partial v_3}{\partial x_3}\end{vmatrix}+\frac{\partial}{\partial x_2}\begin{vmatrix}\frac{\partial v_2}{\partial x_3}&\frac{\partial v_2}{\partial x_1}\\\frac{\partial v_3}{\partial x_3}&\frac{\partial v_3}{\partial x_1}\end{vmatrix}+\frac{\partial}{\partial x_3}\begin{vmatrix}\frac{\partial v_2}{\partial x_1}&\frac{\partial v_2}{\partial x_2}\\\frac{\partial v_3}{\partial x_1}&\frac{\partial v_3}{\partial x_2}\end{vmatrix}\\&=0\,,\end{aligned}$$

after integration by parts, it follows that

$$\int_D\sum_{i=1}^3\frac{\partial v_1}{\partial x_i}(\mathrm{adj}_2(A+\nabla v))_i^1=0\,.$$

This proves the conclusion.

2. Rank one convex $\not\Longrightarrow$ quasi-convex. There is a counterexample due to Sverak [Sv 1].

4.3.2 Morrey Theorem

The importance of the notion of quasi-convexity is due to the following:

Theorem 4.3.8 *Suppose that $f : M^{n\times N} \to \mathbb{R}^1$ is continuous and quasi-convex. If the following growth condition holds:*

$$(C_p) \quad \begin{aligned} &|f(A)| \leqslant \alpha(1+|A|), \quad as \ \ p=1\,, \\ &-\alpha(1+|A|^q) \leqslant f(A) \leqslant \alpha(1+|A|^p), \ \ as \ \ 1\leqslant q<p<\infty\,, \\ &|f(A)| \leqslant \eta(|A|), as \ p=+\infty\,, \end{aligned}$$

where η is continuous and increasing, and $\alpha > 0$, $\forall A \in M^{n\times N}$, then for every bounded open domain $\Omega \subset \mathbb{R}^n$,

$$J(u) = \int_\Omega f(\nabla u)dx$$

is s.w.l.s.c. in $W^{1,p}(\Omega, \mathbb{R}^N)$ (s.w.l.s.c. in $W^{1,\infty}(\Omega, \mathbb{R}^N)$).*

Before going to the proof, we need the piecewise affine function approximation in Sobolev spaces.

Definition 4.3.9 *(Triangulation) Let $\Omega \subset R^n$ be a bounded domain with piecewise affine boundary. A triangulation $\tilde{\tau}$ of Ω is a collection of finitely many n-simplices $\{K_i | i = 1, 2, \ldots, I\}$, such that $\forall i \neq j, K_i \cap K_j$ is either empty or equal to a p-simplex, $0 \le p \le n-1$, and $\Omega = \bigcup_{0\le i\le I} K_i$.*

We call $h_{\tilde{\tau}} = \max\{\mathrm{diam}(K_i)|i = 1, \ldots, n\}$ the mesh size of $\tilde{\tau}$.

To an n-simplex, $K = \{p_0, p_1, \ldots, p_n\}$, (in which, $p_i \in R^n, i = 0, 1, \ldots, n$, and $\{p_i - p_0 | i = 1, 2, \ldots, n\}$ are linearly independent). We define $n+1$ affine functions $\{\lambda_0, \lambda_1, \ldots, \lambda_n\}$ such that

$$\lambda_i(p_j) = \delta_{ij}, \quad i, j = 0, 1, \ldots, n\,.$$

$\forall v \in C(K)$, we define a piecewise affine function by the following interpolation formula:

$$\tilde{v}_K(x) = \sum_{i=0}^n v(p_i)\lambda_i(x)\,. \tag{4.27}$$

One shows that $\forall v \in C^2(K), \forall x \in K$,

$$\|\nabla v - \nabla \tilde{v}_K\|_\infty \le \frac{n^2(n+1)}{2}\frac{h_K^2}{\rho_K}\|v\|_{C^2(K)}\,, \tag{4.28}$$

where $h_K = \mathrm{diam}\, K$, and $\rho_K = \sup\{2R|B_R(x) \subset K, x \in K\}$.

In fact, let $p_i = (p_i^1, \ldots, p_i^n), i = 0, 1, \ldots, n$. We consider the functions $v_j(x) = x^j, j = 1, 2, \ldots, n$, where $x = (x^1, \ldots, x^n)$, and $v_0(x) \equiv 1$ respectively. We obtain:

$$x^j = \sum_0^n p_i^j \lambda_i(x), \ j = 1, 2, \ldots, n, \text{ and } \sum_0^n \lambda_i(x) = 1\,.$$

Differentiating these formulas, it follows that

$$\sum_{i=0}^{n} p_i^j \frac{\partial \lambda_i}{\partial x^k} = \delta_k^j, \ j,k = 1,2,\ldots,n,$$

$$\sum_{i=0}^{n} \frac{\partial \lambda_i}{\partial x^k} = 0, \quad k = 1,2,\ldots,n\,.$$

Thus by Taylor expansion,

$$\begin{aligned}\frac{\partial v_K(x)}{\partial x^k} &= \sum_{i=0}^{n} v(p_i)\frac{\partial \lambda_i(x)}{\partial x^k}\\ &= \sum_{i=0}^{n}\left[v(x) + \nabla v(x)(p_i - x) + \frac{1}{2}\nabla^2 v(\xi_x)(p_i - x)^2\right]\frac{\partial \lambda_i(x)}{\partial x^k}\\ &= \frac{\partial v(x)}{\partial x^k} + \sum_{i=0}^{n} R_i(x)\frac{\partial \lambda_i(x)}{\partial x^k}\,,\end{aligned}$$

where $\xi_x \in K$, and $R_i(x) = \frac{1}{2}\nabla^2 v(\xi_x)(p_i - x)^2$. Since

$$|R_i(x)| \leq \frac{n^2}{2}\|v\|_{C^2} h_K^2 \ \forall i\,,$$

and $|\nabla \lambda_i| \leq \frac{1}{\rho_K}$. We obtain the desired estimate (4.28).

A triangulation $\tilde{\tau}$ is called regular, if $\exists$ a constant $C > 0$, such that $\frac{h_K}{\rho_K} \leq C, \forall K \in \tilde{\tau}$.

Lemma 4.3.10 *Suppose that $\Omega \subset R^n$ is a bounded domain with piecewise affine boundary. Then for $\forall u \in W^{1,p}(\Omega, R^N), \forall \epsilon > 0, \exists$ a piecewise affine function v on Ω, satisfying:*

1. $\|u - v\|_{1,p} < \epsilon$,
2. $|\nabla v(x)| \leq |\nabla u(x)|$, *and* $|v(x)| \leq |u(x)|$ *a.e.*

Proof. With the aid of the extension operator and the mollifier, we have a function $\tilde{u} \in C^2(\overline{\Omega}, R^N)$, such that

1. $\|\tilde{u} - u\|_{1,p} < \frac{\epsilon}{3}$,
2. $|\tilde{u}(x)| \leq |u(x)|, |\nabla \tilde{u}(x)| \leq |\nabla u(x)|$ a.e.,
3. $\|\tilde{u}\|_{C^2(\overline{\Omega})} \leq C\|u\|_{1,p}$,

where $C > 0$ is a constant depending on Ω.

To any regular triangulation $\tilde{\tau}$ of Ω, we define a linear operator π from $W^{1,p}(\Omega, R^N)$ into itself as follows:

$$(\pi u)(x) = \tilde{u}_K(x) \ \forall x \in K, \forall K \in \tilde{\tau}\,,$$

where $\tilde{u}_K$ is defined as in equation (4.27).

From $|(\pi u)(x)| \le |u(x)|$, and $|\nabla(\pi u)(x)| \le |\nabla u(x)|$ *a.e.*, it follows that $\|\pi\| \le 1$. On account of (4.28) $\|\tilde{u} - \pi\tilde{u}\|_{1,p}$ is small as $h_{\tilde{\tau}} > 0$ is small. Therefore,

$$\|u - \pi u\|_{1,p} \le \|u - \tilde{u}\|_{1,p} + \|\tilde{u} - \pi\tilde{u}\|_{1,p} + \|\pi\tilde{u} - \pi u\|_{1,p} < \epsilon\,.$$

Set $v = \pi u$. This is what we need. □

Now, let us return to the proof of Theorem 4.3.8 for the case $p = \infty$. Other cases are similar, so are omitted.

Proof. Suppose $u \in W^{1,\infty}(\Omega, R^N)$ and $u_j \stackrel{*}{\rightharpoonup} u$ in $W^{1,\infty}$, we shall prove that

$$\liminf \int_\Omega f(\nabla u_j) \ge \int_\Omega f(\nabla u)\,.$$

We may assume $|\nabla u_j|, |\nabla u| \le C, M = \sup\{f(\xi) | |\xi| \le 3C+1\}$. $\forall \epsilon > 0 \exists \delta > 0$ such that $|f(\xi) - f(\xi')| \le \frac{\epsilon}{6m(\Omega)}$ as $|\xi - \xi'| \le \delta$ and $|\xi|, |\xi'| \le 3C$. One chooses $\Omega_1 \subset \overline{\Omega_1} \subset \Omega$, in which $\partial\Omega_1$ is piecewise affine with $m(\Omega \backslash \Omega_1) < \frac{\epsilon}{8M}$. There exists a regular triangulation of $\Omega_1 : \tau_k = \{D_1^k, \ldots, D_{I_k}^k\}, \exists v_k \in W^{1,\infty}(\Omega, R^N)$ such that v_k is a piecewise affine function on Ω_1 with $v_k = u$ on $\partial\Omega$, and $\|v_k - u\|_{1,\infty} < \epsilon$. Let $v_k = A_i^k\ \forall x \in D_i^k$. This can be realized by considering a domain $\Omega_2 : \overline{\Omega_1} \subset \Omega_2 \subset \overline{\Omega_2} \subset \Omega$, with a piecewise affine boundary $\partial\Omega_2$, which is parallel to $\partial\Omega_1$. We define v_k in Ω_1 according to Lemma 4.3.10, $v_k(x) = u(x),\ \forall x \in \Omega \backslash \Omega_2$, and linearly interpolate $u|_{\partial\Omega_2}$ and $v_k|_{\partial\Omega_1}$ in $\Omega_2 \backslash \Omega_1$.

According to Lemma 4.3.10, we may choose τ_k with $h_{\tau_k} \to 0$ such that $\nabla v_k \to \nabla u$ in measure.

Let $u_{jk} = v_k + u_j - u$, then $u_{jk} \stackrel{*}{\rightharpoonup} v_k$ in $W^{1,\infty}$ as $j \to \infty$; and $|\nabla u_{jk}| \le 3C$. One has $k_0 \in \mathcal{N}$, as $k > k_0$,

$$\int_\Omega |f(\nabla v_k) - f(\nabla u)| < \frac{\epsilon}{6} \tag{4.29}$$

and

$$\int_\Omega |f(\nabla u_{jk}) - f(\nabla u_j)| < \frac{\epsilon}{6}\,. \tag{4.30}$$

Fixing $k > k_0$, let p_i^k be the center of $D_i^k, \forall r \in (0,1)$; we consider a similar subsimplex $C_i^k = r(D_i^k - \{p_i^k\}) + p_i^k, i = 1, 2, \ldots, I_k$, and $\Omega_1' = \bigcup_1^{I_k} C_i^k$. Define a smooth function ρ on Ω_1, such that $\rho(x) = 1\ \forall x \in \Omega_1', \rho(x) = 0\ \forall x \in \bigcup_1^{I_k} \partial D_i^k$, and $|\nabla\rho| \le C(\tilde{\tau_k}, r)$, a constant depending on $\tilde{\tau_k}$ and r.

Define $v_{jk} = v_k + \rho(u_{jk} - v_k) = v_k + \rho(u_j - u)$; we have

$$\begin{aligned}
\int_{\Omega_1} f(\nabla u_{jk}) &= \int_{\Omega_1'} f(\nabla v_{jk}) + \int_{\Omega_1 \backslash \Omega_1'} f(\nabla u_{jk}) \\
&= \int_{\Omega_1} f(\nabla v_{jk}) + \int_{\Omega_1 \backslash \Omega_1'} [f(\nabla u_{jk}) - f(\nabla v_{jk})] \\
&= A + B + C\,.
\end{aligned}$$

On account of the quasi-convexity of f; we have

$$A = \int_{\Omega_1} f(\nabla v_{jk}) = \sum_{i=1}^{I_k} \int_{D_i^k} f(A_i^k + \nabla[\rho(u_{jk} - u)])$$
$$\geq \sum_{i=1}^{I_k} f(A_i^k) m(D_i^k)$$
$$= \int_{\Omega_1} f(\nabla v_k) .$$

It is easily seen that

$$B = \int_{\Omega_1 \backslash \Omega_1'} f(\nabla u_{jk}) \geq -M(1 - r^n) m(\Omega_1) .$$

As to C, from $\|u_j - u\|_\infty \to 0$ as $j \to \infty$, we obtain

$$\|\nabla v_{jk}\|_\infty \leq \|\nabla v_k\|_\infty + \|\nabla u_j\|_\infty + \|\nabla u\|_\infty + \|\nabla \rho\|_\infty \|u_j - u\|_\infty \leq 3C + 1 ,$$

as $j > j_0(\tilde{\tau}_k, r)$. Thus

$$\int_{\Omega_1 \backslash \Omega_1'} f(\nabla v_{jk}) \leq M(1 - r^n) m(\Omega_1) .$$

Putting them together, we have

$$\int_{\Omega_1} f(\nabla u_{jk}) \geq \int_{\Omega_1} f(\nabla v_k) - 2Mm(\Omega_1)(1 - r^n), \tag{4.31}$$

as $j > j_0(k, r)$.

Combining (4.29), (4.30) and (4.31), and suitably choosing $r \in (0, 1)$, we have

$$\int_\Omega f(\nabla u_j) \geq \int_{\Omega \backslash \Omega_1} f(\nabla u_{jk}) + \int_{\Omega_1} f(\nabla u_{jk}) - \frac{\epsilon}{6}$$
$$\geq \int_\Omega f(\nabla v_k) + \int_{\Omega \backslash \Omega_1} [f(\nabla u_{jk}) - f(\nabla v_k)] - \frac{\epsilon}{6} - 2M(1 - r^n) m(\Omega)$$
$$\geq \int_\Omega f(\nabla u) - \epsilon ,$$

as $j > j_0(k, r)$.

Since $\epsilon > 0$ is arbitrarily small, we have proved the s.w*.l.s.c. of J. □

Remark 4.3.11 *Theorem 4.3.8 is initially due to Morrey under some additional conditions. It was refined by Meyers, and greatly improved by Acerbi and Fusco [AF] and Marcellini [Mar]. In the final version, f can be a Caratheodory function: $\overline{\Omega} \times \mathbb{R}^N \times \mathbb{M}^{n \times N} \to \mathbb{R}^1$, and (C_p) is replaced by the following:*

$$|f(x,u,A)| \leqslant \alpha(1+|u|+|A|) \quad as\ p=1\,,$$

$$-\alpha(1+|u|^r+|A|^q) \leqslant f(x,u,A) \leqslant \alpha(1+|u|^r+|A|^p)\ as\ 1<p\leqslant n\,,$$

where $1 \leqslant q < p$, *and* $r \in [1, \frac{np}{n-p})$,

$$-\alpha(1+|A|^q) \leqslant f(x,u,A) \leqslant \alpha(1+|A|^p)\ as\ n<p<\infty\,,$$

where $1 \leq q < p$, *and*

$$f(x,u,A) \leqslant \eta(x,|u|,|A|)\ as\ p=\infty\,,$$

where η *is an increasing function in each of its argument.*

Theorem 4.3.12 *Suppose that* $f : \mathbb{M}^{n\times N} \to \mathbb{R}^1$ *is continuous, quasi-convex and satisfying*

$$c|A|^p \leqslant f(A) \leqslant C(1+|A|^p) \quad for\ C>c>0\,,$$

for $1<p<\infty$. *Then for* $\forall v \in W^{1,p}(\Omega, R^N)$, *the functional*

$$J(u) = \int_\Omega f(\nabla u)$$

achieves its minimum on $E = W^{1,p}_v := \{u \in W^{1,p}(\Omega,R^N)|\ u|_{\partial\Omega} = v|_{\partial\Omega}\}$.

Proof. In fact, J is s.w.l.s.c. and coercive, and E is weakly closed. □

4.3.3 Nonlinear Elasticity

The body of a given elastic material occupies a bounded domain Ω in $\mathbb{R}^3$; it is called the reference configuration. External force gives the material a deformation $u : \overline{\Omega} \to \mathbb{R}^3$. The gradient of u, $\nabla u = (\frac{\partial u^i}{\partial x_\alpha})_{1\leq i,\alpha\leq 3}$ is called the deformation gradient. To the hyper-elastic materials, a stored energy density function $W : \overline{\Omega} \times \mathbb{M}^{3\times 3}_+ \to \mathbb{R}^1$ is introduced with certain symmetric properties (e.g., it is left invariant under rotations and right invariant under certain isotropic groups), where $\mathbb{M}^{3\times 3}_+$ denotes all 3×3 matrices with positive determinants. The elastic energy reads as

$$I(u) = \int_\Omega W(x,\nabla u)dx\,.$$

The basic assumption of the variational approach to this problem is that the observed deformations correspond to minimizers of the elastic energy.

After the reductions from these invariances, the stored energy density is, in fact, dependent on the determinant, the minors and the eigenvalues of the deformation gradient ∇u only.

A function $f : \mathbb{M}^{n\times N} \to \mathbb{R}^1 \cup \{+\infty\}$ is called poly-convex if there exists $g : \mathbb{R}^{\tau(n,N)} \to \mathbb{R}^1$ convex such that

$$f(A) = g(T(A))$$

where

$$\tau(n, N) = \sum_{s=1}^{n\wedge N} \frac{n!N!}{(s!)^2(n-s)!(N-s)!},$$

and $T : \mathbb{M}^{n\times N} \to \mathbb{R}^{\tau(n,N)}$ is of the form:

$$T(A) = (A, \mathrm{adj}_2 A, \ldots, \mathrm{adj}_{n\wedge N} A),$$

and $\mathrm{adj}_s A$ denotes the matrix of all $s \times s$ minors of A.

Examples. Let $A = (a_{ij}) \in \mathbb{M}^{n\times n}$, then

$$\| A \|^2 = \sum_{i,j=1}^{n} |a_{ij}|^2 ,$$

$$\det(A) ,$$

and

$$\| \mathrm{adj}_2 A \|^p \quad p \geq 1 ,$$

are poly-convex functions.

What are the relationships between poly-convexity and other convexities? We claim that

$$\text{convex} \Longrightarrow \text{poly-convex} \Longrightarrow \text{quasi-convex}.$$

In fact, the first implication is trivial.

To verify the second implication, let us recall

$$\frac{1}{m(D)} \int_D \det(B + \nabla v) = \det(B),\ \forall B \in \mathbb{M}^{n\times n},\ \forall D \subset \Omega,\ \forall v \in C_0^\infty(D, R^n) .$$

It implies that $\forall A \in \mathbb{M}^{n\times n}$,

$$\frac{1}{m(D)} \int_D \mathrm{adj}_s(A + \nabla v) = \mathrm{adj}_s(A) \quad 2 \leqslant s \leqslant n ,$$

and then

$$\frac{1}{m(D)} \int_D T(A + \nabla v) = T(A) .$$

Thus if f is poly-convex, and $f(A) = g(T(A))$, where g is convex, then by Jessen's inequality

$$\begin{aligned} \frac{1}{m(D)} \int_D f(A + \nabla v) &= \frac{1}{m(D)} \int_D g(T(A + \nabla v)) \\ &\geqslant g\left(\frac{1}{m(D)} \int_D T(A + \nabla v)\right) \end{aligned}$$

$$= g(T(A))$$
$$= f(A)\ ,$$

i.e., f is quasi-convex.

In summary, we have the following implications:

$$\text{convex} \implies \text{poly-convex} \implies \text{quasi-convex} \implies \text{rank-one convex}\ .$$

One may ask: Are these implications strict?

The following example is due to Dacorogna and Marcellini [DM]: $A \in \mathbb{M}^{2\times 2}$, $f(A) = \| A \|^4 - \gamma \| A \|^2 \det A$, then

$$\begin{aligned} &f \text{ is convex} \Longleftrightarrow |\gamma| \leqslant \frac{4}{3}\sqrt{2}\ , \\ &f \text{ is poly-convex} \Longleftrightarrow |\gamma| \leqslant 2\ , \\ &f \text{ is quasi-convex} \Longleftrightarrow |\gamma| \leqslant 2+\varepsilon, \text{ for some } \varepsilon > 0\ , \\ &f \text{ is rank-one convex} \Longleftrightarrow |\gamma| \leqslant \frac{4}{\sqrt{3}}\ . \end{aligned}$$

It is not known whether $2+\varepsilon = \frac{4}{\sqrt{3}}$. In addition to the counterexample due to Sverak [Sv 1], all the reverse implications are false.

4.4 Relaxation and Young Measure

Not all variational integrands are quasi-convex. In this case a minimizing sequence may not converge to a minimizer, there may even be no minimizer!

For example, let $\Omega = (0,1)$, we consider the functional

$$J(u) = \int_0^1 [u^2(x) + (u'(x)^2 - 1)^2]dx \tag{4.32}$$

on the Sobolev space $W_0^{1,4}(\Omega)$. The integrand is not quasi-convex. It is easily seen that $\inf J = 0$. Indeed, on the one hand, we define a sequence of sawtooth functions:

$$u_n(x) = \begin{cases} x - \frac{k}{n} & \text{if } x \in [\frac{k}{n}, \frac{2k+1}{2n}]\ , \\ -x + \frac{k+1}{n} & \text{if } x \in [\frac{2k+1}{n}, \frac{k+1}{n}]\ . \end{cases}$$

From $|u_n'(x)| = 1$ a.e., and $|u_n| \leqslant \frac{1}{2n}$, we conclude that

$$J(u_n) \leqslant \frac{1}{4n^2} \to 0\ .$$

On the other hand, $J \geqslant 0$. Then $\{u_n\}$ is a minimizing sequence with $\inf J = 0$.

However, $u_n \rightharpoonup \theta$ in $W_0^{1,4}(\Omega)$, because

$$\int_0^1 u_n' \cdot \varphi = -\int_0^1 u_n \cdot \varphi' \to 0, \ \forall \varphi \in C_0^\infty(\Omega), \text{ as } n \to 0 \, .$$

But $J(\theta) = 1$.

Moreover, the functional does not have a minimizer in $W_0^{1,4}(\Omega)$. Because, if $u_0 \in W_0^{1,4}(\Omega)$ achieves $J(u_0) = 0$, then $u_0(x) = 0$, a.e., and $u_0'(x) = 0$, a.e., which imply that $J(u_0) = 1$. This is impossible.

Nevertheless, to this kind of problem, the minimizing sequences must provide some useful information in understanding the variational problems. There are several ways to studying these minimizing sequences:

(1) Relaxation method: Introduce a relaxed functional with a quasi-convex integrand. It shares a common minimizing sequence with J.
(2) Young measure: Extend the working Sobolev space to a measure space so that the minimizing sequences converge in some sense to a measure.
(3) Change the working space to something other than the Sobolev space $W^{1,p}(\Omega, \mathbb{R}^N)$.

4.4.1 Relaxations

In the first aspect, let us recall the relationship between s.w*.l.s.c. and the quasi-convexity. We are inspired to study various convex envelopes for non-convex functions.

Definition 4.4.1 *Let $f : \mathbb{M}^{n\times N} \to \mathbb{R}^1 \cup \{+\infty\}$, we call*

$$\begin{aligned} Cf &= \sup\{g \leqslant f |\ g \ \textit{convex}\} , \\ Pf &= \sup\{g \leqslant f |\ g \ \textit{poly-convex}\} , \\ Qf &= \sup\{g \leqslant f |\ g \ \textit{quasi-convex}\} , \\ Rf &= \sup\{g \leqslant f |\ g \ \textit{rank-one convex}\} , \end{aligned}$$

the convex, poly-convex, quasi-convex and rank-one convex envelope of f, respectively.

Obviously,

(1) Cf, Pf, Qf and Rf are convex, poly-convex, quasi-convex and rank-one convex functions, respectively.
(2) $Cf \leqslant Pf \leqslant Qf \leqslant Rf \leqslant f$.

In particular, if $n = 1$ or $N = 1$, then $Cf = Pf = Qf = Rf$. By definition, $f \geqslant Cf \geqslant f^{**}$, the biconjugate function of f. But one can show that if f is proper, then $Cf = f^{**}$. Since Qf is quasi-convex, in addition to some growth condition, the associated functional

$$\widetilde{J}(u) = \int_\Omega (Qf)(x, u(x), \nabla u(x)) \tag{4.33}$$

is s.w*.l.s.c. in a certain Sovolev space $W^{1,p}(\Omega, \mathbb{R}^N), 1 < p \leq \infty$. The following theorem is very basic in this aspect.

Theorem 4.4.2 *Suppose that $f : \mathbb{M}^{n\times N} \to \mathbb{R}^1$ is measurable and locally integrable, and satisfies*

$$C_2|A|^p \leqslant f(A) \leqslant C_1(1 + |A|^p) \tag{4.34}$$

for some $C_1 > C_2 > 0$ and $p \in (1, \infty)$. For $\forall v \in W^{1,p}(\Omega, \mathbb{R}^N)$, define $W_v^{1,p} = \{u \in W^{1,p}(\Omega, \mathbb{R}^N)|\ u|_{\partial\Omega} = v|_{\partial\Omega}\}$, then

$$\inf\{J(u)|\ u \in W_v^{1,p}\} = \inf\{\widetilde{J}(u)|\ u \in W_v^{1,p}\}\ .$$

Moreover, a function $u_0 \in W_v^{1,p}$ is a minimizer of $\widetilde{J}$ if and only if it is a cluster point (with respect to the weak topology of $W^{1,p}$) of a minimizing sequence for J.

Before going to the proof of Theorem 4.4.2, we give a characterization of the quasi-convex envelope.

Lemma 4.4.3 *If $f : \mathbb{M}^{n\times N} \to \mathbb{R}^1$ is locally bounded and Borel measurable, then*

$$(Qf)(A) = \inf\left\{\frac{1}{m(D)}\int_D f(A + \nabla\varphi(x))dx|\ \varphi \in W_0^{1,\infty}(D, \mathbb{R}^N)\right\}, \tag{4.35}$$

where $D \subset \mathbb{R}^n$ is a bounded domain with $m(\partial D) = 0$.

Proof. We denote the RHS of (4.35) by $(qf)(A, D)$. By definition, $f \leq g$ implies that $qf \leq qg$; and if f is quasi-convex, then $qf = f$.

We want to show:

(1) $(qf)(A, D)$ is independent of D, then we denote it by $(qf)(A)$.
(2) $(qf)(A)$ is quasi-convex.

We assume these two conclusions at this moment, then by the definition of the quasi-convex envelope,

$$(qf)(A) \geqslant (q(Qf))(A) = (Qf)(A)\ \ \forall A \in M^{n\times N}\ .$$

From (2), qf is quasi-convex and $qf \leqslant f$. Since Qf is the largest quasi-convex function $\leqslant f$, it follows that $qf \leqslant Qf$. Therefore

$$qf = Qf\ .$$

1. Verification of (1).

One observes that for any two bounded domains D_1 and D_2, if after translation and scaling, D_1 is transformed into D_2 (denoted by $D_1 \sim D_2$), then

$$(qf)(A, D_1) = (qf)(A, D_2) .$$

This proves that $(qf)(A, D)$ are equal for all $D \sim D_1$. Moreover, if $D_1 \subset D_2$, then $(qf)(A, D_2) \leq (qf)(A, D_1)$.

Now for any two bounded domains D_1 and D_2 in $\mathbb{R}^n$, we may assume that D_1 is the unit cube. And for any $\varepsilon > 0$, we find $\{E_j | \sim D_1, \ j = 1, 2, \ldots, m\}$, with $E_i \cap E_j = \emptyset \ i \neq j$ and $m(D_2 \backslash \bigcup_{j=1}^{m} E_j) < \varepsilon$. On each E_j one may find a piecewise affine function $\varphi_j^\varepsilon \in W_0^{1,\infty}(E_j, \mathbb{R}^n)$, such that

$$\int_{E_j} f(A + \nabla \varphi_j^\varepsilon(x)) dx \leqslant (\varepsilon + (qf)(A, D_1)) m(E_j) ,$$

and let

$$\varphi^\varepsilon(x) = \begin{cases} \varphi_j^\varepsilon(x) & x \in E_j \\ 0 & x \in D_2 \backslash \bigcup_{j=1}^{m} E_j . \end{cases}$$

Thus

$$\begin{aligned} \int_{D_2} f(A + \nabla \varphi^\varepsilon(x)) dx &\leqslant \sum_{j=1}^{m} \int_{E_j} f(A + \nabla \varphi_j^\varepsilon(x)) dx + \int_{D_2 \backslash \bigcup_{j=1}^{m} E_j} f(A) dx. \\ &\leqslant (\varepsilon + (qf)(A, D_1)) \sum_{j=1}^{m} m(E_j) + f(A) m \left(D_2 \backslash \bigcup_{j=1}^{m} E_j \right) . \end{aligned}$$

Since $\varepsilon > 0$ is arbitrary, it follows that

$$(qf)(A, D_2) \leqslant (qf)(A, D_1) .$$

The positions of D_1 and D_2 are symmetric, thus

$$(qf)(A, D_1) = (qf)(A, D_2) ,$$

i.e., $(qf)(A, D)$ is independent of D, and will be written as $(qf)(A)$.

2. Verification of (2).

First, we verify:

$$\frac{1}{m(D)} \int_D (qf)(A + \nabla \psi(x)) dx \geqslant (qf)(A) ,$$

for all piecewise affine $\psi \in W_0^{1,\infty}(D, \mathbb{R}^N)$, where D is the unit cube. In fact, let $D = \bigcup_{i=1}^{m} D_i$, where $D_i, i = 1, \ldots, m$ are simplexes with $D_j^\circ \cap D_i^\circ = \emptyset \ \forall i \neq j$

and $\nabla\psi|_{D_i} = B_i\ \forall i$. By the definition of qf, for $\forall\varepsilon > 0$, $\exists\varphi_i \in W_0^{1,\infty}(D_i, \mathbb{R}^N)$ such that

$$(qf)(A+\nabla\psi(x)) = (qf)(A+B_i) \geqslant \frac{1}{m(D_i)}\int_{D_i} f(A+\nabla\psi+\nabla\varphi_i)-\varepsilon,\ \forall x \in D_i\ .$$

One extends φ_i to be zero outside D_i, and define $\varphi = \psi + \sum_{i=1}^m \varphi_i$. Then $\varphi \in W_0^{1,\infty}(D, \mathbb{R}^N)$, and

$$\begin{aligned}\int_D (qf)(A+\nabla\psi(x))dx &= \sum_{i=1}^m \int_{D_i} (qf)(A+B_i)\\ &\geqslant \sum_{i=1}^m \int_{D_i} f(A+\nabla\varphi) - \varepsilon m(D)\\ &= \int_D f(A+\nabla\varphi) - \varepsilon m(D)\\ &\geqslant m(D)[(qf)(A) - \varepsilon]\ .\end{aligned}$$

Since $\varepsilon > 0$ is arbitrary, it follows that

$$\begin{aligned}\inf\Bigg\{\frac{1}{m(D)}\int_D (qf)(A+\nabla\psi)dx|\ &\psi \in W_0^{1,\infty}(D,\mathbb{R}^N)\\ \text{piecewise linear}\Bigg\} &\geqslant (qf)(A)\ .\end{aligned} \tag{4.36}$$

By the same procedure as in Sect. 4.3, we conclude that (qf) is rank-one convex. Therefore qf is continuous, and in fact, is locally Lipschitzian (Theorem 2.2.1, Sect. 2.2). Since piecewise affine functions are dense in $W_0^{1,\infty}(D, \mathbb{R}^N)$, (4.36) is extended to all $\psi \in W_0^{1,\infty}(D, \mathbb{R}^N)$. Thus qf is quasi-convex. □

Proof of Theorem 4.4.2.

Proof. 1. Since $Qf \leqslant f$, $\inf\{\widetilde{J}\} \leqslant \inf\{J\}$. We only want to show that any minimizing sequence of $\widetilde{J}$ is indeed a minimizing sequence of J. Thus it is sufficient to prove that $\forall u \in W_v^{1,\infty}(\Omega, \mathbb{R}^N)\ \exists\{u_l\}_1^\infty \subset W_v^{1,\infty}(\Omega, \mathbb{R}^N)$ such that

$$J(u_l) \to \widetilde{J}(u) \quad \text{as } l \to \infty\ .$$

Due to the assumption (4.34) on f and the density of piecewise affine functions in $W_0^{1,\infty}(\Omega, \mathbb{R}^N)$, we may assume that u is piecewise affine in a smaller domain $\Omega_0 \subset \overline{\Omega}_0 \subset \Omega$. By a standard argument, it is sufficient to replace Ω by Ω_0.

2. Let $\Omega_0 = \bigcup\limits_{i=1}^m D_i$, where $D_i, i = 1, \ldots, m$ are simplexes with $D_i^\circ \cap D_j^\circ = \emptyset\ i \neq j$ and let $\nabla u|_{D_i} = B_i\ \forall i$. Then we choose small cubes $\{C_i^{(j)}\}_{j=1}^{p_i} \subset D_i$ with $C_i^{(j)\circ} \cap C_i^{(k)\circ} = \emptyset$ as $j \neq k$, and

$$\int_{D_i \setminus \bigcup_{j=1}^{p_i} C_i^{(j)}} (1 + |\nabla u(x)|)^p dx \leqslant \frac{\varepsilon}{mC_1} .$$

We are going to construct a sequence u_l on each cube $C = C_i^{(j)}$. By Lemma 4.4.3, $\exists \varphi_l \in W_0^{1,\infty}(C, \mathbb{R}^N)$ such that for $\forall x \in C$

$$(Qf)(\nabla u(x)) \leqslant \frac{1}{m(C)} \int_C f(\nabla u(x) + \nabla \varphi_l(x))dx \leqslant (Qf)(\nabla u(x)) + \frac{1}{l} . \quad (4.37)$$

Since C is a cube and $\varphi_l \in W_0^{1,\infty}(C, \mathbb{R}^N)$, we may extend φ_l periodically to $\mathbb{R}^N$. Let

$$u_l(x) = u(x) + \frac{1}{l}\varphi_l(lx), \quad \forall x \in C .$$

Then $u_l|_{\partial C} = u|_{\partial C}$, and

$$\begin{aligned} \int_C f(\nabla u_l(x))dx &= \int_C f(B_i + \nabla \varphi_l(lx))dx \\ &= \frac{1}{l^n} \int_{lC} f(B_i + \nabla \varphi_l(y))dy \\ &= \int_C f(B_i + \nabla \varphi_l(y))dy . \end{aligned} \quad (4.38)$$

Combining (4.37) and (4.38), we obtain

$$\int_{C_i^{(j)}} f(\nabla u_l(x))dx \to (Qf)(B_i)m(C_i^{(j)}) ,$$

as $l \to \infty$.

3. "Putting them together" we define

$$u_l(x) = \begin{cases} u & x \in \Omega_0 \setminus \bigcup_{i=1}^{m} \bigcup_{j=1}^{p_i} C_i^{(j)} , \\ u_l(x) & x \in \bigcup_{i=1}^{m} \bigcup_{j=1}^{p_i} C_i^{(j)} . \end{cases}$$

It follows that

$$\begin{aligned} \int_{\Omega_0} f(\nabla u_l(x))dx &= \int_{\Omega_0 \setminus \bigcup_{i=1}^{m}\bigcup_{j=1}^{p_i} C_i^{(j)}} f(\nabla u(x))dx \\ &\quad + \sum_{i=1}^{m} \sum_{j=1}^{p_i} m(C_i^{(j)})(Qf)(B_i) + o(1) \\ &= \int_{\Omega_0} (Qf)(\nabla u(x))dx + o(1) . \end{aligned}$$

After a diagonal process, we obtain

$$J(u_l) \to \widetilde{J}(u) \quad \text{as } l \to \infty \, .$$

4. If u_0 is a minimizer of $\widetilde{J}$, then by the above argument, we obtain a minimizing sequence $\{u_l\}$ of J, with $u_l|_{\partial\Omega} = u_0|_{\partial\Omega}$.

Combining the first inequality of (4.34) with the second inequality of (4.37), we obtain

$$\begin{aligned} C_2 \int_\Omega |\nabla u_l|^p &\leqslant \int_\Omega f(\nabla u_l) \\ &\leqslant \widetilde{J}(u_0) + m(\Omega) \\ &\leqslant J(u_0) + m(\Omega) \, . \end{aligned}$$

Therefore there exists a weakly convergent subsequence $\{u_{l_i}\}$ in $W_v^{1,p}$.

Recalling the construction of $\{u_l\}$, we have $u_l(x) \to u_0(x)$, a.e. Therefore, $u_{l_i} \rightharpoonup u_0$ in $W_v^{1,p}$.

Conversely, if $\{u_l\}$ is a minimizing sequence of J, then $\widetilde{J}(u_l) \leqslant J(u_l) \to \inf J = \inf \widetilde{J}$, i.e., $\{u_l\}$ is also a minimizing sequence of $\widetilde{J}$. Again by the first inequality of (4.34), there exists a subsequence $\{u_{l_i}\}$ that weakly converges to u_0 in $W_0^{1,p}$. Since $\widetilde{J}$ is s.w.l.s.c., u_0 is a minimizer of $\widetilde{J}$. □

Now, let us return to the example we met in the beginning of this section. The function

$$f(u, \xi) = u^2 + (\xi^2 - 1)^2$$

is obviously not convex in ξ. Since in the case where $N = 1$, convex = quasi-convex, the convex envelope of $f(u, \xi)$ is the following:

$$Qf(u, \xi) = \begin{cases} u^2 + (\xi^2 - 1)^2 & \text{if } |\xi| \geqslant 1 \\ 0 & \text{if } |\xi| < 1 \, , \end{cases}$$

and then

$$\widetilde{J}(u) = \int_0^1 Qf(u(x), u'(x))dx \, .$$

In fact, the minimizing sequence introduced previously weakly converges to 0 in $W^{1,4}([0,1])$, and $u = 0$ is exactly the minimizer of $\widetilde{J}$.

Since the initial functional J is in the lack of weakly lower semi-continuity, we then minimize another functional $\widetilde{J}$ instead, which is s.w.l.s.c. and shares a common minimizing sequence with the original one. This passage is called relaxation. The solution of $\widetilde{J}$ is called a generalized solution of J. Noticing that the representation of the quasi-convex envelope in Lemma 4.4.3 can be seen as an average over fine scale oscillations, we give a physical interpretation of this phenomenon: J represents the microscopic energy, while $\widetilde{J}$ is the macroscopic. There are models in elastic crystals, in which this phenomenon occurs.

The quasi convex envelope of a non-convex function is not easy to compute. People have developed various methods in approaching the computations. Readers who are interested in this problem are referred to the book of

Dacorogna [Dac]. In particular, to the interesting example on the two well problem: $K = \{A, -A\}$, the quasi convex envelope of the square distance function $Qdist(P, K)^2$ has been expressed explicitly by K. Zhang [Zh 2].

4.4.2 Young Measure

In a variational problem, there may be many minimizing sequences. If minimizers exist, then the minimizers describe the common feature of these sequences. Otherwise, the sequences do not converge to a minimizer as an element of the given function space. In his study of generalized solutions of optimal control problems, L. C. Young [Yo] introduced a family of measures in certain measure space as a replacement of the minimizer. The idea is as follow: Let $\Omega \subset \mathbb{R}^n$ be a measurable set, and let $z_j : \Omega \to \mathbb{R}^N, j = 1, 2, \ldots,$, be a sequence of measurable functions. To each z_j, we relate a family of probability measures $\nu_x^j = \delta_{z_j(x)}$ on $\mathbb{R}^N, \forall x \in \Omega$. Instead of the convergence of z_j, we consider the convergence of the family of probability measures: $\{\nu_x^j\}$; i.e., whether there exists a family of measures ν_x on $\mathbb{R}^N$, $\forall x \in \Omega$, such that for $\forall f \in C_0(\mathbb{R}^N)$, the space of continuous functions with compact support,

$$\int_{\mathbb{R}^N} f(y) d\nu_x^j(y) \to \int_{\mathbb{R}^N} f(y) d\nu_x(y) ,$$

as $j \to \infty$. The family of measures $\{\nu_x | x \in \Omega\}$ is a Young measure.

However, the function z_j is well defined in a.e. sense, what is the meaning of the measure: $\delta_{z_j(x)}$? What is the exact meaning of the family of measures $\{\nu_x | x \in \Omega\}$?

In order to clarify the meaning, we appeal to the slicing measures. Let μ be a finite nonnegative Radon measure on $\mathbb{R}^{n+N}$. Denote by σ the canonical projection of μ onto $\mathbb{R}^n$: $\sigma(E) = \mu(E \times \mathbb{R}^N)$ $\forall$ Borel set $E \subset \mathbb{R}^n$.

Lemma 4.4.4 *For $\sigma -$ a.e. point $x \in \mathbb{R}^n, \exists$ a Radon probability measure ν_x on $\mathbb{R}^N$ such that for each bounded continuous function f,*

1. $x \mapsto \int_{\mathbb{R}^N} f(x, y) d\nu_x(y)$ *is $\sigma -$ measurable,*
2. $\int_{\mathbb{R}^n \times \mathbb{R}^N} f(x, y) d\mu(x, y) = \int_{\mathbb{R}^n} (\int_{\mathbb{R}^N} f(x, y) d\nu_x(y)) d\sigma(x)$.

Proof. Recall that the dual space of $C_0(\mathbb{R}^N)$, the space of continuous functions with compact support, is the signed measure space $\mathcal{M}(\mathbb{R}^N)$. The family of measures ν_x on $\mathbb{R}^N, \forall x \in \mathbb{R}^n$, should be determined with the aid of functions on $C_0(\mathbb{R}^N)$.

Since $C_0(\mathbb{R}^N)$ is separable, there is a countable dense subset $\{g_k\}_1^\infty$. Now, $\forall$ Borel sets $E \subset \mathbb{R}^n$, we define

$$\gamma_k(E) = \int_{E \times \mathbb{R}^N} g_k(y) d\mu(x, y), \quad \forall k .$$

Obviously, γ_k is absolutely continuous with respect to σ. There exists a σ-null set $\mathcal{N}$ such that the derivatives exist for $\forall x \notin \mathcal{N}$:

$$D_\sigma \gamma_k(x) = \lim_{r\to 0} \frac{\gamma_k(B_r(x))}{\sigma(B_r(x))}, \quad \forall k\ , \tag{4.39}$$

and then $x \mapsto D_\sigma\gamma_k(x)$ are bounded and σ measurable.

Moreover, for all Borel sets $E \subset \mathbb{R}^n$,

$$\int_{E\times\mathbb{R}^N} g_k(y)d\mu(x,y) = \gamma_k(E) = \int_E D_\sigma\gamma_k(x)d\sigma(x),\ \forall k\ . \tag{4.40}$$

Since $\{g_k\}_1^\infty$ is dense in $C_0(\mathbb{R}^N), \forall g \in C_0(\mathbb{R}^N)$, we have a subsequence $g_{k_j} \to g$ uniformly on $\mathbb{R}^N$. Thus for $\forall x \notin \mathcal{N}, \Gamma_x(g) = \lim_{j\to\infty} D_\sigma\gamma_{k_j}(x)$ exists, and

$$|\Gamma_x(g)| \le C_\mu \|g\|_{C_0(\mathbb{R}^N)}\ , \tag{4.41}$$

where $C_\mu = \mu(\mathbb{R}^n \times \mathbb{R}^N)$. Since

$$|\gamma_k(B_r(x)) - \gamma_l(B_r(x))| \le \|g_k - g_l\|_{C_0(\mathbb{R}^N)}\sigma(B_r(x))\ ,$$

the limit $\Gamma_x(g)$ does not depend on the choice of the subsequence converging to g. Also, the map: $g \mapsto \Gamma_x(g)$ is linear and positive. Provided by the Riesz representation theorem, $\exists \nu_x \in \mathcal{M}(\mathbb{R}^N)$, which is nonnegative, such that for $\forall g \in C_0(\mathbb{R}^N)$:

$$\Gamma_x(g) = \int_{\mathbb{R}^N} g(y)d\nu_x(y),\ \forall x \notin \mathcal{N}\ .$$

By taking limits, we see $\nu_x(\mathbb{R}^N) = \Gamma_x(1) = 1$. Therefore, it is a probability measure.

As a function of x on $R^N\backslash\mathcal{N}, x \mapsto \Gamma_x(g)$ is bounded and σ-measurable, because it is the pointwise limit of a sequence of σ-measurable functions $D_\sigma\gamma_{k_j}(x)$. Therefore, the conclusion (1) is obtained by approximation.

Following (4.40), by taking limits we obtain

$$\int_{E\times\mathbb{R}^N} g(y)d\mu(x,y) = \int_E \Gamma_x(g)d\sigma(x) = \int_E \left(\int_{\mathbb{R}^N} g(y)d\nu_x(y)\right) d\sigma(x)\ . \tag{4.42}$$

From (4.42) $\forall h \in C_0(\mathbb{R}^n)$, we have

$$\int_{\mathbb{R}^n\times\mathbb{R}^N} g(y)h(x)d\mu(x,y) = \int_{\mathbb{R}^n} \left(\int_{\mathbb{R}^N} g(y)d\nu_x(y)\right) d\sigma(x)\ . \tag{4.43}$$

Since $C_0(\mathbb{R}^n)\bigotimes C_0(\mathbb{R}^N)$ is dense in $C_0(\mathbb{R}^{n+N})$, (2) is proved by approximation. □

Remark 4.4.5 *By standard measure theoretic argument, one can show that the statements of Lemma 4.4.4 hold, if the bounded continuous function f is replaced by $f \in L^1(\mathbb{R}^n \times \mathbb{R}^N, \mu)$. Moreover, we have*

$$f(x,\cdot) \in L^1(\mathbb{R}^N, \nu_x) \text{ for a.e., } x \in \mathbb{R}^n\ ,$$

and the function $x \mapsto \int_{\mathbb{R}^N} f(x,y)d\nu_x(y)$ is in $L^1(\mathbb{R}^n, \sigma)$.

A map $\nu : \mathbb{R}^n \to \mathbb{M}(\mathbb{R}^N)$ is called weakly $^*\sigma$-measurable, if the functions $x \mapsto \int_{\mathbb{R}^N} f(y) d\nu_x(y)$ are σ-measurable for all $f \in C_0(\mathbb{R}^N)$ on $\mathbb{R}^n$.

In the context of the duality between $\mathcal{M}(\mathbb{R}^N)$ and $C_0(\mathbb{R}^N)$, the map $\nu : \mathbb{R}^n \to \mathcal{M}(\mathbb{R}^N)$, obtained in Lemma 4.4.4 is w^* σ-measurable.

After the preparations, we are able to identify an $L^\infty(\Omega, \mathbb{R}^N)$ function with a w^*-measurable map $\nu : \Omega \mapsto \mathcal{M}(\mathbb{R}^N)$.

Suppose that $\Omega \subset \mathbb{R}^n$ is a Lebesgue measurable set with a finite Lebesgue measure. For $\forall z \in L^\infty(\Omega, \mathbb{R}^N)$, one defines a measure μ on $\mathbb{R}^n \times \mathbb{R}^N$, provided by the duality:

$$\int_{\mathbb{R}^n \times \mathbb{R}^N} f(x,y) d\mu(x,y) = \int_\Omega f(x, z(x)) dx, \quad \forall f \in C_0(\mathbb{R}^n \times \mathbb{R}^N), \tag{4.44}$$

and then a family of measures on $\mathbb{R}^N$: $\delta_{z(x)}$ for a.e. $x \in \Omega$. Since the right-hand side of (4.44) is linear and positive in f, and

$$\left| \int_\Omega f(x, z(x)) dx \right| \le m(\Omega) \|f\|_{C_0},$$

the existence of μ follows from the Riesz representation theorem.

Obviously, the canonical projection of μ onto R^n is the Lebesgue measure restricted on the measurable set Ω, and the w^*-measurable map $\nu : \Omega \to \mathcal{M}(\mathbb{R}^N)$ satisfies $\nu_x = \delta_{z(x)}$ for a.e. $x \in \Omega$.

Now we turn to the fundamental theorem for Young measures.

Theorem 4.4.6 *Let $\Omega \subset \mathbb{R}^n$ be a bounded open domain and let $\{z_k\} \subset L^\infty(\Omega, \mathbb{R}^N)$ be a bounded sequence. Then there exist a subsequence $\{z_{k_j}\}$ and a w^*-measurable map $\nu : \Omega \to \mathcal{M}(\mathbb{R}^N)$ such that*

1. *ν_x is a Radon probability measure for a.e. $x \in \Omega$.*
2. *For $\forall g \in C_0(R^N)$, one has*

$$g(z_{k_j}) \stackrel{*}{\rightharpoonup} \overline{g}(x) := \langle \nu_x, g \rangle = \int_{\mathbb{R}^N} g(y) d\nu_x(y), \text{ in } L^\infty(\Omega).$$

3. *Let $K \subset \mathbb{R}^N$ be closed, and let $\mathrm{dist}(z_{k_j}, K) \to 0$ in measure as $j \to \infty$. Then $\mathrm{supp}\, \nu_x \subset K$ for a.e. $x \in \Omega$.*

Proof. According to (4.44), $\forall z_k, \exists \mu_k$ on $\mathbb{R}^n \times \mathbb{R}^N$, for which the canonical projection onto R^n is the Lebesgue measure restricted to Ω. Since $\mu_k(\mathbb{R}^n \times \mathbb{R}^N) \le \mathcal{L}^n(\Omega)$, there exist a measure μ on $\mathbb{R}^n \times \mathbb{R}^N$ and a subsequence such that $\mu_{k_j} \stackrel{*}{\rightharpoonup} \mu$, according to the Banach–Alaoglu theorem.

We claim that the canonical projection σ of μ into R^n is the Lebesgue measure $\mathcal{L}^n$ restricted on Ω. On one hand, if $U \subset \Omega$ is open, then by the w^*-convergence:

$$\sigma(U) = \mu(U \times \mathbb{R}^N) \le \liminf \mu_{k_j}(U \times \mathbb{R}^N) = \mathcal{L}^n(U).$$

According to the Radon–Nikodym theorem, there exists $a \in L^1(\Omega, \mathcal{L}^n)$ such that $\sigma = a\mathcal{L}^n$.

On the other hand, for any compact subset $K \subset \Omega$, we show that $\sigma(K) \geq \mathcal{L}^n(K)$. Since $\{z_k\}$ is bounded in $L^\infty(\Omega, \mathbb{R}^N), \exists R > 0$, such that $\text{supp}(\mu_{k_j})$, and then $\text{supp}(\mu)$ are contained in $\Omega \times B_R(\theta). \forall \epsilon > 0, \exists f \in C_0(\mathbb{R}^n \times \mathbb{R}^N)$, with $f = 1$ on $K \times B_R(\theta)$ and $f \geq 0$, such that

$$\int_{\mathbb{R}^n \times \mathbb{R}^N} f(x,y) d\mu(x,y) < \mu(K \times B_R(\theta)) + \epsilon .$$

Thus,

$$\begin{aligned}
\sigma(K) &= \mu(K \times \mathbb{R}^N) \\
&= \mu(K \times \overline{B_R(\theta)}) \\
&> \int_{\mathbb{R}^n \times \mathbb{R}^N} f(x,y) d\mu(x,y) - \epsilon \\
&= \lim_{j \to \infty} \int_{\mathbb{R}^n \times \mathbb{R}^N} f(x,y) d\mu_{k_j}(x,y) - \epsilon \\
&> \mathcal{L}^n(K) - \epsilon .
\end{aligned}$$

Since $\epsilon > 0$ is arbitrary, we have proved that

$$\sigma(K) \geq \mathcal{L}^n(K) . \tag{4.45}$$

Therefore $a = 1$, a.e., and σ is the restriction of the Lebesgue measure on Ω. The first conclusion follows directly from Lemma 4.4.4.

We turn to proving the second conclusion. $\forall h \in C(\overline{\Omega})$, and $\forall g \in C_0(\mathbb{R}^N), \exists$ an extension of $h, \tilde{h} \in C_0(\mathbb{R}^n)$, from (4.44), we have

$$\begin{aligned}
&\int_{\mathbb{R}^{n+N}} \tilde{h}(x) g(y) d\mu_{k_j}(x,y) \\
&\quad = \int_\Omega h(x) g(z_{k_j}(x)) dx .
\end{aligned}$$

By taking limits, and from (4.43), we have

$$\begin{aligned}
\int_\Omega h(x) g(z_{k_j}(x)) dx &\to \int_{\mathbb{R}^n \times \mathbb{R}^N} \tilde{h}(x) g(y) d\mu(x,y) \\
&= \int_\Omega h(x) \left(\int_{\mathbb{R}^N} g(y) d\nu_x(y) \right) dx .
\end{aligned}$$

Since $C(\overline{\Omega})$ is dense in $L^1(\Omega)$ and $g(z_{k_j})$ is bounded in $L^\infty(\Omega)$, the above equality implies that

$$g(z_{k_j}) \overset{*}{\rightharpoonup} \int_{\mathbb{R}^N} g(y) d\nu(y) =< \nu, g > \text{ in } L^\infty(\Omega, \mathbb{R}^N) ,$$

i.e., (2) holds.

We prove the third conclusion. $\forall g \in C_0(R^N \backslash K), \forall \epsilon > 0, \exists C_\epsilon > 0$, such that

$$|g(y)| \le \epsilon + C_\epsilon \text{dist}(y, K), \ \forall y \in \mathbb{R}^N ,$$

which implies that

$$|g(z_k(x))| \le \epsilon + C_\epsilon \text{dist}(z_k(x), K), \ \text{a.e.,} \ x \in \Omega .$$

From $\text{dist}(z_{k_j}(x), K) \to 0$ in measure, we have

$$|\langle \nu_x, g \rangle| \le \epsilon, \quad \text{a.e. } x \in \Omega$$

Since $\epsilon > 0$ is arbitrary, $\text{supp}(\nu_x) \subset K$, a.e. $x \in \Omega$. □

Definition 4.4.7 *The w^*-measurable map $\nu : \Omega \to \mathcal{M}(\mathbb{R}^N)$, defined in Theorem 4.4.6, is called a Young measure generated by the sequence $\{z_{k_j}\} \subset L^\infty(\Omega, \mathbb{R}^N)$.*

Remark 4.4.8 *The conclusion (2) has a probability interpretation: In the limit, $g(z_{k_j})$ takes the value $g(y)$ with probability $\nu_x(y)$ at x. Therefore the Young measure can be used to describe the local phase proportions in an infinitesimally fine mixture.*

Corollary 4.4.9 *Assume that ν is the Young measure associated with the sequence $\{z_k\}$ and that $z_k \to z$ in measure. Then $\nu_x = \delta_{z(x)}$* a.e.

Proof. $\forall g \in C_0(\mathbb{R}^N), g(z_k) \to g(z)$ in measure. But by Theorem 4.4.6, $g(z_k) \overset{*}{\rightharpoonup} \overline{g} = \langle \nu, g \rangle$. Therefore,

$$\langle \nu_x, g \rangle = g(z(x)) = \langle \delta_{z(x)}, g \rangle, \ \text{a.e} .$$

□

Remark 4.4.10 *In Theorem 4.4.6, the L^∞-bounded sequence $\{z_k\} \subset L^\infty(\Omega, \mathbb{R}^N)$ can be replaced by the L^1 bounded sequence $\{z_k\} \subset L^1(\Omega, \mathbb{R}^N)$. In this case, ν_x is $\mathcal{L}^n \lfloor \Omega$ a.e. a probability measure with*

$$\int_\Omega \int_{\mathbb{R}^N} |y| d\nu_x(y) dx \le \liminf_{k \to \infty} \|z_k\|_{L^1} .$$

Again we claim that σ is the restriction of the Lebesgue measure on Ω. In fact, only (4.45) should be modified: $\forall R > 0$

$$\begin{aligned} \sigma(K) &\ge \mu(K \times \overline{B_R(\theta)}) \\ &\ge \limsup_{k \to \infty} \mu_k(K \times \overline{B_R(\theta)}) \\ &\ge \limsup_{k \to \infty} \mathcal{L}^n(\{x \in K \,|\, |z_k(x)| \le R\}) \end{aligned}$$

$$\geq \mathcal{L}^n(K) - \frac{1}{R}\sup_k \|z_k\|_{L^1} .$$

As $R \to \infty$, it follows that $\sigma(K) \geq \mathcal{L}^n(K)$. Finally, we use Remark 4.4.5, and choose a sequence of positive functions $f_k(x,y) \uparrow |y|$ pointwisely; the monotone convergence theorem implies the conclusion.

Corollary 4.4.11 *Assume that $\{z_k\} \subset L^\infty(\Omega, \mathbb{R}^N)$, satisfies $z_k \to z$, a.e., and that $\{w_k\} \subset L^\infty(\Omega, \mathbb{R}^M)$ generates the Young measure ν. Then $\{z_k, w_k\} : \Omega \to \mathbb{R}^{N+M}$ generates the Young measure $\delta_{z(x)} \bigotimes \nu_x, x \in \Omega$.*

Proof. $\forall \varphi \in C_0(\mathbb{R}^N), \forall \psi \in C_0(\mathbb{R}^M), \forall \eta \in L^1(\Omega)$, by definition and the Lebesgue's dominance theorem, we have

$$\begin{aligned} \varphi(z_k) &\to \varphi(z), \text{ a.e. }, \\ \eta\varphi(z_k) &\to \eta\varphi(z) \text{ in } L^1(\Omega) , \\ \psi(w_k) &\overset{*}{\rightharpoonup} \overline{\psi} = \langle \nu, \psi \rangle \text{ in } L^\infty(\Omega) . \end{aligned}$$

This implies that

$$\int_\Omega \eta(\varphi \bigotimes \psi)(z_k, w_k)dx = \int_\Omega \eta\varphi(z_k)\psi(w_k)dx \to \int_\Omega \eta(x)\varphi(z(x))\langle \nu_x, \psi\rangle dx .$$

Then,

$$(\varphi \bigotimes \psi)(z_k, w_k) \overset{*}{\rightharpoonup} \langle \delta_z \bigotimes \nu, \varphi \bigotimes \psi \rangle \text{ in } L^\infty(\Omega) .$$

Since $C_0(\mathbb{R}^N) \bigotimes C_0(\mathbb{R}^M)$ is dense in $C_0(\mathbb{R}^{N+M})$, it follows that

$$f(z_k, w_k) \overset{*}{\rightharpoonup} \langle \delta_z \bigotimes \nu, f \rangle \text{ in } L^\infty(\Omega) ,$$

for all $f \in C_0(\mathbb{R}^{N+M})$. □

Example 4.4.12 *Let φ be the function defined in Corollary 4.3.3, and $\varphi_k(x) = \varphi(kx), \forall k \in \mathcal{N}$. It is known that*

$$\varphi_k \overset{*}{\rightharpoonup} \int_0^1 \varphi(x)dx = \lambda\alpha + (1-\lambda)\beta .$$

Similarly, $\forall f \in C(\mathbb{R}^1)$, we have

$$f \circ \varphi_k \overset{*}{\rightharpoonup} \int_0^1 f(\varphi(x))dx = \lambda f(\alpha) + (1-\lambda)f(\beta) .$$

Therefore the sequence $\{\varphi_k\}$ is associated with a Young measure: $\nu = \lambda\delta_\alpha + (1-\lambda)\delta_\beta$; i.e., $\nu_x = \lambda\delta(x-\alpha) + (1-\lambda)\delta(x-\beta), \ \forall x \in \mathbb{R}^1$.

Example 4.4.13 *Let $\{u_n\}$ be the sequence of sawtooth functions:*

$$u_n(x) = \begin{cases} x - \frac{k}{n} & \text{if } x \in [\frac{k}{n}, \frac{2k+1}{n}] \\ -x + \frac{k+1}{n} & \text{if } x \in [\frac{2k+1}{n}, \frac{k+1}{n}] \,, \end{cases}$$

which is a minimizing sequence of the functional J (see equation 4.32). Let

$$z_n(x) = u_n'(x) = \begin{cases} 1 & \text{if } x \in [\frac{k}{n}, \frac{2k+1}{n}] \,, \\ -1 & \text{if } x \in [\frac{2k+1}{n}] \,. \end{cases}$$

Let ν be the Young measure associated with to the sequence z_k.

Conclusion: $\nu = \frac{1}{2}(\delta_{-1} + \delta_{+1})$.

In fact,

$$g(z_n(x)) \overset{*}{\rightharpoonup} \langle \nu_x, g \rangle \,,$$

for all $g \in C_0(\mathbb{R}^1)$. Let us take $g(y) = \min\{(y^2 - 1)^2, 1\}$. Since $J(u_n) \to 0$, $\langle \nu_x, g \rangle = 0$ a.e. This implies that $\text{supp}(\nu_x) \subset \{-1, 1\}$, i.e., $\nu_x = \lambda(x)\delta_{-1} + (1 - \lambda(x))\delta_{+1}$.

Also, $z_n \overset{*}{\rightharpoonup} 0$ in $L^\infty(\mathbb{R}^1)$ and the relation $g_1(z_n(x)) \overset{*}{\rightharpoonup} \langle \nu_x, g_1 \rangle$ in $L^\infty(\mathbb{R}^1)$, holds for all $g_1 \in C_0(\mathbb{R}^1)$ with $g_1(y) = y$, as $|y| < 2$. Thus

$$\langle \lambda(x)\delta_{-1} + (1 - \lambda(x))\delta_{+1}, g_1 \rangle = 0 \,.$$

This implies that $1 - 2\lambda(x) = 0$, or $\lambda(x) = \frac{1}{2}$.

Example 4.4.14 *Let $B, C \in M^{n \times N}$ satisfy $rank(B - C) = 1$. Assume $\exists \lambda \in (0, 1)$, such that $\lambda B + (1 - \lambda)C = 0$. It is known that without loss of generality, one may assume $B = (1 - \lambda)a \bigotimes e_1$, and $C = -\lambda a \bigotimes e_1$ for some $a \in \mathbb{R}^N$ and $e_1 \in \mathbb{R}^n$. Let*

$$\varphi(t) = \begin{cases} (1 - \lambda)t & \text{if } t \in [0, \lambda] \,, \\ -\lambda(t - 1) & \text{if } t \in [\lambda, 1] \,. \end{cases}$$

$\forall x = (x_1, \ldots, x_n) \in D := [0, 1]^n$, let

$$u_k(x) = ak^{-1}\varphi(kx_1), \quad k = 1, 2, \ldots, \quad \text{and } z_k = \nabla u_k \,.$$

Then $\{z_k\}$ is associated with the Young measure $\nu = \lambda\delta_B + (1 - \lambda)\delta_C$, where δ_B and δ_C are the probability measures concentrated at the matrices B and C, respectively.

In fact,

$$z_k(x) = \begin{cases} B & \text{if } \{kx_1\} \in (0, \lambda) \,, \\ C & \text{if } \{kx_1\} \in (\lambda, 1) \,, \end{cases}$$

so is $\text{dist}(z_k, \{B, C\}) = 0$. By Theorem 4.4.6, the probability measure ν satisfies $\text{supp}(\nu_x) \subset \{B, C\}$ a.e. Therefore, $\exists \mu(x) \in [0, 1]$ such that $\nu_x =$

$\mu(x)\delta_B + (1-\mu(x))\delta_C$, and $\forall g \in C_0(M^{n\times N}), g(z_{k_j}) \stackrel{*}{\rightharpoonup} \langle \nu, g\rangle$ for some subsequence $\{z_{k_j}\}$. In particular, we take $g(y) = y$ in a ball centered at θ, containing $a \bigotimes e_1$. Again, from $z_k \stackrel{*}{\rightharpoonup} 0$, we obtain $\mu(x)B + (1-\mu(x))C = 0$, i.e., $\mu(x) = \lambda$. Therefore, $\nu = \lambda\delta_B + (1-\lambda)\delta_C$.

A w*-measurable map $\nu : \Omega \to \mathcal{M}(M^{n\times N})$ is called a gradient Young measure, if $\exists\{u_j\} \subset W^{1,\infty}(\Omega, R^N)$ such that

$$u_j \stackrel{*}{\rightharpoonup} u \text{ in } W^{1,\infty}(\Omega, R^N), \text{ and } \delta_{\nabla u_j} \stackrel{*}{\rightharpoonup} \nu\,.$$

A central problem for the gradient Young measure is the following: Given a set $K \subset M^{n\times N}$, how do we characterize all $W^{1,\infty}$ gradient Young measures ν such that $\mathrm{supp}(\nu_x) \subset K$ for a.e. $x \in \Omega$?

Indeed, letting ν be the Young measure generated by $\{u_j\}$, the range of $\{\nabla u_j\}$ must be contained in a big ball $B_R(\theta)$ of $M^{n\times N}$. Then by taking $K = \overline{B}_R(\theta)$, and according to (3) of Theorem 4.4.6, it follows that:

(1). There exists a compact set $K \subset M^{n\times N}$ such that $\mathrm{supp}(\nu_x) \subset K$, a.e.

If we choose $g(y) = y$ on $B_R(\theta)$ and $g \in C_0(M^{n\times N})$, then $g(\nabla u_j) = \nabla u_j \stackrel{*}{\rightharpoonup} \nabla u$ in $L^\infty(\Omega, M^{n\times N})$. From the second conclusion of Theorem 4.4.6, we have $g(\nabla u_j) \stackrel{*}{\rightharpoonup} \langle \nu, g\rangle$ in $L^\infty(\Omega, M^{n\times N})$. Since g is arbitrary outside the ball $B_R(\theta)$, we may simply write $\langle \nu, g\rangle$ as $\langle \nu, \mathrm{id}\rangle$. Thus,

(2) $\langle \nu_x, \mathrm{id}\rangle = \nabla u(x)$ a.e.

Finally, assume that $g : M^{n\times N} \to \mathbb{R}^1$ is continuous and quasi-convex, then by the Morrey theorem, for all open sets $U \subset \Omega$, we have

$$\liminf_{k\to\infty} \int_U g(\nabla u_k(x))dx \geq \int_U g(\nabla u(x))dx\,.$$

Note that

$$g(\nabla u_k(x)) \stackrel{*}{\rightharpoonup} \langle \nu_x, g\rangle, \text{ and } g(\nabla u(x)) = g(\langle \nu_x, \mathrm{id}\rangle)\,;$$

we obtain

$$\int_U \langle \nu_x, g\rangle dx \geq \int_U g(\langle \nu_x, \mathrm{id}\rangle)dx\,.$$

Since $U \subset \Omega$ is arbitrary, we arrive at

(3) $\langle \nu_x, g\rangle \geq g(\langle \nu_x, \mathrm{id}\rangle)$ a.e.

In fact, all these propositions together characterize the $W^{1,\infty}$ gradient Young measure, i.e., the converse is also true. Namely, one has the following:

Theorem 4.4.15 *A w^*-measurable map $\nu : \Omega \to \mathcal{M}(M^{n\times N})$ is a gradient Young measure if and only if $\nu_x \geq 0$ a.e., $\exists$ a compact set $K \subset M^{n\times N}$ and $\exists u \in W^{1,\infty}(\Omega, \mathbb{R}^N)$ such that*

1. $\mathrm{supp}(\nu_x) \subset K$, *a.e.*
2. $\langle \nu_x, \mathrm{id}\rangle = \nabla u(x)$, *a.e.*
3. $\langle \nu_x, g\rangle \geq g(\langle \nu_x, id\rangle)$ *a.e., $\forall$ continuous quasi-convex g: $M^{n\times N} \to \mathbb{R}^1$.*

The sufficient part of the proof can be found in [KP 1].

Remark 4.4.16 *In the modeling of microstructure, the reason people use $W^{1,\infty}$ sequences approaching a compact set K is based on a truncation lemma due to Kewei Zhang [Zh 1]: Let K be a compact set in $M^{n\times N}$, and let $\|K\|_\infty = \sup\{|A| \mid A \in K\}$. Let $\Omega \subset R^n$ be a bounded domain. If $\{u_j\} \subset W^{1,1}_{loc}(\Omega, R^N)$ satisfies $\mathrm{dist}\,(\nabla u_j, K) \to 0$ in $L^1(\Omega)$, and $u_j \to u$ in $L^1_{loc}(\Omega)$, then there exists a sequence $\{v_j\} \subset W^{1,1}_{loc}(\Omega, R^N)$ such that $|\nabla v_j| \le c_{n,N}\|K\|_\infty$, $|\{x \in \Omega \,|\, u_j(x) \neq v_j(x)\}| \to 0$, and $v_j = u$ near $\partial\Omega$.*

As an application, we extend that Morrey theorem to the case where the integrand is a function depending on $(x, u, \nabla u)$.

Corollary 4.4.17 *Suppose that $f : \Omega \times \mathbb{R}^N \times M^{n\times N} \to \mathbb{R}^1$ is a nonnegative Caratheodory function, and that $u_j \rightharpoonup u$ in $W^{1,\infty}(\Omega, \mathbb{R}^N)$ generates a gradient Young measure ν. Then*

$$\liminf_{j\to\infty} \int_\Omega f(x, u_j(x), \nabla u_j(x))dx \ge \int_\Omega \left(\int_{\mathcal{M}^{n\times N}} f(x, u(x), \xi) d\nu_x(\xi) \right) dx$$

If further, $\forall (x, u) \in \Omega \times R^N, \xi \to f(x, u, \xi)$ is quasi-convex, then

$$\liminf_{j\to\infty} \int_\Omega f(x, u_j(x), \nabla u_j(x))dx \ge \int_\Omega f(x, u(x), \nabla u(x))dx \ ,$$

i.e., the functional $J(u) = \int_\Omega f(x, u(x), \nabla u(x))dx$ is s.w.l.s.c. on $W^{1,\infty}(\Omega, \mathbb{R}^N)$.*

Proof. According to Corollary 4.4.13, the sequence $\{z_j = (u_j, \nabla u_j)\}$ generates the Young measure $\delta_u \bigotimes \nu$.

Since the range of $\{z_j\}$ are contained in a big ball $B_R(\theta) \subset \mathbb{R}^N \times M^{n\times N}$, we may assume that $f = 0$ outside $B_R(\theta)$.

Since f is a Caratheodory function, by the Scorza–Dragoni–Vainberg theorem (a version of the Luzin theorem with parameters, see also Vainberg [Va], there exists an increasing sequence of compact sets $C_k \subset \Omega$ such that $m(\Omega \backslash C_k) \to 0$, and $f|_{C_k \times \mathbb{R}^N \times \mathcal{M}^{n\times N}}$ is continuous. Let $g_k(x, \cdot) = \chi_{C_k}(x) f(x, \cdot) : \Omega \to C_0(\mathbb{R}^N \times \mathcal{M}^{n\times N})$. We obtain

$$\begin{aligned} \int_\Omega f(x, u_j(x), \nabla u_j(x))dx &\ge \int_\Omega g_k(x, u_j(x), \nabla u_j(x))dx \\ &\to \int_\Omega \langle \delta_{u(x)} \bigotimes \nu_x, g_k(x, \cdot)\rangle dx, \text{ as } j \to \infty \ . \end{aligned}$$

But,

$$\begin{aligned} \text{RHS} &= \int_{C_k} \langle \delta_{u(x)} \bigotimes \nu_x, f(x, \cdot, \cdot)\rangle dx \\ &= \int_{C_k} \left(\int_{M^{n\times N}} f(x, u(x), \xi) d\nu_x(\xi) \right) dx \end{aligned}$$

$$\to \int_\Omega \left(\int_{M^{n\times N}} f(x, u(x), \xi) d\nu_x(\xi) \right) dx, \text{ as } k \to \infty \,.$$

This proves the first conclusion.

If further, f is quasi-convex in ξ, then from the conclusions (2) and (3) of Theorem 4.4.15, we have

$$\int_{\mathcal{M}^{n\times N}} f(x, u(x), \xi) d\nu_x(\xi) \geq f(x, u(x), \nabla u(x)), \text{ a.e. } x \in \Omega \,.$$

The s.w*.l.s.c. of J is proved. □

Remark 4.4.18 *A gradient Young measure ν, if it is constant up to a null set, is called homogeneous. This is a useful notion in studying oscillations of sequences because it isolates the oscillation near a point in a domain. Examples 4.4.13 and 4.4.14 are homogeneous.*

4.5 Other Function Spaces

We have seen that the w^*-compactness plays an important role in the calculus of variations. That is why we always choose Sobolev spaces $W^{m,p}, 1 < p \leq \infty$ as working spaces. The case $p = 1$ is different, even if to the Lebesgue space L^1, the Banach–Alaoglu theorem cannot be applied. In fact, the positive delta-type sequence does not weakly * converge to any L^1 function. In the extremal case, we study some other function spaces as replacements. In this section, the BV space, the Hardy space and some other related spaces are introduced in the applications to variational problems.

4.5.1 BV Space

We study the minimal surface problem. Let $\Omega \subset \mathbb{R}^n$ be a bounded open domain; a hypersurface $u \in C^1(\overline{\Omega}, \mathbb{R}^1)$ with prescribed boundary condition has the area:

$$A(u) = \int_\Omega \sqrt{(1 + |\nabla u|^2)} \,.$$

This leads to the following variational problem: Min$\{A(u)\}$ in the above set of hypersurfaces. One might expect that the Sobolev space $W^{1,1}$ would be a working space for the direct method. Unfortunately, the closed ball in this space is not weakly closed, and so a minimizing sequence may fail to converge. We are introduced to the space of bounded variations, which is widely used in the so-called free discontinuous boundary problems, in which the boundary of the domain is to be determined in the extremum problems.

Recall that a continuous linear functional defined on the space of continuous functions on $[0, 1]$ can be represented by a BV function u as follows:

$$L(\phi) = \int_0^1 \phi(t)du(t), \ \forall \phi \in C([0,1]) . \tag{4.46}$$

If $\phi \in C_0^1([0,1])$, then

$$L(\phi) = -\int_0^1 u(t)\phi'(t)dt . \tag{4.47}$$

We shall extend this relation to higher dimensional space. Let $\Omega \subset \mathbb{R}^n$ be an open domain, and let $C_0(\Omega)$ be the space of continuous functions of compact support with respect to Ω.

First, we note that L is a linear continuous functional on $C_0(\Omega)$ if and only if there exists a Radon measure μ and a μ-measurable function ν on Ω with $|\nu(x)| = 1$, a.e., such that

$$L(\phi) = \int_\Omega \phi(x)\nu(x)d\mu(x), \ \phi \in C_0(\Omega) .$$

Second, for each compact subset $K \subset \Omega$, let $L_K = L|_{C(K)}$, then

$$\|L_K\| := \sup\{L(f) \mid \|f\|_{C(K)} \leq 1\} = \mu(K) .$$

Motivated by (4.46), and (4.47), one defines the counterpart of a BV function u in higher dimensional spaces, as being a function $u \in L^1(\Omega)$ such that the functional

$$L(\phi) := \int_\Omega u \,\mathrm{div}\phi$$

is defined for all $\phi \in C_0^1(\Omega, \mathbb{R}^n)$, and can be extended continuously to $C_0(\Omega, \mathbb{R}^n)$. Then by Riesz representation theorem, there exist Radon measures ρ_i, and ρ_i measurable functions $\tau_i(x)$ on Ω, with $|\tau_i(x)| \leq 1, \rho_i-$ a.e., $i = 1, \ldots, n$, and let $\mu = \Sigma_{i=1}^n \rho_i, \nu_i = \tau_i \frac{d\rho_i}{d\mu}, i = 1, \ldots, n.$ such that

$$\int_\Omega u \,\mathrm{div}\phi = -\int_\Omega \Sigma_1^n \phi_i \tau_i d\rho_i = -\int_\Omega (\phi \cdot \nu)d\mu , \tag{4.48}$$

where $\nu(x) = (\nu_1(x), \ldots, \nu_n(x))$ with $\|\nu(x)\|_\infty \leq 1 \, \mu-$ a.e. We define $\|\cdot\| := \|\cdot\|_{R^n}$.

Definition 4.5.1 *Let $\Omega \subset \mathbb{R}^n$ be open. A function $u \in L^1(\Omega)$ is said to be of bounded variations, if the total variation of u on Ω is:*

$$\|Du\|(\Omega) := \sup\left\{\int_\Omega u \,\mathrm{div}\phi \mid \phi \in C_0^1(\Omega, \mathbb{R}^n), \|\phi(x)\| \leq 1, \ for\, a.e.\, x \in \Omega\right\} < \infty .$$

The space $BV(\Omega)$ consists of all bounded variation functions on Ω with norm:

$$\|u\|_{BV} = \|u\|_{L^1} + \|Du\|(\Omega) .$$

Since $\|.\|_{R^n}$ and $\|.\|_\infty$ are equivalent, for $u \in BV(\Omega)$, the total variation $\|Du\|$ can be regarded as a measure μ:

$$\frac{1}{\sqrt{n}}\mu(\Omega) \le \|Du\|(\Omega) \le \sqrt{n}\mu(\Omega) . \tag{4.49}$$

Moreover, the distributional derivative Du makes sense, it is related to the Radon measure μ and vector measurable function $\nu(x)$.

Example 4.5.2 *If $u \in W^{1,1}(\Omega)$, then $u \in BV(\Omega)$, and*

$$\|u\|_{BV} = \|u\|_{W^{1,1}} .$$

In fact, we shall verify that $\|Du\|(\Omega) = \int_\Omega |\nabla u| dx$. On one hand,

$$|\int_\Omega u \,\mathrm{div}\phi \, dx| = |\int_\Omega \nabla u \cdot \phi \, dx| \le \int_\Omega |\nabla u| dx$$

for all $\phi \in C_0^1(\Omega, R^n)$ with $\|\phi(x)\| \le 1 \forall x \in \Omega$, it follows that $\|Du\|(\Omega) \le \int_\Omega |\nabla u| dx$.

On the other hand, $\forall u \in W^{1,1}(\Omega), \forall \epsilon > 0, \exists \phi_\epsilon \in C_0^1(\Omega, \mathbb{R}^n)$ with $\|\phi_\epsilon(x)\| \le 1$, $\forall x \in \Omega$ satisfying:

$$\int_\Omega |\nabla u| \, dx \le \int_\Omega \nabla u \cdot \phi_\epsilon \, dx + \epsilon = \int_\Omega u \cdot \mathrm{div}\phi_\epsilon dx + \epsilon \le \|Du\|(\Omega) + \epsilon .$$

Since $\epsilon > 0$ is arbitrary, we obtain $\int_\Omega |\nabla u| dx \le \|Du\|(\Omega)$.

Example 4.5.3 *Let $S \subset \mathbb{R}^n$ be a C^∞ compact $(n-1)$-dimensional hypersurface with the induced metric, and let H^{n-1} be the $(n-1)$-dimensional Hausdorff measure in $\mathbb{R}^n$. The area of S is $H^{n-1}(S)$. Let Ω be the body bounded by S, and let χ_Ω be the characteristic function of Ω. Then $\chi_\Omega \in BV(R^n)$, and*

$$\|D\chi_\Omega\|(\mathbb{R}^n) = H^{n-1}(S) .$$

Indeed, by the Gauss formula, $\forall \phi \in C_0^1(\mathbb{R}^n, \mathbb{R}^n)$

$$\int_{R^n} \chi_\Omega \mathrm{div}\phi dx = \int_\Omega \mathrm{div}\phi dx = \int_S \mathsf{n}(x) \cdot \phi(x) d\mathsf{H}^{n-1} ,$$

where $\mathsf{n}(x)$ is the unit exterior normal. Thus

$$\|D\chi_\Omega\|(\mathbb{R}^n) \le \mathsf{H}^{n-1}(S) .$$

On the other hand, one extends n to be a C^∞ vector field V over $\mathbb{R}^n$ with $\|V(x)\| \le 1 \; \forall x \in \mathbb{R}^n$. This can be done by a partition of unity. Then, $\forall \rho \in C_0^\infty(\mathbb{R}^n, \mathbb{R}^1)$ with $|\rho(x)| \le 1$, $\forall x \in \mathbb{R}^n$, let $\phi = \rho V$, we have

$$\int_{R^n} \chi_\Omega \mathrm{div}\phi \, dx = \int_S \rho d\mathsf{H}^{n-1} .$$

Thus,

$$\begin{aligned}
&\|D\chi_\Omega\|(\mathbb{R}^n) \\
&= sup\left\{\int_{\mathbb{R}^n} \chi_\Omega \mathrm{div}\phi dx \mid \phi \in C_0^\infty(\mathbb{R}^n, \mathbb{R}^n), \text{ with } \|\phi(x)\| \leq 1, \ \forall x \in \mathbb{R}^n\right\} \\
&\geq sup\left\{\int_S \rho d\mathsf{H}^{n-1} \mid \rho \in C_0^\infty(\mathbb{R}^n, \mathbb{R}^1), |\rho(x)| \leq 1, \ \forall x \in \mathbb{R}^n\right\} \\
&= \mathsf{H}^{n-1}(S) .
\end{aligned}$$

These two examples show that $W^{1,1}(\Omega)$ is strictly contained in $BV(\Omega)$, since for $n = 1$, the characteristic function $\chi_{[0,1]} \in BV(\mathbb{R}^1)$ but not in $W^{1,1}$.

This leads us to extend the definition of the co-dimensional one area to the boundary of more general domains.

Definition 4.5.4 *Let E be a Borel set in an open domain $\Omega \subset \mathbb{R}^n$. We call*

$$\|\partial E\|(\Omega) = \|D\chi_E\|(\Omega)$$

the perimeter of E in Ω.

As a function space with the norm $\|u\|_{BV} = \|u\|_{L^1} + \|Du\|(\Omega)$, $BV(\Omega)$ is a Banach space. Only the completeness remains to be verified.

Lemma 4.5.5 *(Lower semi-continuity) If $\{u_j\} \subset BV(\Omega)$, and $u_j \to u$ in L^1; then for every open $U \subset \Omega$*

$$\|Du\|(U) \leq \liminf_{j\to\infty} \|Du_j\|(U) . \tag{4.50}$$

If further, $\sup\{\|Du_j\|(\Omega) \mid j \in \mathcal{N}\} < \infty$, *then $u \in BV(\Omega)$.*

Proof. $\forall \phi \in C_0^1(U, \mathbb{R}^n)$ with $\|\phi(x)\| \leq 1$, one has

$$\int_U u \, \mathrm{div}\phi \, dx = \lim_{j\to\infty} \int_U u_j \, \mathrm{div}\phi \, dx \leq \liminf_{j\to\infty} \|Du_j\|(U) .$$

(4.50) is proved. □

Theorem 4.5.6 *$BV(\Omega)$ is complete.*

Proof. For a Cauchy sequence $\{u_j\}$ in the BV norm, it is obvious that $u_j \to u$ in L^1. By the previous lemma, $\|Du\|(\Omega) < \infty$, and then $u \in BV(\Omega)$. It remains to show that $\|D(u_j - u)\|(\Omega) \to 0$. Again, from the lower semi-continuity lemma, $\forall \epsilon > 0, \exists j_0 \in \mathcal{N}$ such that

$$\|D(u_j - u)\|(\Omega) \leq \liminf_{k\to\infty} \|D(u_j - u_k)\|(\Omega) < \epsilon, \text{ as } j \geq j_0 .$$

□

Now we consider the possibility of C^∞ approximation of the BV functions. Since the $W^{1,1}$ norm equals the BV norm for C^1 functions, and C^∞ is dense in $W^{1,1}$, we can only have:

Theorem 4.5.7 *Let Ω be an open domain of $\mathbb{R}^n$. Then for $\forall u \in BV(\Omega)$, $\exists u_j \in BV(\Omega) \cap C^\infty(\Omega)$ such that*

1. *$u_j \to u$ in $L^1(\Omega)$,*
2. *$\|Du_j\| \to \|Du\|$ in the sense of Radon measure.*

In particular, $\|Du_j\|(\Omega) \to \|Du\|(\Omega)$.

We omit the proof, but refer to Giusti [Gi], p. 14.

Theorem 4.5.8 *(Compactness) Let Ω be a bounded open domain of $\mathbb{R}^n$. Any sequence $\{u_j\} \subset BV(\Omega)$ with $\|u_j\|_{BV} \le M < \infty$ possesses a convergent subsequence in the L^1 norm, and the limit $u \in BV(\Omega)$, with $\|u\|_{BV} \le M$.*

Proof. We take a sequence $\{v_j\} \subset BV(\Omega) \cap C^\infty(\Omega)$ such that $\|u_j - v_j\|_{L^1} < \frac{1}{j}$, and $\|Dv_j\|(\Omega) \le M + \frac{1}{j}$. From Example 4.5.2, we have $\|v_j\|_{W^{1,1}} = \|Dv_j\|_{BV}$. According to the Rellich–Kondrachev compactness theorem, there is a subsequence v_{j_k} L^1-converges to u. From the lower semi-continuity lemma, $\|Du\|(\Omega) \le M$, and $u \in BV(\Omega)$. Obviously, $u_{j_k} \to u$ in L^1. □

Besides the above preparations, in order to study the minimal surface problem stated at the beginning of the subsection we have to define the trace of BV functions. The following theorem can be found in Giusti [Gi] and Evans and Gariepy [EG]:

Theorem 4.5.9 *(Trace) Let $\Omega \subset \mathbb{R}^n$ be a bounded domain with Lipschitzian boundary $\partial\Omega$. Then the trace operator $T : BV(\Omega) \to L^1(\partial\Omega, H^{n-1})$ is bounded, and we have*

$$\int_\Omega u \operatorname{div} \phi = -\int_\Omega (\phi \cdot \nu) d\mu + \int_{\partial\Omega} Tu(\phi \cdot \mathsf{n}) dH^{n-1}, \ \forall \phi \in C_0^1(\Omega, \mathbb{R}^n),$$

where μ is the Radon measure, ν is the μ-measurable vector function with $\|\nu(x)\| \le 1$, a.e. with respect to μ, and n is the unit normal vector field over $\partial\Omega$, which is well defined almost everywhere.

We omit the proof and turn to the existence of the minimal surface problem.

Theorem 4.5.10 *(Nonparametric minimal surface) Let $\Omega \subset \mathbb{R}^n$ be a bounded Lipschitzian domain, and let $u_0 \in BV(\Omega)$. Then there exists a minimizer of the problem:*

$$\operatorname{Min}\{A(u) \,|\, u \in BV(\Omega),\, Tu = Tu_0\},$$

where

$$A(u) = \int_\Omega \sqrt{1 + |Du|^2}$$

$$:= \sup\left\{ \int_\Omega (\phi_0 + u\mathrm{div}\phi)dx \mid (\phi_0, \phi) \in C_0^1(\Omega, \mathbb{R}^{1+n}) \right.$$
$$\left. with\ |\phi_0(x)| + \|\phi(x)\| \leq 1\, \forall x \in \Omega \right\} .$$

Proof. One considers the set: $X_M := \{u \in BV(\Omega) \,|\, Tu = Tu_0$, and $\|Du\|(\Omega) \leq M\}$, for any $M > 0$. For sufficiently large $M > 0$, $X_M \neq \emptyset$ since we have

$$\|Du\|(\Omega) \leq A(u) \leq \|Du\|(\Omega) + \mathrm{mes}(\Omega) .$$

Combining with Theorems 4.5.7 and 4.5.8, X_M is compact in L^1-topology. By the same proof as for Lemma 4.5.5, A is l.s.c. on X_M in L^1-topology. The proof follows from the general principle of the direct method, i.e., Theorem 4.2.1. □

The problem stated above is called the nonparametric minimal surface problem, there is another formulation called parametric minimal surface problem:

Again, let Ω be a bounded open domain in R^n, and let L be a set of finite perimeters. Define

$$\mathsf{F} = \{F \text{ is Borel measurable} \,|\, F\backslash\Omega = L\backslash\Omega\} .$$

Theorem 4.5.11 *(De Giorgi) The problem*

$$\mathrm{Min}\{\|D\chi_F\|(R^n) \mid F \in \mathsf{F}\}$$

has a solution.

Proof. Since Ω is bounded, $\exists R > 0$ such that $\Omega \subset B_R(\theta)$. If $F\backslash\Omega = L\backslash\Omega$, then $F\backslash B_R(\theta) = L\backslash B_R(\theta)$. It follows that

$$\|D\chi_F\|(R^n) = \|D\chi_F\|(B_R(\theta)) + \|D\chi_L\|(R^n\backslash B_R(\theta)) .$$

Thus the problem is reduced to

$$\mathrm{Min}\{\|D\chi_F\|(B_R(\theta)) \,|\, F \in \Gamma\} .$$

The functional $J(u) := \|Du\|(B_R(\theta))$ is bounded from below on the set $X := \{u = \chi_F \mid \|Du\|(B_R(\theta)) \leq \|\partial\chi_L\|(B_R(\theta))\}$.

We shall verify that X is a compact subset under L^1-topology. According to Theorem 4.5.8, it remains to verify that any limit point u of X is a characteristic function of a measurable set of finite perimeter. In fact, it is the a.e. limit of a sequence of X.

Since J is l.s.c. on X under L^1-topology, there exists a minimizer u. □

Although we have proved the existence of minimal surfaces, the solutions we obtained are very weak. Thus, in order to verify that the weak solutions are geometric minimal surfaces, the difficulties lie in the regularity, cf E. Giusti [EG] and De Giorgi [DG 1].

4.5.2 Hardy Space and BMO Space

It is known that a bounded sequence in L^1 space may not have a w*-convergent subsequence. Therefore L^1 is not a good space for the variational method on which we work. Comparing with the L^p space $p \in (1, \infty)$, the latter has many important properties in analysis: the uniform convexity, the reflexivity, the smoothness of the norm, the boundedness of singular integral operators in Calderon–Zygmund theory, as well as the L^p estimates in the elliptic theory, which L^1 space does not have. The Hardy space H^1 is in some sense a replacement of L^1. It is defined as follows:

Definition 4.5.12 *(Hardy space)*

$$\boldsymbol{H}^1(\mathbb{R}^N) = \{f \in L^1(\mathbb{R}^N) \mid R_j f \in L^1(\mathbb{R}^N),\ j = 1, 2, \ldots, N\} .$$

where $R_j = \partial_j(-\Delta)^{-\frac{1}{2}}$ *is the Riesz transform,* $j = 1, 2, \ldots, N$. *The norm is defined to be*

$$\|f\|_{\boldsymbol{H}^1} = \|f\|_{L^1} + \sum_1^N \|R_j f\|_{L^1} .$$

There are many equivalent characterizations of the Hardy space. The following is one of them. $\forall h \in C_0^\infty(\mathbb{R}^N)$, with $h \geq 0$, and $\int h = 1$. $\forall t > 0$ let $h_t(x) = \frac{1}{t^N} h(\frac{x}{t})$. For every distribution f, one defines

$$f^*(x) = \sup_{t>0} |(h_t * f)(x)| .$$

If $f^* \in L^1$, then from the Lebesgue dominance theorem, we have $f \in L^1$ and $\|f\|_{L^1} \leq \|f^*\|_{L^1}$. By definition f^* depends on h, but it is proved that $f \in H^1$ if and only if $f^* \in L^1$ for any such h, and

$$\|f\| = \|f^*\|_{L^1}$$

is an equivalent norm. $\mathbf{H}^1$ is a Banach space.

In order to study the $\mathbf{H}^1$ space, we need some knowledge on harmonic analysis, in particular, the notion of Hardy-Littlewood maximal functions: $\forall f \in L^1_{\text{loc}}$, the function

$$M(f)(x) = \sup_{r>0} \frac{1}{|B_r(x)|} \int_{B_r(x)} |f(y)| dy$$

is called the maximal function of f. It possesses the following properties:

1. $M(cf) = |c| M(f), \forall c \in \mathbb{R}^1$, and $M(f_1 + f_2) \leq M(f_1) + M(f_2), \forall f_1, f_2 \in L^1_{\text{loc}}$.
2. $|f(x)| \leq M(f)(x)$ a.e.
3. If $M(f) \in L^1(\mathbb{R}^n)$, then $f = 0$ a.e.

Proof. If $|f| > 0$ on a positive measure set E, then we may assume that $\exists R > 0, \exists \epsilon > 0$ such that $E \subset B_R(\theta), |f(x)| \geq \epsilon, \forall x \in E$. Thus,

$$M(f)(x) \geq C_N \epsilon |E|(|x| + R)^{-N} ,$$

where C_N is a constant depending on N only. It contradicts $M(f) \in L^1(\mathbb{R}^N)$. □

4. The maximal function operator is of weak $(1,1)$ type, i.e., there exists a constant $C > 0$ such that

$$|E_\lambda| \leq \frac{C}{\lambda}\|f\|_{L^1} \ \forall \lambda > 0, f \in L^1(\mathbb{R}^N) ,$$

where $E_\lambda = \{x \in \mathbb{R}^N \mid M(f)(x) > \lambda\}$.

Proof. For $\forall x \in E_\lambda, \exists r_x > 0$ such that

$$\int_{B_{r_x}(x)} |f| dx \geq \lambda |B_{r_x}(x)| .$$

Thus, $\{B_{r_x}(x) \mid x \in E_\lambda\}$ is a covering of E_λ. According to the Vitali covering theorem, there exist a family of countable disjoint balls $\{B_{r_{x_j}}(x_j)\}$ and a constant c_N such that

$$|E_\lambda| \leq c_N \sum_j |B_{r_{x_j}}(x_j)| \leq c_N \lambda^{-1} \sum_j \int_{B_{r_{x_j}}(x_j)} |f| dx \leq c_N \lambda^{-1} \|f\|_{L^1} .$$

□

5. For $p \in (1, \infty)$, if $f \in L^p(\mathbb{R}^N)$, then $M(f) \in L^p(\mathbb{R}^N)$ and there exists a constant $C_p > 0$ such that

$$\|M(f)\|_{L^p} \leq C_p \|f\|_{L^p} .$$

Proof. Since M is of (∞, ∞) type, in combination with property 4, the conclusion follows from the interpolation theory. □

6. (Kolmogorov's inequality) For any $\delta \in (0,1)$, there is a constant $C = C_\delta > 0$ such that

$$\int_E M(f)^\delta dx \leq \frac{1}{1-\delta} C^\delta |E|^{1-\delta} \|f\|_{L^1}^\delta, \ \forall \text{ measurable } E \subset \mathbb{R}^N .$$

Proof. In fact from property 4,

$$\begin{aligned}\int_E M(f)^\delta dx &= \delta \int_0^\infty |\{x \mid M(f)(x) > \lambda\} \cap E| \lambda^{\delta-1} d\lambda \\ &= \delta \left[\int_0^t + \int_t^\infty\right] |\{x \mid M(f)(x) > \lambda\} \cap E| \lambda^{\delta-1} d\lambda\end{aligned}$$

$$\leq |E|t^\delta + C\delta \int_t^\infty \lambda^{\delta-2} d\lambda \|f\|_{L^1}$$
$$\leq |E|t^\delta + \frac{C\delta}{1-\delta} t^{\delta-1} \|f\|_{L^1} /;,$$

where C is the weak $(1,1)$ norm of $M(f)$. Setting $t = \frac{C\|f\|_{L^1}}{|E|}$, we obtain the desired inequality. □

We investigate the behavior of $\mathbf{H}^1$ functions.

1. There exists a constant $C_1 > 0$ such that $f^* \leq C_1 M(f)$. If $f \geq 0$, then $M(f) \leq C_2 f^*$ for some constant C_2. Thus if $f \geq 0$ and $f \in \mathbf{H}^1$, then $f = 0$.
2. If $f \in \mathbf{H}^1$, then $\int_{\mathbb{R}^N} f dx = 0$.

Proof. Let $\phi \in C^\infty(\mathbb{R}^N)$ with $\text{supp}\phi \subset B_1(\theta)$ and $\phi(\theta) = 1$. $\forall x \in \mathbb{R}^N, \forall s \geq |x|$, one defines $h(y) = c\phi(s^{-1}x - 2y)$, where c is the normalized constant such that $\int_{\mathbb{R}^N} h(y)dy = 1$. Then

$$\begin{aligned} f^*(x) &= \sup_{t>0} |(h_t * f)(x)| \\ &\geq (2s)^{-N} \left| \int_{\mathbb{R}^N} h\left(\frac{x-y}{2s}\right) f(y)dy \right| \\ &= (2s)^{-N} c \left| \int_{\mathbb{R}^N} \phi\left(\frac{y}{s}\right) f(y)dy \right| , \end{aligned}$$

since the last integral tends to $\int_{\mathbb{R}^N} f(y)dy$ as $s \to \infty$. If $\int_{\mathbb{R}^N} f(y)dy \neq 0$, then we would have $\liminf_{|x|\to\infty} f^*(x)|x|^N > 0$, which contradicts $f^* \in L^1(\mathbb{R}^N)$. □

3. If $f \in L^\infty(\mathbb{R}^N)$ with compact $\text{supp}f$ and $\int_{\mathbb{R}^N} f(x)dx = 0$, then $f \in \mathbf{H}^1$.

   Proof. Assume $\text{supp}f \subset B_R(\theta)$. Since

$$\begin{aligned} |(h_t * f)(x)| &= \left| \int_{\mathbb{R}^N} [h_t(x-y) - h_t(\theta)] f(y)dy \right| \\ &= \left| \int_{\mathbb{R}^N} \int_0^1 (x-y) \cdot \nabla h_t(s(x-y)) ds f(y)dy \right| \\ &\leq C \int_{|x-y|\leq t,\, |y|\leq R} t^{-(N+1)} dy \|f\|_\infty , \end{aligned}$$

   therefore $f^*(x) \leq CR|x|^{-N-1}\|f\|_\infty$. The conclusion follows. □

The dual space of $\mathbf{H}^1$ is the BMO space, which is defined as follows: $\forall f \in L^1_{loc}(\mathbb{R}^N), \forall x \in \mathbb{R}^N, \forall r > 0$, let

$$\overline{f}_r(x) = \frac{1}{|B_r(x)|} \int_{B_r(x)} f(y)dy ,$$

and define

Definition 4.5.13 *(BMO space)*

$$\|f\|_{\text{BMO}} = \sup_{x\in R^N} \sup_{r>0} \frac{1}{|B_r(x)|} \int_{B_r(x)} |f(y) - \overline{f}_r(x)| dy \, .$$

$\text{BMO}(\mathbb{R}^N)$ *is the space of functions such that* $\|f\|_{\text{BMO}} < \infty$.

The quantity $\|f\|_{\text{BMO}}$ is a semi-norm; in fact, $\|f\|_{\text{BMO}} = 0$ if and only if $f =$ constant. After modulo constants, BMO is a Banach space.

There is an equivalent BMO semi-norm:

$$\|f\|' = \sup_{x\in \mathbb{R}^N} \sup_{r>0} \inf_{c} \frac{1}{|B_r(x)|} \int_{B_r(x)} |f(y) - c| dy \, .$$

In fact, on one hand it is obvious $\|f\|' \leq \|f\|_{\text{BMO}}$, and on the other hand,

$$⨍_{B_r(x)} |f(y) - \overline{f}_r(x)| dy \leq ⨍_{B_r(x)} |f(y) - c| dy + |\overline{f}_r(x) - c| \leq 2\|f\|' \, ,$$

where $⨍$ denotes the average.

The following simple properties are easily seen from the definition.

1. $L^\infty \subset \text{BMO} \subset L^1_{\text{loc}}$.
2. The function $\log|x| \in \text{BMO}$. In fact, the scaling transformation: $\forall \lambda > 0, f(x) \to f(\lambda x)$ maps BMO functions to BMO functions, and preserves the semi-norm. Under the scaling, $\log|x|$ is changed by adding a constant. The verification is then reduced to

$$\int_B |\log|x|| dx \leq C, \text{ or } \int_B |\log|x| - \log|x_0|| dx \leq C \, ,$$

   where B is unit ball centered at x_0, The first inequality is true for $|x_0| \leq 1$ and the second for $|x_0| \geq 1$. These hold by elementary calculation.
3. If $f \in \text{BMO}$, then $|f| \in \text{BMO}$, and then $f_\pm \in \text{BMO}$.
   Let $⨍_\Omega = \frac{1}{|\Omega|} \int_\Omega$ denote the average; one has

$$⨍_{B_r(x)} ||f(y)| - |c|| dy \leq ⨍_{B_r(x)} |f(y) - c| dy \, ,$$

   which implies that $\| |f| \|' \leq \|f\|'$.
   In the case where (x, r) is not specified, we define $Q = B_r(x)$ and $m_Q(f) = \overline{f}_r(x)$.
4. If $f \in L^1_{loc}(\mathbb{R}^N)$ and if there exists $C > 0$ such that $m_Q(f) - \text{essinf}_Q f \leq C \; \forall Q$, then $f \in \text{BMO}$ and $\|f\|_{\text{BMO}} \leq 2C$.

   Proof. Since $\int_Q [f(y) - m_Q(f)] dy = 0$,

$$\begin{aligned} ⨍_Q |f(y) - m_Q(f)| dy &= 2⨍_Q \max[m_Q(f) - f(y), 0] dy \\ &\leq 2⨍_Q [m_Q(f) - \text{essinf}_Q f] dy \leq 2C \, . \end{aligned}$$

   □

5. Let $f \in L^1_{\text{loc}}(\mathbb{R}^N)$ with $f \geq 0$. If there exists a constant $C > 0$ such that $m_Q(f) \leq C\text{essinf}_Q f$, $\forall\, Q$, then $\log f \in \text{BMO}$.

   Proof. Taking logarithms and using Jessen's inequality, we have

   $$m_Q(\log f) = ⨍_Q \log f dy \leq \log ⨍_Q f dy = \log m_Q(f) \leq \log C + \text{essinf}_Q \log f \,.$$

   The conclusion follows from property 4. □

6. Assume $f \in L^1_{\text{loc}}(\mathbb{R}^N)$ and that $M(f)(x) \neq \infty$ a.e. $\forall\, \delta \in (0,1)$, set $w = M(f)^\delta$. Then there exists a constant $C = C_\delta > 0$ independent of Q, such that $m_Q(w) \leq C\text{inf}_Q w$, $\forall\, Q$.

   Proof. Fixing Q, we may assume $Q = B_R(\theta)$ without loss of generality. Decompose $f = f_1 + f_2$, where $f_1 = \chi_{Q'} \cdot f$ with $Q' = 2Q$. Let $w_i = M(f_i)^\delta, i = 1, 2$. Since $M(f) \leq M(f_1) + M(f_2)$, there is a constant c depending on δ, such that $w \leq c(w_1 + w_2)$. Now we apply Kolmogorov's inequality to w_1:

   $$\begin{aligned} m_Q(w_1) = ⨍_Q M(f_1)^\delta(y) dy &\leq \frac{C_\delta}{|Q|}|Q|^{1-\delta}\|f_1\|^\delta_{L^1} \\ &= 2^N C_\delta m_{Q'}(f)^\delta \\ &\leq C_1 \text{inf}_{Q'} w \leq C_1 \text{inf}_Q w \end{aligned}$$

   where C_1 is independent of Q. Next we turn to estimating w_2. It remains to verify that there exists a constant $C_2 > 0$ such that

   $$M(f_2)(x) \leq C_2 M(f_2)(y), \;\; \forall\, x, y \in Q \,.$$

   Indeed, it implies that $w_2(x) \leq C_2 \text{inf}_Q w_2 \leq C_2 \text{inf}_Q w \;\forall x \in Q$, then $m_Q(w_2) \leq C_2 \text{inf}_Q w$. Combining these two estimates together, we obtain

   $$m_Q(w) \leq c(m_Q(w_1) + m_Q(w_2)) \leq C\text{inf}_Q w \,.$$

   Now we return to estimate $M(f_2)$. Notice that for any $r > 0$, only when $B_r(x) \cap (Q')^c \neq \emptyset$, i.e., $r \geq R$, $\int_{B_r(x)} |f_2(z)| dz$ can be nonzero. From

   $$⨍_{B_r(x)} |f_2(z)| dz \leq \frac{|B_{r+R}(y)|}{B_r(x)} ⨍_{B_{r+R}(y)} |f_2(z)| dz \leq 2^N M(f_2)(y) \,,$$

   it follows that $M(f_2)(x) \leq 2^N M(f_2)(y)$. □

Combining properties 5 and 6, we obtain:

7. If $f \in L^1_{\text{loc}}(\mathbb{R}^N)$ and $M(f)(x) \neq \infty$ a.e., then $\log M(f) \in \text{BMO}$ and $\|\log M(f)\|_{\text{BMO}} \leq c_N$, a constant depending on N only.

Definition 4.5.14 *(VMO space) The space* $\mathrm{VMO}(\mathbb{R}^N)$ *is the closure of* $C_0^\infty(\mathbb{R}^N)$ *in* BMO.

It is proved (see Stein [Ste 2]): that $(\mathrm{VMO})^* = \mathbf{H}^1$, $(\mathbf{H}^1)^* = \mathrm{BMO}$, in the following sense, $\forall f \in \mathrm{BMO}$, the following integral makes sense:

$$l(g) = \int f \cdot g dx, \ \ \forall g \in \mathbf{H}^1_\alpha := \left\{ g \in L^\infty \,|\, \mathrm{supp}\,(g) \text{ is compact, and } \int g = 0 \right\}$$

(see property 3 of $\mathbf{H}^1$ functions), and has a unique bounded extension to $\mathbf{H}^1$:

$$|l(g)| \le C \|f\|_{\mathrm{BMO}} \|g\|_{\mathbf{H}^1}$$

for some constant $C > 0$.

Conversely, $\forall l \in (\mathbf{H}^1)^*, \exists f \in \mathrm{BMO}$, such that $l(g) = \int f \cdot g \ \forall g \in \mathbf{H}^1_\alpha$, with $\|f\|_{\mathrm{BMO}} \le C \|l\|_{(\mathbf{H}^1)^*}$.

Since $\mathbf{H}^1$ is the dual of a separable Banach space VMO, the weak-* topology on $\mathbf{H}^1$ is well defined, and then the Banach–Alaoglu theorem is applicable. Namely:

If $\{f_j\}$ is a bounded sequence in $\mathbf{H}^1$, then there is a subsequence $\{f_j'\}$, which converges in the distribution sense to $f \in \mathbf{H}^1$, with $\|f\|_{\mathbf{H}^1} \le \liminf_{j\to\infty} \|f_j\|_{\mathbf{H}^1}$.

4.5.3 Compensation Compactness

If we intend to use the space $\mathbf{H}^1$ in variational problems, then the prices we have to pay are: (1) To verify the $\mathbf{H}^1$ boundedness of nonlinear quantities appearing in the integrand of the functional along a minimizing sequence, and (2) such nonlinear quantities are sequentially weakly * l.s.c.

There are few examples of typical nonlinear quantities, which are in the space L^1 with some additional compensation conditions so that they fall into $\mathbf{H}^1$ and preserve the w^*-continuity.

Example 1. (div-curl) Suppose $\{E_n\} \subset L^p(\mathbb{R}^N, \mathbb{R}^N), \mathrm{div} E_n = 0$ in the distribution sense, and $\{B_n\} \subset L^{p'}(\mathbb{R}^N, \mathbb{R}^N), \mathrm{curl} B_n = 0$ in the distribution sense, where $p \in (1, \infty), \frac{1}{p} + \frac{1}{p'} = 1$. Assume that both $\{E_n\}, \{B_n\}$ are bounded in their own spaces respectively, then $E_n \cdot B_n$ is bounded in $L^1(\mathbb{R}^N, \mathbb{R}^1)$. The compensation conditions $\mathrm{div} E_n = 0$, $\mathrm{curl} B_n = 0$ play a role in the weak*-convergence.

Example 2. (Jacobian) Suppose $u \in L^q_{\mathrm{loc}}(\mathbb{R}^N, \mathbb{R}^N)$, for some $q \in (1, \infty)$, in which $\nabla u \in L^N(\mathbb{R}^N, \mathbb{R}^{N\times N})$. Then $\det(\nabla u) \in L^1(\mathbb{R}^N, \mathbb{R}^1)$. The special structure of the Jacobian is also a kind of compensation condition.

First we verify that the compensation condition implises the nonlinear quantities are indeed in H^1, and then these nonlinear quantities are w^* continuous along weakly convergent sequences.

Theorem 4.5.15 *Suppose that* $E \in L^p(\mathbb{R}^N, \mathbb{R}^N), B \in L^{p'}(\mathbb{R}^N, \mathbb{R}^N)$, *with* $p \in (1, \infty), \frac{1}{p} + \frac{1}{p'} = 1$. *If* $\operatorname{curl} B = 0$ *and* $\operatorname{div} E = 0$, *then* $E \cdot B \in \boldsymbol{H}^1$ *and* $\exists C > 0$ *a constant such that* $\|E \cdot B\|_{\boldsymbol{H}^1} \leq C\|E\|_p \|B\|_{p'}$.

Proof. Taking α, β satisfying $\frac{1}{\alpha} + \frac{1}{\beta} = 1 + \frac{1}{N}, \alpha \in (1, p), \beta \in (1, p'), \forall h \in C_0^\infty$ nonnegative, with $\int h = 0$, we claim that

$$|\{h_t * (E \cdot B)\}(x)| \leq C(⨍_{B_t(x)} |E|^\alpha)^{\frac{1}{\alpha}} (⨍_{B_t(x)} |B|^\beta)^{\frac{1}{\beta}} ,$$

where $⨍_B = \frac{1}{|B|} \int_B$ denotes the average.

Indeed, From $\operatorname{curl} B = 0$, we have $\omega \in W^{1,p'}(\mathbb{R}^N, \mathbb{R}^1)$ such that $\nabla \omega = B$. $\operatorname{div} E = 0$ yields

$$\operatorname{div}(\omega E) = \omega \operatorname{div} E + E \cdot \nabla \omega = E \cdot B .$$

By suitably choosing h such that $\operatorname{supp}(h) \subset B_1(\theta)$, we have

$$\begin{aligned} |h_t * (E \cdot B)(x)| &= \left| \int \frac{1}{t^{N+1}} \nabla h \left(\frac{x-y}{t} \right) E(y) \omega(y) dy \right| \\ &= \left| \int \frac{1}{t^{N+1}} \nabla h \left(\frac{x-y}{t} \right) E(y) \{\omega - ⨍_{B_t(x)} \omega\} dy \right| \\ &\leq C(⨍_{B_t(x)} |E|^\alpha)^{\frac{1}{\alpha}} \left(⨍_{B_t(x)} \left[\frac{1}{t} |\omega - ⨍_{B_t(x)} \omega| \right]^{\alpha'} \right)^{\frac{1}{\alpha'}} \\ &\leq C(⨍_{B_t(x)} |E|^\alpha)^{\frac{1}{\alpha}} (⨍_{B_t(x)} |\nabla \omega|^\beta)^{\frac{1}{\beta}} \end{aligned}$$

from the Sobolev–Poincaré inequality.

With the aid of the maximal function of $f \in L^1_{\text{loc}}(\mathbb{R}^n)$:

$$M(f)(x) = \sup_{r>0} \frac{1}{m(B_r(x))} \int_{B_r(x)} |f(y)| dy .$$

The above inequality can be rewritten as

$$\sup_{t>0} h_t * (E \cdot B)(x) \leq C(M(|E|^\alpha)(x))^{\frac{1}{\alpha}} (M(|B|^\beta)(x))^{\frac{1}{\beta}} .$$

Again by Holder inequality and the L^p boundedness of the maximal function, we have

$$\begin{aligned} \|E \cdot B\|_{\mathbf{H}^1} &= \int \sup_{t>0} |h_t * (E \cdot B)(x)| dx \\ &\leq C \left(\int (M(|E|^\alpha))^{\frac{p}{\alpha}} dx \right)^{\frac{1}{p}} \left(\int (M(|B|^\beta) \right)^{\frac{p'}{\beta}} dx)^{\frac{1}{p'}} \\ &\leq C\|E\|_p \|B\|_{p'} . \end{aligned}$$

□

The second example can be reduced to the first one, since

$$\det(\nabla u) = \sum_i \partial_i u^1 A_i^1 = \nabla u^1 \cdot \sigma \,,$$

where $u = (u^1, u^2, \dots, u^N)$, and $\sigma = (A_1^1, A_2^1, \dots, A_N^1)$ is the N-vector consisting of all $N-1$ minors of $\det(\nabla u)$ with respect to the first row. Define $E = \sigma$ and $B = \nabla u^1$, then $E \in L^{\frac{N}{N-1}}(\mathbb{R}^N, \mathbb{R}^N)$, satisfying $\mathrm{div} E = 0$, and $B \in L^N(\mathbb{R}^N, \mathbb{R}^N)$, satisfying $\mathrm{curl} B = 0$.

According to Theorem 4.5.15, $\det(\nabla u) \in \mathbf{H}^1$ and there exists a constant $C > 0$ such that

$$\|\det(\nabla u)\|_{\mathbf{H}^1} \le C \|\nabla u\|_{L^N}^N \,.$$

Apart from the w* compactness, compensation compactness theory identifies classes of such nonlinear quantities, which are sequentially weakly * continuous. Generally speaking, weak convergence may not preserve nonlinear quantities, e.g., if $u_n \rightharpoonup u, v_n \rightharpoonup v$, then we cannot conclude that $u_n v_n \rightharpoonup uv$. This can be seen from the example $u_n = v_n = \sin(n\pi x)$ in $L^2[0,1]$. Both u_n and $v_n \rightharpoonup 0$, but $u_n v_n = \frac{1}{2}(1 - \cos(2n\pi x)) \rightharpoonup \frac{1}{2}$. However, luckily we have:

Theorem 4.5.16 *Suppose that $\{E_j\}, \{B_j\}$ are two bounded sequences in $L^2(\mathbb{R}^N)$ satisfying $\mathrm{div}(E_j) = 0$ and $\mathrm{curl}(B_j) = 0$ in the distribution sense. If further, $E_j \rightharpoonup E, B_j \rightharpoonup B$ in $L^2(\mathbb{R}^N)$, then $E_j \cdot B_j \rightharpoonup E \cdot B$ in the distribution sense, and then $E_j \cdot B_j \stackrel{*}{\rightharpoonup} E \cdot B$ in $\boldsymbol{H}^1$.*

Proof. If B is curl free, we claim that there exists a distribution w such that $\nabla w = B$. This can be shown by Fourier transformations. Let $\hat{B}(\xi)$ be the Fourier transform of B, since $\mathrm{curl} B = 0$ implies that $\xi_i \hat{B}_k - \xi_k \hat{B}_i = 0, \ \forall \ i,k = 1,2,\dots,N$. Let w be the Fourier inverse transform of $\frac{\sum_1^N \xi_k \hat{B}_k}{|\xi|^2}$. Obviously, $\nabla w = B$. Moreover, if B has a bounded support: $\mathrm{supp} B \subset B_R(\theta)^\circ$, then we have $w \in L^2$ and $\mathrm{supp} w \subset B_R(\theta)^\circ$.

Now let w_j be the distribution such that $B_j = \nabla w_j$. For a given $\phi \in C_0^\infty(\mathbb{R}^N)$, we want to prove that

$$\int_{\mathbb{R}^N} E_j \cdot B_j \phi dx \to \int_{\mathbb{R}^N} E \cdot B \phi dx \text{ as } j \to \infty \,.$$

In this case, we may assume all the supports $\mathrm{supp}\, w_j \subset B_R(\theta)^\circ \; j = 1,2,\dots,$ for some $R > 0$. Thus $\{w_j\} \subset H^1(B_R(\theta))$ and is bounded. Modulo a subsequence, we may assume $w_j \to w$ in $L^2(B_R(\theta))$. Thus

$$\begin{aligned} \langle E_j \cdot B_j \,, \phi \rangle &= \langle \mathrm{div}(w_j E_j) \,, \phi \rangle \\ &= -\langle E_j \,, w_j(\nabla \phi) \rangle \\ &\to -\langle E \,, w(\nabla \phi) \rangle \\ &= \langle \mathrm{div}(wE) \,, \phi \rangle = \langle E \cdot B \,, \phi \rangle \end{aligned}$$

where in taking limits, we used $E_j \rightharpoonup E$ and $w_j \to w$ in $L^2(B_R(\theta))$.

Since C_0^∞ is dense in VMO space, $E_j \cdot B_j \stackrel{*}{\rightharpoonup} E \cdot B$ in $\mathbf{H}^1$. □

Remark 4.5.17 *The above theorem has the following generalization: Let $\Omega \subset \mathbb{R}^N$ be an open domain. Assume that $\{E_j\}, \{B_j\}$ are two bounded sequences in $L^2(\Omega, \mathbb{R}^N)$ such that*

1. *$\{\operatorname{div} E_j\}$ lies in a compact subset of $W^{-1,2}(\Omega)$,*
2. *$\{\operatorname{curl} B_j\}$ lies in a compact subset of $W^{-1,2}(\Omega, \mathbb{M}^{N\times N})$.*

If further, $E_j \rightharpoonup E$ and $B_j \rightharpoonup B$ in $L^2(\Omega, \mathbb{R}^N)$, then $E_j \cdot B_j \rightharpoonup E \cdot B$ in the distribution sense.

Similarly, one has:

Theorem 4.5.18 *Assume that $\{u_j\}$ is bounded in $W^{1,N}(\mathbb{R}^N, \mathbb{R}^N)$, and $u_j \rightharpoonup u$. Then $\det(\nabla u_j) \rightharpoonup \det(\nabla u)$ in the distribution sense, so is $\operatorname{div}(\nabla u_j) \overset{*}{\rightharpoonup}$ $\operatorname{div}(\nabla u)$ in $\boldsymbol{H}^1(R^n)$.*

Proof. Again we use the previous special decomposition:

$$\det(\nabla u) = \sum_k \partial_k u^1 A_k^1 = \nabla u^1 \cdot \sigma = \nabla(u^1 \cdot \sigma) ,$$

where $\sigma = (A_1^1, A_2^1, \ldots, A_N^1)$. Since $\operatorname{div}\sigma = 0$, for $\forall \phi \in C_0^\infty(\mathbb{R}^N)$, one has

$$\langle \det(\nabla u_j) , \phi \rangle = -\langle u_j^1 \cdot \sigma_j , \nabla \phi \rangle \to -\langle u^1 \cdot \sigma , \nabla \phi \rangle = \langle \det(\nabla u) , \phi \rangle .$$

□

One can define a Hardy space on a domain Ω in $\mathbb{R}^N$. A distribution f on Ω is said to be in $\mathbf{H}^1_{\rm loc}(\Omega)$ if for each compact set $K \subset \Omega$, there is an $\epsilon > 0$ so that

$$\int_K (\sup_{0<t<\epsilon} |h_t * f(x)|) dx < \infty ,$$

where h is defined as before, and $h_t * f(x)$ is defined for h as long as $t < \operatorname{dist}(x, \partial\Omega)$.

An equivalent definition is as follows:

$f \in \mathbf{H}^1_{\rm loc}(\Omega)$ if and only if $\forall \phi \in C_0^\infty(\Omega)$ with $\int_\Omega \phi \neq 0$, there is a constant c such that $\phi(f - c) \in \mathbf{H}^1(\mathbb{R}^N)$.

4.5.4 Applications to the Calculus of Variations

We shall now show how the Hardy space is applied to variational problems.

Recall that in a compact space X with Radon measure μ, a sequence $\{f_j\} \subset L^1(X, \mathcal{B}, \mu)$ is weakly precompact, i.e., there exists a weakly convergent subsequence, if and only if $\{f_j\}$ is bounded in L^1 norm and is uniformly absolutely continuous, see for instance [DS], Theorem IV.8.9. For an L^1-bounded sequence $\{f_j\}$ without the latter condition, we cannot say whether it is weakly convergent in L^1; this can be seen from the δ-type sequences. However, Chacon introduced the following:

Lemma 4.5.19 *(Bitting lemma) Let $\Omega \subset \mathbb{R}^N$ be a bounded open domain, and let $\{u_j\} \subset L^1(\Omega)$ be bounded. Then there exist a subsequence and decreasing Borel measurable subsets $\{E_k\} \subset \Omega$ such that $|E_k| \to 0$ and that $\chi_{\Omega \setminus E_k}|u_j|$ is uniformly absolutely continuous for all $k \in \mathcal{N}$.*

Proof. Without loss of generality, we may assume that $\{u_j\}$ are nonnegative. Let $\{\nu_j\}$ be the associated Young measures in $\mathbb{R}^1$, which w^*-converges in $\mathbb{R}^N \times \mathbb{R}^1$ to $\mathcal{L}^N \lfloor \Omega \bigotimes \nu_x$. $\forall R > 0$, we define $g_R(y) = y$ if $0 \leq y \leq R$, $= 2R - y$ if $R \leq y \leq 2R$, $= 0$ if $y > 2R$ or $y < 0$. Obviously, $g_R \in C_0(\mathbb{R}^1)$, and $|y\chi_{[0,R]}(y)| \leq g_R(y) \leq |y|$. According to Remark 4.4.10, we have

$$\begin{aligned}\limsup_{j\to\infty} \int_{\{u_j \leq R\}} u_j dx &\leq \limsup_{j\to\infty} \int_{\Omega\times\mathbb{R}^1} g_R(y) d\nu_j(x,y) \\ &= \int_\Omega \int_{\mathbb{R}^1} g_R(y) d\nu_x(y) dx \\ &\leq \int_\Omega \int_{\mathbb{R}^1} |y| d\nu_x(y) dx < \infty \,.\end{aligned}$$

We choose sequences $n_j \geq 2^j, k_j \geq j$ such that

$$\limsup_{j\to\infty} \int_{\{v_j \leq n_j\}} v_j dx \leq \int_\Omega \int_{\mathbb{R}^1} |y| d\nu_x(y) dx, \tag{4.51}$$

where $v_j = u_{k_j}$. $\forall k \in \mathcal{N}$, let $E_k = \bigcup_{j\geq k} \{v_j > n_j\}$. We shall verify:

1. $|E_k| \to 0$ as $k \to \infty$.
2. For $\forall k \in \mathcal{N}$, $\{v_j \chi_{F_k}\}$ is uniformly absolutely continuous in j, where $F_k = \Omega \setminus E_k$.

By definition we have the estimates:

$$|E_k| \leq \sum_{j\geq k} |\{v_j > n_j\}| \leq \sup_j \|v_j\|_{L^1} \sum_{j\geq k} \frac{1}{n_j} \leq \frac{1}{2^{k-1}} \sup_j \|v_j\|_{L^1} \,.$$

Thus (1) is verified.

We claim that

$$\limsup_{R\to\infty} \limsup_{j\to\infty} \int_{F_k \cap \{v_j > R\}} v_j dx = 0 \,. \tag{4.52}$$

If it is proved, then for $\forall \epsilon > 0$, let $R_0 > 0, j_0 > 0$ be such that $\int_{F_k \cap \{v_j > R_0\}} v_j < \frac{\epsilon}{2}$ as $j > j_0$. For any $\mathcal{L}^N$ measurable set U with $|U| < \frac{\epsilon}{2R_0}$, we have

$$\int_{U\cap F_k} v_j dx \leq \int_{F_k \cap \{v_j > R_0\}} v_j dx + \int_{U \cap F_k \cap \{v_j \leq R_0\}} v_j dx \leq \frac{\epsilon}{2} + R_0 |U| < \epsilon \,,$$

for $\forall j \geq j_0$, i.e., v_j is uniformly absolutely continuous on F_k $\forall k$.

Now we return to proving (4.52). Since for $\forall$ Borel sets $E \in \mathcal{B}(\Omega), \forall \phi \in C_0(\mathbb{R}^1)$,

$$\lim_{j\to\infty} \int_{E\cap\{v_j\le n_j\}} \phi(v_j(x))dx = \int_E \int_{\mathbb{R}^1} \phi(y)d\nu_x(y)dx \ ,$$

Choosing $\phi_j \uparrow |y|$, it follows that

$$\liminf_{j\to\infty} \int_{E\cap\{v_j\le n_j\}} v_j(x)dx \ge \int_E \int_{\mathbb{R}^1} |y|d\nu_x(y)dx \ , \tag{4.53}$$

because for $j \ge k, F_k \subset \{v_j \le n_j\}, F_k \cup (\{v_j \le n_j\} \cap E_k) = \{v_j \le n_j\}$. Combining (4.51) and (4.53) with $E = E_k$, we have

$$\limsup_{j\to\infty} \int_{F_k} v_j(x)dx \le \int_{F_k} \int_{\mathbb{R}^1} |y|d\nu_x(y)dx \ . \tag{4.54}$$

Thus,

$$\begin{aligned}
&\limsup_{R\to\infty} \limsup_{j\to\infty} \int_{F_k\cap\{v_j>R\}} v_j(x)dx \\
&= \limsup_{R\to\infty} \limsup_{j\to\infty} \left(\int_{F_k} - \int_{F_k\cap\{v_j\le R\}} \right) v_j(x)dx \\
&\le \limsup_{R\to\infty} \limsup_{j\to\infty} \left(\int_{F_k} \int_{\mathbb{R}^1} |y|d\nu_x(y)dx - \int_{F_k} g_R(v_j(x)) \Big) dx \right) \\
&= \limsup_{R\to\infty} \int_{F_k} \int_{\mathbb{R}^1} (|y| - g_R(y))d\nu_x(y)dx = 0 \ .
\end{aligned}$$

$\square$

Definition 4.5.20 *Let $\{u_j\} \subset L^1(\Omega)$ be bounded; we say that u_j converges to $u \in L^1(\Omega)$ in the sense of the bitting lemma, if for all $\epsilon > 0$ there exists a measurable subset $E \subset \Omega$ such that $|E| < \epsilon$ and $u_j \rightharpoonup u$ in $L^1(\Omega\backslash E)$.*

What is the relationship between $\mathbf{H}^1 - w^*$ convergence and the convergence in the sense of the bitting lemma?

Lemma 4.5.21 *Let $\{u_j\} \overset{*}{\rightharpoonup} u$ in $\mathbf{H}^1(\mathbb{R}^N)$. For each $R > 0$, let $v \in L^1(B_R(\theta))$ be such that there exists a subsequence n_j for which u_{n_j} converges to v in the sense of the bitting lemma in $B_R(\theta)$, then $u = v$ a.e. on $B_R(\theta)$.*

Proof. We want to show that for $\forall \phi \in C_0^\infty(\mathbb{R}^N)$, $\int u\phi dx = \int v\phi dx$. Assume $\mathrm{supp}\phi \subset B_R(\theta)$ for suitable $R > 0, v_j = u_{n_j}\ \forall j$. By the assumption, for all $\epsilon > 0$ there exists E such that $|E| < \epsilon$ and $v_j \rightharpoonup v$ on $L^1(\Omega\backslash E)$. For all $\lambda > 0$ define $w_\lambda = (1+\lambda \log M(\chi_E))_+$, where $M(f)$ is the maximal function of f. Provided by the propositions (3) and (7) of BMO, $w_\lambda \in \mathrm{BMO}$, and$\|\log M(\chi_E)\| \le c_N$, and then $\|w_\lambda\|_{\mathrm{BMO}} \le C\lambda$, where C is a constant independent of λ and ϵ.

Since $\chi_E \le M(\chi_E) \le 1$ a.e., it follows $\chi_E \le w_\lambda \le 1$ a.e. According to the proposition (4) of the maximal function, we have

$$|\{w_\lambda > 0\}| = \left|\left\{M(\chi_E) > \exp\left(-\frac{1}{\lambda}\right)\right\}\right| \le C\exp\left(\frac{1}{\lambda}\right)\|\chi_E\|_{L^1} \le C\varepsilon\exp\left(\frac{1}{\lambda}\right),$$

and then

$$\begin{aligned}\int_{\mathbb{R}^N} u_{n_j}\phi dx &= \int_{\mathbb{R}^N} v\phi dx \\ &\quad + \int_{\mathbb{R}^N} u_{n_j}\phi w_\lambda dx + \int_{\mathbb{R}^N}(u_{n_j}-v)\phi(1-w_\lambda)dx - \int_{\mathbb{R}^N} v\phi w_\lambda dx \\ &= I_1 + I_2 + I_3 + I_4\,,\end{aligned}$$

where

$$\begin{aligned} I_1 &= \int_{\mathbb{R}^N} v\phi dx, \\ I_2 &\le \|u_{n_j}\|_{\mathbf{H}^1}\|w_\lambda\phi\|_{\mathrm{BMO}} \le C\|w_\lambda\phi\|_{\mathrm{BMO}}, \\ I_3 &= \int_{\Omega\backslash E}(u_{n_j}-v)(1-w_\lambda)\phi dx \to 0, \text{ as } j\to\infty, \\ I_4 &\le \int_{\{w_\lambda>0\}}|v||\phi|dx \le \|\phi\|_{L^\infty}\int_{B_R(\theta)\cap\{w_\lambda>0\}}|v|\,. \end{aligned}$$

The limit of I_3 is due to the assumption that u_j converges weakly to v in $L^1(\Omega\backslash E)$. We turn to estimate I_2. $\forall r > 0, \forall x \in \mathbb{R}^N$, let $Q = B_r(x)$, we have

$$\begin{aligned} f_Q|\phi w_\lambda - f_Q\phi w_\lambda|dz &\le \|\phi\|_{L^\infty}\|w_\lambda\|_{\mathrm{BMO}} + \frac{1}{|Q|}f_Q dz\int_{\{w_\lambda>0\}}|\phi(z)-\phi(y)|dy \\ &\le \begin{cases} C(\lambda + R|\{w_\lambda>0\}|) \text{ if } r<R\,, \\ C(\lambda + \frac{1}{R^N}|\{w_\lambda>0\}|) \text{ if } r\ge R\,. \end{cases}\end{aligned}$$

$\forall\eta > 0$ fixing $\lambda > 0$ such that $C\lambda < \frac{\eta}{2}$ at first, and then we choose $\epsilon > 0$ so small such that the summation of I_4 and $C(R + R^{-N})|\{w_\lambda > 0\}| < \frac{\eta}{2}$, we obtain the desired conclusion. □

Combining Lemmas 4.5.19 and 4.5.21, we have:

Theorem 4.5.22 *Assume that $u_j \overset{*}{\rightharpoonup} u$ in $\mathbf{H}^1(\mathbb{R}^N)$. Then for all $R > 0$ there exists a subsequence $\{u_{n_j}\}$ converging to u on $B_R(\theta)$ in the sense of the bitting lemma.*

As an application to the calculus of variations, we study the polyconvex functionals in Sect. 4.3.3. Recall that a function $f : \Omega\times M^{n\times n} \to \mathbb{R}^1$ is called polyconvex, if $f(x, A) = g(x, T(A))$, where $g : \Omega\times\mathbb{R}^{\tau_n} \to \mathbb{R}^1$ is a Caratheodory function, for a.e. $x \in \Omega, \xi \mapsto g(x,\xi)$ is convex, $T = \{\mathrm{adj}_s\}_1^n : M^{n\times n} \to \mathbb{R}^{\tau_n}$, and

$$\tau_n = \sum_1^n \left(\frac{n!}{s!(n-s)!}\right)^2 .$$

Theorem 4.5.23 *Let Ω be an open bounded domain in $\mathbb{R}^n$; $f(x,A) = g(x,T(A))$ is polyconvex, satisfying*

$$C|A|^n \le g(x,T(A)) \ \forall A \in M^{n\times n}, C > 0 .$$

Then the functional

$$J(u) = \int_\Omega f(x,\nabla u)dx$$

has a minimizer in $W_0^{1,n}(\Omega,\mathbb{R}^n)$.

Proof. Let $c = \inf_{u\in W_0^{1,n}} J(u)$, and $\{u_j\}$ be a minimizing sequence, i.e., $J(u_j) \to c$. By the coercive condition, $\|\nabla u_j\|_{L^n(\Omega,M^{n\times n})}$ is bounded, and then $\|u_j\|_{W^{1,n}(\Omega,\mathbb{R}^n)}$ is bounded. Modulo a subsequence, we have $u_j \rightharpoonup u$ in $W^{1,n}(\Omega,\mathbb{R}^n)$, $\mathrm{adj}_s(\nabla u_j) \rightharpoonup \mathrm{adj}_s(\nabla u)$, in $L^{\frac{n}{s}}(\Omega,\mathbb{R}^{(\frac{n!}{s!(n-s)!})^2}), 1 \le s < n$. Since we can extend u_j to be zero outside Ω, provided by Theorem 4.5.18, we have $\det(\nabla u_j) \ {}^*\!\rightharpoonup \det(\nabla u)$ in $\mathbf{H}^1(\mathbb{R}^n)$.

For $\forall\epsilon > 0$, we find $\delta > 0$ such that if $|E| < \delta$ then $\int_E g(x,T(u(x)))dx < \epsilon$. According to Theorem 4.5.22, there exists a measurable set $E \subset \Omega$ such that $|E| < \delta$ and $\det(\nabla(u_j)) \rightharpoonup \det(\nabla u)$ in $L^1(\Omega\backslash E)$. Obviously, we also have $\mathrm{adj}_s(\nabla u_j) \rightharpoonup \mathrm{adj}_s(\nabla u), 1 \le s < n$ in $L^1(\Omega\backslash E,\mathbb{R}^{(\frac{n!}{s!(n-s)!})^2})$. Since a.e., $x \in \Omega, \xi \mapsto g(x,\xi)$ is convex, we have

$$\begin{aligned}\liminf_{j\to\infty}\int_\Omega g(x,T(\nabla u_j(x)))dx &\ge \liminf_{j\to\infty}\int_{\Omega\backslash E} g(x,T(\nabla u_j(x)))dx\\ &\ge \int_{\Omega\backslash E} g(x,T(\nabla u(x)))dx\\ &\ge \int_\Omega g(x,T(\nabla u(x)))dx - \epsilon .\end{aligned}$$

Since $\epsilon > 0$ is arbitrary, we obtain

$$\liminf_{j\to\infty}\int_\Omega g(x,T(\nabla u_j(x)))dx \ge \int_\Omega g(x,T(\nabla u(x)))dx .$$

Thus $u \in W_0^{1,n}(\Omega,\mathbb{R}^n)$ is a minimizer. □

Comparing this result with Theorem 4.3.12, the growth condition for polyconvex functions is dropped. It is useful in the elasticity theory.

Another advantage in using Hardy space is on regularity. As we mentioned above, the Calderon–Zygmund theory can be applied to the Hardy space. Once some nonlinear expression of the minimizer is proved in $\mathbf{H}^1$, the Euler

Lagrange equation is applied to obtain an a priori estimate, and gain better regularity of the solution, cf. Evans [Ev 2], Helein [Hel 1].

The compensation compactness method has been successfully applied to elasticity and hyperbolic systems by Ball [Bal 1], Di Perna [Di], Tartar [Tar 1] etc.

4.6 Free Discontinuous Problems

4.6.1 Γ-convergence

Γ-convergence has been studied extensively by De Giorgi's school. The idea is to use a family of functionals J_ϵ, depending on a parameter ϵ and approaching the given functional J in a variational problem, such that the minimizers v_ϵ of J_ϵ have a limit point v which is a minimizer of J.

Definition 4.6.1 *Let X be a metric space and let $F_n : X \to [0,\infty]$ be a sequence of functionals. We say that F_n Γ-converges to F on X as $n \to \infty$, written as $\Gamma - \lim_{n\to\infty} F_n = F$, if*

(1) $\forall u \in X, \forall \{u_n\}$ *such that* $u_n \to u$ *in* X *one has* $\liminf_{n\to\infty} F_n(u_n) \geq F(u)$,
(2) $\forall u \in X, \exists \{u_n\}$ *such that* $u_n \to u$ *in* X *and* $\limsup_{n\to\infty} F_n(u_n) \leq F(u)$.

The importance of the notion lies in the following theorems.

Theorem 4.6.2 *If $F = \Gamma - \lim_{n\to\infty} F_n$, then F is l.s.c.*

Proof. We prove by contradiction. If F is not l.s.c., then $\exists x_n \to x$, such that $F(x) > \lim_{n\to\infty} F(x_n)$. However, by definition, $\forall n, \exists \{x_{nm}\} \subset X, x_{nm} \to x_n$, satisfying $F_n(x_{nm}) \to F(x_n)$. We may assume that $\lim_{n\to\infty} F(x_n) > -\infty$ and $F(x) < \infty$. Let $\delta = \frac{1}{4}(F(x) - \lim_{n\to\infty} F(x_n)) > 0$, then, $\forall n \in \mathcal{N}, \exists m_n \in \mathcal{N}$ such that

$$|F_n(x_{nm_n}) - F(x_n)| < \delta,\ m_n \to \infty, \text{ and } x_{nm_n} \to x\ . \tag{4.55}$$

On one hand, by Γ-convergence,

$$F(x) \leq \liminf_{n\to\infty} F_n(x_{nm_n})\ .$$

On the other hand, by the definition of δ, $F_n(x_{nm_n}) > F(x) - \delta$ for n large, we have

$$F(x_n) < F(x) - 3\delta\ . \tag{4.56}$$

It follows that $F_n(x_{nm_n}) - F(x_n) > 2\delta$, which contradicts (4.55). □

Theorem 4.6.3 *Suppose that $\Gamma - \lim_{n\to\infty} F_n = F$ and that v_n minimizes F_n over X $\forall n \in \mathcal{N}$. If v is a limiting point of v_n, then v is a minimizer of F and $F(v) = \liminf_{n\to\infty} F_n(v_n)$.*

Proof. By (2) of the definition of Γ-convergence, for every $u \in X$, there exists $\{u_n\} \subset X$ such that $u_n \to u$, and

$$\lim_{n\to\infty} F_n(u_n) = F(u) .$$

Since v_n is a minimizer of F_n, we have $F_n(v_n) \le F_n(u_n)$. Therefore (1) of the definition of Γ-convergence yields

$$F(u) \ge \limsup_{n\to\infty} F_n(u_n) \ge \liminf_{n\to\infty} F_n(v_n) \ge F(v) , \tag{4.57}$$

as $v_n \to v$ in X. This proves the theorem. □

We note that Γ-convergence is different from pointwise convergence.

Example 4.6.4

$$F_n(x) = \begin{cases} 1 \text{ if } x \ge \frac{1}{n} , \\ nx \text{ if } x \in [-\frac{1}{n}, \frac{1}{n}] , \\ -1 \text{ if } x \le -\frac{1}{n} \end{cases}$$

then

$$\Gamma - \lim_{n\to\infty} F_n(x) = \begin{cases} 1, & \text{if } x > 0 , \\ -1 & \text{if } x \le 0 . \end{cases}$$

But pointwisely,

$$F_n(x) \to \text{sgn}(x) .$$

The difference is at $x = 0$.

Example 4.6.5 $\Gamma - \lim_{n\to\infty} \sin nx = -1$.

4.6.2 A Phase Transition Problem

We present here an example due to Cahn and Hilliard on phase transitions showing how Γ-convergence is applied to variational problems.

In a container $\Omega \subset R^3$ filled with two immiscible and incompressible fluids, the two fluids arrange themselves in order to minimize the area of the interface which separates these two phases. Let u be a function, which takes the value 0 on the set occupied by the first fluid, and the value 1 on the set occupied by the second. Let $V = \int u$ be the total volume of the second fluid. Obviously one has $0 < V < m(\Omega)$. Let $S(u)$ be the interface, i.e., the singular set of u. The equilibrium configuration is obtained by minimizing the energy $F(u) := \sigma \mathsf{H}^2(S(u))$, where σ is the surface tension between the two fluids, and H^2 is the two dimensional Hausdorff measure.

On the macroscopic level, we allow for a mixture of two fluids; let $u : \Omega \to [0, 1]$ be the average density of the second fluid, and let W be a double-well potential, i.e., a continuous positive function that vanishes only at 0 and 1.

Cahn and Hilliard established a model: $\forall \epsilon > 0$, we consider a minimizer u_ϵ of the following energy functional:

$$E_\epsilon(u) = \epsilon \int_\Omega |\nabla u|^2 + \frac{1}{\epsilon} \int_\Omega W(u), \text{ with } V = \int_\Omega u \,. \tag{4.58}$$

Under the volume constraint u_ϵ cannot be a constant; but as $\epsilon \to 0$, for a.e. $x \in \Omega, u_\epsilon(x)$ must tend to either 0 or 1. Does u_ϵ tend to a minimizer of F? The following Modica–Mortola theorem gives an answer to this question. Before going to the proof, we need a few lemmas.

Lemma 4.6.6 *Suppose that* $u_m \to u$ *in a metric space* (X, d) *satisfying* $\limsup_{m\to\infty} F(u_m) \le F(u)$, *and that* $\forall m, \exists u_{m,n} \to u_m$ as $n \to \infty$ *such that* $\limsup_{n\to\infty} F_n(u_{m,n}) \le F(u_m)$. *Then* $\exists m_n \to \infty$, *such that for* $v_n = u_{m_n n}$ *we have* $v_n \to u$ *and* $\limsup_{n\to\infty} F_n(v_n) \le F(u)$.

Proof. We may assume $d(u, u_m) < \frac{1}{m}$. Define

$$m_n = \max\left\{ m \,|,\; d(u_{mn}, u) < \frac{1}{m}, F_m(u_{mn}) \le F(u_m) + \frac{1}{m} \right\} .$$

It is easy to verify that $m_n \to \infty$. We set $v_n = u_{m_n n}$, then v_n satisfies the requirements. □

The following lemma is just the one-dimensional version of the general theorem.

Lemma 4.6.7

$$\min\left\{ \int_{\mathbb{R}^1} [|\gamma'(t)|^2 + W(\gamma(t))]dt \mid \gamma \in C^1(R^1), \gamma(-\infty) = 0, \gamma(\infty) = 1 \right\}$$
$$= 2\int_0^1 \sqrt{W(t)}dt \,.$$

Proof. We write the functional to be minimized as $J(\gamma)$, and the right-hand integral as σ. According to the Cauchy–Schwarz inequality,

$$J(\gamma) \ge 2\int_{\mathbb{R}^1} \sqrt{W(\gamma(t))}\gamma'(t)dt = 2\int_0^1 \sqrt{W(s)}ds = \sigma \,.$$

The equality holds if and only if $\gamma' = \sqrt{W(\gamma)}$. Since $\sqrt{W}$ is continuous and positive between the two zeroes 0 and 1, the first-order equation has a global solution γ on $\mathbb{R}^1$, with $\gamma(-\infty) = 0, \gamma(\infty) = 1$. □

Let $U \subset \mathbb{R}^n$ be an open set with nonempty boundary ∂U. One defines the signed distance function as follows:

$$d(x) = \begin{cases} \text{dist}(x, \partial U), & \text{if } x \in U, \\ -\text{dist}(x, \partial U), & \text{if } x \notin U. \end{cases}$$

Thus, d is Lipschitzian continuous with Lipschitz constant 1.

Lemma 4.6.8 *Assume that $\partial U \in C^2$ is bounded. $\forall t \in \mathbb{R}^1$ we set $\Sigma_t = \{x \in \mathbb{R}^n \mid d(x) = t\}$. Then $\exists \epsilon > 0$ such that for $|s| < \epsilon$, Σ_s is a hypersurface of C^2 class and $\lim_{s\to 0} H^{n-1}(\Sigma_s) = H^{n-1}(\partial U)$. Moreover,*

$$|\nabla d(x)| = 1 .$$

Proof. $\forall x_0 \in \partial U$, let $\mathsf{n}(x_0)$ be the unit outer normal vector of ∂U at x_0, and let $T_{x_0}(\partial U)$ be the tangent hyperplane at x_0. Rotate the coordinates such that the first $n-1$ coordinate axes lie on $T_{x_0}(\partial U)$ and the x_n coordinate axis points in the direction $-\mathsf{n}(x_0)$. In a neighborhood $V(x_0)$ of x_0, ∂U can be represented by $x_n = f(x')$ where $x' = (x_1, \ldots, x_{n-1})$ with $f(x_0') = 0, \nabla f(x_0') = \theta$.

There exists a tubular neighborhood Ξ of ∂U, such that $\forall x \in \Xi$, there is a unique representation $\xi = \pi(x) \in \partial U$ such that $x = \pi(x) - \mathsf{n}(x)d(x)$, where $\mathsf{n}(x) = \mathsf{n}(\pi(x))$, and that ξ is the unique point satisfying $|x - \xi| = d(x)$ is a neighborhood of x. There exists $\epsilon > 0$ such that $\Sigma_s \subset \Xi, \forall |s| < \epsilon$. We claim that:

$$|\nabla d(x)| = 1, \text{ and } \nabla d(x) = -\mathsf{n}(\xi) .$$

Thus $\forall x \in \Sigma_s, s = d(x)$, we have a local representation: $x = F(x', s) := (x', f(x')) - \mathsf{n}(x', f(x'))s$. By the use of the implicit function theorem, one can verify that F is invertible as $\epsilon > 0$ is small. This implies that $d(x) = d(\xi - s\mathsf{n}(\xi)) = s$. Therefore, $\nabla d(x)\mathsf{n}(\xi) = -1$. Since d has Lipschitz constant 1, the claim is proved.

Using the Gauss–Green formula, we have

$$\int_{0<d(x)<s} \operatorname{div}(\nabla d(x)) = \int_{\Sigma_0} \nabla d(x)\mathsf{n}(x) d\mathsf{H}^{n-1} + \int_{\Sigma_s} \nabla d(x)\mathsf{n}(x) d\mathsf{H}^{n-1} .$$

As $s \to 0$, the left-hand side tends to zero, and we obtain

$$\mathsf{H}^{n-1}(\Sigma_s) \to \mathsf{H}^{n-1}(\Sigma_0) = \mathsf{H}^{n-1}(\partial U) .$$

□

Theorem 4.6.9 *(Modica–Mortola) Assume that $W \in C([0,1])$ is nonnegative and equals zero at 0 and 1 only. Let $X = \{u : \Omega \to [0,1] \mid u$ is measurable and $\int_\Omega u = V\}$ endowed with the L^1 norm. $\forall \epsilon > 0$, let*

$$F_\epsilon(u) := \begin{cases} E_\epsilon(u) = \epsilon \int_\Omega |\nabla u|^2 + \frac{1}{\epsilon}\int_\Omega W(u) \text{ if } u \in W^{1,2} \cap X , \\ +\infty, \quad \textit{otherwise} , \end{cases} \tag{4.59}$$

and

$$F(u) = \begin{cases} \sigma \|D\chi_E\|(\Omega) \text{ if } u = \chi_E , \\ +\infty, \quad \textit{otherwise} , \end{cases} \tag{4.60}$$

where $\sigma = 2\int_0^1 \sqrt{W(t)}dt$, and E is a set of finite perimetrics. Then $\Gamma - \lim F_\epsilon = F$ in X, and every sequence of minimizers of F_ϵ is subconvergent in X to a minimizer of F.

Proof. 1. We claim that $\forall u \in X$ and for every $u_{\epsilon_n} \to u$ in X with $\epsilon_n \to 0$, if $\exists M > 0$ such that

$$\epsilon_n \int |\nabla u_{\epsilon_n}|^2 + \frac{1}{\epsilon_n} \int W(u_{\epsilon_n}) \leq M ,$$

then $\int W(u) \leq \liminf_{n\to\infty} \int W(u_{\epsilon_n}) = 0$, provided by Fatou's lemma. From the double well assumption on W $u = \chi_E$ for some Borel set E.

2. We begin with the verification of (1) of the definition of Γ-convergence. We may assume without loss of generality that $\liminf_{\epsilon\to 0} F_\epsilon(u_\epsilon)$ is bounded. According to step 1, $u = \chi_E$ for some Borel set E. From the Cauchy–Schwarz inequality and the l.s.c. of the total variations, we have

$$\begin{aligned}
\liminf_{\epsilon\to 0} F_\epsilon(u_\epsilon) &= \liminf_{\epsilon\to 0} \int_\Omega \left(\epsilon |\nabla u_\epsilon|^2 + \frac{1}{\epsilon} W(u_\epsilon) \right) dx \\
&\geq \liminf_{\epsilon\to 0} 2 \int_\Omega \sqrt{W(u_\epsilon)} |\nabla u_\epsilon| dx \\
&= \liminf_{\epsilon\to 0} \int_\Omega |\nabla(\Psi(u_\epsilon))| dx \\
&= \liminf_{\epsilon\to 0} \|D\Psi(u_\epsilon)\|(\Omega) \\
&\geq \|D\Psi(u)\|(\Omega) ,
\end{aligned}$$

where Ψ is the primitive of $2\sqrt{W}$ satisfying $\Psi(0) = 0$. One has $\Psi(u) = \Psi(\chi_E) = \Psi(1)\chi_E = \sigma\chi_E$. Thus, $\|D\Psi(u)\| = F(u)$, and then (1) is verified.

3. We now verify (2) of the definition of Γ-convergence. By the definition of F, it is sufficient to verify (2) for $u = \chi_E$. Suitably modifying the approximation theorem of BV functions (Theorem 4.5.7), we find approximate functions of the form:

$$u_j = \sum_i \chi_{\Omega_i^j}, \text{ with } \|u_j - u\|_{L^1} \to 0 ,$$

where $\Omega_i^j, i = 1, 2, \ldots, j = 1, 2, \ldots$, are disjoint bounded open sets with smooth boundary $\partial\Omega_i^j$. On account of Lemma 4.6.6, the verification is reduced to verifying (2) for characteristic functions of open sets with bounded smooth boundaries.

Let $u = \chi_\Omega$, let d be the signed distance function of $\partial\Omega$, and let γ be a minimizer obtained in Lemma 4.6.7. $\forall \epsilon > 0$, we define $u_\epsilon(x) = \gamma(\frac{d(x)}{\epsilon})$. From Lemma 4.6.8 and the co-area formula, we have

$$\begin{aligned}
F_\epsilon(u_\epsilon) &= \frac{1}{\epsilon} \int_\Omega \left[\left| \gamma' \left(\frac{d}{\epsilon} \right) \right|^2 + W \left(\gamma \left(\frac{d}{\epsilon} \right) \right) \right] dx \\
&= \frac{1}{\epsilon} \int_{\mathbb{R}^1} g(t) [\gamma'(t/\epsilon)^2 + W(\gamma(t/\epsilon))] dt \\
&= \int_{\mathbb{R}^1} g(\epsilon t) [\gamma'(t)^2 + W(\gamma(t))] dt .
\end{aligned}$$

where $g(t) = \mathsf{H}^{n-1}(\Sigma_t)$. Lemma 4.6.8 yields $g(\epsilon t) \to \mathbf{H}^{n-1}(\partial\Omega)$ as $\epsilon \to 0$. Thus

$$\limsup F_\epsilon(u_\epsilon) \leq \|D\chi_E\|(\Omega) \int_{R^1} [|\gamma'(t)|^2 + W(\gamma(t))]dt = F(u) .$$

4. Finally we verify that if u_ϵ is a minimizer of F_ϵ, then $\{u_\epsilon\}$ is precompact.

In fact, one may assume that $F_\epsilon(u_\epsilon)$ is bounded, and then by step 1, so is $\|D\Psi(u_\epsilon)\|(\Omega)$. Thus $|\Psi(u_\epsilon)\|_{BV} = \|\Psi(u_\epsilon)\|_{L^1} + \|D\Psi(u_\epsilon)\|(\Omega)$ is bounded. According to the compactness property of BV functions, it is precompact in $L^1(\Omega)$. Since Ψ admits a continuous inverse, u_ϵ is precompact in L^1. □

4.6.3 Segmentation and Mumford–Shah Problem

In image segmentation, we want to detect the edge of an image from a picture.

Given a bounded domain $\Omega \subset \mathbb{R}^2$ and an image represented by a function $g \in L^2(\Omega)$, find a closed set $K \subset \overline{\Omega}$ with finite one-dimensional Hausdorff measure $\mathsf{H}^1(K)$, and a function $u \in H^1(\Omega\backslash K)$, which minimize the following cost functional:

$$E(K,u) = \int_{\Omega\backslash K} |\nabla u|^2 dx + \mu \int_{\Omega\backslash K} |u-g|^2 dx + \lambda \mathsf{H}^1(K) ,$$

where $\lambda,\ \mu > 0$ are parameters. This is the Mumford–Shah problem.

The original image g is thereby estimated by u which is smooth in Ω except for the discontinuous set K (in fact, in $H^1(\Omega\backslash K)$), and the latter is the estimated edge.

This is a variational problem with unknown (K,u). Obviously, the difficulty lies in K, which is an object we have never met.

Tentatively, suppose K is given. Since E is quadratic in u, so is convex and coercive, we obtain a minimizer $u(K)$ with $m_{\Omega\backslash K} := E(K,u(K))$. The problem is then reduced to minimizing the functional:

$$J(K) = m_{\Omega\backslash K} + \lambda \mathsf{H}^1(K) . \tag{4.61}$$

Since the discontinuous set K is only assumed to be Hausdorff measurable, the problem is more complicated and requires further and deeper prerequisites. For pedagogical purposes, we follow Nordström [No] and restrict ourselves to assuming that only the union of finitely many C^1-curves γ is the admissible candidate for K.

Given $\xi \in C^1([0,1],\mathbb{R}^2)$ the image of ξ is a C^1-curve (may be self intersection). Thus the union of $N - C^1$-curves γ is defined by $\xi = (\xi_1,\ldots,\xi_N) \in C^1([0,1],\mathbb{R}^2)^N$, with

$$\gamma = \xi[0,1] := \bigcup_{n=1}^{N} \xi_n[0,1] .$$

Let us recall the notion of a domain with minimally smooth boundary (cf. Stein [Ste 1]).

An open set $\mathcal{O} \subset \mathbb{R}^n$ is said to have a minimally smooth boundary if $\exists \varepsilon > 0$, $\exists$ an integer N, a positive $M > 0$, and a sequence of open sets $U_1, U_2, \ldots$, such that

(i) If $x \in \partial\mathcal{O}$, then $B_\varepsilon(x) \subset U_i$ for some i,
(ii) No point of $\mathbb{R}^n$ is contained in more than N of U_i's,
(iii) $\forall i$, $\exists$ a special Lipschitz domain D_i whose bound does not exceed M so that

$$U_i \cap \mathcal{O} = U_i \cap D_i \, .$$

For example, if $\partial\mathcal{O} \in C^1$, then $\partial\mathcal{O}$ is minimally smooth.

The introduction of this notion is due to the following extension theorem:

Theorem 4.6.10 *Let $\mathcal{O} \subset \mathbb{R}^n$ be a domain with minimally smooth boundary. Then $\exists$ a linear operator T mapping functions on $\mathcal{O}$ to functions on $\mathbb{R}^n$ such that*

(1) $Tu|_{\mathcal{O}} = u$,
(2) $\| Tu \|_{W^{l,p}(\mathbb{R}^n)} \leqslant C_{l,p} \| u \|_{W^{l,p}(\mathcal{O})}$ for some constant $C_{l,p}, \forall l \in \mathcal{N}, p \in [1,\infty]$.

Definition 4.6.11 *An $N - C^1$-curve γ is said to be an admissible image segmentation of Ω if $\forall$ connected components G of $\Omega\backslash\gamma$, $\forall \varepsilon > 0$, $\exists$ a domain with minimally smooth boundary $\mathcal{O}_\varepsilon$ such that*

(1) $\overline{\mathcal{O}}_\varepsilon \subset G$,
(2) $m(G - \mathcal{O}_\varepsilon) < \varepsilon$.

We denote by E_N the set of all $N - C^1$-curve admissible image segmentations of Ω endowed with the $C^1([0,1], \mathbb{R}^2)^N$ topology.

Thus for $\forall \gamma \in E_N$,

$$\mathsf{H}^1(\gamma) = \int_0^1 |\dot{\xi}(t)| dt := \sum_{n=1}^N \int_0^1 |\dot{\xi}_n(t)| dt, \quad \text{where } \gamma = \xi[0,1] = \bigcup_{n=1}^N \xi_n[0,1] \, ,$$

i.e., the length of the piecewise C^1-curve γ.

Now, let us make some preparations. To simplify the notations, we assume $\lambda = \mu = 1$.

For any bounded open domain G, we define

$$I_G(u) = \int_G |\nabla u|^2 + \int_G |u - g|^2 \quad \text{on } H^1(G) \, .$$

It is known that I_G has an unique minimizer u_G. Let

$$m_G = I_G(u_G) = \min_{u \in H^1(G)} I_G \, .$$

By simple calculations,

$$m_G = \int_G |g|^2 - g \cdot u_G \,.$$

Moreover, we have the following properties:

(1) $I'_G(u_G) = 0$, i.e.,

$$\int_G \nabla u_G \nabla \varphi + \int_G (u_G - g)\varphi = 0 \quad \forall \varphi \in H^1(G)\,.$$

(2) $\int_G |\nabla u_G|^2 + |u_G|^2 \leqslant \int_G |g|^2$.

Lemma 4.6.12 *Let G be a connected component of $\Omega \backslash \gamma$ where γ is an admissible image segmentation of Ω. Then for $\forall \varepsilon > 0$, $\exists$ a domain with minimally smooth boundary $\mathcal{O}$ such that*

$$|m_{\mathcal{O}} - m_G| < \varepsilon \,.$$

Proof. By the definition of the admissibility for $\gamma, \forall \epsilon > 0, \exists$ a domain with minimally smooth boundary $\mathcal{O} \subset \overline{\mathcal{O}} \subset G$ such that $m(G \backslash \mathcal{O}) < \epsilon$. Letting $u_{\mathcal{O}}$ be the minimizer of $I_{\mathcal{O}}$, we have

$$\begin{aligned} m_G - m_{\mathcal{O}} &= \int_G (|g|^2 - g u_G) - \int_{\mathcal{O}} (|g|^2 - g u_{\mathcal{O}}) \\ &= \int_{G \backslash \mathcal{O}} (|g|^2 - g u_G) + \int_{\mathcal{O}} g(u_{\mathcal{O}} - u_G)\,. \end{aligned}$$

Noticing that

$$\begin{aligned} &\int_{G \backslash \mathcal{O}} \nabla u_G \nabla \widetilde{u}_{\mathcal{O}} + \int_{G \backslash \mathcal{O}} u_G \widetilde{u}_{\mathcal{O}} \\ &= \int_G g \widetilde{u}_{\mathcal{O}} - \int_{\mathcal{O}} g u_G \\ &= \int_{G \backslash \mathcal{O}} g \widetilde{u}_{\mathcal{O}} + \int_{\mathcal{O}} g(u_{\mathcal{O}} - u_G)\,, \end{aligned}$$

where $\widetilde{u}_{\mathcal{O}} = T u_{\mathcal{O}}$, we have

$$\int_{\mathcal{O}} g(u_{\mathcal{O}} - u_G) = \int_{G \backslash \mathcal{O}} (\nabla u_G \nabla \widetilde{u}_{\mathcal{O}} + u_G \widetilde{u}_{\mathcal{O}}) - \int_{G \backslash \mathcal{O}} g \widetilde{u}_{\mathcal{O}}\,.$$

Applying the Cauchy–Schwarz inequality, the first integral $\leq \|u_G\|_{H^1(G \backslash \mathcal{O})} \|\widetilde{u}_{\mathcal{O}}\|_{H^1}$. By (2) and Theorem 4.6.10 $\exists C_0 > 0$ such that

$$\|\widetilde{u}_{\mathcal{O}}\|_{H^1} \leq C_0 \|u_{\mathcal{O}}\|_{H^1(\mathcal{O})} \leq C_0 \|g\|_{L^2(\mathcal{O})} \leq C_0 \|g\|_{L^2(G)}\,.$$

Therefore $\exists C > 0$ a constant such that

$$|m_G - m_{\mathcal{O}}| \leqslant C \parallel g \parallel_{L^2(G)} \left(\int_{G\backslash\mathcal{O}} |g|^2 + \|u_G\|^2_{H^1(G\backslash\mathcal{O})} \right)^{\frac{1}{2}} .$$

Since $m(G\backslash\mathcal{O})$ can be sufficiently small, the conclusion follows. □

Lemma 4.6.13 *The function J restricted on E_N is l.s.c.*

Proof. The second term $\mathcal{H}^1(\gamma)$ of J is trivially continuous in C^1-topology, so is l.s.c.

It is sufficient to verify the l.s.c. of $m_{\Omega\backslash\gamma}$ with respect to γ, i.e., for $\forall\gamma_0 \in E_N$, $\forall\varepsilon > 0$, $\exists\delta > 0$ such that dist $(\gamma, \gamma_0) < \delta$ implies that $m_{\Omega\backslash\gamma} > m_{\Omega\backslash\gamma_0} - \varepsilon$.

Let $\mathcal{E}$ be the collection of all connected components of $\Omega\backslash\gamma_0$, then $\mathcal{E}$ is countable, say $\mathcal{E} = \{G_i\}_1^\infty$. From

$$\sum_{i=1}^{\infty} \int_{G_i} |g|^2 = \int_{\Omega\backslash\gamma} |g|^2 \leqslant \parallel g \parallel^2_{L^2(\Omega)} ,$$

there exists $K > 0$ such that

$$\sum_{i=K+1}^{\infty} \int_{G_i} |g|^2 < \frac{\varepsilon}{2} . \tag{4.62}$$

Now we focus our study on $\{G_i\}_{1\leqslant i\leqslant K}$. According to Lemma 4.6.12, $\exists\mathcal{O}_i$ with minimally smooth boundary satisfying. $\overline{\mathcal{O}}_i \subset G_i$ and $|m_{\mathcal{O}_i} - m_{G_i}| \leqslant \frac{\varepsilon}{2K}$ $\forall i = 1, 2, \ldots, K$.

Let $\delta > 0$ be such that dist $(\overline{\mathcal{O}}_i, \partial G_i) \geqslant \delta > 0$; we choose a $\frac{\delta}{2}$ neighborhood of γ_0, then $\bigcup_{i=1}^{K} \mathcal{O}_i \subset \Omega\backslash\gamma$ as dist $(\gamma, \gamma_0) < \frac{\delta}{2}$, and we have

$$\begin{aligned} m_{\Omega\backslash\gamma} = I_{\Omega\backslash\gamma}(u_{\Omega\backslash\gamma}) &\geqslant \sum_{i=1}^{K} I_{\mathcal{O}_i}(u_{\Omega\backslash\gamma}) \\ &\geqslant \sum_{i=1}^{K} m_{\mathcal{O}_i} \\ &\geqslant \frac{-\varepsilon}{2} + \sum_{i=1}^{K} m_{G_i} . \end{aligned}$$

Let $u(x) = u_{G_i}$ as $x \in G_i, i = 1, 2, \ldots, K$ and $u(x) = 0$ as $x \in \mathcal{E}\backslash\bigcup_{i=1}^{K} G_i$, then $\Omega\backslash(\gamma_0 \cup \bigcup_{i=1}^{K} G_i) = \bigcup_{i=K+1}^{\infty} G_i$, we obtain from inequality (4.62)

$$m_{\Omega\backslash\gamma_0} \leq I_{\Omega\backslash\gamma_0}(u) = \sum_{i=1}^{K} I_{G_i}(u_{G_i}) + I_{\Omega\backslash(\gamma_0\cup\bigcup_{i=1}^{k} G_i)}(0)$$

$$\leq \sum_{i=1}^{K} m_{G_i} + \frac{\varepsilon}{2} .$$

This proves that

$$m_{\Omega\backslash\gamma} > m_{\Omega\backslash\gamma_0} - \varepsilon .$$

□

Now for $\forall N \in \mathcal{N}, \forall \alpha \in (0,1), \forall \rho > 0, \forall \omega > 0$, let us define

$$\mathcal{C}_N(\alpha,\rho,\omega) = \Big\{ \gamma = \xi[0,1] \in E_N \mid \| \xi \|_{C^1} + \sup_{s,t\in[0,1],s\neq t} \frac{|\dot{\xi(t)} - \dot{\xi(s)}|}{|t-s|^\alpha} \leq \rho, \int_0^1 \| \dot{\xi}(t) \| \, dt \geq \omega \Big\}$$

endowed with the $C^1[0,1]^N$ topology, and let

$$\mathcal{C} = \{\emptyset\} \cup \bigcup_{N=1}^{\infty} \mathcal{C}_N(\alpha,\rho,\omega) .$$

We arrive at:

Theorem 4.6.14 $\exists \gamma_0 \in \mathcal{C}, \exists u_0 \in H^1(\Omega\backslash\gamma_0)$, *such that*

$$E(\gamma_0, u_0) = \mathrm{Min}_{(\gamma,u)\in(\mathcal{C}\times H^1(\Omega\backslash\gamma))} E(\gamma, u) .$$

Proof. The lower bound $\omega > 0$ forces us to minimize E on $C_N(\alpha,\rho,\omega)$ for some finite N, and the latter is a closed compact set in $C^1[0,1]^N$. Our conclusion follows from the general principle of calculus of variations. □

The simplified model of Mumford–Shah problem is restricted to the space of possible discontinuous sets consisting of those corresponding to line drawings.

The Mumford–Shah functional has been studied by De Giorgi and Ambrosio [DG 2], [Amb], [DA]. In the variational problem, the working space is an SBV space, a subspace of $BV(\Omega)$, which consists of those BV functions in which the Cantor part of their derivatives vanishes. The distributional derivative of a BV function u can be decomposed into three parts: $Du = D^a u + D^j u + D^c u$, where D^a, D^j, and D^c are the absolutely continuous part, jump part, and the Cantor part, resp. The advantage of the subspace SBV (Ω) is in the characterization of its w^*-compactness. The reader is referred to Ambrosio [Amb].

Mumford and Shah conjectured that if (K, u) is an optimal essential pair of J, then K is locally in Ω the union of finitely many $C^{1,1}$ embedded curves, see Morel [Mor].

4.7 Concentration Compactness

The loss of compactness breaks down the standard variational techniques. However, most problems arising in geometry (prescribing scalar curvature problem, Yamabe problem, minimal surfaces, fixed points and intersections in symplectic geometry etc.) and in physics (N-body problem, Yang–Mills equation, nonlinear Schrödinger equations, gravity theory etc.) are in this realm. Semilinear elliptic PDEs in $\mathbb{R}^n$ and those with the critical Sobolev exponent are two typical examples on this topic. They have been studied extensively in the last two decades.

If $\Omega \subset \mathbb{R}^n$ is a bounded domain, then a bounded sequence $\{u_j\}$ in $H^1(\Omega)$ has a weakly convergent subsequence, and then a strongly convergent subsequence in $L^q(\Omega)$ as $q \in [2, 2^*)$, according to the compactness of the embedding $H^1(\Omega) \hookrightarrow L^q(\Omega)$. This argument is frequently used. (See Example 3 in Sect. 4.2.) However, when Ω is unbounded, the argument does not work. This can be seen by the following example: Let $\Omega = \mathbb{R}^n$, $\{x_j\}$ be a sequence that tends to infinity. Given a nonzero $u_0 \in C_0^\infty$, let $u_j(x) = u_0(x - x_j)$. Then $||u_j||_{H^1} = ||u_0||_{H^1}$ and $|u_j|_q = |u_0|_q$. Therefore, $u_j(x) \to 0$ a.e., and then $u_j \rightharpoonup 0$ in $H^1(R^n)$, but u_j cannot L^q converge to 0. The reason is that the associated measures

$$\mu_j(E) = \int_E |u_j|^2 \ \forall \text{ Lebesgue measurable set } E \subset \mathbb{R}^n \tag{4.63}$$

leak out at infinity.

4.7.1 Concentration Function

Relating to a measure μ, P. Levy introduced the concentration function

$$Q(r) = \sup_{x \in \mathbb{R}^n} \mu(B_r(x)), \ r \geq 0 \,.$$

We have the following:

Lemma 4.7.1 *Suppose that $\{u_j\} \subset H^1(\mathbb{R}^n)$ is a bounded sequence. If $\exists\, R > 0$ such that $\underline{\lim}_{j \to +\infty} Q_j(R) = 0$, where Q_j is the concentration functions with respect to μ_j defined in equation (4.63), then there exists a subsequence $\{u_{j_k}\}$ such that $u_{j_k} \to 0$ in $L^q(\mathbb{R}^n)$, $\forall\, q \in (2, 2^*)$.*

Proof. Due to the interpolation inequality, we have:

$$||u||_{L^q(B_R(x))} \leq ||u||^{1-\theta}_{L^2(B_R(x))} ||u||^{\theta}_{L^{2^*}(B_R(x))}$$

$\forall\, x \in \mathbb{R}^n$, where $\theta = \frac{(q-2)n}{2q}$.

In the case where $\theta q \geq 2$, we have $q \geq 2 + \dfrac{4}{n}$ and a constant $C > 0$ such that

$$\int_{B_R(x)} |u|^q \le C||u||^{(1-\theta)q}_{L^2(B_R(x))}||u||^{\theta q-2}_{H^1(B_R(x))}||u||^2_{H^1(B_R(x))}$$

$$\le C(\sup_{x\in\mathbb{R}^n} ||u||^{(1-\theta)q}_{L^2(B_R(x))})||u||^{\theta q-2}_{H^1(\mathbb{R}^n)} \int_{B_R(x)} (|u|^2 + |\nabla u|^2)$$

from the Sobolev embedding theorem. We find a locally finite covering, i.e., $\exists \{x_k\}$ and $\exists \ell \in \mathbb{N}$ such that $\forall x \in \mathbb{R}^n$, $\exists$ at most ℓ balls of $B_R(x_k)$ to which x belongs. Then

$$\int_{\mathbb{R}^n} |u|^q \le \sum_k \int_{B_R(x_k)} |u|^q$$

$$\le \ell C||u||^{\theta q}_{H^1(\mathbb{R}^n)} \left(\sup_{x\in\mathbb{R}^n} ||u||^{(1-\theta)q}_{L^2(B_R(x))}\right) .$$

By the assumptions that $\|u_j\|_{H^1}$ is bounded and modulo a subsequence $\sup_{x\in\mathbb{R}^n} \int_{B_R(x)} |u_j|^2 \to 0$, it follows that $\int_{\mathbb{R}^n} |u_j|^q \to 0$.

In the case where $\theta q < 2$, we have $2 < q < 2 + \dfrac{4}{n} := q_0$. Since $\theta q_0 = 2$, $||u_{j_k}||_{L^{q_0}(\mathbb{R}^n)} \to 0$. Letting $\frac{1}{q} = \frac{\lambda}{2} + \frac{(1-\lambda)}{q_0}$, we obtain, by interpolation, that

$$||u_{j_k}||_{L^q(\mathbb{R}^n)} \le |u_{j_k}|_2^{\lambda} |u_{j_k}|_{q_0}^{1-\lambda} \to 0 .$$

□

The lemma shows that the concentration function plays an important role in the compactness argument. It will be seen by the following:

Corollary 4.7.2 *(Strauss) Let $H^1_r(R^n)$ be the subspace of $H^1(R^n)$ consisting of radial symmetric functions. The embedding $H^1_r(R^n) \hookrightarrow L^p(R^n), 2 < p < 2^*, n \ge 2$ is compact.*

Proof. $\forall x \in R^n, \forall R > 0$, let $m(x, R)$ be the largest number of disjoint balls with radius R and the centers lie on the same sphere with radius $|x|$ centered at θ. It is easily seen that $m(x, R) \to \infty$ as $\|x\| \to \infty$. By definition $\forall u \in L^2(R^n), \forall r > 0$,

$$\int_{B_r(x)} |u|^2 \le m(x, r)^{-1}\|u\|^2_{L^2} .$$

If $\{u_j\}$ is a bounded sequence in $H^1_r(R^n)$, then $\forall \epsilon > 0$, $\exists R > 0$ such that

$$\sup\left\{\int_{B_r(x)} |u_j|^2 \mid |x| \ge R\right\} < \epsilon .$$

We may assume $u_j \rightharpoonup 0$ in $H^1(R^n)$; then by the Rellich theorem, after a subsequence $\int_{B_{R+r}(\theta)} |u_{n_j}|^2 \to 0$. It follows that

$$\sup\left\{\int_{B_r(x)} |u_{n_j}|^2 \mid |x| \le R\right\} \to 0\,.$$

According to Lemma 4.7.1, $u_{n_j} \to 0$ in $L^p(R^n)$. □

If μ is a probability measure on $\mathbb{R}^n$, then the concentration function Q with respect to μ is nonnegative, nondecreasing and $\lim\limits_{r\to\infty} Q(r) = 1$.

Now we carefully study various cases of the behavior of a sequence of concentration functions.

Let $\{Q_j\}$ be a sequence of concentration functions associated with probability measures $\{\mu_j\}$, then they consist of a bounded set in $\mathrm{BV}[0,\infty)$. By Helly's lemma, after a subsequence, we again denote it by $\{Q_j\}$, there exists a nonnegative nondecreasing function $Q \in \mathrm{BV}[0,\infty)$ such that $Q_j(r) \to Q(r)$ a.e.

One may assume that $Q(r-0) = Q(r)$ and that $Q(r) \le \underline{\lim} Q_j(r)$. Let us consider the limit: $\lambda = \lim\limits_{r\to\infty} Q(r) \in [0,1]$.

If $\lambda = 0$, then $\lim\limits_{j\to\infty} Q_j(r) = 0$, $\forall\, r$.

If $\lambda = 1$, then $\forall\, \epsilon > 0$, $\exists\, R > 0$, $\exists\, \{x_j\} \subset \mathbb{R}^n$ such that $\mu_j(B_R(x_j)) \ge 1-\epsilon$. In fact, by definition, $\exists\, R_0 > 0$ s.t. $Q(R_0) > \frac{1}{2}$. Let $x_j \in \mathbb{R}^n$ satisfying $Q_j(R_0) \le \mu_j(B_{R_0}(x_j)) + \frac{1}{j}$. For $\epsilon \in (0, \frac{1}{2})$, we choose $R_1 > 0$ such that $Q(R_1) > 1-\epsilon$, and let y_j satisfy $Q_j(R_1) \le \mu_j(B_{R_1}(y_j)) + \frac{1}{j}$. Then

$$\mu_j(B_{R_1}(y_j)) + \mu_j(B_{R_0}(x_j)) \ge Q_j(R_1) + Q_j(R_0) - \frac{2}{j} > 1, \text{ for } j \text{ large}\,.$$

Therefore $B_{R_1}(y_j) \cap B_{R_0}(x_j) \ne \emptyset$, which implies $B_{R_1}(y_j) \subset B_{2R_1+R_0}(x_j)$ and $\mu_j(B_{2R_1+R_0}(x_j)) \ge \mu_j(B_{R_1}(y_j)) \ge Q_j(R_1) - \frac{1}{j} > 1-\epsilon-\frac{1}{j}$ for j large. This is the conclusion.

For $\lambda \in (0,1)$, $\forall\, \epsilon > 0$, $\exists\, R_0 > 0$, $\exists\, \{x_j\}$ such that $Q_j(R_0) \ge \mu_j(B_{R_0}(x_j)) \ge \lambda - \epsilon$ for j large. Also, we may find a sequence $R_j \to \infty$ such that

$$Q_j(R_0) \le Q_j(R_j) < \lambda + \epsilon \text{ for } j \text{ large}\,.$$

Thus $\lambda - \epsilon < \mu_j(B_{R_0}(x_j)) \le \mu_j(B_{R_j}(x_j)) \le \lambda + \epsilon$. Now for any given $R > R_0$, we may assume $R_j > R$ for all j and let $\mu_j^1 = \mu_j|_{B_{R_0}(x_j)}$ and $\mu_j^2 = \mu_j|_{\mathbb{R}^n \setminus B_{R_j}(x_j)}$. Then $0 \le \mu_j^1 + \mu_j^2 \le \mu_j$, supp $\mu_j^1 \subset B_{R_0}(x_j)$, supp $\mu_j^2 \subset \mathbb{R}^n \setminus B_{R_j}(x_j) \subset \mathbb{R}^n \setminus B_R(x_j)$, and so

$$\begin{aligned}
&|\lambda - \mu_j^1(\mathbb{R}^n)| + |(1-\lambda) - \mu_j^2(\mathbb{R}^n)| \\
&\le |\lambda - \mu_j(B_{R_0}(x_j))| + |\lambda - \mu_j(B_{R_j}(x_j)| < 2\epsilon\,.
\end{aligned}$$

In summary, we have:

Theorem 4.7.3 *(Concentration compactness principle) Suppose that* $\{\mu_j\} \subset M(\mathbb{R}^n)$ *is a sequence of probability measures. Then one of the following three conclusions holds:*

1. (Compactness) $\exists\ \{x_j\} \subset \mathbb{R}^n$ *such that* $\forall\ \epsilon > 0$, $\exists\ R > 0$ *with* $\mu_j(B_R(x_j)) \geq 1 - \epsilon$ *for all* j.
2. (Vanishing) $\forall\ R > 0$, $\lim_{j\to\infty}(\sup_{x\in\mathbb{R}^n} \mu_j(B_R(x))) = 0$.
3. (Dichotomy) $\exists\ \lambda \in (0,1)$ *such that* $\forall\ \epsilon > 0$, $\exists\ R > 0$ *and* $\exists\ \{x_j\}$ *with the property: Given* $R' > R$, *there exist positive measures* μ_j^1 *and* μ_j^2 *such that* $0 \leq \mu_j^1 + \mu_j^2 \leq \mu_j$, $\text{supp } \mu_j^1 \subset B_R(x_j)$, $\text{supp } \mu_j^2 \subset \mathbb{R}^n \setminus B_{R'}(x_j)$, *and* $\limsup_{j\to\infty}\{|\lambda - \mu_j^1(\mathbb{R}^n)| + |(1-\lambda) - \mu_j^2(\mathbb{R}^n)|\} \leq \epsilon$.

We shall present an example showing how this principle is applied.

In contrast with Example 3 in Sect. 4.2, we consider the same equation but on the whole space $\mathbb{R}^n$: Find a nontrivial solution $u \in H^1(\mathbb{R}^n)$ satisfying

$$-\triangle u + u = |u|^{q-2}u, \ \ 2 < q < 2^* \ .$$

If we follow the steps in the proof of that example, then it can be stated as a minimizing problem for the functional:

$$J(u) = \int_{\mathbb{R}^n} (|\nabla u|^2 + |u|^2)$$

subject to the constraint:

$$M = \left\{ u \in H^1(\mathbb{R}^n) \mid \int_{\mathbb{R}^n} |u|^q = 1 \right\} .$$

It is important to note that both J and M are translation invariant.

From the translation invariance, we intend to prove that, for any minimizing sequence $\{u_j\}$, after suitable translation, the new minimizing sequence $\{v_j\}$ is L^q subconvergent. Then the limit $v_0 \in M$.

Let

$$S_q = \inf_M J \ .$$

In light of the concentration compactness principle, let $\mu_j = |u_j|^q\ dx$. If one can exclude Cases 2 and 3, then $\exists \{x_j\} \subset R^n$, $\forall\ \epsilon > 0$, $\exists\ R > 0$, such that

$$1 = \int_{\mathbb{R}^n} |u_j(x - x_j)|^q\ dx \geq \mu_j(B_R(x_j)) \geq 1 - \epsilon \ .$$

Let $v_j(x) = u_j(x - x_j)$. Again $\{v_j\}$ is a minimizing sequence, $\|v_j\|_{H^1(R^n)}$ is bounded, after a subsequence $v_j \rightharpoonup v_0$, and then $v_j \to v_0$ strongly in $L^q(B_R(\theta))$.

Then we have $1 \geq \int_{B_R(0)} |v_0(x)|^q \geq 1 - \epsilon$. Therefore, $\int_{\mathbb{R}^n} |v_0|^q = 1$, i.e., $v_0 \in M$. Moreover, we have $v_j \to v_0$ in $L^q(R^n)$ strongly.

• To exclude the vanishing case: Suppose we had $R > 0$ such that

$$\liminf_{j\to\infty} \sup_{x\in\mathbb{R}^n} \int_{B_R(x)} |u_j|^q = 0 .$$

Then $\liminf_{j\to\infty} \sup_{x\in\mathbb{R}^n} \int_{B_R(x)} |u_j|^2 = 0$. According to the Lemma 4.7.1, $u_j \to 0$ in $L^p(\mathbb{R}^n)$ $\forall$ $p \in (2, 2^*)$. In particular, when we take $p = q$, it contradicts $|u_j|_q = 1$.

• To exclude the dichotomy case: Suppose $\exists \lambda \in (0,1)$ such that $\forall$ $\epsilon > 0$, $\exists$ $R > 0$, $\exists$ $\{x_j\}$, and positive measures μ_j^1 and μ_j^2 such that $0 \leq \mu_j^1 + \mu_j^2 \leq \mu_j$, supp $\mu_j^1 \subset B_R(x_j)$, supp $\mu_j^2 \subset \mathbb{R}^n \setminus B_{2R}(x_j)$, and $\overline{\lim}_{j\to\infty}\{|\lambda - \mu_j^1(\mathbb{R}^n)| + |(1-\lambda) - \mu_j^2(\mathbb{R}^n)|\} \leq \epsilon$.

Choosing $\epsilon_j \to 0$, $\exists$ $R_j > 0$, after a subsequence, we have supp $\mu_j^1 \subset B_{R_j}(x_j)$, supp $\mu_j^2 \subset \mathbb{R}^n \setminus B_{2R_j}(x_j)$, and $\overline{\lim}_{j\to\infty}\{|\lambda - \mu_j^1(\mathbb{R}^n)| + |(1-\lambda) - \mu_j^2(\mathbb{R}^n)|\} = 0$.

In fact, we may assume $R_j \to \infty$. Let $\phi \in C_0^\infty(B_2(\theta))$ such that $\phi = 1$ in $B_1(\theta)$ and let $\phi_j(x) = \phi(\frac{x - x_j}{R_j})$. We write $v_j = u_j\phi_j + u_j(1-\phi_j)$. Then $J(u_j) = J(u_j\phi_j) + J(u_j(1-\phi_j)) + B_j$, where B_j is the interaction term:

$$B_j = 2\int \nabla(u_j\phi_j)\nabla(u_j(1-\phi_j)) + u_j^2\phi_j(1-\phi_j) .$$

By simple estimation $\forall\epsilon > 0$,

$$B_j \geq -\epsilon - \frac{1}{\epsilon}E_j ,$$

where

$$E_j = 2\int_{B_{2R_j}(x_j)\setminus B_{R_j}(x_j)} |u_j\nabla\phi_j|^2 \leq \frac{C}{R_j^2}\int_{B_{2R_j}(x_j)\setminus B_{R_j}(x_j)} |u_j|^2 .$$

Fixing $\epsilon > 0$ arbitrarily small, since $E_j \to 0$ as $j \to \infty$, we have

$$\begin{aligned} J(u_j) &\geq S_q(|u_j\phi_j|_q^2 + |u_j(1-\phi_j)|_q^2) - \epsilon - \frac{1}{\epsilon}E_j \\ &\geq S_q(\mu_j^1(B_{R_j}(x_j))^{\frac{2}{q}} + \mu_j^2(\mathbb{R}^n \setminus B_{2R_j}(x_j))^{\frac{2}{q}}) - \epsilon - \frac{1}{\epsilon}E_j \\ &\geq S_q(\lambda^{\frac{2}{q}} + (1-\lambda)^{\frac{2}{q}}) - \epsilon + o(1) . \end{aligned}$$

But $\lambda^{\frac{2}{q}} + (1-\lambda)^{\frac{2}{q}} > 1$ as $\lambda \in (0,1)$. This is a contradiction.

Namely we obtain:

Theorem 4.7.4 *(P.L. Lions) For any minimizing sequence $\{u_j\} \subset H^1(\mathbb{R}^n)$ of the functional $J(u) = \int_{\mathbb{R}^n}(|\nabla u|^2 + |u|^2)$ subject to $|u|_q = 1$ for $q \in (2, 2^*)$,* *there exists a sequence* $\{x_j\} \subset \mathbb{R}^n$ *such that* $v_j(x) = u_j(x - x_j)$ L^q*-strongly converges to a minimizer of* J.

Example 1. Assume $a \in C(R^n)$ satisfies $a(x) \to a_\infty$ as $|x| \to \infty$. One studies the nontrivial solution $u \in H^1(R^n)$ of the equation:

$$-\Delta u + a(x)u = |u|^{q-2}u, \quad 2 < q < 2^* .$$

Define

$$I(u) = \int_{R^n} (|\nabla u|^2 + a(x)|u|^2), \quad I_\infty(u) = \int_{R^n} (|\nabla u|^2 + a_\infty |u|^2) ,$$

and

$$M(u) := |u|_q^q = \int_{R^n} |u|^q .$$

We define

$$S = \inf\{I(u) \mid M(u) = 1\} \text{ and } S_\infty = \inf\{I_\infty(u) \mid M(u) = 1\} ,$$

and conclude: If $S < S_\infty$, then there is a nontrivial solution of the above equation.

Before going to the proof, we need a lemma (see [BL]):

Lemma 4.7.5 *(Brezis–Lieb) Suppose that* $\Omega \subset \mathbb{R}^n$ *and* $\{u_j\} \subset L^p(\Omega)$, $p \in [1, \infty)$ *If* $\{u_j\}$ *is bounded in* $L^p(\Omega)$, *and* $u_j \to u$ *a.e.,* on Ω. *Then*

$$\lim_{j\to\infty} (\|u_j\|_p^p - \|u_j - u\|_p^p) = \|u\|_p^p .$$

Proof. According to Fatou's lemma, $|u|_p^p \le \liminf_{j\to\infty} \|u_j\|_p^p < \infty$. We begin with an elementary inequality: $\forall \epsilon > 0$, there exists $C_\epsilon > 0$ such that $\forall x, y \in \mathbb{R}^1$,

$$||x + y|^p - |x|^p| \le \epsilon |x|^p + C_\epsilon |y|^p .$$

Let

$$v_{j,\epsilon} = (|\,|u_j|^p - |u_j - u|^p - |u|^p\,| - \epsilon |u_j - u|^p)_+ ,$$

then $v_{j,\epsilon} \le (1 + C_\epsilon)|u|^p$, and $v_{j,\epsilon} \to 0$ a.e.

Then we can use Lebesgue's dominance theorem to conclude: $\int_\Omega v_{j,\epsilon} \to 0$ as $j \to \infty$. Since

$$|\,|u_j|^p - |u_j - u|^p - |u|^p\,| \le v_{j,\epsilon} + \epsilon |u_j - u|^p ,$$

we have

$$\limsup_{j\to\infty} \int_\Omega |\,|u_j|^p - ||u_j - u|^p - |u|^p\,| \le C\epsilon ,$$

where $C = \sup\{\|u_j - u\|_p^p < \infty$. Letting $\epsilon \to 0$, the lemma is proved. □

Now we turn to the proof of the conclusion in Example 1. In fact, by the previous discussion, after a subsequence, any

minimizing sequence $\{u_j\} \subset H^1(R^n) \cap M^{-1}(1)$ is weakly H^1 convergent to some $u_0 \in H^1(R^n)$, so it is sufficient to show that $u_0 \in M^{-1}(1)$. Let $v_j = u_j - u_0$, we have $v_j \rightharpoonup 0$, in $H^1(R^n)$ and $v_j \to 0$ in L^2_{loc}, and then $v_j \to 0$ a.e. Since we have:

1. $I(u_j) = I(u_0) + I(v_j) + o(1)$.
2. $|u_j|_q^q = |u_0|_q^q + |v_j|_q^q + o(1)$ (from Lemma (4.7.6))
3. $\int a(x)|v_j|^2 = a_\infty \int |v_j|^2 + o(1)$,
   (from the estimate

$$\int |a(x) - a_\infty||v_j(x)|^2 \le \int_{|x|\ge R} |a(x) - a_\infty||v_j|^2 + C \int_{|x|\le R} |v_j|^2 ,$$

for suitably chosen $R > 0$, where $C = 2\,\text{Max}\{a(x)\}$).

It follows that $I(v_j) = I_\infty(v_j) + o(1)$. Let $|u_0|_q^q = \lambda$. Obviously $\lambda \in [0,1]$; we are going to exclude the case that $\lambda \in [0,1)$. Suppose not, either $\lambda \in (0,1)$, we would have

$$\begin{aligned} S &\ge S|u_0|_q^2 + S_\infty |v_j|_q^2 + o(1) \\ &= S\lambda^{\frac{2}{q}} + S_\infty (1-\lambda)^{\frac{2}{q}} + o(1) \\ &\ge S(\lambda^{\frac{2}{q}} + (1-\lambda)^{\frac{2}{q}}) + o(1) > S , \end{aligned}$$

or, $\lambda = 0$, then $S > S_\infty$. These are all impossible.

4.7.2 The Critical Sobolev Exponent and the Best Constants

The following best constant plays an important role in many variational problems arising in geometry and analysis:

$$S = \inf_{u \in D^{1,2} \setminus \{\theta\}} \frac{\int_{\mathbb{R}^n} |\nabla u|^2}{(\int_{\mathbb{R}^n} |u|^{2^*})^{\frac{2}{2^*}}} ,$$

where $D^{1,2}$ is the closure of $C_0^\infty(\mathbb{R}^n)$ under the norm: $\|u\| = (\int_{\mathbb{R}^n} |\nabla u|^2)^{\frac{1}{2}}$.

We intend to figure out the precise value of S and reduce it to a variational problem. Let

$$I(u) = \|u\|^2 := \int_{\mathbb{R}^n} |\nabla u|^2 ,$$

and

$$M(u) = |u|_{2^*}^{2^*} = \int_{\mathbb{R}^n} |u(x)|^{2^*} dx .$$

We study the minimization problem:

$$\text{Min}\{I(u) \mid M(u) = 1\} .$$

Both I and M are not only translation invariant, but also invariant under the scaling transformation:

$$T_\theta : u(x) \to \theta^{-\frac{n-2}{2}} u\left(\frac{x}{\theta}\right), \quad \forall \theta > 0 .$$

By the same idea as in the previous subsection, one may use the concentration compactness principle to show that after translation and scaling, the minimizing sequence is again convergent in $L^{2^*}(\mathbb{R}^n)$, and then a minimizer does exist. Let u_0 be a minimizer; obviously $|u_0|$ is also. Therefore we may assume the minimizer u_0 is nonnegative. The Euler–Lagrange equation of the variational problem reads as

$$-\Delta u = \lambda |u|^{2^*-2} u \quad \text{in } R^n, \tag{4.64}$$

where $\lambda > 0$ is the Lagrange multiplier. After suitably adjusting a factor one may assume $\lambda = 1$. By a moving plane argument due to Gidas, Ni and Nirenberg, [GNN] (see also Yanyan Li [Li 2]), u_0 is radially symmetric: $u_0(x) = g(|x|)$, for some nonnegative function on R^1_+. Moreover, g is nonincreasing. Plugging g into equation (4.64), it becomes an ODE:

$$-(r^{n-1} g')' = S r^{n-1} g^{2^*-1} .$$

However, we shall present here a more direct proof due to Lieb.

Lemma 4.7.6 *If the problem* $\mathrm{Min}\{I(u) \mid M(u) = 1\}$ *possesses a minimizer, then it has a radially symmetric and nonincreasing minimizer* u_0*; i.e., there is a nonincreasing function* g *defined on* $\mathbb{R}^1_+$ *such that* $u_0(x) = g(|x|)$.

Proof. We may restrict ourselves to nonnegative minimizers, because $\| |u| \| \le \|u\|$, and $\| |u| \|_q = \|u\|_q$. Let u^* be the Schwarz rearrangement of u (see Lemma 4.2.5), then u^* is nonincreasing, radial symmetric and satisfies $m(u_t) = m(u^*_t)$, $\forall t \in \mathbb{R}^1$, where $u_t = \{x \in \mathbb{R}^n \mid u(x) \ge t\}$. Thus,

$$|u|_q^q = \int_{-\infty}^{\infty} |t|^q dm(u_t) = \int_{-\infty}^{\infty} |t|^q dm(u^*_t) = |u^*|_q^q .$$

According to the Faber–Krahn inequality, it follows that

$$\int_{\mathbb{R}^n} |\nabla u^*|^2 \le \int_{\mathbb{R}^n} |\nabla u|^2 .$$

□

Combining Corollary 4.7.2 with Lemma 4.7.6, one proves that the following equation possesses a positive solution in $H^1(R^n)$ directly:

$$-\Delta u + u = |u|^{q-2} u \quad \text{for } 2 < q < 2^* .$$

Lemma 4.7.7 *The minimizer of the problem in lemma 4.7.6 is achieved.*

Proof. 1. By changing variables, let $F(\alpha t) = e^{\alpha t} g(e^t)$, where $\alpha = \frac{n}{2} - 1$, and $r = e^t$; we have

$$\frac{1}{\alpha}\int_{-\infty}^{\infty} F^{2^*} dt = \int_0^{\infty} g(r)^{2^*} r^{n-1} dr, \ \alpha \int_{-\infty}^{\infty} (F' - F)^2 dt = \int_0^{\infty} g'(r)^2 r^{n-1} dr \, .$$

Setting $f(g) = \int_0^\infty g'(r)^2 r^{n-1} dr$, $\phi(g) = \int_0^\infty g(r)^{2^*} r^{n-1} dr$, and $E_0 = \{g \in H^1_{\text{loc}}(R^1_+) \,|\, g \geq 0, \phi(g) = \omega_{n-1}^{-1}\}$, $E_0' = \{g \in E_0 \,|\, g \text{ is nonincreasing}\}$, where $\omega_{n-1} = \frac{n\pi^{\frac{n}{2}}}{\Gamma(\frac{n}{2}+1)}$ is the area of the unit sphere S^{n-1}, we have:

$$\begin{aligned} \text{Min}\{I(u) \,|\, M(u) = 1\} < \text{Min}\{f(u) \,|\, u \in E_0\} &\leq \text{Min}\{f(u) \,|\, u \in E_0'\} \\ &\leq \text{Min}\{I(u) \,|\, M(u) = 1\} \, . \end{aligned}$$

Our problem is reduced to find:

$$\text{Min}\{f(u) \,|\, u \in E_0'\} \, . \tag{4.65}$$

2. Let $f_1(F) = \int_{-\infty}^\infty (F'^2 + F^2) dt, \phi_1(F) = \int_{-\infty}^\infty F^{2^*} dt$, and $E_1 = \{F \in H^1(R^1) \,|\, \phi_1(F) = \alpha\omega_{n-1}^{-1}\}$. Then the problem (4.65) is equivalent to the problem:

$$\text{Min}\{f_1(F) \,|\, F \in E_1\} \, .$$

In fact, on the one hand, if F solves the latter, then $F(t) \to 0$ as $t \to \pm\infty$, One has

$$\int_{-\infty}^{\infty} (F' - F)^2 dt = \int_{-\infty}^{\infty} (F'^2 + F^2) dt \, . \tag{4.66}$$

On the other hand, functions with compact supports consist of a dense subset of E_0, and for g with compact support $F(t)$ is zero as t becomes large. Moreover, $F(t) \leq g(0)e^t$, and then $F(t) \to 0$ as $t \to \pm\infty$. Again equation (4.66) holds.

3. Let $E_1' = \{F \in E_1 \,|\, F \text{ is even and nonincreasing on } R^1_+\}$. Again, by symmetric rearrangement, we have

$$\text{Min}\{f_1(F) \,|\, F \in E_1\} = \text{Min}\{f_1(F) \,|\, F \in E_1'\} \, .$$

Let $\{F_j\}$ be a minimizing sequence in E_1', the sequence is weakly convergent to some $F_0 \in E_1'$. We are going to verify that $|F_0|_{2^*}^{2^*} = \frac{\alpha}{\omega_{n-1}}$.

Since $F_j \to F_0$, a.e., it is sufficient to find a dominant function in $L^{2^*}(\mathbb{R}^1_+)$. From $F_j(t) \to 0$ as $t \to \infty$, we have

$$F_j(t)^2 = -2\int_t^\infty F_j' F_j dt \leq \int_0^\infty (F_j'^2 + F_j^2) dt \leq C \, .$$

Since F_j is nonincreasing on $[0, \infty)$, and

$$tF_j(t)^2 \leq \int_0^t F_j^2 \leq C \, ,$$

the function $M(t) = \text{Min}\{1, t^{-\frac{1}{2}}\}$ is a dominant function for F_j in L^{2^*} $(\mathbb{R}^1_+)$. □

Now we can find the minimizer explicitly. Following the above notations, we conclude that the function F_0 satisfies the ordinary differential equation:

$$\begin{cases} -F'' + F = S\alpha^{-2}F^{2^*-1} , \\ F'(0) = 0 , \\ F(\infty) = 0 . \end{cases}$$

where S is the best constant, $\alpha = \frac{n}{2} - 1$. The unique solution (see Lieb [Lieb], Aubin [Au 1] and Talenti [Tal]) reads as

$$F_0 = \left(\sqrt{\frac{n(n-2)}{4S}}\operatorname{sech}(\alpha^{-1}t)\right)^{\alpha} .$$

or

$$g_0(r) = r^{-\alpha}F_0(\alpha \log r) = \left(\sqrt{\frac{n(n-2)}{S}}\frac{1}{1+r^2}\right)^{\frac{n}{2}-1} .$$

By direct computation, we obtain

$$S = n(n-2)\pi\left(\frac{\Gamma(n)}{\Gamma\left(\frac{n}{2}\right)}\right)^{-\frac{2}{n}} .$$

Therefore we have proved:

Theorem 4.7.8 *The functions*

$$u(x) = \left[\frac{(n(n-2)\theta)}{(\theta^2 + |x-y|^2)^2}\right]^{\frac{n-2}{4}} , \ \forall \theta > 0, \ \forall\, y \in \mathbb{R}^n \tag{4.67}$$

are minimizers of the problem $S = \mathrm{Min}\{\|u\|_2^2 \,|\, |u|_{2^*} = 1\}$.

In fact, all minimizers for S are of the form (4.67), see Caffarelli, Gidas, and Spruck [CGS]. A simple proof via the moving plane method [GNN] was given by W. Chen and C. Li, Yanyan Li, see [ChL 2], [Li 2].

Example 2. Find a nontrivial solution $u \in H_0^1(\Omega)$ of the following equation, where $\Omega \subset \mathbb{R}^n$ is a bounded domain:

$$-\Delta u - \lambda u = |u|^{2^*-2}u \text{ in } \Omega . \tag{4.68}$$

where we assume $\lambda < \lambda_1$, the first eigenvalue of $-\Delta$. Again, the minimization problem $\mathrm{Min}\{I_\lambda(u) \,|\, |u|_{2^*} = 1\}$ is considered, where

$$I_\lambda(u) = \|u\|^2 - \lambda|u|_2^2 .$$

Let

$$S_\lambda(\Omega) = \inf\{I_\lambda(u) \,|\, u \in H_0^1(\Omega), |u|_{2^*} = 1\} .$$

Lemma 4.7.9

$$S_0(\Omega) = S \ .$$

Proof. 1. For $\Omega_1 \subset \Omega_2$, by definition, $S_0(\Omega_2) \le S_0(\Omega_1)$.

2. For $\forall R > 0$, let $\Omega_j = B_{jR}(\theta), j = 1, 2$. We claim that $S_0(\Omega_1) = S_0(\Omega_2)$ This is due to the scaling invariance of the functional $S_0(\Omega)$.

3. By the translation invariance, we conclude that for all $\Omega, S_0(\Omega) = S_0(B_1(\theta))$.

This proves the lemma. □

Following the steps in Example 1, there exists a minimizing sequence $\{u_j\}$ with $|u_j|_{2^*} = 1$ and $u_j \rightharpoonup u_0$ in $H_0^1(\Omega)$. Let $v_j = u_j - u_0$, we have.

1. $I_\lambda(u_j) = I_\lambda(u_0) + I_\lambda(v_j) + o(1)$.
2. $|u_j|_{2^*}^{2^*} = |u_0|_{2^*}^{2^*} + |v_j|_{2^*}^{2^*} + o(1)$.

Let $|u_0|_{2^*}^{2^*} = \mu \in [0, 1]$. It remains to verify that $\mu = 1$. We shall exclude the cases $\mu = 0$, and $\mu \in (0, 1)$. In fact, if $\mu \in (0, 1)$, then

$$\begin{aligned} S_\lambda &\ge S_\lambda(|u_0|_{2^*}^2 + |v_j|_{2^*}^2) + o(1) \\ &= S_\lambda(\mu^{\frac{2}{2^*}} + (1-\mu)^{\frac{2}{2^*}}) + o(1) \\ &> S_\lambda + o(1) \ . \end{aligned}$$

This is impossible. If $\mu = 0$, then $u_0 = \theta$, i.e., $u_j \rightharpoonup \theta$ in $H_0^1(\Omega)$, and then $|u_j|_2 \to 0$. Thus

$$\begin{aligned} I_\lambda(u_j) &= |\nabla u_j|_2^2 + o(1) \\ &\ge S|u_j|_{2^*}^{\frac{2}{2^*}} + o(1) = S + o(1) \ . \end{aligned}$$

Now, we need the following:

Lemma 4.7.10 *(Brezis–Nirenberg [BN 1] For $n \ge 4, \forall \lambda > 0$,*

$$S_\lambda(\Omega) := \inf_{u \in H_0^1(\Omega) \setminus \{\theta\}} \frac{I_\lambda(u)}{|u|_{2^*}^2} < S \ . \tag{4.69}$$

Proof. One may assume $\theta \in \Omega$. Let $\varphi \in C_0^\infty(\Omega)$, and $\varphi(x) = 1$ in a small neighborhood of θ. For $\varepsilon > 0$, we define

$$v_\varepsilon(x) = \frac{\varphi(x)}{(\varepsilon^2 + |x|^2)^{\frac{n}{2}-1}} \ .$$

Then

$$\nabla v_\varepsilon(x) = \frac{\nabla \varphi(x)}{(\varepsilon^2 + |x|^2)^{\frac{n}{2}-1}} - \frac{(n-2)\varphi x}{(\varepsilon^2 + |x|^2)^{\frac{n}{2}}} \ .$$

We have constants $K_1, K_2, K_3 > 0$ such that

$$\|v_\varepsilon\|_2^2 = (n-2)^2 \int_\Omega \frac{|x|^2 dx}{(\varepsilon^2+|x|^2)^n} + O(1) = (n-2)^2 \int_{R^n} \frac{|x|^2 dx}{(\varepsilon^2+|x|^2)^n} + O(1)$$
$$= \frac{K_1}{\varepsilon^{n-2}} + O(1) ,$$

$$|v|_{2^*}^{2^*} = \int_\Omega \frac{|\varphi|^{2^*} dx}{(\varepsilon^2+|x|^2)^n} = \int_{R^n} \frac{dx}{(\varepsilon^2+|x|^2)^n} + O(1) = \frac{K_2}{\varepsilon^n} + O(1) ,$$

and

$$|v_\varepsilon|_2^2 = \int_\Omega \frac{|\varphi|^2 dx}{(\varepsilon^2+|x|^2)^{n-2}} = \int_{R^n} \frac{dx}{(\varepsilon^2+|x|^2)^{n-2}} + O(1) = \frac{K_3}{\varepsilon^{n-4}} + O(1) ,$$

as $n \geq 5$, and

$$|v_\varepsilon|_2^2 = K_3 |\log \varepsilon| + O(1) ,$$

as $n = 4$.

In summary, we have

$$S_\lambda \leq \begin{cases} \frac{K_1 - \lambda K_3 \varepsilon^2}{K_2^{1-\frac{2}{n}}} & \text{if } n \geq 5 , \\ \frac{K_1 - \lambda K_3 \varepsilon^2 \log|\varepsilon|}{K_2^{\frac{1}{2}}} & \text{if } n = 4 . \end{cases}$$

Since $K_3 > 0$ and $\frac{K_1}{K_2^{1-\frac{2}{n}}} = S$ for small $\varepsilon > 0, S_\lambda < S$. □

Provided by Lemma 4.7.10, the vanishing case is excluded. We have proved the existence of a nontrivial solution.

Remark 4.7.11 *In fact, the restriction $\lambda < \lambda_1$ is not necessary. Provided by minimax methods, which we shall study in the next subsection, one can remove this restriction. Also the case $n = 3$ has been discussed, see Brezis–Nirenberg [BN 1].*

Remark 4.7.12 *(Yamabe problem) Let (M, g_0) be a connected compact n-dimensional Riemannian manifold. One asks: Does there exist a metric g pointwise conformal to g_0 such that the scalar curvature R with respect to g is a constant?*

Set $g = u^{\frac{4}{n-2}} g_0$, where $u \in C^\infty(M)$ with $u > 0$. It is reduced to the following PDE:

$$-\frac{4(n-1)}{n-2} \Delta_{g_0} u + R_{g_0} u = R u^{2^*-1} ,$$

where Δ_{g_0} is the Laplacian Beltrami operator, and R_{g_0} is the scalar curvature with respect to g_0. Then it is transferred to the variational problem:

$$\lambda(M) = \min_{u \in H^1(M)} \frac{\int_M |\nabla_{g_0} u|^2 + \frac{n-2}{4(n-1)} R_{g_0} u^2}{(\int_M |u|^{2^*})^{\frac{2}{2^*}}} .$$

This is a problem very similar to Example 2. Again, $\lambda(M) \leq S$. If $\lambda(M) < S$, then the problem is solvable by the previous method. However, it is easily seen that if $(M, g_0) = (S^n, \hat{g}_0)$, where $\hat{g}_0$ is the canonical metric, then $\lambda(S^n) = S$. The Yamabe problem has been solved by Aubin [Au 1] and Schoen [Sco]. They proved:

Theorem 4.7.13 *For $n \geq 3$, if (M, g_0) is a compact connected Riemannian manifold, which is not conformally equivalent to $(S^n, \hat{g}_0)$, then $\lambda(M) < S$. Then Yamabe problem is solvable.*

4.8 Minimax Methods

4.8.1 Ekeland Variational Principle

It is well known that the direct method does not work in the lack of compactness (i.e., the coerciveness). Without coerciveness, only approximate minimizers can be found. Let us recall the Ekeland variational principle, which we have derived in Chap. 2 as an equivalence of Caristi fixed-point theorem. Due to the importance of this principle, we present here a direct proof.

Theorem 4.8.1 *(Ekeland) Let (X, d) be a complete metric space, and let $f : X \to \mathbb{R}^1 \cup \{+\infty\}$, but $f \not\equiv +\infty$. If f is bounded from below and l.s.c., and if $\exists\, \varepsilon > 0$, $\exists x_\varepsilon \in X$ satisfying $f(x_\varepsilon) < \inf_X f + \varepsilon$. Then $\exists\, y_\varepsilon \in X$ such that*

1. $f(y_\varepsilon) \leqslant f(x_\varepsilon)$,
2. $d(x_\varepsilon, y_\varepsilon) \leqslant 1$,
3. $f(x) > f(y_\varepsilon) - \varepsilon d(y_\varepsilon, x), \qquad \forall x \neq y_\varepsilon$.

Proof. Note that the required y_ε is the minimum of the function $f(x) + \varepsilon d(y_\varepsilon, x)$, while the new function contains y_ε itself. Accordingly, we define a sequence approaching y_ε.

Choose $u_0 = x_\varepsilon$. Suppose that u_n is already chosen. Set

$$S_n = \{w \in X \mid f(w) \leqslant f(u_n) - \varepsilon d(w, u_n)\} .$$

Obviously $S_n \neq \varnothing$. We choose $u_{n+1} \in S_n$ satisfying

$$f(u_{n+1}) - \inf_{S_n} f \leqslant \frac{1}{2}\left[f(u_n) - \inf_{S_n} f\right] , \tag{4.70}$$

$n = 0, 1, 2, \ldots$. We want to show that $\{u_n\}$ is a Cauchy sequence. Indeed,

$$\varepsilon d(u_m, u_n) \leqslant f(u_n) - f(u_m), \qquad \forall m > n . \tag{4.71}$$

Since f is bounded below, $f(u_n) - f(u_m) \to 0$ as $m, n \to \infty$, it follows $u_n \to u^* \in X$. From the l.s.c. of f and (4.70), we have

$$f(u^*) \leqslant \varliminf_{n\to\infty} f(u_n) \leqslant \varliminf_{n\to\infty} \inf_{S_n} f \, . \tag{4.72}$$

We are going to verify that $y_\varepsilon = u^*$ satisfies the conclusions 1–3. In fact, since from (4.71) $f(u_n)$ is nonincreasing, conclusion 1 is trivially true. Again by (4.71), we have

$$\begin{aligned} \varepsilon d(x_\varepsilon, y_\varepsilon) &= \varepsilon d(u_0, u^*) \\ &\leqslant f(x_\varepsilon) - f(u^*) \\ &\leqslant f(x_\varepsilon) - \inf_{S_n} f \\ &< \varepsilon \, . \end{aligned}$$

Thus, conclusion 2 follows. Conclusion 3 is proved by contradiction. If $y_\varepsilon = u^*$ does not satisfy conclusion 3, then $\exists\, w \neq u^*$ such that

$$f(w) \leqslant f(u^*) - \varepsilon d(u^*, w) \, . \tag{4.73}$$

By combining (4.71) with (4.73), it follows that

$$f(w) \leqslant f(u_n) - \varepsilon d(u_n, w), \qquad \forall n \geqslant 0 \, ,$$

i.e., $w \in \bigcap_{n=1}^{\infty} S_n$. Thus, by (4.72), $f(u^*) \leqslant f(w)$. It contradicts (4.73). □

Corollary 4.8.2 *Let (X, d) be a complete metric space, and let $f : X \to \mathbb{R}^1_+ \cup \{+\infty\}$ be proper, bounded from below and l.s.c. Then for $\forall \varepsilon > 0, \exists\, y_\varepsilon \in X$ such that $f(x) > f(y_\varepsilon) - \varepsilon d(x, y_\varepsilon)$, $\forall x \neq y_\varepsilon$.*

Let us introduce the following notion, which plays an important role in the calculus of variations in the large.

Definition 4.8.3 *Let X be a Banach space, and $f \in C^1(X, \mathbb{R}^1)$. If any sequence $\{x_j\}_1^\infty \subset X$ along which $f(x_j) \to c$ and $f'(x_j) \to \theta$ implies a convergent subsequence. Then f is said to satisfy the $(PS)_c$ condition. If f satisfies the $(PS)_c$ condition, for $\forall c$, then it is said to satisfy the (PS) condition.*

Let us return to the minimizing problem:

Corollary 4.8.4 *Let X be a Banach space, and let $f \in C^1(X, \mathbb{R}^1)$ be bounded from below. If $(PS)_c$ holds with $c = \inf_X f$, then f possesses a minimum.*

Proof. According to Theorem 4.8.1, $\forall n \geqslant 0 \quad \exists\, x_n \in X$ satisfying

$$\begin{cases} f(x) > f(x_n) - \frac{1}{n}\|x - x_n\| \\ f(x_n) < c + \frac{1}{n} \end{cases}$$

which implies that $f(x_n) \to c$ and $f'(x_n) \to \theta$. Therefore we obtain a convergent subsequence $x_{n_j} \to x^*$, according to $(PS)_c$. So is

$$f(x^*) = c = \inf_X f \, .$$

4.8.2 Minimax Principle

In this section, we are concerned with those critical points that are saddle points but not minima. The following geometric intuition suggests a minimax consideration.

A valley is surrounded by mountains. Starting from a point x_1 on the ground outside these mountains, we intend to get into a place x_0 in the valley. A path l one would like to take, is a path along which the highest point is lower than those at neighboring paths.

The highest point on this path is indeed a saddle point of the height function.

Theorem 4.8.5 *(Ambrosetti–Rabinowitz [AR]) Let X be a Banach space, and $f \in C^1(X, \mathbb{R}^1)$. Suppose that $\Omega \subset X$ is an open set, $x_0 \in \Omega$, and $x_1 \notin \overline{\Omega}$. Set*

$$\Gamma = \{l \in C([0,1], X) \,\big|\, l(i) = x_i, i = 0, 1\} \ ,$$

and

$$c = \inf_{l \in \Gamma} \max_{t \in [0,1]} f \circ l(t) \ .$$

If

1. *$\alpha = \inf_{x \in \partial\Omega} f(x) > \max\{f(x_0), f(x_1)\}$,*
2. *$(PS)_c$ holds for f;*

then $c \geqslant \alpha$ is a critical point.

We shall not prove this theorem at this moment, but introduce a more general notion.

Definition 4.8.6 *Let X be a Banach space. Let $Q \subset X$ be a compact manifold with boundary ∂Q and let $S \subset X$ be a closed subset of X. ∂Q is said linking with S, if*

1. *$\partial Q \cap S = \varnothing$,*
2. *$\forall \varphi : Q \to X$ continuous with $\varphi|_{\partial\Omega} = \mathrm{id}\,|_{\partial\Omega}$, we have $\varphi(Q) \cap S \neq \varnothing$.*

Example 4.8.7 *(Mountain pass) Let Ω, x_0, x_1 be as in Theorem 4.8.5. Set $Q =$ the segment $\{\lambda x_0 + (1-\lambda)x_1 \mid \lambda \in [0,1]\}$, and $S = \partial\Omega$. Then, $\partial Q = \{x_0, x_1\}$ and S link.*

Example 4.8.8 *Let X_1 be a finite-dimensional linear subspace of the Banach space X, and let X_2 be its complement: $X = X_1 + X_2$. Let $Q = B_R \cap X_1, S = X_2$, where B_R is the ball with radius $R > 0$ centered at θ. Then ∂Q and S link.*

Indeed, $\forall \varphi : Q \to X$ continuous with $\varphi\big|_{\partial Q} = \mathrm{id}\,\big|_{\partial\Omega}$, we want to show: $\varphi(Q) \cap X_2 \neq \varnothing$.

Let P be the projection onto X_1. It is sufficient to show that $P\circ\varphi : Q \to X_1$ has a zero.

Obviously

$$\deg(P \circ \varphi, Q, \theta) = \deg(\mathrm{id}, Q, \theta) = 1 \,.$$

The conclusion follows from the Brouwer degree theory.

Theorem 4.8.9 *Let X be a Banach space, and let $f \in C^1(X, \mathbb{R}^1)$. Assume that $Q \subset X$ is a compact manifold with boundary ∂Q which links with a closed subset $S \subset X$. Set*

$$\Gamma = \{\varphi \in C(Q, X) \,|\, \varphi|_{\partial Q} = \mathrm{id}\,|_{\partial Q}\} \,,$$

and

$$c = \inf_{\varphi\in\Gamma} \max_{\xi\in Q} f \circ \varphi(\xi) \,.$$

If $\exists\, \alpha < \beta$ such that

$$\sup_{x\in\partial Q} f(x) \leqslant \alpha < \beta \leqslant \inf_{x\in S} f(x) \,,$$

and if $(PS)_c$ holds, then $c\,(\geqslant \beta)$ is a critical value.

Proof. Let d be the distance on $C(Q, X)$. Then (Γ, d) is a metric space. Let

$$J(\varphi) = \max_{\xi\in Q} f \circ \varphi(\xi) \,.$$

Invoke the assumption that ∂Q and S link, $J \geqslant \inf\limits_S f$. Moreover, J is locally Lipschitzian, since

$$\begin{aligned} J(\varphi_1) - J(\varphi_2) &\leqslant \max_{\xi\in Q}[f \circ \varphi_1(\xi) - f \circ \varphi_2(\xi)] \\ &\leqslant \max_{\substack{\theta\in[0,1] \\ \xi\in Q}} \|f'(\theta\varphi_1(\xi) + (1-\theta)\varphi_2(\xi))\| d(\varphi_1, \varphi_2) \,. \end{aligned}$$

It follows that

$$\big|J(\varphi_1) - J(\varphi_2)\big| \leqslant C d(\varphi_1, \varphi_2) \,,$$

where C is a constant depending on φ_1, φ_2. We apply the Ekeland variational principle to J, and obtain a sequence $\{\varphi_n\} \subset \Gamma$ satisfying

$$c \leqslant J(\varphi_n) < c + \frac{1}{n} \,, \tag{4.74}$$

and

$$J(\varphi) \geqslant J(\varphi_n) - \frac{1}{n} d(\varphi, \varphi_n) \,, \tag{4.75}$$

$n = 1, 2, 3, \ldots$.

Set

$$\mathcal{M}(\varphi) = \{\xi \in Q \,|\, f \circ \varphi(\xi) = J(\varphi)\} \,.$$

Obviously $\mathcal{M}(\varphi)$ is compact. We claim that $\mathcal{M}(\varphi) \subset \mathrm{int}(Q)$. Indeed, since ∂Q and S link, if $\exists \xi_0 \in \mathcal{M}(\varphi) \cap \partial Q$, then

$$f \circ \varphi(\xi_0) = \max_{\xi \in Q} f \circ \varphi(\xi) \geqslant \inf_S f \geqslant \beta \,.$$

But,

$$f \circ \varphi(\xi_0) = f(\xi_0) \leqslant \sup_{x \in \partial Q} f(x) \leqslant \alpha \,.$$

This is a contradiction.

Set

$$\Gamma_0 = \{\psi \in C(Q, X) \,|\, \psi|_{\partial Q} = \theta\} \,.$$

It is a linear closed subspace of X. Let $\|\cdot\|$ be the norm of $C(Q, X)$. $\forall h \in \Gamma_0$, with $\|h\| = 1, \forall \lambda_j \downarrow 0, \forall \xi_j \in \mathcal{M}(\varphi_n + \lambda_j h)$, we have

$$\lambda_j^{-1}[\, f \circ (\varphi_n + \lambda_j h)(\xi_j) - f \circ \varphi_n(\xi_j)\,] \geqslant -\frac{1}{n} \,, \tag{4.76}$$

from (4.75). Since $\{\xi_j\} \subset Q$, we obtain a convergent subsequence $\xi_j \to \eta_n^* \in \mathcal{M}(\varphi_n)$, which depends on φ_n, λ_j and h. After taking limits, we have

$$\langle f' \circ \varphi_n(\eta_n^*), h(\eta_n^*)\rangle \geqslant -\frac{1}{n} \,. \tag{4.77}$$

We want to show that $\exists \, \eta_n \in \mathcal{M}(\varphi_n)$ such that

$$\langle f' \circ \varphi_n(\eta_n), u\rangle \geqslant -\frac{1}{n} \,, \tag{4.78}$$

$\forall u \in X$ with $\|u\| = 1$. If not, $\forall \eta \in \mathcal{M}(\varphi_n), \exists \, v_\eta \in X$ with $\|v_\eta\| = 1$, satisfying

$$\langle f' \circ \varphi_n(\eta), v_\eta\rangle < -\frac{1}{n} \,,$$

then there exists a neighborhood of η, $O_\eta \subset \mathrm{int}(Q)$ such that

$$\langle f' \circ \varphi_n(\xi), v_\eta\rangle < -\frac{1}{n}, \qquad \forall \xi \in O_\eta \,.$$

Since $\mathcal{M}(\varphi_n)$ is compact, there is a finite covering. Let m be the least number of covering: $\bigcup_{i=1}^m O_{\eta_i} \supset \mathcal{M}(\varphi_n)$. We obtain the associate $\{v_{\eta_i}\}_1^m$, $\|v_{\eta_i}\| = 1$ satisfying

$$\langle f' \circ \varphi_n(\xi), v_{\eta_i}\rangle < -\frac{1}{n} \qquad \forall \xi \in O_{\eta_i} \,,$$

$i = 1, 2, \ldots, m$. Construct a partition of unity subject to $\{O_{\eta_i}\}_1^m : 0 \leqslant \varrho_i \leqslant 1, \sup \varrho_i \subset O_{\eta_i}, i = 1, \ldots, m$, and

$$\sum_{i=1}^m \varrho_i(\xi) \equiv 1, \qquad \forall \xi \in \mathcal{M}(\varphi_n) \,.$$

We set

$$v = v(\xi) = \sum_{i=1}^{m} \varrho_i(\xi) v_{\eta_i} \, .$$

Thus $v \in \Gamma_0$ and $\|v\| \leqslant 1$. But, $\exists \xi^* \in \mathcal{M}(\varphi_n)$ such that there is only one i_0 satisfying $\xi^* \in O_{\eta_{i_0}}$. Therefore $\|v\| = 1$. We obtain

$$\langle f' \circ \varphi_n(\xi), v(\xi) \rangle < -\frac{1}{n}, \ \forall \xi \in \mathcal{M}(\varphi_n) \, .$$

This contradicts (4.77). Thus (4.78) holds.

Setting $x_n = \varphi_n(\eta_n)$, we have

$$f'(x_n) \to \theta \, .$$

Combining with (4.74), we have also

$$f(x_n) \to c \, .$$

The $(PS)_c$ condition implies that c is a critical value. □

The above proof is based on Shi [Shi] and Aubin and Ekeland [AE].

The following example due to Brezis and Nirenberg [BN 3] asserts that the $(PS)_c$ condition is crucial in Theorem 4.5.5.

Example. The function $\varphi(x, y) = x^2 + (1 - x)^3 y^2$ defined on R^2 does have a mountain surround $(0, 0)$: $\varphi(x, y) \geq c > 0$ on the circle $x^2 + y^2 = \frac{1}{4}$, $\varphi(0, 0) = 0$, and $\varphi(4, 1) = -11$. But by direct computation it has only one critical point: $(0, 0)$.

4.8.3 Applications

The mountain pass lemma and the related minimax principles are widely used in the study of differential equations. We are satisfied to give few examples.

Example 1. We study the following forced resonance problem.

Given $h \in L^1([0, \pi])$ satisfying

$$\int_0^{\pi} h(t) \sin t dt = 0 \, . \tag{4.79}$$

Assume that $g \in C(\mathbb{R}^1)$ is T-periodic, $T > 0$, and satisfies:

$$\int_0^{T} g(t) dt = 0 \, . \tag{4.80}$$

Find a solution of the nonlinear BVP:

$$\begin{cases} \ddot{x}(t) + x(t) = h(t) + g(x(t)) & t \in (0, \pi) \\ x(0) = x(\pi) = 0 \, . \end{cases} \tag{4.81}$$

First we introduce a generalized Riemann-Lebesgue lemma.

Lemma 4.8.10 *Under the above assumptions of g, if $u_n \to u$ in $C([0,\pi])$, $\alpha_n \to \infty$, $u \in C^1([0,\pi])$ and $k \in L^1([0,\pi])$, then*

$$\lim_{n\to\infty} \int_0^\pi g(u_n(t) + \alpha_n \sin t)k(t)dt \to 0 \, .$$

Proof. Since the sequence is bounded in $C([0,\pi])$, and since $g(u_n(t) + \alpha_n \sin nt) - g(u(t) + \alpha_n \sin t)$ uniformly converges to 0, it is sufficient to prove that

$$\lim_{n\to\infty} \int_0^\pi g(u(t) + \alpha_n \sin t) \cdot \chi_E(t)dt = 0 \, ,$$

$\forall E = (a,b) \subset (0,\pi)$, where χ_E is the characteristic function of E. We may assume $\alpha_n \to +\infty$. We write $v_n(t) = \alpha_n^{-1}u(t) + \sin t$, and let $G(t)$ be a primitive of g. Then G is T-periodic, according to (4.80).

At first, we assume $\frac{\pi}{2} \notin E$. Then for n large, $s = v_n(t)$ is strictly increasing (decreasing), if $b < \frac{\pi}{2}$ (or $a > \frac{\pi}{2}$ resp.). We have $t = v_n^{-1}(s)$.

Let $\delta = \mathrm{dist}\,(E, \frac{\pi}{2})$, $\exists$ a constant C_δ depending on $\delta > 0$, such that $\frac{1}{|v_n'(v_n^{-1}(s))|} \leqslant C_\delta$ a.e. By changing variables,

$$\begin{aligned}\left|\int_a^b g(\alpha_n v_n(t))dt\right| &= \left|\int_{v_n(a)}^{v_n(b)} g(\alpha_n s)\frac{ds}{v_n'(v_n^{-1}(s))}\right| \\ &\leqslant \frac{C_\delta}{\alpha_n}|G(\alpha_n v_n(b)) - G(\alpha_n v_n(a))| \, . \end{aligned} \tag{4.82}$$

Since G is continuous and periodic, the RHS of the above inequality tends to zero as $n \to \infty$.

Second, we consider the case $\frac{\pi}{2} \in E$. Since g is bounded, provided by the absolute continuity of the integral of k, $\forall \varepsilon > 0$, $\exists \delta > 0$ such that

$$\left|\int_{\pi/2-\delta}^{\pi/2+\delta} k(t)g(\alpha_n v_n(t))dt\right| < \frac{\varepsilon}{2} \, . \tag{4.83}$$

Combining (4.82) with (4.83), the conclusion follows. □

Let us reformulate (4.81) in a variational version. Again let G be a primitive of g satisfying $\int_0^T G(t)dt = 0$. Setting $X = H_0^1([0,\pi])$,

$$I(u) = \int_0^\pi \left[\frac{1}{2}(\dot{u}^2 - u^2) + hu\right] dt, \;\; N(u) = \int_0^\pi G(u(t))dt \, ,$$

and

$$J(u) = I(u) + N(u) \, .$$

The functionals I and J are not coercive. In fact, letting $u_n(t) = n \sin t$, we have $I(u_n) = 0$, and from $|G(t)| \leqslant C$, a constant, we have $|N(u_n)| \leqslant C\pi$, and then $|J(u_n)| \leqslant C\pi$.

Now, let us decompose X into $H_1 \oplus \text{span}\{e_1\}$, where $e_1 = \sin t$, and $H_1 = e_1^\perp$, i.e., $\forall u \in H_0^1([0,\pi])$ one has the orthogonal decomposition:

$$u = u_1 + \alpha e_1 ,$$

where

$$\alpha = \int_0^\pi u(t) \sin t dt .$$

Then we have:

(1) I is s.w.l.s.c. and coercive on H_1. Moreover, from (4.79), $I(u) = I(u_1)$.
(2) N is bounded and weakly continuous on X, and according to Lemma 4.8.10, $N(u_1 + \alpha e_1) \to 0$ as $\alpha \to \infty$, $\forall u_1 \in H_1$.
(3) J is bounded from below on X and coercive on H_1.

Let

$$m^* = \inf_{u \in X} J(u) .$$

Lemma 4.8.11 *If m^* is not a minimum of J, then $\exists \{u_{1j}\} \subset H_1$ and $\alpha_j \to \infty$ such that $J(u_{1j} + \alpha_j e_1) \to m^*$. Moreover, $u_{1j} \rightharpoonup \overline{u}_1$ in H_1, which is a solution of $\ddot{x} + x = h$, and $m^* = I(\overline{u}_1) = \inf I(u_1),\ u_1 \in H_1$.*

Proof. 1. We choose a minimizing sequence $\{u_j\}$ of J, and decompose it on $H_1 \oplus \text{span}\{e_1\}$:

$$u_j = u_{1j} + \alpha_j e_1 .$$

We claim that $\alpha_j \to \infty$. Indeed, invoke the coerciveness of I on H_1 and

$$I(u_{1j}) = I(u_j) = J(u_j) - N(u_j) < m^* + 1 + C\pi ,$$

$\{u_{1j}\}$ is bounded in H_1. Thus after a subsequence $u_{1j} \rightharpoonup \overline{u}_1$ in H_1, which implies $u_{1j} \to \overline{u}_1$ in $C([0,\pi])$.

If $\{\alpha_j\}$ is bounded, then after a subsequence $\alpha_j \to \overline{\alpha}$, we have

$$N(u_{1j} + \alpha_j e_1) \to N(\overline{u}) ,$$

where $\overline{u} = \overline{u}_1 + \overline{\alpha} e_1$. We have

$$\liminf_{j\to\infty} I(u_j) = \lim_{j\to\infty} I(u_{1j}) \geqslant I(\overline{u}_1) = I(\overline{u}) ,$$

and

$$m^* = \lim_{j\to\infty} J(u_j) = \lim_{j\to\infty} (I(u_j) + N(u_j)) \geqslant I(\overline{u}) + N(\overline{u}) = J(\overline{u}) .$$

Therefore, $\bar{u}$ is a minimizer of J. This is a contradiction. The claim is proved.

2. $\forall \alpha \in \mathbb{R}^1, \forall u_1 \in H_1$, one has

$$m^* \leq J(u_1 + \alpha e_1) = I(u_1) + N(u_1 + \alpha e_1) .$$

Letting $\alpha \to \infty$, provided by Lemma 4.8.10, $N(u_1 + \alpha e_1) \to 0$, thus

$$m^* \leqslant I(u_1) = I(\bar{u}_1), \quad \forall u_1 \in H_1 . \tag{4.84}$$

But to the minimizing sequence $\{u_j\}$ we have

$$\begin{aligned} I(\overline{u}_1) &\leqslant \lim_{j\to\infty} I(u_{1j}) \\ &= \lim_{j\to\infty} (J(u_j) - N(u_{1j} + \alpha_j e)) \\ &= \lim_{j\to\infty} J(u_j) = m^* . \end{aligned} \tag{4.85}$$

Combining (4.84) and (4.85), we conclude that $m^* = I(\overline{u}_1)$ is the minimum of I on H_1, and $\overline{u}_1$ is the associated minimizer, which satisfies $I'(\overline{u}_1) = 0$, i.e., $\overline{u}_1$ is a solution of

$$\begin{cases} \ddot{x} + x = h , \\ x(0) = x(\pi) = 0 . \end{cases}$$

□

Lemma 4.8.12 *$\forall c > m^*$, J satisfies $(PS)_c$.*

Proof. Assume that $\{u_j\}$ is a $(PS)_c$ sequence of J, i.e.,

$$J(u_j) \to c \text{ and } J'(u_j) \to \theta \text{ in } X .$$

From the coerciveness of I on H_1, and

$$I(u_{1j}) = J(u_j) - N(u_{1j} + \alpha_j e_1) \leqslant c + 1 + c\pi ,$$

after a subsequence, we have $u_{1j} \rightharpoonup \overline{u}_1$ in H_1.

We are going to verify that after a subsequence $\alpha_j \to \overline{\alpha}$. For otherwise, $\alpha_j \to \infty$, we would have $g(u_j) \overset{*}{\rightharpoonup} \theta$ in $L^\infty([0,\pi])$, according to Lemma 4.8.10. Noticing that the differential operator $(\frac{d}{dt})^2 + 1$ has a compact inverse $\mathbb{K}$ on H_1, from $J'(u_j) \to \theta$, we have

$$u_{1j} = \mathbb{K}h + \mathbb{K}Pg(u_j) + o(1) , \tag{4.86}$$

where P is the orthogonal projection from X onto H_1. Thus $u_{1j} \to \mathbb{K}h = \overline{u}_1$ in H_1 strongly. It follows from 2 in the proof of Lemma 4.8.11,

$$\begin{aligned} c &= \lim_{j\to\infty} J(u_j) \\ &= \lim_{j\to\infty} (I(u_{1j}) + N(u_{1j} + \alpha_j e_1)) \\ &= \lim_{j\to\infty} I(u_{1j}) \\ &= I(\overline{u}_1) \\ &= m^* . \end{aligned}$$

This is a contradiction. Therefore $\alpha_j \to \overline{\alpha}$. Substituting in (4.86), we obtain a convergent sequence:

$$u_j = u_{1j} + \alpha_j e_1 \to \mathbb{K}h + \mathbb{K}Pg(\overline{u}_1 + \overline{\alpha}e_1) + \overline{\alpha}e_1 \, .$$

□

Conclusion. *The problem (4.81) has at least one solution.*

Proof. If m^* is attainable at u^*, then u^* is a minimizer, which is a solution of the Euler Lagrange equation (4.81) for J.

Otherwise, let

$$b = \inf_{u_1 \in H_1} J(u_1) \, .$$

Note that $J|_{H_1}$ is s.w.l.s.c. and coercive, $\exists u_1^* \in H_1$ such that $b = J(u_1^*)$. Therefore $b > m^*$. Since

$$\lim_{\alpha \to \pm\infty} J(\overline{u}_1 + \alpha e_1) = I(\overline{u}_1) = m^* \, ,$$

we may choose $a > 0$ large enough such that

$$J(\overline{u}_1 \pm a e_1) < b \, .$$

Define

$$\Gamma = \left\{ l \in C([0,1], X) |\, l\left(\frac{1}{2} \pm \frac{1}{2}\right) = \overline{u}_1 \pm a e_1 \right\} \, ,$$

and

$$c = \inf_{l \in \Gamma} \max_{t \in [0,1]} J(l(t)) \, .$$

Since for $\forall l \in \Gamma$, $l[0,1] \cap H_1 \neq \emptyset$, we have $c \geqslant b > m^*$. Thus $(PS)_c$ holds. According to the mountain pass theorem there is a critical point u^* with $J(u^*) = c$. Again u^* solves (4.81). □

Example 2. We present here an application to the superlinear elliptic boundary problem, see Ambrosetti and Rabinowitz [AR].

Let $\Omega \subset \mathbb{R}^n$ be a bounded domain with smooth boundary. Given a function $f \in C(\overline{\Omega} \times \mathbb{R}^1)$ satisfying the following conditions:

$[F_1]$ There exist positive constants C_1, C_2 and $\alpha \in (1, \frac{n+2}{n-2})$ (when $n \geq 3$) such that

$$|f(x,t)| \leq C_1 + C_2 |t|^\alpha \, .$$

$[F_2]$ $\exists \theta \in (0, \frac{1}{2}), \exists$ a constant $M > 0$ such that

$$0 < F(x,t) := \int_0^t F(x,s) ds \leq \theta t f(x,t), \ \text{ as } |t| \geq M \, .$$

[F_3]

$$\limsup_{t\to 0} \frac{f(x,t)}{t} < \lambda_1, \text{ uniformly in } x \in \overline{\Omega} .$$

[F_4]

$$\liminf_{t\to\infty} \frac{f(x,t)}{t} > \lambda_1, \text{ uniformly in } x \in \overline{\Omega} .$$

where λ_1 is the first eigenvalue of the Laplacian $-\Delta$ on Ω with Dirichlet boundary conditions.

Theorem 4.8.13 *Under the assumptions F_1, F_2, F_3, F_4, the problem*

$$\begin{cases} -\Delta u = f(x,u), & x \in \Omega , \\ \quad u = 0, & x \in \partial\Omega , \end{cases} \tag{4.87}$$

has a nontrivial solution.

Proof. We consider the functional on $H_0^1(\Omega)$:

$$J(u) = \int_\Omega \left[\frac{1}{2}|\nabla u|^2 - F(x,u)\right] dx . \tag{4.88}$$

From the growth condition F_1, J is C^1.

Noticing $J(\theta) = 0$, we claim that there is a mountain surrounding θ and $\exists \alpha > 0, \exists \rho > 0$ and $u_0 \notin B_\rho(\theta)$ with $J(u_0) = 0$. Indeed, from F_3, we have $\epsilon > 0, \delta > 0$ such that

$$\frac{f(x,t)}{t} \le \lambda_1 - \epsilon, \text{ as } 0 < |t| < \delta ,$$

which implies that

$$F(x,t) \le \frac{1}{2}(\lambda_1 - \epsilon)t^2 \text{ as } |t| < \delta .$$

Combining with F_1, we have a constant C_3 such that

$$F(x,t) \le \frac{1}{2}(\lambda_1 - \epsilon)t^2 + C_3|t|^p ,$$

where $p = \alpha + 1$, thus

$$\int_\Omega F(x,u(x))dx \le \frac{1}{2}\left(1 - \frac{\epsilon}{\lambda_1}\right) \|u\|_{H^1}^2 + C_4\|u\|_{L^p}^p .$$

This proves that

$$J(u) \ge \frac{\epsilon}{2\lambda_1}\|u\|_{H^1}^2 - C_4\|u\|_{L^p}^p .$$

Therefore $\exists \alpha > 0, \exists \rho > 0$ such that $J|_{\partial B_\rho(\theta)} \ge \alpha$.

Now, we take $u_0 = s\phi_1$, where $\phi_1 > 0$ is the normalized first eigenfunction, and $s > 0$ is to be determined. Setting

$$g(s) = J(s\phi_1) = \frac{\lambda_1}{2}s^2 - \int_\Omega F(x, s\phi_1(x))dx \;,$$

From F_4, $g(s) \to -\infty$ as $s \to +\infty$. Therefore there exists $s_0 > 0$ satisfying $g(s_0) = 0$, set $s = s_0, u_0$ is as required.

It remains to verify the Palais-Smale condition. Suppose that a sequence $\{u_j\} \subset H_0^1(\Omega)$ satisfies $J(u_j) \to c, J'(u_j) \to \theta$; we shall show that it possesses a convergent subsequence. Firstly, $\{u_j\}$ is bounded. In fact, $\exists C_3 > 0$ such that

$$-C_3 \leq \frac{1}{2}\|u_j\|_{H^1}^2 - \int_\Omega F(x, u_j(x))dx \leq C_3 \;.$$

According to F_2,

$$-\theta \int_\Omega u_j(x)f(x, u_j(x))dx \leq - \int_\Omega F(x, u_j(x))dx + C_4. \;,$$

it follows that

$$\left(\frac{1}{2} - \theta\right) \|u_j\|_{H^1}^2 + \theta J'(u_j)(u_j) \leq C_5 \;.$$

Since $J'(u_j) \to \theta$, this proves that $\{u_j\}$ is bounded in $H_0^1(\Omega)$.

Next, we show the existence of a convergent subsequence. Noticing that $p < \frac{2n}{n-2}$, from the Sobolev embedding theorem $H^1(\Omega) \hookrightarrow L^p(\Omega)$ is bounded, so that $\|u_j\|_{L^p}$ is bounded, and then after a subsequence, $\{f(x, u_j(x))\}$ is weakly convergent in $L^{p'}(\Omega)$, $\frac{1}{p} + \frac{1}{p'} = 1$. However, $(-\Delta)^{-1} : L^{p'}(\Omega) \to H_0^1(\Omega)$ is compact. Therefore, $(-\Delta)^{-1}f(\cdot, u_j)$ is strongly convergent in $H_0^1(\Omega)$. Since

$$J'(u_j) = u_j - (-\Delta)^{-1}f(\cdot, u_j) \to \theta \;,$$

u_j converges strongly. □

It is interesting to compare the assumptions F_1 to F_4 with those in Sect. 3.6, Example 4. They are very similar. But the differences lie as follows: The partial differential operator in the cited example is not necessarily a divergent form, in which case the variational method cannot be applied; while in the above example, the solution is not necessarily positive. Moreover, one may even change the assumption F_3 (see below); in these cases the a priori estimate, which the degree theoretic argument relies on, either fails or holds under other assumptions.

Example 3. (Ni [Ni]) We make the conditions

$[F_3']$

$$f(x, t) = o(t) \text{ as } t \to 0 \text{ uniformlyin } x \in \overline{\Omega} \;.$$

$[F_4']$

$$tf(x, t) \geq 0, \;\forall (x, t) \in \overline{\Omega} \times \mathbb{R}^1 \;.$$

Theorem 4.8.14 *Under the assumptions F_1, F_2, F_3', F_4', assume $a \in C(\overline{\Omega})$ is positive. Then the following equation possesses a nontrivial solution:*

$$\begin{cases} -\Delta u(x) = a(x)u(x) + f(x, u(x)), & x \in \Omega\,, \\ \qquad\qquad u(x) = 0. & x \in \partial\Omega\,. \end{cases} \tag{4.89}$$

Proof. We consider the functional on $H_0^1(\Omega)$:

$$J(u) = \int_\Omega \left\{ \frac{1}{2}[|\nabla u|^2 - au^2] - F(x,u) \right\} dx\,,$$

For which the Euler–Lagrange equation is (4.89). In the same manner we verify the Palais–Smale condition. We claim that there are linking sets, which separate values of J.

Indeed, let $0 < \lambda_1 < \lambda_2 \leq \cdots \leq \lambda_k \leq 1 < \lambda_{k+1} \leq \cdots$ be the eigenvalues and $\{\phi_1, \ldots, \phi_k, \phi_{k+1}, \ldots\}$ be the eigenfunctions of the eigenvalue problem:

$$-\Delta u = \lambda a \cdot u,\;\; u \in H_0^1(\Omega)\,.$$

Taking $E_k = \operatorname{span}\{\phi_1, \ldots, \phi_k\}$, we have the orthogonal decomposition: $H_0^1(\Omega) = E_k \bigoplus \hat{E}$.

On $\hat{E}$, we have

$$\int_\Omega |\nabla u|^2 dx \geq \lambda_{k+1} \int_\Omega au^2 dx\,,$$

therefore

$$J(u) \geq \int_\Omega \left\{ \frac{1}{2}\left(1 - \frac{1}{\lambda_{k+1}}\right) |\nabla u|^2 - F(x,u) \right\} dx\,.$$

From F_3',

$$\int_\Omega F(x, u(x))dx = o(\|u\|^2) \;\text{ as }\; \|u\| \to 0\,.$$

Therefore there exist $\rho > 0, \alpha > 0$ such that

$$J(u) \geq \alpha \;\text{ as }\; u \in \partial B_\rho(\theta) \cap \hat{E}\,.$$

From $F_4', F(x,t) \geq 0\;\; \forall (x,t) \in \overline{\Omega} \times \mathbb{R}^1$, it implies that

$$J(u) \leq \frac{1}{2} \int_\Omega [|\nabla u|^2 - a(x)u^2]dx \leq 0,\;\; \forall u \in E_k\,.$$

From F_2, we have

$$F(x,t) \geq C_1|t|^{\frac{1}{\theta}} - C_2\,,$$

where C_1, C_2 are positive constants. Since $\theta < \frac{1}{2}$,

$$J(u) \leq (\lambda_{k+1} - 1) \int_\Omega au^2 dx - C_1 \int_\Omega |u|^{\frac{1}{\theta}} dx + C_2|\Omega| \to -\infty\,,$$

as $u \in E_{k+1} := \text{span}\{\phi_1, \ldots, \phi_k, \phi_{k+1}\}$ and $\|u\|_{H^1} \to \infty$. There exists $R > \rho$ such that

$$J(u) \leq 0 \text{ as } u \in \partial B_R(\theta) \cap E_{k+1} \,.$$

Setting $S = \partial B_\rho(\theta) \cap \hat{E}$ and $\partial Q = (B_R(\theta) \cap E_k) \cup (\partial B_R(\theta) \cap \{(x_1, \ldots, x_{k+1}) \in E_{k+1} \,|\, x_{k+1} \geq 0\})$. Then S and ∂Q link, and they separate values of J.

Theorem 4.8.9 yields the existence of a critical value $c \geq \alpha > 0$, which implies the existence of a nontrivial solution of equation (4.89). □

5

Topological and Variational Methods

This chapter is devoted to the critical point theory and Conley index theory.

A typical problem from geometry in the calculus of variations is to find geodesics between two points q_0, q_1 on a given Riemannian manifold (M, g). A path on M connecting q_0 and q_1 is denoted by $\gamma : [0,1] \to M$ with $\gamma(i) = q_i, i = 0, 1$. Let

$$N = \{\gamma \in C^{1-0}([0,1], M) \,|\, \gamma(i) = q_i,\ i = 0, 1\}\ ,$$

and define the energy functional:

$$I(\gamma) = \int_0^1 \left|\left|\frac{d\gamma}{dt}\right|\right|^2 dt\ ,$$

where $||\cdot||^2$ is the scalar product induced by the metric g, or in the local coordinates:

$$\left|\left|\frac{d\gamma}{dt}\right|\right|^2 = \Sigma_{i,j} g_{ij}(\gamma(t)) \frac{d\gamma}{dt}\frac{d\gamma^j}{dt}\ ,$$

and $\gamma = (\gamma^1, \ldots, \gamma^p), p = \dim \mathrm{M}$. Then the Euler–Lagrange equation reads as the geodesic equation:

$$\left(\frac{d}{dt}\right)^2 \gamma^i = \Sigma_{j,k} \Gamma^i_{jk}(\gamma) \frac{d\gamma^j}{dt}\frac{d\gamma^k}{dt}, \quad i = 1, 2, \ldots, p\ .$$

Geometrically, a curve satisfying the geodesic equation has the feature that tangents along it are parallel. Generally speaking, there are many geodesics joining q_0 and q_1, which are critical points of the energy functional. However, besides those minimal in length, how do we reach the others?

Influenced by the pioneering work of Birkhoff (1917) [Bi] on closed geodesics, two global methods have appeared: The minimax method and the Morse theory. Both are based on topological arguments.

The minimax method, as an outgrowth of the max-min characterization of the eigenvalues of Laplacian with Dirichlet data, was successfully developed by Ljusternik and Schnirelmann (1934)[LjS 1]. The method provided a proof of the existence of at least three closed geodesics on a closed surface with genus zero, see Ljusternik and Schnirelmann [Lj, LjS 1], Ballmann [Bm], Klinenberg [Kl].

In the 1950s Krasnosel'ski [Kr 1] introduced the notion of genus, which is a topological index related to the group $\mathbf{Z}_2$, in the study of the nonlinear eigenvalue problem for a class of integral equations.

An important step towards the recent development of critical point theory is due to Palais (1966) [Pa 1, Pa 2]. Two major contributions are: (1) The Ljusternik Schnirelmann theory was extended to infinite-dimensional manifolds. (2) Homotopically stable families were used in the minimax principle.

The reviva of the study of the minimax method began with the works of Ambrosetti and Rabinowitz (1974)[AR] and Rabinowitz [Ra 3]. The first paper deals with functionals unbounded from below, and provides many applications in the study of nonlinear elliptic equations, Hamiltonian systems, and problems from geometry and mathematical physics. The second paper provides a proof of the existence of a closed orbit of an autonomous Hamiltonian system on a star-shaped hypersurface in R^{2n} (independently, see Weinstein [Wei 1]). This is the first global result on this problem and is a breakthrough to the Weinstein conjecture.

In parallell, in the 1930s, Morse developed a theory which reveals the relationship between the critical points of a nondegenerate function and the topology of the underlying compact manifold. Although the nondegeneracy condition and the compactness assumption in his theory do not meet in most variational problems, he succeeded in applying his theory in the study of closed geodesics and unstable minimal surfaces, see Morse [Mo], Milnor [Mi 1], Bott[Bo 1], Morse and Tompkins [MT], Schiffman [Sch]. Moreover, Morse theory became a basic tool in computing the homology of compact manifolds. The work of Smale on the solution of the Poincaré conjecture for $n \geq 5$ pushes the development of the theory to a new peak, see Smale [Sm 1], Milnor [Mi 3].

In the 1950s and 1960s Rothe [Ro 1–3], Palais [Pa 3], and Smale [Sm 2] extended the Morse theory to infinite-dimensional manifolds by using the Palais–Smale condition. Later, Marino and Prodi [MP 1, MP 2] and Gromoll and Meyer [GM] endeavored in weakening the restriction of the nondegeneracy. In applying Morse theory to differential equations, most functionals are unbounded from below, and the nondegenerate assumption should be removed; to this end, critical groups for isolated critical points are introduced and the relative homology is involved, by which the interconnection with the minimax principle has been revealed, see K. C. Chang [Ch 4,5,7].

Conley extended the Morse theory to flows on a compact space without a variational structure. The Conley index for isolated invariant sets with respect to the flow is a homotopy invariant. A Morse decomposition is also extended. With the aid of Conley's theory, Floer established his homology theory, which is an important tool in the symplectic geometry.

In this chapter we shall introduce these theories as tools in the study of multiple solutions for differential equations.

Morse theory is introduced in Sect. 5.1, we shall present it in a way comparing with the Leray–Schauder degree. After the basic theorems, we shall focus on the computation of critical groups for isolated critical points, in particular, those obtained by various minimax methods.

Section 5.2 is on the minimax principles, which include the Ljusternik–Schnirelmann category theory, cap length estimates, Krasnosel'ski's genus and other index theories. Various extensions can be found in Rabinowitz [Ra 4], [Ra 5]. The central result is the multiplicity theorem.

Section 5.3 is deals with an application to the Weinstein conjecture, see Viterbo [Vi 1].

We return to the prescribed Gauss curvature problem on S^2 in Sect. 5.4. We shall apply the Morse theory with boundary conditions to attack the problem.

We introduce the definitions of the isolated invariant set, the index pair, and the Conley index in metric space without compactness. The fundamental properties of the Conley index, in particular, the homotopy invariance and the Morse decomposition are studied. Examples and the relationship with the Morse theory are also presented.

5.1 Morse Theory

5.1.1 Introduction

Morse revealed a deep relation between the critical points of any nondegenerate function and the topology of the underlying compact manifold M.

Let $f \in C^1(M)$. $\forall a \in R^1$, the set $f_a = \{x \in M | f(x) \leq a\}$ is called a level set. We denote the set of all critical points by K. Here the terminology "critical point" is coincident with that for general differential mappings, which was introduced in Sect. 1.3.5, i.e., $K = \{x \in M \mid f'(x) = \theta\}$. A real value $\{c\}$ is called critical if $f^{-1}(c) \cap K \neq \emptyset$; otherwise, it is called regular. According to the Sard–Smale theorem, the set of critical values is of the first category.

A C^2 function f is called nondegenerate if it has only nondegenerate critical points. A critical point p is called nondegenerate if the Hessian $f''(p)$ at this point has a bounded inverse ("boundedness" is assumed if M is an infinite-dimensional Hilbert–Riemannian manifold). As an application of the Sard–Smale theorem, the set of nondegenerate functions is dense in $C^2(M)$. To a nondegenerate critical point p, we call the dimension of the subspace of negative eigenvectors of $f''(p)$ the Morse index, and denote it by $\text{ind}(f''(p))$.

The basic idea in the Morse theory is to relate the local behavior of a nondegenerate critical point with the variations of the topological structures of the level sets f_a. More precisely, if $f^{-1} \cap K = \{p\}$ and $\text{ind}(f''(p)) = j$, then

$$f_{c+\epsilon} \sim f_{c-\epsilon} \cup B^j ,$$

where B^j is the j-ball, for $\epsilon > 0$ small, "$\cup$" means "attached by" and $\sim$ means homotopical equivalence.

Since M is compact, the nondegenerate function f can only have finitely many critical points $\{p_1, \ldots, p_l\}$. Then the space M is homotopically equivalent to the space by gluing finitely many balls with dimensions $\{\text{ind}(f''(p_i))|i = 1, \ldots, l\}$ according to their levels.

The above consideration is based on the following Morse lemma. We assign a Riemannian structure M.

Lemma 5.1.1 *(Morse lemma) Suppose that $f \in C^2(M, R^1)$ and that p is a nondegenerate critical point; then there exists a neighborhood U of p and a local diffeomorphism $\Phi : U \to T_p(M)$ with $\Phi(p) = \theta$, such that*

$$f \circ \Phi^{-1}(\xi) = f(p) + \frac{1}{2}(f''(p)\xi, \xi)_p \quad \forall \xi \in \Phi(U) ,$$

where $(,)_p$ is the Riemannian structure at p.

From the homology point of view, for a pair of topological spaces (A, B) with $B \subset A$, let $H_q(A, B, G)$ be the qth relative homology group with coefficient group G, $q = 0, 1, 2, \ldots$. They are homotopically invariant.

According to the Morse lemma, after a linear homeomorphism, roughly speaking, we may assume that f is a quadratic function in a neighborhood $B_\epsilon(\theta)$ of θ in $T_p(M)$ of the form:

$$f(x) = c + \frac{1}{2}(\| x_+ \|^2 - \| x_- \|^2) ,$$

where $x = x_+ + x_- \in H_\pm$, $\text{dim} H_- = j$, and $R^n = H_+ \oplus H_-$ is an orthogonal decomposition, where $n = \text{dim} M$.

To characterize homologically the nondegenerate critical point p with Morse index j, we have

$$H_q(f_{c+\epsilon}, f_{c-\epsilon}, G) = H_q(B^j, G) = \delta_{qj} G \quad \forall q \in \mathbb{N} .$$

Since p satisfies $f'(p) = \theta$, it is a nondegenerate zero of the mapping $f' : M \longrightarrow T^*(M)$, the index of the Brouwer degree at p is well defined and

$$i(f', p) = \text{sgn}\det(f''(p)) = (-1)^j = \sum_{q=0}^{n} (-1)^q \text{rank} H_q(B^j, G) .$$

If further, p is the only critical point in $f^{-1}[c - \epsilon, c + \epsilon]$, then as a local characterization of f at p, $H_q(f_{c+\epsilon}, f_{c-\epsilon}, G)$, $q = 0, 1, \ldots, n$, is better than the Brouwer index.

We are encouraged by using the homotopy invariance $\{H_q(f_b, f_a; G)\}_{q=0}^n$ (for $a < b$) as a replacement for $\deg(f', f^{-1}(a, b), \theta)$.

Unfortunately there is no additivity (but only the subadditivity, see later). In the degree theory, the excision property and then the Kronecker existence, which makes it useful in the study of fixed points, are based on the additivity; while for relative homology groups, the excision property is related to a deformation argument. In this context, we shall establish the counterpart of the Kronecker existence theorem by deformation as follows:

Theorem 5.1.2 *(Nontrivial interval theorem) If $\exists q \in \mathbb{N}$ and $\exists a < b$ such that $H_q(f_b, f_a; G)$ is nontrivial, then $K \cap f^{-1}[a,b] \neq \emptyset$.*

One proves the theorem by contradiction, i.e., if there is no critical value in the interval $[a,b]$, then $f_b \sim f_a$. In fact, a homotopy between f_b and f_a can be easily constructed by the negative gradient flow:

$$\dot{x}(t) = -f(x(t)) \quad \forall x \in f^{-1}[a,b]\ .$$

Noticing that along the flow line f is strictly decreasing and that $f^{-1}[a,b]$ is compact, the proof is left to readers as an exercise.

We shall present Morse theory in this book as a topological tool in the study of the existence and the multiplicity of solutions of certain nonlinear differential equations with variational structures.

To our purpose the two assumptions: the compactness of manifolds and the nondegeneracy of functions, are too restricted. In dealing with infinite-dimensional manifolds, a certain kind of compactness is assumed on the function f, e.g., the Palais–Smale condition. Under this condition, deformation theorems are derived, which implies the noncritical interval theorem. Let us use singular relative homology groups with an Abelian coefficient group G, $H_*(X,Y;G)$ to describe the topological difference between the topological spaces X and Y, with $Y \subset X$. Namely, if $H_*(f_b, f_a; G)$ is not trivial, then $K \cap f^{-1}[a,b] \neq \emptyset$. This is the main result in Sect. 5.1.2.

Instead of the nondegeneracy, we study a certain kind of isolatedness of critical points. A series of critical groups is introduced to replace the Morse index in describing the local behavior of isolated critical points. The basic properties and computations of critical groups are studied, e.g., the shifting theorem and that for mountain pass points. The local theory is described in Sect. 5.1.3.

The Morse relation, which is the subadditivity of topological invariances $H_q(f_b, f_a; G), q = 0, 1, 2, \ldots$, links, on one hand, the global invariants: $H_*(f_b, f_a; G), * = 0, 1, \ldots$, and on the other hand, the local invariants: the critical groups for isolated critical points in $f^{-1}[a,b]$. It is applied to the estimation of the critical groups for isolated critical points obtained by minimax principles. It is also used to set up the relationship between the Leray–Schauder index and the critical groups for an isolated zero of a potential compact vector field, by which, we see that for a potential compact vector field, critical groups provide more information than the Leray–Schauder index. This is the contents of Sect. 5.1.4.

5.1.2 Deformation Theorem

The following terminologies are used:

Let X be a topological space. A deformation of X is a continuous map $\eta : X \times [0,1] \to X$ such that $\eta(\cdot, 0) = \mathrm{id}$.

Definition 5.1.3 *(Deformation retract) Given a topological pair $Y \subset X$. A continuous map $r : X \to Y$ is called a deformation retract, if $r \circ i = id_Y$ and $i \circ r \sim id_X$, where $i : Y \to X$ is the injection. In this case Y is called a deformation retraction of X.*

Definition 5.1.4 *(Strong deformation retract) A deformation retract r is called a strong deformation retract, if there exists a deformation $\eta : X \times [0,1] \to X$, such that $\eta(\cdot, t)|_Y = id_Y \ \forall t \in [0,1]$ and $\eta(\cdot, 1) = i \circ r$. Then Y is called a strong deformation retraction of X.*

According to the homotopy invariance of the relative singular homology group $H_*(X, Y; G)$ (where G is the coefficient group), if Y is a strong deformation retraction of X, then $H_*(X, Y; G) = 0$.

In the following, M is assumed to be a Banach–Finsler manifold with a Finsler structure $\| \cdot \|$. But for pedagogical reasons, all of the following theorems are proved only on Banach spaces. Readers who are familiar with the background material on infinite-dimensional manifolds may complete the proofs themselves or may refer to [Ch 1].

As we have seen in Sect. 5.1.1, the deformation is constructed by negative gradient flow on Riemannian manifolds. However, if the C^1 function f is defined on a Banach space M, $f'(x) \in M^*$, the dual space of M, then the gradient flow does not make sense. We introduce:

Definition 5.1.5 *(Pseudo-gradient vector field) Let M be a Banach–Finsler manifold and $f \in C^1(M, R^1)$. Let K be the critical set of f. A vector field X over M is called a pseudo-gradient if*

(1) $\| X(p) \| < 2|f'(p)|$ and
(2) $\langle f'(p), X(p) \rangle > |f'(p)|^2, \ \forall p \in \widetilde{M} := M \backslash K$,

where $\langle , \rangle$ is the duality between $T^*(M)$ and $T(M)$, $\| \cdot \|$ and $| \cdot |$ stand for the Finsler structures on $T(M)$ and $T^*(M)$, respectively.

Our main results of this section are the following:

Theorem 5.1.6 *(Noncritical interval theorem) If $f \in C^1(M, R^1)$ satisfies $(PS)_c \ \forall c \in [a,b]$ and if $K \cap f^{-1}[a,b] = \emptyset$, then f_a is a strong deformation retraction of f_b.*

As a direct consequence, Theorem 5.1.2 is extended to Banach–Finsler manifolds for C^1 functions satisfying the $(PS)_c$ condition $\forall c \in [a,b]$.

Theorem 5.1.7 *(Second deformation theorem) If $f \in C^1(M, R^1)$ satisfies $(PS)_c \ \forall c \in [a,b]$ and M is C^2. If $K \cap f^{-1}(a,b] = \emptyset$ and the connected components of $K \cap f^{-1}(a)$ are only isolated points, then f_a is a strong deformation retraction of f_b.*

We need the following lemmas:

Lemma 5.1.8 *(The existence of a pseudo-gradient vector field) $\forall f \in C^1(M, R^1)$, there exists a continuous pseudo-gradient vector field of f on $\widetilde{M} = M \backslash K$.*

Proof. $\forall p_0 \in \widetilde{M}, \exists \xi(p_0) \in T_{p_0}(M)$ such that $\| \xi(p_0) \| = 1$ and $\langle f'(p_0), \xi(p_0) \rangle > \frac{2}{3}|f'(p_0)|$. Setting $X_{p_0} = \frac{3}{2}|f'(p_0)|\xi(p_0)$, we have

$$\| X_{p_0} \| < 2|f'(p_0)|, \text{ and}$$

$$\langle f'(p_0), X_{p_0} \rangle > |f'(p)|^2 \quad \forall p \in V_{p_0}$$

Since $\widetilde{M}$ is metrizable, it is paracompact. There is a locally finite partition of unity $\{\eta_\beta | \beta \in B\}$, with supp $\eta_\beta \subset V_{p_0}$ for some $p_0 = p_0(\beta) \in \widetilde{M}$. Let

$$X(p) = \sum_{\beta \in B} \eta_\beta(p) X_{p_0(\beta)} .$$

This is a required pseudo-gradient vector field. □

Lemma 5.1.9 *If f satisfies $(PS)_c$ $\forall c \in [a, b]$, and if $K \cap f^{-1}[a, b] = \emptyset$, then $\exists \epsilon, \delta > 0$ such that*

$$|f'(x)| \geq \epsilon \quad \forall x \in f^{-1}[a - \delta, b + \delta] .$$

Proof. If not, then there exists a sequence $\{x_n\} \subset f^{-1}[a - \frac{1}{n}, b + \frac{1}{n}]$ such that $|f'(x_n)| < \frac{1}{n}$. We find a subsequence $\{x'_n\}$ such that $f(x'_n) \to c \in [a, b]$ and $f'(x'_n) \to \theta$. According to $(\mathrm{PS})_c$, we find $x^* \in f^{-1}(c) \cap K$. This is a contradiction. □

Proof. (Proof of the nontrivial interval theorem) The deformation is constructed by the flow, which deforms the level sets:

$$\begin{cases} \dot{\sigma}(t) = -X(\sigma(t)) \\ \sigma(0) = x_0 \in f^{-1}[a, b] . \end{cases}$$

1. It is a decreasing flow:

$$\frac{d}{dt} f(\sigma(t)) = \langle f'(\sigma(t)), \dot{\sigma}(t) \rangle \leq -|f'(\sigma(t))|^2 .$$

2. The flow does not stop until it arrives at f_a. Indeed, if the maximal existence time for the initial data x_0 is T_{x_0} and $f(\sigma(T_{x_0}) - 0) > a$, from Lemma 5.1.9:

$$a - b \leq f(\sigma(t)) - f(x_0) = \int_0^t \langle f'(\sigma(\iota)), \dot{\sigma}(\iota) \rangle d\iota < -\epsilon^2 t \ \ \forall t < T_{x_0} ,$$

then $T_{x_0} < \frac{b-a}{\epsilon^2}$, and

$$d(\sigma(t_2), \sigma(t_1)) \leq \int_{t_1}^{t_2} \| \dot{\sigma}(t) \| \, dt \leq 2 \int_{t_1}^{t_2} |f'(\sigma(t))| dt$$
$$\leq 2\sqrt{(b-a)(t_2 - t_1)}, \forall t_1, t_2 \in (0, T_{x_0}) \, .$$

It implies that $\sigma(T_{x_0})$ exists and is not a critical point. Therefore the flow is extendible beyond T_{x_0}. It contradicts the maximality of T_{x_0}.

3. The arrival time T_x is a continuous function of x. Indeed, $t = T_x$ solves the equation $f(\sigma(t)) = a$. Since

$$\frac{d}{dt} f(\sigma(t))|_{t=T_x} = \langle f'(\sigma(T_x)), \dot{\sigma}(T_x) \rangle < -\epsilon^2 \, ,$$

the continuity follows from the implicit function theorem.

4. Define $\eta : f_b \times [0,1] \to f_b$ by

$$\eta(x,t) = \begin{cases} \sigma(x, T_x t) & x \in f^{-1}[a,b] \\ x & x \in f_a \, , \end{cases}$$

where $\sigma(x,t)$ denotes the flow with initial data x.

This is a strong deformation retract. □

Proof. (Proof of the second deformation theorem) We define the pseudo-gradient flow as before:

$$\begin{cases} \dot{\sigma}(t,x) = -\frac{X(\sigma(t,x))}{\|X(\sigma(t,x))\|^2} \\ \sigma(0,x) = x \in f^{-1}[a,b] \backslash K_b \, . \end{cases}$$

In the same manner, we show that $\forall x \in f^{-1}(a,b] \backslash K_b$, there exists the arrival time $T_x > 0$ such that $\lim_{t \to T_x - 0} f \circ \sigma(t,x) = a$.

1. We claim that $\lim_{t \to T_x - 0} \sigma(t,x)$ does exist and then

$$f(\sigma(T_x - 0, x)) = a \, .$$

Indeed, $(\mathrm{PS})_a$ implies that K_a is compact. Either one of the following cases holds:

$$\text{(a)} \inf_{t \in [0, T_x)} \operatorname{dist}(\sigma(t,x), K_a) > 0 \, ,$$
$$\text{(b)} \inf_{t \in [0, T_x)} \operatorname{dist}(\sigma(t,x), K_a) = 0 \, .$$

In case (a), again by $(\mathrm{PS})_c$ $\forall c \in [a,b]$, $\exists \alpha > 0$ such that

$$\inf_{t \in [0, T_x)} \| f'(\sigma(t,x)) \| \geq \alpha \, .$$

Thus

$$\begin{aligned}\operatorname{dist}(\sigma(t_2,x),\sigma(t_1,x)) &\le \int_{t_1}^{t_2} \left|\left|\frac{d\sigma}{dt}\right|\right| dt\\ &\le \int_{t_1}^{t_2} \frac{dt}{\| X(\sigma(t,x)) \|} \le \int_{t_1}^{t_2} \frac{dt}{\| f'(\sigma(t,x)) \|} \le \frac{|t_2-t_1|}{\alpha} .\end{aligned}$$

Since T_x is finite, $\lim_{t\to T_x-0} \sigma(t,x) = z.$

In case (b), we shall prove that $\exists z \in K_a$ such that

$$\lim_{t\to T_x-0} \sigma(t,x) = z .$$

First, we claim that

$$\lim_{t\to T_x-0} \operatorname{dist}(\sigma(t,x),K_a) = 0 . \tag{5.1}$$

If not, $\exists \epsilon_0 > 0, \exists t_i \to T_x - 0$, such that

$$\operatorname{dist}(\sigma(t_i,x),K_a) \ge \epsilon_0 .$$

By the assumption (b), $\exists t_i' \to T_x - 0$ such that

$$\lim_{i\to\infty} \operatorname{dist}(\sigma(t_i',x),K_a) = 0 .$$

Thus we have two sequences $t_i^* < t_i^{**}$ both converging to T_x such that

$$\begin{aligned}\operatorname{dist}(\sigma(t_i^*,x),K_a) &= \frac{\epsilon_0}{2} ,\\ \operatorname{dist}(\sigma(t_i^{**},x),K_a) &= \epsilon_0 ,\end{aligned}$$

and

$$\sigma(t,x) \in \overline{(K_a)}_{\epsilon_0} \backslash (K_a)^{\circ}_{\frac{\epsilon_0}{2}} \ \ \forall t \in [t_i^*, t_i^{**}] ,$$

where $(K_a)_\delta$ denotes the δ-neighborhood of K_a.

Again by $(\mathrm{PS})_c$ $\forall c \in [a,b]$, we have

$$\inf_{t\in[t_i^*,t_i^{**}]} \| f'(\sigma(t,x)) \| \ge \alpha > 0 .$$

Therefore

$$\begin{aligned}\frac{\epsilon_0}{2} &\le \operatorname{dist}(\sigma(t_i^{**},x),\sigma(t_i^*,x))\\ &\le \int_{t_t^*}^{t_i^{**}} \| \frac{d\sigma}{dt} \| \, dt\\ &\le \frac{1}{\alpha}|t_i^{**} - t_i^*| \to 0 .\end{aligned}$$

This is a contradiction, (5.1) is proved.

It follows that

$$\lim_{t \to T_x - 0} f'(\sigma(t,x)) = \theta ,$$

from the compactness of K_a.

The $(\text{PS})_a$ condition then implies that the limit set A of the orbit $\{\sigma(t,x) | t \in [0, T_x)\}$ is nonempty and that, for each sequence $t_i \to t_{x-0}$, there exists a subsequence $\widetilde{t_i}$ such that $\sigma(\widetilde{t_i}, x)$ is convergent.

Next we prove that A is a compact connected subset of K_a, and then a single point, according to the assumption that the connected components of K_a are isolated points. The compactness of A is obvious. We only want to prove the connectedness. If not, $\exists$ open subsets O and O' such that

$$O \cap O' = \emptyset, \;\; A = (O \cap A) \cup (O' \cap A) ,$$

and $O \cap A \neq \emptyset, \;\; O' \cap A \neq \emptyset$.

Choosing $z \in O \cap A, \;\; z' \in O' \cap A, \;\; \exists t_i \to T_x - 0, \;\; t_i' \to T_x - 0$ satisfying $\sigma(t_i, x) \to z, \;\; \sigma(t_i', x) \to z'$. For large i, we have

$$\sigma(t_i, x) \in O, \;\; \sigma(t_i', x) \in O' ,$$

so $\exists t_i^* \in [t_i, t_i']$ (or $[t_i', t_i]$) such that

$$\sigma(t_i^*, x) \notin O \cup O' .$$

Denote the limit of $\sigma(t_i^*, x)$ by z^*. (It exists because of $(\text{PS})_a$.) So $z^* \in A$, but $z^* \notin O \cup O'$.

This is a contradiction. We conclude that $A = \{z\} \subset K_a$. In both cases, claim 1 is proved.

2. We shall prove the continuity of the function T_x.

As in the proof of Theorem 5.1.6, if $\sigma(T_{x_0} - 0, x_0) \notin K_a$, then the function T_x is continuous at x_0. So we restrict ourselves to the case $z = \sigma(T_{x_0} - 0, x_0) \in K_a$.

If T_x is not continuous at such a x_0, then $\exists \epsilon_0 > 0, \exists x_n \to x_0$ such that

$$|T_{x_n} - T_{x_0}| \geq \epsilon_0 ,$$

so there exists a subsequence

$$\text{either } T_{x_n} \leq T_{x_0} - \epsilon_0, \text{ or } T_{x_n} \geq T_{x_0} + \epsilon_0 .$$

Since

$$\begin{aligned} f \circ \sigma(T_x - \epsilon, x) - f \circ \sigma(t, x) &= \int_t^{T_x - \epsilon} f'(\sigma(t,x)) dt \\ &\leq -\frac{1}{4}(T_x - \epsilon - t) , \end{aligned}$$

we have

$$f \circ \sigma(t,x) \geq a + \frac{1}{4}(T_x - \epsilon - t)\,.$$

But for any fixed $\epsilon > 0$, according to the continuous dependence of the initial data ODE,

$$\mathrm{dist}(\sigma(T_{x_0} - \epsilon, x_n), \sigma(T_{x_0} - \epsilon, x_0)) \to 0\,.$$

If $T_{x_n} \geq T_{x_0} + \epsilon_0$, then we have

$$\begin{aligned} f \circ \sigma(T_{x_0} - \epsilon, x - 0) &= \lim_{n\to\infty} f \circ \sigma(T_{x_0} - \epsilon, x_n) \\ &\geq \lim_{n\to\infty} \left[a + \frac{1}{4}(T_{x_n} - T_{x_0} + \epsilon)\right] \\ &\geq a + \frac{1}{4}(\epsilon_0 + \epsilon)\,. \end{aligned}$$

Letting $\epsilon \to 0$, we obtain

$$a \geq a + \frac{\epsilon_0}{4}\,.$$

This is a contradiction.

Similarly, we prove that $T_{x_n} \leq T_{x_0} - \epsilon_0$ is impossible. Therefore, T_x is continuous.

3. Finally, we define the deformation retract as follows:

$$\eta(t,x) = \begin{cases} x & \text{if } (t,x) \in [0,1] \times f_a\,, \\ \sigma(T_x t, x) & \text{if } (t,x) \in [0,1) \times (f_b \backslash (f_a \cup K_b))\,, \\ \sigma(T_x - 0, x) & \text{if } (t,x) \in \{1\} \times (f_b \backslash (f_a \cup K_b))\,. \end{cases}$$

Claim. Only the continuity of η has to be verified. Four cases are distinguished:

(a) $(t,x) \in [0,1] \times \overset{\circ}{f}_a$,
(b) $(t,x) \in [0,1) \times (f^{-1}(a,b] \backslash K_b)$,
(c) $(t,x) \in \{1\} \times (f^{-1}(a,b] \backslash K_b)$,
(d) $(t,x) \in [0,1] \times f^{-1}(a)$.

Only cases (c) and (d) have to be verified. Since their proofs are similar, we only give the verification for (c).

If η is discontinuous at $(1, x_0)$, then $\exists \epsilon > 0$, $\exists t_n \to T_{x_0} - 0$, and $\exists x_n \to x_0$ such that

$$\mathrm{dist}(\sigma(t_n, x_n), \sigma(T_{x_0} - 0, x_0)) \geq \epsilon\,.$$

Let $z = \sigma(T_{x_0} - 0, x_0) (\in K_a)$, and let

$$F_1 = \{z\},\ F_2 = (M \backslash \overset{\circ}{B}_\epsilon(z)) \cap K_a\,.$$

Both F_1 and F_2 are compact subsets of K_a. Provided by the assumption of K_a, and Lemma 3.5.2, we have compact subsets $K_1,\ K_2 \subset K_a$ such that

$K_1 \cap K_2 = \emptyset,\ F_i \subset K_i,\ i = 1, 2$, and $K_1 \cup K_2 = K_a$. Obviously, we may take $K_1 \subset \overset{\circ}{B}_\epsilon(z)$. Let $N = K_2 \cup (M \backslash \overset{\circ}{B}_\epsilon(z))$, then

$$\alpha = \text{dist}(N, K_1) > 0 .$$

The continuity of the flow as well as of the arrival time T_x implies that $\exists \delta > 0$ such that

$$\text{dist}(\sigma(t, x_0), \sigma(T_{x_0} - 0, x_0)) < \frac{\alpha}{8}$$

if $t \in [T_{x_0} - \delta, T_{x_0}]$, and $\exists \delta_1 > 0$ such that

$$T_x > T_{x_0} - \delta$$

if $x \in B_\delta(x_0)$.

For $t \in [T_{x_0} - \delta, T_{x_0}) \cap [T_{x_0} - \delta, T_x),\ x \in B_{\delta_1}(x_0),\ \exists t \in (0, \delta)$ such that

$$\text{dist}(\sigma(t, x), \sigma(t, x_0)) < \frac{\alpha}{8} .$$

In summary, for such a t, and for any $x \in B_{\delta_t}(x_0)$,

$$\sigma(t, x) \in (K_1)_{\frac{\alpha}{4}},\ \text{the } \frac{\alpha}{4} \text{ neighborhood of } K_1 .$$

For large $n,\ x_n \in B_{\delta_t}(x_0)$, let t'_n be such a t, satisfying

$$\sigma(t'_n, x_n) \in (K_1)_{\frac{\alpha}{4}} .$$

Reducing δ, and repeating the above procedure, we obtain subsequences $\{x_n, t'_n, t_n\}$ such that

$$\begin{cases} t'_n, t_n \to T_{x_0} - 0 \\ \sigma(t'_n, x_n) \in (K_1)_{\frac{\alpha}{4}} \\ \sigma(t_n, x_n) \notin B_\epsilon(z) . \end{cases}$$

We may assume $t'_n < t_n$; then we have $\overline{t'_n},\ \overline{t_n}$ such that

$$t'_n \le \overline{t'_n} < \overline{t_n} \le t_n$$

and

$$\begin{aligned} &\sigma(\overline{t'_n}, x_n) \in \partial[(K_1)_{\frac{\alpha}{4}}] \\ &\sigma(\overline{t_n}, x_n) \in \partial[(N)_{\frac{\alpha}{4}}] \\ &\sigma(t, x_n) \notin (K_1)_{\frac{\alpha}{4}} \cup (N)_{\frac{\alpha}{4}}\ \forall t \in [\overline{t'_n}, \overline{t_n}] , \end{aligned}$$

$n = 1, 2, \ldots$.

According to $(\text{PS})_c$, $\forall c \in [a, b]$

$$\beta = \inf \{\| f'(x) \| \,|x \in f^{-1}[a, b] \backslash ((K_1)_{\frac{\alpha}{4}} \cup (N)_{\frac{\alpha}{4}})\} > 0 .$$

Therefore

$$\frac{\alpha}{2} \le \text{dist}(\sigma(\overline{t'_n}, x_n), \sigma(\overline{t_n}, x_n)) \le \int_{\overline{t'_n}}^{\overline{t_n}} \| \frac{d\sigma}{dt} \| \,dt \le \frac{1}{\beta} |\overline{t_n} - \overline{t'_n}| \to 0 .$$

This is a contradiction. The proof is completed. □

Remark 5.1.10 *A flow on a metric space M is a continuous map $\eta : M \times R^1 \to M$ possessing the following properties:*

(i) $\eta(x,0) = x \quad \forall x \in M,$
(ii) $\eta(\eta(x,t_1),t_2) = \eta(x,t_1+t_2) \quad \forall t_1,t_2 \in R^1, \ \forall x \in M.$

A set $A \subset M$ is said to be (positively) invariant with respect to η, if

$$\eta(x,t) \in A \quad \forall x \in A, \ \forall t \in R^1 \ (\forall t \in R^1_+) \,.$$

By definition, for any subset A, $\widetilde{A} = \underset{t\in R^1}{\cup} \eta(A,t)$ $(A^+ = \underset{t\in R^1_+}{\cup} \eta(A,t))$ is the smallest η- (positively) invariant set containing A.

With these terminologies, it is easy to extend Theorems 5.1.6 and 5.1.7 to pseudo-gradient flow invariant sets. Namely, let η be a pseudo-gradient flow for $f \in C^1(M,R^1)$ satisfying $(\mathrm{PS})_c$ $\forall c \in [a,b]$, and let A be a η-positively invariant set.

If $K \cap f^{-1}(a,b] \cap A = \emptyset$, and the connected components of $K \cap f^{-1}(a) \cap A$ are only isolated points, then $f_a \cap A$ is a strong deformation retraction of $f_b \cap A$.

5.1.3 Critical Groups

Definition 5.1.11 *Let f be a C^1 function defined on M, let p be an isolated critical point f, and let $c = f(p)$.*

$$C_q(f,p) = H_q(f_c \cap U, (f_c \backslash \{p\}) \cap U; G)$$

is called the qth critical group of f at $p, q = 0,1,2,\ldots,$ where U is an isolated neighborhood of p, i.e., $K \cap U = \{p\}$.

According to the excision property of the singular homology theory, the critical groups are well defined, i.e., they do not depend on the special choice of U.

From the definition, we have:

Example 1. If p is an isolated minimum point of f, then

$$C_q(f,p) = \delta_{q0} \cdot G \,.$$

Under some additional conditions, the converse of the above statement is true. Namely: If $f \in C^{2-0}(M,R^1)$ satisfies (PS), and if p is an isolated critical point, which is not a local minimum point, then

$$C_0(f,p) = 0 \,.$$

See [Ch 5].

Example 2. If M is n-dimensional, and p is an isolated local maximum point of f, then

$$C_q(f,p) = \delta_{qn} \cdot G \, .$$

In the following, we assume that M is a Hilbert–Riemannian manifold. The Morse lemma, which is a special case of the following more general decomposition theorem – the splitting theorem – proved later, is a cornerstone in studying the local behavior of a nondegenerate critical point.

Definition 5.1.12 *Let p be a nondegenerate critical point of f, we call the dimension of the negative space corresponding to the spectral decomposition of $f''(p)$, the Morse index of p, and denote it by $ind(f,p)$.*

Example 3. Suppose that $f \in C^2(M, R^1)$ and p is a nondegenerate critical point of f with Morse index j; then

$$C_q(f,p) = \delta_{qj} \cdot G \, .$$

Proof. According to the Morse lemma, after a linear homeomorphism, we may assume that f is a quadratic function on a Hilbert space H of the form:

$$f(x) = \frac{1}{2}(\| x_+ \|^2 - \| x_- \|^2) \, ,$$

where $x = x_+ + x_-$, $x_\pm \in H_\pm$, and $H = H_+ \oplus H_-$ is an orthogonal decomposition.

Let $U = B_\epsilon$ be the ϵ-ball centered at θ:

$$B_\epsilon \cap f_0 = \{x \in B_\epsilon | \; \| x_+ \| \le \| x_- \| \} \, .$$

Define

$$\eta(x,t) = x_- + t x_+ \quad \forall (x,t) \in (B_\epsilon \cap f_0) \times [0,1] \, .$$

It is a strong deformation retract from $(B_\epsilon \cap f_0, B_\epsilon \cap (f_0 \backslash \{\theta\}))$ to $(H_- \cap B_\epsilon, (H_- \backslash \{\theta\}) \cap B_\epsilon)$. Thus

$$\begin{aligned} C_q(f,q) &\cong H_q(H_- \cap B_\epsilon, (H_- \backslash \{\theta\}) \cap B_\epsilon) \\ &\cong H_q(B^j, S^{j-1}) = \delta_{qj} \cdot G \, , \end{aligned}$$

if $j < +\infty$. Nevertheless, for $j = +\infty$, since S^∞ is contractible, again, we have

$$C_q(f,p) \cong 0 \, .$$

The conclusion is proved. □

Theorem 5.1.13 *(Splitting) Suppose that U is a neighborhood of θ in a Hilbert space H, and that $f \in C^2(U, R^1)$. Assume that θ is the only critical point of f, and that $A = f''(\theta)$ is a Fredholm operator. Then there exist a ball $B \subset U$ centered at θ, an origin-preserving local homeomorphism φ defined on B, and a C^1 map $h : B \cap N \to N^\perp$, where $N = ker(A)$ such that*

$$f \circ \varphi(z+y) = \frac{1}{2}(Az, z) + f(y + h(y)) \quad \forall x \in B\,,$$

where $y \in N$ *and* $z \in N^\perp$, *with* $x = y + z$.

Proof. • Decomposing H into $N \oplus N^\perp$, let P be the orthogonal projection onto $N^\perp$. We solve the equation for fixing $y \in B \cap N$:

$$Pf'(y+z) = \theta_1 \quad (\theta_1 \text{ and } \theta_2 \text{ are the origins in } N^\perp \text{ and } N \text{ resp.})\,.$$

Since $f'(\theta_1+\theta_2) = \theta_1$, and $f''(\theta_1+\theta_2) = A$, by the implicit function theorem, there exist a ball B and $h : B \cap N \to N^\perp$ satisfying

$$Pf'(y + h(y)) = \theta_1 \quad \forall y \in B \cap N\,.$$

• Setting $u = z - h(y)$, and letting

$$\begin{cases} F(u,y) = f(z+y) - f(h(y)+y)\,, \\ F_2(u) = \frac{1}{2}(Au, u)\,, \end{cases}$$

we have

$$\begin{aligned} F(\theta_1, y) &= 0\,, \\ F'_u(\theta_1, y) = Pf'(y+h(y)) &= \theta_1\,, \\ F''_u(\theta_1,\theta_2) = Pf''(\theta) &= A|_{N^\perp}\,. \end{aligned}$$

Define $\xi : (u,y) \to u_0 \in F_2^{-1} \circ F(u,y) \cap \{\eta(u,t) | |t| <\| u \|\}$, where η is the flow defined by the following ODE:

$$\begin{cases} \dot\eta(s) = -A\eta(s)/ \parallel A\eta(s) \parallel\,, \\ \eta(0) = u\,. \end{cases}$$

Claim: η is well defined for $|t| <\| u \|$.

Indeed, $\| \eta(s) - u \| \le |s|$, we have $\| \eta(u,t) \| \ge \| u \| - |t|$.

Noticing that $\eta(u,t) \in N^\perp$, the denominator of the vector field does not vanish for $|t| <\| u \|$.

If there exists a unique $\bar t(u,y) \in (0, \| u \|)$ such that

$$F_2(\eta(u, \bar t(u,y))) = F(u,y)\,,$$

for $u \ne \theta_1$, we set

$$\xi(u,y) = \begin{cases} \theta_1, & u = \theta_1\,, \\ \eta(u, \bar t(u,y)), & u \ne \theta_1\,, \end{cases}$$

then $(\xi(u,y), y) \longmapsto (z,y)$ defines a local homeomorphism φ.

Indeed, from the implicit function theorem, $\bar t(u,y)$ is continuous provided that

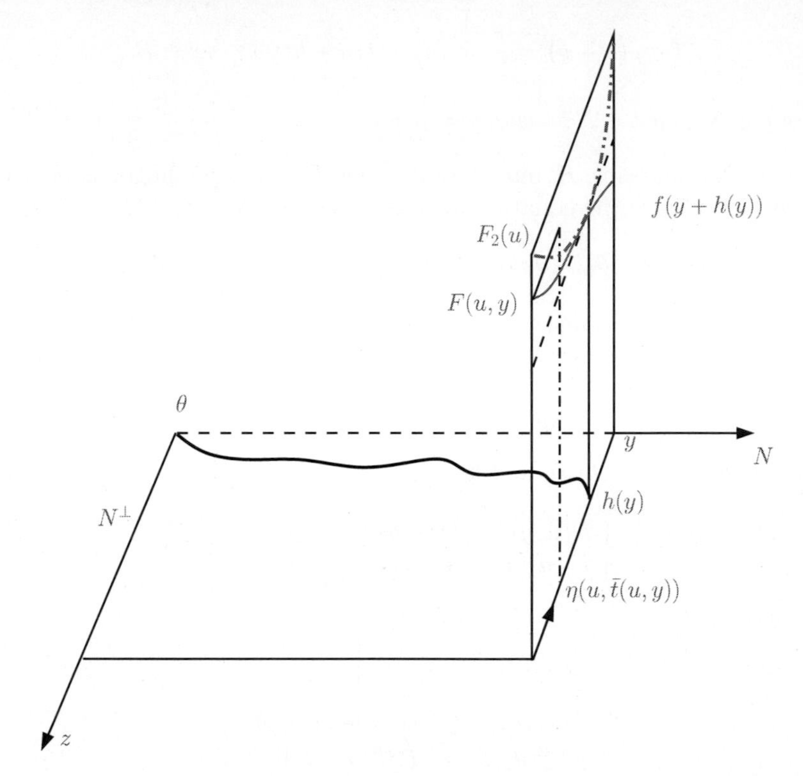

Fig. 5.1.

$$\frac{\partial}{\partial t} F_2 \circ \eta(u, \bar{t}) = - \parallel A\eta(u, \bar{t}) \parallel \neq 0 \ ,$$

if $u \neq \theta_1$.

It is easy to verify that φ is a local homeomorphism.

• We verify the existence and the uniqueness of $\bar{t}(u, y)$ in a few steps. Let $B_1 = B \cap N^{\perp}$ and $B_2 = B \cap N$.

1. For any $\epsilon > 0$, there exists suitable B, such that

$$\begin{aligned} |F(u, y) - F_2(u)| &= |F(u, y) - F(\theta_1, y) - (F'_u(\theta_1, y), u) - F_2(u)| \\ &= | \int_0^1 (1 - t)((F''_u(tu, y) - F''_u(\theta_1, \theta_2))u, u)dt| \\ &< \epsilon \parallel u \parallel^2 , \end{aligned}$$

for $(u, y) \in B_1 \times B_2$.

2.

$$\begin{aligned} |F_2(\eta(u, t)) - F_2(u)| &= \left| \int_0^t \frac{d}{ds} F_2(\eta(u, s))ds \right| \\ &= \left| \int_0^t (F'_2(\eta), \dot{\eta})ds \right| \end{aligned}$$

$$\begin{aligned} &= \int_0^{|t|} \| A\eta(u,s) \| \, ds \\ &\geq C \int_0^{|t|} \| \eta(u,s) \| \, ds \\ &\geq C \left(\| u \| \, |t| - \frac{t^2}{2} \right) , \end{aligned}$$

where $C > 0$ is a constant determined by the spectrum of A.

$F_2(\eta(u,t))$, as a function of t, is strictly decreasing on $(- \| u \|, \| u \|)$, and $F_2(\eta(u,-t)) > F(u,y) > F_2(\eta(u,t))$ holds for

$$\left(1 - \sqrt{1 - \frac{2\epsilon}{C}} \right) \| u \| \leq t \leq \| u \| .$$

We conclude the existence and the uniqueness of the function $\bar{t}(u,y)$, with

$$|\bar{t}(u,y)| \leq \left(1 - \sqrt{1 - \frac{2\epsilon}{C}} \right) \| u \| .$$

□

Remark 5.1.14 *In the case $N = \{0\}$, the Morse lemma is a consequence of the above theorem.* It remains to verify that φ is a diffeomophism. We omit the verification, in fact, a weaker version is sufficient to prove Example 3.

Remark 5.1.15 *In the study of nonlinear PDEs, sometimes we work on Banach spaces rather than Hilbert spaces. We note that the above theorem can be applied under the following assumptions:*

Let X be a Banach space, embedded continuously into a Hilbert space H as a dense linear subspace. Assume:

$(A1)$ There are Banach spaces as follows:

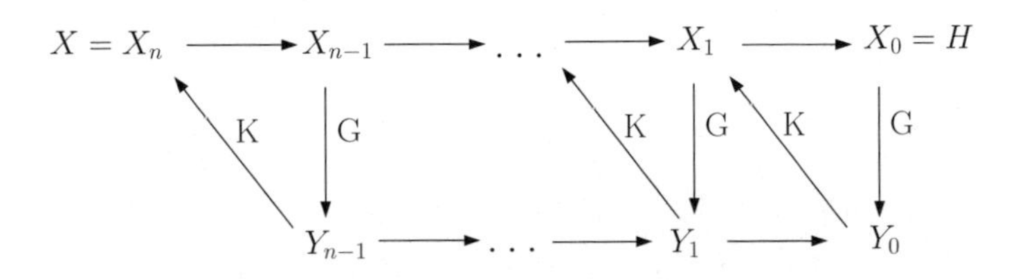

Fig. 5.2.

$(A2)$ $f \in C^2(H, \mathbb{R}^1)$ has a derivative of the form

$$f'(x) = x - K \cdot G(x) ,$$

where $K : Y_i \to X_{i+1}$ is linear and continuous, and $G : X_i \to Y_i$ is continuous, $i = 0, 1, \ldots, n-1$.

Let $A = I - K \cdot G'(\theta)$, and let H_*, $* = -, 0, +$, be the negative, null, and positive spaces resp., according to the spectral decomposition of A.

$(A3)$ $H_0 \subset X$.

A map is called regular, if it maps X into X.

We claim that the splitting theorem remains true for the Banach space X if assumptions $(A_1), (A_2)$ and (A_3) hold.

To this end, we continue to use the notations φ, ξ, η, h and B of Theorem 5.1.13, and prove that φ is regular. It is sufficient to show that the map $\psi : (u, y) \longmapsto (\xi(u, y), y)$ is regular, and that $\mathrm{Im}\, h \subset X$.

Indeed, the flow η is a reparametrization of the following flow ζ:

$$\begin{cases} \dot{\zeta}(s) = -\zeta(s) + K \cdot G'(\theta)\zeta(s) \\ \zeta(0) = u \,. \end{cases}$$

Or equivalently,

$$\zeta(t) = e^{-t}u + K \cdot G'(\theta) \int_0^t e^{t-s}\zeta(s)ds \,.$$

By a bootstrap procedure, we have $\zeta \in X$ whenever $u \in X \cap B$. This shows the regularity of ψ.

Again, as an element in H, $z = h(y)$ satisfies the equation:

$$Pf'(y + z) = \theta \,,$$

i.e.,

$$h(y) = -PK \cdot G(y + h(y)) + (I - P)h(y) \,.$$

Noticing that $(I - P)H = H_0 \subset X$, P is regular. By the same reason, $PX_i \subset X_i$, $i = 0, 1, \ldots, n$. After a bootstrap iteration, it follows that $\mathrm{Im}\, h \subset X$.

Theorem 5.1.16 *Under the assumptions of the splitting theorem,* $(A1)$, $(A2)$ *and* $(A3)$, *we have*

$$C_q(f, \theta) = C_q(f|_X, \theta) \,.$$

Proof. We use φ and B as in the splitting theorem, and choose $U = \varphi(B)$. Define a deformation $\eta : U \times [0, 1] \to U$ as follows:

$$\eta(x, t) = \varphi(tx_+ + x_-) \,,$$

where $x_+ + x_- = \varphi^{-1}(x)$, and $x_+ \in H_+$, $x_- \in H_- \oplus H_0$. Then η is a deformation retract $f_c \cap U \to f_c \cap U_-$, where $U_- = \varphi(B \cap (H_- \oplus H_0))$. Therefore

$$C_*(f, \theta) = H_*(f_c \cap U, f_c \cap U \backslash \{\theta\})$$

$$= H_*(f_c \cap U_-, f_c \cap U_- \backslash \{\theta\}) ,$$

where $c = f(\theta)$.

Since φ is regular, we also have

$$C_*(\hat{f}, \theta) = H_*(\hat{f}_c \cap U_-, \hat{f}_c \cap U_- \backslash \{\theta\}) ,$$

where $\hat{f} = f|_X$. However, $f_c \cap U_- = \hat{f}_c \cap U_-$ from $(A3)$. This proves our conclusion. □

As a consequence of the splitting theorem, we have:

Theorem 5.1.17 *(Shifting) Assume that the Morse index of f at an isolated critical point p is $j < +\infty$. Under the assumptions of the splitting theorem, we have*

$$C_q(f, p) = C_{q-j}(\widetilde{f}, \theta) ,$$

where $\widetilde{f}(y) = f(y + h(y))$ as in the splitting theorem.

Proof. We may assume that f is of the form $f(z + y) = f_1(z) + f(y + h(y))$, where $f_1(z) = \frac{1}{2}(\| z_+ \|^2 - \| z_- \|^2)$, $z = z_+ + z_-$, $(z_+, z_-) \in H_+ \oplus H_- = N^\perp$, and $y \in N$, where the notations are inherited from the splitting theorem. Since $\dim N < \infty$, it is easy to construct a new function f_2 such that f_2 equals $f(y + h(y))$ in a neighborhood of θ in N, and f_2 satisfies the (PS) condition.

Let U_i be an isolated neighborhood of θ_i for f_i in the space $Y_i, i = 1, 2$, where $Y_1 = N^\perp$ and $Y_2 = N$. Suppose $U_i \subset f_i^{-1}[-\epsilon, \epsilon]$, $i = 1, 2$, for some $\epsilon > 0$. Let $W_i = \widetilde{U}_i \cap f_i^{-1}[-\epsilon, \epsilon]$, then W_i is again an isolated neighborhood of θ_i for f_i, where $\widetilde{U}_i$ is the smallest η-invariant set containing U_i, $i = 1, 2$. We have a strong deformation retract ψ deforming $(f_2)_\varepsilon$ to $(f_2)_0$, according to the second deformation theorem. One constructs a deformation as follows:

$$\varphi(z, y, t) = z_- + (1 - t)z_+ + \psi(y, t) .$$

Accordingly,

$$\begin{aligned}
C_*(f, p) &\cong H_*(f_0 \cap (W_1 \times W_2), (f_0 \cap (W_1 \times W_2) \backslash \{(\theta_1, \theta_2)\}); G) \\
&\cong H_*(((f_1)_0) \cap W_1) \times ((f_2)_0 \cap W_2) , \\
&\quad ((f_1)_0) \cap W_1) \times ((f_2)_0 \cap W_2) \backslash \{(\theta_1, \theta_2)\}; G) \\
&\cong H_*((f_1)_0 \cap W_1, (f_1)_0 \cap W_1 \backslash \{\theta_1\}; G) \\
&\quad \otimes H_*((f_2)_0 \cap W_2, (f_2)_0 \cap W_2 \backslash \{\theta_2\}; G) \\
&\cong C_*(f_1, \theta_1) \otimes C_*(f_2, \theta_2)
\end{aligned}$$

Provided by the Kunneth formula, where we omit the verification of the pair $((((f_1)_0 \cap W_1) \backslash \{\theta_1\}) \times ((f_2)_0 \cap W_2),\ ((f_1)_0 \cap W_1) \times ((f_2)_0 \cap W_2) \backslash \{\theta_2\})$ is an excision couple.

By the use of Example 3, we obtain

$$C_q(f, p) = C_{q-j}(f_2, \theta) .$$

Since $f_2(y) = f(y + h(y)) = \widetilde{f}(y)$, this proves our conclusion. □

Corollary 5.1.18 *Under the assumption of the shifting theorem with* $\dim N = k$, *if* p *is:*

(1) a local minimum point of $\widetilde{f}$, *then* $C_q(f,p) = \delta_{qj} \cdot G$;
(2) a local maximum point of $\widetilde{f}$, *then* $C_q(f,p) = \delta_{q(j+k)} \cdot G$;
(3) neither a local minimum point nor a local maximum point, then

$$C_q(f,p) = 0 \text{ for } q \leq j \text{ and } q \geq j+k \ .$$

Definition 5.1.19 *(Mountain pass point) An isolated critical point* p *of* f *is called a mountain pass point if* $C_1(f,p) \neq 0$.

Theorem 5.1.20 *Suppose that* $f \in C^2(M, R^1)$ *has a mountain pass point* p, *and that* $f''(p)$ *is a Fredholm operator satisfying the condition:*

$$(\Phi) \quad f''(p) \geq 0 \text{ and } 0 \in \sigma(f''(p)) \Longrightarrow \dim\ker f''(p) = 1 \ .$$

Then

$$C_q(f,p) = G \cdot \delta_{q1} \ .$$

Proof. Let $j = \operatorname{ind}(f''(p))$.

If p is nondegenerate, then by Example 3: $C_q(f,p) = \delta_{qj} \cdot G$; we have $j = 1$. The conclusion follows.

Otherwise, from the shifting theorem, $C_q(f,p) = C_{q-j}(\widetilde{f},p)$; we obtain $j \leq 1$.

In the case where $j = 1$, $C_0(\widetilde{f},\theta) \neq 0$. θ is a local minimum of f, from Example 1, $C_q(f,p) = C_{q-1}(\widetilde{f},\theta) = \delta_{q1} \cdot G$.

In the case where $j = 0$, $C_1(\widetilde{f},\theta) \neq 0$. Now, $\dim\ker(f''(p)) = 1$, θ is a local maximum. From Example 2, $C_q(f,p) = C_q(\widetilde{f},\theta) = \delta_{q1} \cdot G$.

The proof is complete. □

5.1.4 Global Theory

For a function $f \in C^1(M, R^1)$ satisfying the (PS)_c $\forall c \in [a,b]$, where a, b are regular values, we know from the nontrivial interval theorem that the nontriviality of $H_*(f_b, f_a; G)$ implies the existence of a critical point in $f^{-1}(a,b)$. In this subsection we shall show that the topological invariance $H_*(f_b, f_a; G)$ possesses the homotopy invariance and the subadditivity as follows.

Theorem 5.1.21 *Suppose that* $\{f_\sigma \in C^1(M, R^1) | \sigma \in [0,1]\}$ *is a family of functions satisfying the (PS) condition. Assume that* $a(\sigma)$ *and* $b(\sigma)$ *are two continuous functions defined on* $[0,1]$ *with* $a(\sigma) < b(\sigma)$, *and that both* $a(\sigma)$ *and* $b(\sigma)$ *are regular values of* f_σ, $\forall \sigma \in [0,1]$. *Assume that* $\sigma \mapsto f_\sigma$ *is continuous in* $C^1(M)$ *topology. Then the homology group* $H_*((f_\sigma)_{b(\sigma)}, (f_\sigma)_{a(\sigma)}; G)$ *is independent of* σ.

Proof. For simplifying notations, $\forall \sigma_0, \sigma_1 \in [0,1]$, we write f_{σ_i}, $a(\sigma_i)$, $b(\sigma_i)$, and $K(f_{\sigma_i}) \cap f_{\sigma_i}^{-1}(a(\sigma_i), b(\sigma_i))$ as f_i, a_i, b_i and K_i, respectively, $i = 0, 1$.

As $|\sigma_1 - \sigma_0|$ is small, we have $c < d$ such that

$$f_0(K_0) \subset (c,d) \subset [c,d] \subset (a_0, b_0) \cap (a_1, b_1) .$$

The first inclusion is due to the (PS) condition for f_0.

By the continuity of f_0, we have $\delta > 0$ such that

$$f_0((K_0)_\delta) \subset (c,d) ,$$

where $(K_0)_\delta$ is the δ-neighborhood of K_0. Again by the (PS) condition of f_0, $\exists \epsilon = \epsilon(\delta) > 0$ such that

$$\| f_0'(x) \| \geq \epsilon \quad \forall x \in f_0^{-1}[a_0, b_0] \backslash (K_0)_\delta .$$

As $|\sigma_1 - \sigma_0|$ is small, we have $K_1 \subset (K_0)_\delta$, $f_1((K_0)_\delta) \subset (c,d)$ and $f_1^{-1}[c,d] \subset f_0^{-1}(a,b)$. Thus $f_i(K_j) \subset (c,d)$, $i,j = 0,1$.

Again as $|\sigma_1 - \sigma_0|$ is small, we can construct a pseudo-gradient vector field for f_1, which coincides with that of f_0 in $((f_1)_{b_1} \cap (f_0)_{b_0}) \backslash (K_0)_\delta$. Thus the pseudo-gradient flow defines strong deformation retracts $((f_1)_{b_1}, (f_0)_c \cap (f_1)_c) \to ((f_0)_d \cap (f_1)_d, (f_0)_c \cap (f_1)_c)$ and $(f_1)_c \to (f_0)_c \cap (f_1)_c$, according to the nontrivial interval theorem. In the same manner we have strong deformation retracts $((f_0)_{b_0}, (f_0)_c \cap (f_1)_c) \to ((f_0)_d \cap (f_1)_d, (f_0)_c \cap (f_1)_c)$ and $(f_0)_c \to (f_0)_c \cap (f_1)_c$. Therefore the exactness of the homology sequence yields:

$$H_*((f_0)_{b_0}, (f_0)_c; G) = H_*((f_1)_{b_1}, (f_1)_c; G) .$$

Again by the nontrivial interval theorem, there are strong deformation retracts $(f_0)_c \to (f_0)_{a_0}$, $(f_1)_c \to (f_1)_{a_1}$.

We arrive at

$$H_*((f_0)_{b_0}, (f_0)_{a_0}; G) = H_*((f_1)_{b_1}, (f_1)_{a_1}; G) .$$

□

Remark 5.1.22 *One may use other homology theories instead, e.g., the singular cohomology $H^*(A,B;F)$ or Alexander–Spanier cohomology $\bar{H}^*(A,B;F)$, with a coefficient field F, etc. Since they share most common properties but have only a few differences, we use special homology theory for special problems.*

The homotopy invariance $H_*(f_b, f_a; G)$ can be localized on a pseudo-gradient flow invariant set. For details see Chang and Ghoussoub [CG]. We present here a special case, which will be used later.

Let $f \in C^1(M, R^1)$ satisfy the (PS) condition, and let $a < b$ be regular values. Since $K \cap f^{-1}[a,b]$ is compact, it has a bounded neighborhood $U \subset f^{-1}[a,b]$. Let $\widetilde{U}$ be the η-invariant set for a pseudo-gradient flow η for f.

Lemma 5.1.23 *The set $W = \widetilde{U} \cap f^{-1}[a,b]$ is bounded.*

Proof. From the (PS) condition, we have $\epsilon > 0$ such that

$$\| f'(x) \| \geq \epsilon \quad \forall x \in f^{-1}[a,b] \backslash U .$$

$\forall y \in W \backslash U$, there exist $x \in \partial U$ and $t > 0$ (or $t < 0$) such that $y = \eta(x,t)$ and $\eta(x,s) \in \widetilde{U} \backslash U$ $\forall s > 0$ ($\forall s < 0$ resp.). Let X be the pseudo-gradient vector field with respect to η. Thus

$$\begin{aligned} \epsilon^2 t &\leq \int_0^t \| f'(\eta(x,s)) \|^2 \, ds \\ &\leq \int_0^t \langle f'(\eta(x,s)), X(\eta(x,s)) \rangle ds \\ &= -\int_0^t \frac{d}{ds} f(\eta(x,s)) ds = f(x) - f(y) \leq b - a \end{aligned}$$

as $t > 0$. We obtain the estimate $0 < t \leq \frac{b-a}{\epsilon^2}$; similarly, for $t < 0$, by which, we have the estimate

$$\begin{aligned} \| y - x \| = \| \eta(x,t) - x \| &\leq \int_0^t \| \dot{\eta}(x,s) \| \, ds \\ &\leq 2 \left(t \int_0^t \| f'(\eta(x,s)) \|^2 \, ds \right)^{1/2} \\ &\leq 2(t(b-a))^{1/2} \leq \frac{2(b-a)}{\epsilon} . \end{aligned}$$

Therefore W is bounded. □

Since the regular set is open, we have regular values c, d such that $[a,b] \subset (c,d)$. Let $W = \widetilde{U} \cap f^{-1}[c,d]$.

Corollary 5.1.24 *With the above notations, the singular homology groups $H_*(f_b \cap \widetilde{U}, f_a \cap \widetilde{U}; G)$ are invariant under small $C^1(W)$ perturbation of f, i.e., if $\| g - f \|_{C^1(W)}$ is small and both f and g satisfy the (PS) condition, then*

$$H_*(f_b \cap \widetilde{U}, f_a \cap \widetilde{U}; G) = H_*(g_b \cap \widetilde{U}, g_a \cap \widetilde{U}; G) .$$

Corollary 5.1.25 *With the (PS) condition, the critical groups $C_*(f,p)$ for an isolated critical point p are invariant under C^1 topology on any bounded neighborhood of p.*

Now we turn to the subadditivity.

Let S be an integer-valued function on certain pairs of spaces. S is called subadditive if

$$S(X,Z) \leq S(X,Y) + S(Y,Z), \text{ whenever } Z \subset Y \subset X .$$

Example. $\forall q \in \mathbb{N}$, rank $H_q(X, Y; G)$ is subadditive.

In fact, by examining the exact sequence:

$$\cdots \to H_q(Y,Z;G) \xrightarrow{j_*} H_q(X,Z;G) \xrightarrow{i_*} H_q(X,Y;G) \xrightarrow{\partial_*} H_{q-1}(Y,Z;G) \to \cdots$$

we obtain

$$\begin{aligned}\operatorname{rank} H_q(X,Z;G) &= \operatorname{rank} \operatorname{Im} i_* + \operatorname{rank} \ker i_* \\ &= \operatorname{rank} \operatorname{Im} i_* + \operatorname{rank} \operatorname{Im} j_* \\ &\le \operatorname{rank} H_q(X,Y;G) + \operatorname{rank} H_q(Y,Z;G)\,.\end{aligned}$$

Applying this to the triple (f_c, f_b, f_a) with regular values $c > b > a$, we have

$$\operatorname{rank} H_q(f_c, f_a; G) \le \operatorname{rank} H_q(f_b, f_a; G) + \operatorname{rank} H_q(f_c, f_b; G)\,.$$

If further, we write $\epsilon_q(X,Y) = \operatorname{rank} \operatorname{Im} i_*$, $\epsilon_q(x,z) = \operatorname{rank} \operatorname{Im} j_*$ and $\epsilon_{q-1}(Y,Z) = \operatorname{rank} \operatorname{Im} \partial_*$, then

$$\begin{aligned}\operatorname{rank} H_q(X,Y) &= \epsilon_q(X,Y) + \epsilon_{q-1}(Y,Z)\,, \\ \operatorname{rank} H_q(X,Z) &= \epsilon_q(X,Z) + \epsilon_q(X,Y)\,, \\ \operatorname{rank} H_q(Y,Z) &= \epsilon_q(Y,Z) + \epsilon_q(X,Z)\,.\end{aligned}$$

Summing them up, we obtain

$$P(t;X,Y) + P(t;Y,Z) = P(t;X,Z) + (1+t)Q(t;Y,Z)\,,$$

where

$$P(t;X,Y) = \sum_{q=0}^{\infty} t^q \operatorname{rank} H_q(X,Y;G), \text{ and } Q(t;Y,Z) = \sum_{q=0}^{\infty} t^q \epsilon_q(Y,Z)$$

are formal series with nonnegative coefficients.

If there is a multiple $X_0 \subset X_1 \subset X_2 \subset \cdots \subset X_n$, we have

$$\sum_{j=1}^{n} P(t, X_j, X_{j-1}) = P(t, X_n, X_0) + (1+t)Q(t)\,. \tag{5.2}$$

where Q is a formal series with nonnegative coefficients.

Thus, $\forall t \ge 0, P(t;X,Y)$ is subadditive, and $P(-1,X,Y)$ is additive.

We turn to establishing the relationship between the singular relative homology groups $H_*(f_b, f_a, G)$ and the critical groups $C_*(f,p)$. This is the Morse relation.

Suppose that $f \in C^1(M, R^1)$ has only isolated critical values, and that each of them corresponds to a finite number of critical points; say

$$\cdots < c_{-2} < c_{-1} < c_0 < c_1 < c_2 < \cdots$$

are critical values with

$$K \cap f^{-1}(c_i) = \{z_j^i\}_{j=1}^{m_i}, \; i = 0, \pm 1, \pm 2, \dots .$$

One chooses

$$0 < \epsilon_i < \max\{c_{i+1} - c_i, c_i - c_{i-1}\}, \; i = 0, \pm 1, \pm 2, \dots .$$

Definition 5.1.26 *For a pair of regular values $a < b$, we call*

$$M_q(a, b) = \sum_{a < c_i < b} \operatorname{rank} H_q(f_{c_i + \epsilon_i}, f_{c_i - \epsilon_i}; G)$$

the qth Morse type number of the function f on (a, b), $q = 0, 1, 2, \dots$.

For functions satisfying the (PS) condition, according to the nontrivial interval theorem, Morse type numbers are independent of the special choice of $\{\epsilon_i\}$. Moreover, we have:

Theorem 5.1.27 *Assume that $f \in C^1(M, R^1)$ satisfies the (PS) condition, and has an isolated critical value c, with $K \cap f^{-1}(c) = \{z_j\}_{j=1}^m$. Then for sufficiently small $\epsilon > 0$ we have*

$$H_*(f_{c+\epsilon}, f_{c-\epsilon}; G) = \oplus_{j=1}^m C_*(f, z_j) .$$

Proof. By the second deformation theorem and the homotopy invariance of singular homology groups, we have

$$H_*(f_{c+\epsilon}, f_{c-\epsilon}; G) = H_*(f_c, f_{c-\epsilon}; G) ,$$

and

$$H_*(f_c \backslash (K \cap f^{-1}(c)), f_{c-\epsilon}; G) = H_*(f_{c-\epsilon}, f_{c-\epsilon}; G) = 0 .$$

Applying the exactness of singular homology groups to the triple $(f_c, f_c \backslash (K \cap f^{-1}(c)), f_{c-\epsilon})$:

$$\begin{aligned} \cdots \to H_q(f_c \backslash (K \cap f^{-1}(c)), f_{c-\epsilon}) \to H_q(f_c, f_{c-\epsilon}) \\ \to H_q(f_c, f_c \backslash (K \cap f^{-1}(c))) \to H_{q-1}(f_c \backslash (K \cap f^{-1}(c)), f_{c-\epsilon}) \to \cdots \end{aligned}$$

we find

$$0 \to H_q(f_c, f_{c-\epsilon}) \to H_q(f_c, f_c \backslash (K \cap f^{-1}(c))) \to 0 ,$$

i.e.,

$$H_q(f_c, f_{c-\epsilon}) = H_q(f_c, f_c \backslash (K \cap f^{-1}(c))) .$$

By using the excision property, we may decompose the relative singular homology groups into critical groups:

$$\begin{aligned} H_*(f_c, f_c \backslash (K \cap f^{-1}(c))) &= H_*(f_c \cap (\cup_{j=1}^m B_\epsilon(z_j)), f_c \cap (\cup_{j=1}^m (B_\epsilon(z_j) \backslash \{z_j\}))) \\ &= \oplus_{j=1}^m C_*(f, z_j) . \end{aligned}$$

□

Corollary 5.1.28 $M_*(a,b) = \sum_{a<c_i<b} \sum_{j=1}^{m_i} \operatorname{rank} C_*(f, z_j^i)$.

Thus, for a nondegenerate function,the qth Morse type number is the number of critical points with Morse index q.

Define the qth Betti number:

$$\beta_q = \beta_q(a,b) = \operatorname{rank} H_q(f_b, f_a; G), \ q = 0, 1, \dots ,$$

and the two formal series:

$$M(f,a,b;t) = \sum_{q=0}^{\infty} M_q(a,b)t^q,$$
$$P(f,a,b;t) = \sum_{q=0}^{\infty} \beta_q(a,b)t^q.$$

Theorem 5.1.29 *(Morse relation) Suppose that $f \in C^1(M, R^1)$ satisfies $(PS)_c$ $\forall c \in [a,b]$, where a and b are regular values. Assume $(K \cap f^{-1}[a,b])$ is finite. Moreover, if all $M_q(a,b)$ and $\beta_q(a,b)$ are finite, and only finitely many of them are nonzeroes, then*

$$\sum_{q=0}^{\infty}(M_q(a,b) - \beta_q(a,b))t^q = (1+t)Q(t) \ , \tag{5.3}$$

where $Q(t)$ is a formal series with nonnegative coefficients. In particular, $\forall p = 0, 1, 2, \dots ,$

$$\sum_{q=0}^{p}(-1)^{p-q}M_q(a,b) \geq \sum_{q=0}^{p}(-1)^{p-q}\beta_q(a,b) \ . \tag{5.4}$$

More specifically,

$$\sum_{q=0}^{\infty}(-1)^q M_q(a,b) = \sum_{q=0}^{\infty}(-1)^q\beta_q(a,b) \ .$$

Proof. Let $c_1 < c_2 < \cdots < c_n$ be the critical values of f in the interval $[a,b]$. Let us choose regular values $\{d_j\}_{j=0}^n$ such that

$$a = d_0 < c_1 < d_1 < \cdots < d_{j-1} < c_j < d_j < \cdots < c_n < d_n = b \ .$$

Plugging $X_j = f_{d_j}$ into (5.2), $j = 0, 1, 2, \dots, n$, on the one hand, we have

$$\begin{aligned}\sum_{j=1}^{n} P(t; f_{d_j}, f_{d_{j-1}}) &= P(t; f_b, f_a) + (1+t)Q(t) \\ &= P(f,a,b;t) + (1+t)Q(t).\end{aligned}$$

On the other hand,

$$\sum_{j=1}^{n} P(t, f_{d_j}, f_{d_{j-1}}) = \sum_{j=1}^{n}\sum_{q=0}^{\infty} t^q \operatorname{rank} H_q(f_{d_j}, f_{d_{j-1}}; G)$$

$$= \sum_{j=1}^{n}\sum_{q=0}^{\infty} t^q \sum_{i=1}^{m_j} \operatorname{rank} C_q(f, z_i^j)$$

$$= \sum_{q=0}^{\infty} t^q \sum_{j=1}^{n}\sum_{i=1}^{m_j} \operatorname{rank} C_q(f, z_i^j)$$

$$= M(f, a, b; t) \; .$$

(5.4) is proved by setting $t = -1$. Replacing the formal series by finite sums, equation (5.3) is proved similarly. □

Remark 5.1.30 *(5.4) is a sequence of Morse inequalities, and the following equality is called the Morse equality. On the right-hand side, it is related to the Euler characteristic of the pair (X, Y):*

$$\chi(X, Y; G) = \sum_{q=0}^{\infty} (-1)^q \operatorname{rank} H_q(X, Y; G)$$

Remark 5.1.31 *(Morse inequalities under general boundary conditions) Let M be a Hilbert–Riemannian manifold modeled on a Hilbert space H with boundary $\partial M = \Sigma$, which is a smooth oriented submanifold with codimension 1. Let $n(x)$ be the outward normal unit vector of Σ at x.*

$f \in C^1(M, R^1)$ is said to be satisfying the general boundary condition if:

(1) $K \cap \Sigma = \varnothing$, i.e., it has no critical point on Σ;
(2) The restriction $\hat{f} = f|_\Sigma$, as a function on Σ, has only isolated critical points.

Let $\Sigma_- = \{x \in \Sigma \,|\, (f'(x), n(x)) \leq 0\}$, where $(\cdot,\cdot)$ is the inner product of the Hilbert space H.

Suppose that f has only isolated critical points, and let $\{m_0, m_1, \ldots,\}$, and $\{\mu_0, \mu_1, \ldots,\}$ be the Morse type numbers of the function f and $\hat{f}$ on M and Σ_- respectively.

Under the above assumptions, in addition, if we assume the (PS) condition for f on M and for $\hat{f}$ on Σ respectively, then the Morse inequalities are modifid as follows:

$$\Sigma_{q=0}^{\infty}(m_q + \mu_q - \beta_q)t^q = (1+t)Q(t) \; ,$$

where $Q(t)$ is a formal series with nonnegative coefficients.

Readers are referred to K. C. Chang [Ch 5] and Chang and Liu [CL 1].

At the end of this subsection, we study the relationship between the Morse theory and the Leray–Schauder degree theory.

Let H be a Hilbert space, and let $f \in C^2(H, R^1)$. Assume that $f'(x) = x - T(x)$, where T is compact, and that p_0 is an isolated critical point of f, i.e., an isolated zero of the compact vector field f'. Both the Leray–Schauder index

$i(f', p_0)$ and the critical groups $C_*(f, p_0)$ are topological invariants describing the local behavior at p_0 of f'.

Question: What is the relationship between $i(f', p_0)$ and $C_*(f, p_0)$?

Theorem 5.1.32 *Suppose that $f \in C^2(H, R^1)$ satisfies the (PS) condition, and that $f'(x) = x - T(x)$ is a compact vector field. If p_0 is an isolated critical point of f, then*

$$i(f', p_0) = \sum_{q=0}^{\infty} (-1)^q \operatorname{rank} C_q(f, p_0) ,$$

wherever the RHS makes sense.

Proof. 1. If p_0 is nondegenerate, then $C_q(f, p_0) = \delta_{qj} \cdot G$, where j is the Morse index of p_0, i.e., the dimension of the negative space of

$$f''(p_0) = \mathrm{id} - T'(p_0) .$$

Since $T'(p_0)$ is compact, $j < \infty$.

In this case, $i(f', p_0) = (-1)^j$. Thus

$$i(f', p_0) = \sum_{q=0}^{\infty} (-1)^q \operatorname{rank} C_q(f, p_0) .$$

2. p_0 is degenerate. Assume $p_0 = \theta$ and $f(\theta) = 0$. Let U be an isolated bounded neighborhood of θ, and $W = \widetilde{U} \cap f^{-1}[-\gamma, \gamma]$ for small $\gamma > 0$. According to Lemma 5.1.23, it is bounded, and let $\delta > 0$ be sufficiently small such that $B_\delta(\theta) \subset \mathrm{int}(W) \cap f^{-1}[-\gamma/2, \gamma/2]$. We shall define a function $\widetilde{f}$ satisfying the (PS) condition such that:

(1) $|f(x) - \widetilde{f}(x)| < \gamma/2 \ \forall x \in H$,
(2) $f(x) = \widetilde{f}(x) \ \forall x \notin B_\delta$,
(3) $\widetilde{f}$ has only finitely many nondegenerate critical points $\{p_j\}_1^m \subset B_\delta$ in W.

We assume the existence of such a $\widetilde{f}$ at this moment. From part 1, Corollary 5.1.24 and Theorem 5.1.29, we have

$$\begin{aligned} i(f', \theta) &= \deg(f', W, \theta) = \deg(\widetilde{f}', W, \theta) \\ &= \sum_{j=1}^{m} i(\widetilde{f}', p_j) = \sum_{j=1}^{m} \sum_{q=0}^{\infty} (-1)^q \operatorname{rank} C_q(\widetilde{f}, p_j) \\ &= \sum_{q=0}^{\infty} (-1)^q \operatorname{rank} H_q(\widetilde{f}_\gamma \cap W, \widetilde{f}_{-\gamma} \cap W; G) \end{aligned}$$

$$= \sum_{q=0}^{\infty} (-1)^q \operatorname{rank} H_q(f_\gamma \cap W, f_{-\gamma} \cap W; G)$$

$$= \sum_{q=0}^{\infty} (-1)^q \operatorname{rank} C_q(f, \theta) .$$

3. Now, we return to the construction of the function f. Define

$$\widetilde{f}(x) = f(x) + p(\| x \|)(x_0, x)$$

where $p \in C^2([0,\infty), R^1)$ satisfying $0 \le p \le 1$, $|p'(t)| \le \frac{4}{\delta}$, and

$$p(t) = \begin{cases} 1 & t \in [0, \delta/2] \\ 0 & t > \delta , \end{cases}$$

and $x_0 \in H$ will be determined later. Let

$$\beta = \inf \{\| f'(x) \| \, x \in B_\delta \backslash B_{\delta/2}\} ,$$

The (PS) condition of f yields $\beta > 0$. We choose $x_0 \in H$ such that

$$0 < \| x_0 \| < \min\{\beta/6, \gamma/3\} .$$

Then we have

$$|f(x) - \widetilde{f}(x)| < \gamma/3 ,$$
$$\| \widetilde{f}'(x) \| \ge \beta/6 \quad \forall x \in B_\delta \backslash B_{\delta/2} ,$$
$$\widetilde{f}(x) = f(x) \quad \forall x \notin B_\delta .$$

For small $\| x_0 \|$, $\widetilde{f}'$ is a k-set contraction mapping vector field with $k < 1$. Therefore $\deg(\widetilde{f}, W, \theta)$ is well defined. The (PS) condition for $\widetilde{f}$ is verified directly. On account of the Sard–Smale theorem, a suitable x_0 can be chosen such that $\widetilde{f}$ is nondegenerate. □

Remark 5.1.33 *The above theorem shows that for potential compact vector fields, the critical groups provide more information than the Leray–Schauder index for an isolated zero.*

Theorem 5.1.34 *Suppose that $f \in C^1(M, \mathbb{R}^1)$ satisfies the (PS) condition with regular values $a < b$. Let U be a bounded neighborhood of $K \cap f^{-1}[a,b]$, and $W = \widetilde{U} \cap f^{-1}[a,b]$. Assume that M is a Hilbert space and $f'(x) = x - T(x)$ is a compact vector field. Then*

$$\deg(f', W, \theta) = \sum_{q=0}^{\infty} (-1)^q \operatorname{rank} H_q(f_b \cap W, f_a \cap W, G) .$$

Proof. This is a combination of Theorems 5.1.32 and 5.1.29 with an application of the Sard–Smale theorem. □

5.1.5 Applications

We present a few simple examples showing how Morse theory can be applied to the study of the existence and the multiplicity of critical points. More sophisticated examples will be studied later.

1. A Three-Solution Theorem

Theorem 5.1.35 *Let H be a Hilbert space. Suppose that $f \in C^2(H, R^1)$ satisfies the (PS) condition and is bounded from below, and that p_0 is a nondegenerate nonminimum critical point of f with finite index j. Then f has at least three distinct critical points.*

Proof. In fact, from Corollary 4.8.4, there exists a minimizer p_1. One may assume that the minimizer is unique, for otherwise, the proof is done. Thus, we have already two distinct critical points: p_0 and the minimizer p_1, with $C_q(f, p_1) = \delta_{q0}G$ and $C_q(f, p_0) = \delta_{qj}G$, $j \neq 0$. Let $m = f(p_1)$ and $c = f(p_0)$.

If there were no other critical point, then according to noncritical interval theorem, $\forall \epsilon > 0$ $f_{c+\epsilon}$ would be a strong deformation retract of H. Since m is the minimum, $f_{m-\epsilon} = \emptyset$, it follows that $\chi(f_{c+\epsilon}, f_{m-\epsilon}) = \chi(H) = 1$. According to the Morse equality, we would have

$$1 + (-1)^j = 1 .$$

This is a contradiction. $\square$

Example 1. Let $\Omega \subset R^n$ be a bounded domain with smooth boundary; we study the problem:

$$\begin{cases} -\triangle u = g(u) & \text{in } \Omega , \\ u = 0 & \text{on } \partial\Omega . \end{cases} \tag{5.5}$$

Let $\lambda_1 < \lambda_2 \leq \cdots$ be the eigenvalues of $-\triangle$ with the Dirichlet condition, and let $c_i, i = 1, 2, 3$ be various constants. Assume

(g1) $g \in C^1(R^1)$ with $g(0) = 0$.

(g2)

$$|g'(t)| \leq \begin{cases} c_1(1 + |t|^{p-1}),\ p < \frac{n+2}{n-2} & \text{if } n \geq 3 , \\ \text{no restriction} & \text{if } n \leq 2 . \end{cases}$$

(g3) $G(t) := \int_0^t g(s)ds \leq c_2|t|^2 + c_3$, where $c_2 < \lambda_1/2$.

(g4) $\exists i \geq 1$ such that $\lambda_i < g'(0) \leq \lambda_{i+1}$.

Theorem 5.1.36 *The problem (5.5) under assumptions (g1)–(g4) has at least three distinct solutions.*

Proof. Define

$$J(u) = \int_\Omega \frac{1}{2}|\nabla u|^2 - G(u) \text{ on } H_0^1(\Omega) .$$

According to (g1) and (g2), J is C^2. (g3) implies that J is bounded below. $g(0) = 0$ means that θ is a critical point of J. The verification of the (PS) condition follows from (g2) and (g3). In fact, (g3) implies that a (PS) sequence must be bounded in H^1, and (g2) implies that $|g(t)| \le c_4(1+|t|^p)$, the compact embedding is applied.

Since

$$J''(\theta) = \text{id} - g'(0)(-\Delta)^{-1} ,$$

(g4) implies that θ is nondegenerate with finite index $\text{ind}(J''(\theta)) = i \ge 1$. □

2. Bifurcation

We know from Theorem 3.5.1 that on a Banach space, for a compact vector field with parameter $\lambda : F(x,\lambda) = x - \lambda Tx + N(x,\lambda)$, where T is compact, $\| N(x,\lambda) \| = \circ(\| x \|)$ uniformly in λ, the bifurcation point occurs at eigenvalues λ with odd multiplicity. Generally speaking, the odd multiplicity condition cannot be removed. However, if $\forall\, \lambda \in R^1,\ F(\cdot,\lambda)$ is a potential operator, i.e., the differential of a certain functional with a parameter λ, we have:

Theorem 5.1.37 *Suppose that H is a Hilbert space, and that $f \in C^2(H,\mathbb{R}^1)$ satisfies $f'(\theta) = \theta$. If λ_0 is an isolated eigenvalue of the self-adjoint operator $f''(\theta)$ with finite multiplicity, then (θ,λ_0) is a bifurcation point of the equation:*

$$F(u,\lambda) = f'(u) - \lambda u = \theta . \tag{5.6}$$

Let $f(u) = \frac{1}{2}(Lu,u) + g(u)$, where $g(u) = o(\|u\|^2)$, then $L = f''(\theta)$, and $f'(u) = Lu + G(u)$, with $G(u) = g'(u), G(u) = \circ(\| u \|)$ as $\| u \| \to 0$.

The proof depends upon the Lyapunov–Schmidt reduction. Let $X = \ker(L - \lambda_0 I)$, with $\dim X = n$; and let $P, P^\perp$ be the orthogonal projections onto X and $X^\perp$, respectively. Then (5.6) is equivalent to a pair of equations

$$\lambda_0 x + PG(x + x^\perp) = \lambda x , \tag{5.7}$$
$$Lx^\perp + P^\perp G(x + x^\perp) = \lambda x^\perp , \tag{5.8}$$

where $u = x + x^\perp,\ x \in X,\ x^\perp \in X^\perp$. According to the IFT, equation (5.8) is uniquely solvable in a small bounded neighborhood $\mathcal{O}$ of $(\lambda_0,\theta) \in R^1 \times X$, say $x^\perp = \varphi(\lambda,x)$ for $(\lambda,x) \in \mathcal{O}$, where $\varphi \in C^1(\mathcal{O},X^\perp)$. Substituting $x^\perp = \varphi(\lambda,x)$ into (5.7), we have

$$\lambda_0 x + PG(x + \varphi(\lambda,x)) = \lambda x , \tag{5.9}$$

which is again an Euler–Lagrange equation on the finite-dimensional space X. Indeed, let

$$\begin{aligned} J_\lambda(x) &= f(x+\varphi(\lambda,x)) - \frac{\lambda}{2}(\| x \|^2 + \| \varphi(\lambda,x) \|^2) \\ &= \frac{1}{2}(\lambda_0 - \lambda) \| x \|^2 + \frac{1}{2}(L\varphi,\varphi) - \frac{\lambda}{2} \| \varphi \|^2 + g(x+\varphi) \, . \end{aligned}$$

It is easy to verify that (5.9) is the Euler–Lagrange equation for J_λ, and that $\|\varphi(\lambda,x)\| = \circ(\| x \|)$ as $x \to \theta$.

The problem is reduced to finding the critical points of J_λ near $x=\theta$ for fixed λ near λ_0, where $J_\lambda \in C^1(\Omega_1, R^1)$, Ω_1 is a neighborhood of θ in X.

If (θ, λ_0) were not a bifurcation point, then there would be a small neighborhood $\Omega_1 \times (\lambda_0 - \delta, \lambda_0 + \delta)$, $\delta > 0$ such that θ is the unique critical point in Ω_1 for J_λ as $|\lambda - \lambda_0| < \delta$. However, as $\lambda < \lambda_0$, θ is a local minimum point, $C_q(J_\lambda, \theta) = \delta_{q0}G$, but as $\lambda > \lambda_0$, θ is a local maximum point, $C_q(J_\lambda,\theta) = \delta_{qn}G$. This contradicts the invariance of critical groups under C^1-perturbation on bounded neighborhoods of θ.

Note: From the degree point of view, there is no difference between $\lambda > \lambda_0$ and $\lambda < \lambda_0$, except n is odd. Compare with Krasnoselski's theorem (Theorem 3.5.1).

3. Superlinear Elliptic Problem

We turn to the problem studied previously in Sect. 3.6 and Sect. 4.8.

Example 2.

$$\begin{cases} -\Delta u = g(x,u) & \text{in } \Omega \, , \\ u = 0 & \text{on } \partial\Omega \, , \end{cases} \tag{5.10}$$

where $\Omega \subset R^n$ is a bounded domain with smooth boundary. Assume:

(g1) $|g(x,t)| \le C(1+|t|^\alpha)$ $\alpha < \frac{n+2}{n-2}$ if $n \ge 3$.
(g2) $\exists \theta > 2$, $\exists M > 0$ such that $0 < \theta G(x,t) \le t g(x,t)$ $\forall x \in \Omega$, for $|t| \ge M$, where $G(x,t) = \int_0^t g(x,s) ds$.
(g3) $g \in C^1(\overline{\Omega} \times \mathbb{R}^1)$ with $g(x,0) = g_t(x,0) = 0$.

Define

$$J(u) = \int_\Omega \left[\frac{1}{2}|\nabla u|^2 - G(x,u(x)) \right] dx \, .$$

Lemma 5.1.38 *(Z.Q. Wang) Under the assumptions of (g1), (g2) and (g3), there exists a constant $A > 0$, such that*

$$J_a \simeq S^\infty, \text{ the unit sphere in } H_0^1(\Omega)$$

for $-a > A$.

Proof. By (g_2), we have a constant $C > 0$ such that

$$G(x,t) \ge C(|t|^\theta - 1) \;\forall t, |t| \ge M \, .$$

Thus $\forall u \in S^\infty$,

$$J(tu) \to -\infty \text{ as } t \to +\infty .$$

First we want to show that $\exists A > 0$ such that $\forall a < -A$, if $J(tu) \le a$, then $\frac{d}{dt}J(tu) < 0$.

In fact, set

$$A = 2M|\Omega| \max_{(x,t)\in\overline{\Omega}\times[-M,M]} |g(x,t)| + 1 .$$

If $J(tu) = \frac{t^2}{2} - \int_\Omega G(x, tu(x))dx \le a$, then

$$\begin{aligned}
\frac{d}{dt}J(tu) &= (J'(tu), u) \\
&= t - \int_\Omega u(x) \cdot g(x, tu(x))dx \\
&\le \frac{2}{t}\left\{\int_\Omega G(x, tu(x))dx - \frac{1}{2}\int_\Omega tu(x) \cdot g(x, tu(x))dx + a\right\} \\
&\le \frac{2}{t}\left\{\left(\frac{1}{\theta} - \frac{1}{2}\right)\int_{|tu(x)|\ge M} tu(x)g(x, tu(x))dx + (A-1) + a\right\} \\
&\le \frac{2}{t}\left\{\left(\frac{1}{\theta} - \frac{1}{2}\right)C\theta\int_{|tu(x)|\ge M} |t|^\theta|u(x)|^\theta dx - 1\right\} < 0 ,
\end{aligned}$$

as $a < -A$. The implicit function theorem is employed to obtain a unique $T(u) \in C(S^\infty, \mathbb{R}^1)$ such that

$$J(T(u)u) = a \quad \forall u \in S^\infty .$$

Next, we claim that $\| T(u) \|$ possesses a positive lower bound $\epsilon > 0$. In fact, by (g3), $g(x, 0) = g_t'(x, 0) = 0$, $J(tu) = \frac{t^2}{2} - \circ(t^2)$ $\forall u \in S^\infty$ as $|t|$ is small. The conclusion follows.

Finally, let us define a deformation retract $\eta : [0, 1] \times (H\backslash B_\epsilon(\theta)) \to H\backslash B_\epsilon(\theta)$, where $H = H_0^1(\Omega)$ and $B_\epsilon(\theta)$ is the ϵ-ball with center θ, by

$$\eta(s, u) = (1 - s)u + sT(u)u \;\forall u \in H\backslash B_\epsilon(\theta) .$$

This proves that $H\backslash B_\epsilon(\theta) \simeq J_a$, i.e., $J_a \simeq S^\infty$. □

Indeed, J satisfies the (PS) condition (see e.g., Sect. 4.8.3, Example 2). It is easy to give a different proof of the conclusion we obtained previously, i.e., there is a nontrivial solution of (5.10). Indeed, θ is a critical point of J, which is an isolated minimum. Therefore $C_q(J, \theta) = \delta_{q0}G$.

It is known that S^∞ is contractible.

The pair $(H_0^1(\Omega), J_a)$ must be trivial, i.e.,

$$H_q(H_0^1(\Omega), J_a; G) = 0 \;\; \forall q .$$

If there were no other critical points, then the Morse relation would be $1 = 0$. This is impossible.

In fact, from Sect. 4.8, Example 2, we have already known that there is a critical point via the mountain pass lemma, we shall see in the next section, equation (5.10) possesses at least three nontrivial solutions.

5.2 Minimax Principles (Revisited)

The minimax method was initiated by G. Birkhoff in his pioneering work on closed geodesics. It was successfully developed by Ljusternik and Schnirelmann in proving that on a closed surface with genus zero there are at least three closed geodesics. The method has been well developed by M. A. Krasnoselski, P. S. Palais, A. Ambrosetti and P. Rabinowitz etc.

5.2.1 A Minimax Principle

We have studied saddle points by the minimax method in Sect. 4.8. In this section we shall introduce a general minimax principle based on the deformation theorem, which contains various types of this method as special cases.

Definition 5.2.1 *For a given function $f \in C(M, R^1)$ satisfying the (PS) condition on a Banach–Finsler manifold M, for a given $a \in R^1$, let $\Phi_a(f) = \{\phi = \eta(1, \cdot) \,|\, \eta \in C([0,1] \times M, M)$ satisfy $\eta(t,x) = x,\ \forall\,(t,x) \in (\{0\} \times M) \cup ([0,1] \times f_a)\}$. A family of subsets $\mathcal{F}$ of M is called $\Phi_a(f)$-invariant if $\forall A \in \mathcal{F}\ \phi(A) \in \mathcal{F}$, whenever $\phi = \Phi_a(f)$.*

Example 1. (Linking) Let X be a Banach space, let $Q \subset X$ be a compact manifold with boundary ∂Q and let $S \subset X$ be a closed subset of X. Assume ∂Q and S link (see Definition 4.8.6). Let $f \in C^1(X, R^1)$ satisfy the (PS) condition. Assume $a = \max_{\partial Q} f < \inf_S f$. We set the family $\mathcal{F} = \{\psi(Q) | \psi \in C(Q, X)$ with $\psi|_{\partial Q} = \mathrm{id}_{\partial Q}\}$. Then $\mathcal{F}$ is $\Phi_a(f)$-invariant.

In fact, for $A = \psi(Q)$, one has $\phi(A) = \phi \circ \psi(Q), \forall \phi \in \Phi_a(f)$, and then $\phi \circ \psi|_{\partial Q} = id|_{\partial Q}$, so is $\phi(A) \in \mathcal{F}$.

Example 2. (Homology class) Let $f \in C^1(M, R^1)$ satisfy the (PS) condition. For a pair of real numbers $a < b$, if $[\sigma] \in H_q(f_b, f_a; G)$ is a nontrivial q-relative homology class for some $q \in \mathbb{N}$. Set

$$\mathcal{F} = \{|\sigma| | \sigma \in [\sigma] \text{ is a singular } q\text{-closed chain in } (f_b, f_a) \\ \text{with coefficient group } G\} ,$$

where $|\sigma|$ is the support of σ. Then $\mathcal{F}$ is $\Phi_a(f)$-invariant.

In fact, let $\phi = \eta(1, \cdot)$ where $\eta \in C([0,1] \times M, M)$ satisfies $\eta(t, x) = x \forall\,(t,x) \in (\{0\} \times M) \cup ([0,1] \times f_a)$, then $\forall \sigma \in [\sigma], \phi(|\sigma|) = |\phi(\sigma)|$, and $\phi(\sigma) \in [\sigma]$.

Theorem 5.2.2 *(Minimax principle) Suppose that M is a smooth Banach–Finsler manifold and that $f \in C^1(M, R^1)$ satisfies the (PS) condition. Let $\mathcal{F}$ be a $\Phi_a(f)$-invariant family of subsets of M for some $a \in R^1$. Set*

$$c = \inf_{A\in\mathcal{F}} \sup_{x\in A} f(x) . \tag{5.11}$$

If (1) c is finite, (2) $a < c$, then c is a critical value of f.

Proof. It is proved by contradiction. If c is not a critical value, then there exists $\epsilon > 0$ such that $a < c - \epsilon$, $f^{-1}[c-\epsilon, c+\epsilon] \cap K = \emptyset$, provided by the (PS) condition. Taking $A_0 \in \mathcal{F}$ such that $\sup_{x\in A_0} f(x) < c + \epsilon$ i.e., $A_0 \subset f_{c+\epsilon}$. According to the nontrivial interval theorem, $\exists$ a strong deformation retraction $\eta : [0,1] \times M \to M$ such that $\eta(0,\cdot) = \mathrm{id}$, $\eta(t,\cdot)|_{f_{c-\epsilon}} = \mathrm{id}_{f_{c-\epsilon}}\ \forall t \in [0,1]$, and $\eta(1, f_{c+\epsilon}) \subset f_{c-\epsilon}$. Thus $\eta(t,x) = x\, \forall\, (t,x) \in (\{0\}\times M) \cup ([0,1] \times f_a)$, and then $\phi := \eta(1,\cdot) \in \Phi_a(f)$. Since $\mathcal{F}$ is $\Phi_a(f)$-invariant, it follows that

$$\inf_{A\in\mathcal{F}} \sup_{x\in A} f(x) < c - \epsilon .$$

This is a contradiction. □

Thus, in Example 1, Since Q is compact, $c < \infty$. From the S and ∂Q link, we have $c > a$. Theorem 5.2.2 implies Theorem 4.8.9.

Also, applying Theorem 5.2.2 to Example 2,

$$c = \inf_{\sigma\in[\sigma]} \sup_{x\in|\sigma|} f(x) \tag{5.12}$$

is a critical value of f.

In fact, obviously c is finite. We verify that c is a critical value, if not, $c \le a$. $\forall \epsilon > 0, \exists \sigma \in [\sigma]$, with $|\sigma| \subset f_{c+\epsilon}$, according to the second deformation theorem, there exists η deforming $f_{c+\epsilon}$ to f_c. Since $\eta(1,\sigma) \in [\sigma]$ and $|\eta(\sigma)| \subset f_a$, $[\sigma]$ is trivial in (f_b, f_a). A contradiction.

What can we say about the critical point with critical value c defined by equation (5.12)?

Theorem 5.2.3 *Suppose that M is a smooth Banach–Finsler manifold, and that $f \in C^1(M, R^1)$ satisfies $(PS)_d$ $\forall d \in [a,b]$, where a, b are regular values. If $[\sigma]$ is a nontrivial class in $H_q(f_b, f_a; G)$, and c is defined by (5.12). Then $c \in (a,b)$ and $K \cap f^{-1}(c) \neq \emptyset$. Moreover, if c is isolated and $K \cap f^{-1}(c)$ consists of isolated critical points, then $\exists p \in K \cap f^{-1}(c)$ such that*

$$C_q(f,p) \neq 0 .$$

Proof. The first part of the conclusion has been proved. Since c is an isolated critical value, i.e., $\exists \epsilon > 0$ such that $f^{-1}[c-\epsilon, c+\epsilon] \cap K = K \cap f^{-1}(c)$ and $a < c - \epsilon$.

It is sufficient to show $H_q(f_{c-\epsilon}, f_{c+\epsilon}; G) \neq 0$, for then by Theorem 5.1.27, there must be a $p \in K \cap f^{-1}(c)$ such that $C_q(f,p) \neq 0$. Now we prove it by contradiction. If $H_q(f_{c+\epsilon}, f_{c-\epsilon}; G) = 0$, then by the long exact sequence

$$\cdots \longrightarrow H_q(f_{c-\epsilon}, f_a; G) \xrightarrow{i_*} H_q(f_{c+\epsilon}, f_a; G) \xrightarrow{j_*} H_q(f_{c+\epsilon}, f_{c-\epsilon}; G) \longrightarrow \cdots ,$$

we have $\operatorname{Im} i_* = H_q(f_{c+\epsilon}, f_a; G)$. This shows that the class $[\sigma]$ is again non-trivial in $H_q(f_{c-\epsilon}, f_a; G)$. But this contradicts the definition of c. □

Combining Theorem 5.2.3 with the shifting theorem, we have:

Corollary 5.2.4 *Let M be a smooth Hilbert–Riemannian manifold. Suppose that $f \in C^2(M, R^1)$ satisfies the assumptions of Theorem 5.2.3, and that $f''(x)$ is a Fredholm operator $\forall x \in K \cap f^{-1}(c)$. Then $\exists p \in K \cap f^{-1}(c)$ such that* $\operatorname{ind}(f''(p)) \leq q \leq \operatorname{ind}(f''(p)) + \dim\ker f''(p)$.

Return to the mountain pass point, we have:

Corollary 5.2.5 *Suppose that M is a smooth Hilbert–Riemannian manifold and that $f \in C^1(M, R^1)$. If $\exists p_0,\ p_1 \in M$ such that*

$$c = \inf_{l\in\Gamma} \sup_{t\in[0,1]} f \circ l(t) > \max\{f(p_0), f(p_1)\} ,$$

where $\Gamma = \{l \in C([0,1], M) | l(i) = p_i,\ i = 0,1\}$. Assume $(PS)_c$. If $K \cap f_c$ consists of isolated points, then there exists a mountain pass point $p \in K \cap f^{-1}(c)$ (see Definition 5.1.19) i.e.,

$$C_1(f,p) \neq 0 . \tag{5.13}$$

5.2.2 Category and Ljusternik–Schnirelmann Multiplicity Theorem

The notion of category was introduced by Ljusternik and Schnirelmann. It is a topological invariant used in the estimate of the lower bound of the number of critical points.

Definition 5.2.6 *Let M be a topological space, $A \subset M$ be a closed subset. Set*

$$\begin{aligned}\operatorname{cat}_M(A) = \inf\{m \in \mathbb{N} \cup \{+\infty\} \mid &\exists m \text{ contractible closed subsets of } M : \\ &F_1, F_2, \ldots, F_m \text{ such that } A \subset \cup_{i=1}^m F_i\} .\end{aligned}$$

A set F is called contractible (in M) if $\exists \eta : [0,1] \times M \to M$ such that $\eta(0,\cdot) = \mathrm{id}_M$ and $\eta(1,F) =$ one-point set.

Example 1. If C is a closed convex set in a Banach space X, then $\operatorname{cat}_X(C) = 1$. In fact, pick any $p \in C$, and set

$$\eta(t, x) = (1 - t)x + tp .$$

Thus C is contractible.

Example 2. Let S^n be the unit ball in R^{n+1}. Then $\mathrm{cat}_{S^n}(S^n) = 2$.

It is easy to see that $\mathrm{cat}_{S^n}(S^n) \le 2$, because the upper and the lower hemispheres $S^n_\pm = \{(x', x_{n+1}) \in R^n \times R^1 \mid \| x' \|^2 + x^2_{n+1} = 1, \ \pm x_{n+1} \ge 0\}$ are contractible. On the other hand, S^n is not contractible in itself, i.e., $\mathrm{cat}_{S^n}(S^n) > 1$. This can be shown by contradiction. Suppose there were a continuous $\eta : [0,1] \times S^n \to S^n$ such that $\eta(0, \cdot) = \mathrm{id}_{S^n}$ and $\eta(1, S^n) = p \in S^n$. Let $A : B_1^{n+1}(\theta) \to S^n$ be defined by:

$$A(x) = \begin{cases} -p & \text{if}\, x = \theta \,, \\ -\eta(1 - \| x \|, \frac{x}{\|x\|}) & \text{if}\, x \in B_1^{n+1}(\theta)\backslash\{\theta\} \,, \end{cases}$$

then A would be continuous. According to the Brouwer fixed-point theorem, $\exists p_0 \in S^n$ such that $A(p_0) = p_0$, But by the definition of A, $A(p_0) = -p_0$, thus $p_0 = \theta$. This is a contradiction.

Example 3. S^∞ is contractible, so is $\mathrm{cat}_{S^\infty}(S^\infty) = 1$. (see Sect. 3.3).

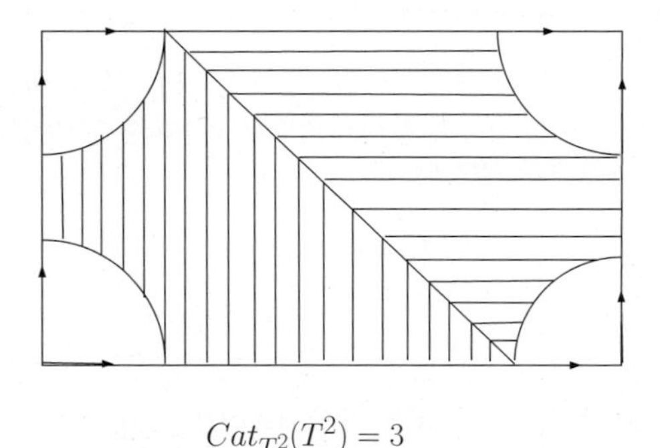

$Cat_{T^2}(T^2) = 3$

Fig. 5.3.

Example 4. Let T^2 be the two-dimensional torus, i.e., $S^1 \times S^1$. Then $\mathrm{cat}_{T^2}(T^2) = 3$. This can be shown by Fig. 5.3.

The following fundamental properties for the category hold:

(1) $\mathrm{cat}_M(A) = 0 \Leftrightarrow A = \emptyset$.
(2) (Monotonicity) $A \subset B \Rightarrow \mathrm{cat}_M(A) \le \mathrm{cat}_M(B)$.
(3) (Subadditivity) $\mathrm{cat}_M(A \cup B) \le \mathrm{cat}_M(A) + \mathrm{cat}_M(B)$.
(4) (Deformation nondecreasing) If $\eta : [0,1] \times M \to M$ is continuous such that $\eta(0, \cdot) = \mathrm{id}_M$, then $\mathrm{cat}_M(A) \le \mathrm{cat}_M(\overline{\eta(1, A)})$.
(5) (Continuity) If A is compact, then there is a closed neighborhood N of A such that $A \subset \mathrm{int}(N)$ and $\mathrm{cat}_M(A) = \mathrm{cat}_M(N)$.
(6) (Normality) $\mathrm{cap}_M(\{p\}) = 1 \ \forall p \in M$.

In fact properties (1), (2), (3) and (6) follow from the definition directly. We shall prove (4) and (5) below.

Proof. Proof of (4). Let $B = \overline{\eta(1, A)}$ and let $\text{cat}_M(B) = m$, i.e., $\exists$ closed contractible sets $F_1, \ldots, F_m$ of M such that $B \subset \cup_{i=1}^m F_i$. Let $G_i = \eta^{-1}(1, F_i)$, then G_i is closed and contractible, because $\exists \xi : [0,1] \times M \to M$ such that $\xi(0, \ldots) = \text{id}_M$ and $\xi(1, F_i) = p_i \in M$. Now let $\varphi = \xi \circ \eta : [0,1] \times M \to M$, we have $\varphi(0, \cdot) = \text{id}_M$ and $\varphi(1, G_i) = p_i$, $i = 1, 2, \ldots, m$. Since $A \subset \cup_{i=1}^m G_i$, property (4) is proved. □

In order to verify (5), we need a result on continuous extensions due to Hanner [Hann] and Palais [Pa 3]:

Let X, Y be metrizable Banach manifolds, and let $A \subset X$ be a closed subset. Then for any continuous map $\phi : (\{0\} \times X) \cup ([0,1] \times A) \to Y$, there is a continuous extension $\widetilde{\phi} : [0,1] \times X \to Y$.

Proof. Since A is compact, $\text{cat}_M(A) < \infty$, i.e., $\exists$ closed contractible sets $F_1, \ldots, F_m$ such that $A \subset \cup_{i=1}^m F_i$. Since $\exists \varphi_i :: [0,1] \times M \to M$ such that $\varphi_i(0, \ldots) = \text{id}_M$ and $\varphi_i(1, F_i) = p_i \in M$, we apply the above continuous extension theorem to ϕ_i and obtain $\tilde{\varphi}_i : [0,1] \times M \to M$ such that $\tilde{\varphi}_i|_{(\{\theta\} \times M) \cup ([0,1] \times F_i)} = \varphi_i|_{(\{\theta\} \times M) \cup ([0,1] \times F_i)}$, $i = 1, 2, \ldots, m$. Let V_i be a closed contractible neighborhood of p_i, then $U_i = \tilde{\varphi}_i^{-1}(1, V_i)$ is closed and contractible, because $\tilde{\varphi}_i^{-1}(0, V_i) = V_i$. Thus $N = \cup_{i=1}^m U_i$ is a closed neighborhood of A, we obtain $m = \text{cat}_M(A) \leq \text{cat}_M(N) \leq m$. □

As a consequence of the additivity and normality, we have

(7) If $\text{cat}_M(A) = m$, then $\sharp A \geq m$, i.e., there are at least m distinct points in A.

The main theorem of this subsection is the following multiplicity theorem:

Theorem 5.2.7 *(Ljusternik–Schnirelman theorem) Let M be a smooth Banach–Finsler manifold. Suppose that $f \in C^1(M, R^1)$ is a function bounded from below, satisfying the (PS) condition. Then f has a least $\text{cat}_M(M)$ critical points.*

This theorem is based on the following version of the deformation theorem:

Theorem 5.2.8 *Let M be a smooth Banach–Finsler manifold. Suppose $f \in C^1(M, R^1)$. Let $N' \subset N$ be two closed neighborhoods satisfying*

$$\text{dist}(N', \partial N) \geq \frac{7}{8}\delta, \ \delta > 0 \, .$$

Suppose that there are positive constants b and $\overline{\epsilon}$, such that

$$\| df(x) \| \geq b \ \forall x \in f_{c+\epsilon} \backslash (f_{c-\epsilon} \cup N') \, ,$$

$$0 < \overline{\epsilon} < 3 \min \left\{ \frac{1}{4}\delta b^2, \frac{1}{8}\delta b \right\} .$$

Then for any $0 < \epsilon < \frac{\overline{\epsilon}}{2}$, there exists $\eta \in C([0,1] \times M, M)$ satisfying

(1) $\eta(0,\cdot) = \text{id}$,
(2) $\eta(t,\cdot)|_{Cf^{-1}[c-\bar{\epsilon},c+\bar{\epsilon}]} = id|_{Cf^{-1}[c-\bar{\epsilon},c+\bar{\epsilon}]}$,
(3) $\eta(t,\cdot) : M \to M$ *is a homeomorphism* $\forall t \in [0,1]$,
(4) $\eta(1, f_{c+\epsilon}\backslash N) \subset f_{c-\epsilon}$,
(5) $f \circ \eta(t,x)$ *is nonincreasing in* t $\forall (t,x) \in [0,1] \times M$.

Proof. Define a smooth function:

$$p(s) = \begin{cases} 0 & \text{for } s \notin [c-\bar{\epsilon}, c+\bar{\epsilon}] \,, \\ 1 & \text{for } s \in [c-\epsilon, c+\epsilon] \,, \end{cases}$$

with $0 \le p(s) \le 1$. Let $A = \overline{M\backslash(N')_{\frac{\delta}{8}}}$, and $B = N'$ be two closed subsets. Let

$$g(x) = \frac{\text{dist}(x,B)}{\text{dist}(x,A) + \text{dist}(x,B)} \,.$$

We see $0 \le g \le 1,\ g = 0$ on N' and $g = 1$ outside $(N')_{\frac{\delta}{8}}$. Define

$$q(s) = \begin{cases} 1 & 0 \le s \le 1 \,, \\ \frac{1}{s} & s \ge 1 \,. \end{cases}$$

Letting X be a pseudo-gradient vector field of f, we define

$$V(x) = -g(x)p(f(x))q(\| X(x) \|)X(x) \,.$$

Thus $V \in C^{1-0}$, and $\| V(x) \| \le 1$. Consider the ODE:

$$\begin{cases} \dot{\sigma}(t) = V(\sigma(t)) \\ \sigma(0) = x_0 \quad \forall x_0 \in M \,. \end{cases}$$

Since V is bounded, the global existence and uniqueness of the flow $\sigma(t)$ on R^1 are known. Let

$$\eta(t,x) = \sigma(t), \text{ with } \sigma(0) = x \,.$$

Then $\eta \in C([0,1] \times M, M)$ satisfies (1), (2), (3) and (5). It remains to verify (4). From (5), it is sufficient to verify (4) for $x \in f_{c+\epsilon}\backslash(f_{c-\epsilon} \cup N)$. More precisely, we shall prove:

$$f \circ \eta\left(\frac{3}{4}\delta, x\right) \le c - \epsilon \ \ \forall x \in f_{c+\epsilon}\backslash(f_{c-\epsilon} \cup N) \,.$$

Indeed, if not, $\exists x \in f^{-1}[c-\epsilon, c+\epsilon]\backslash N$ such that

$$c - \epsilon < f \circ \eta(t,x) \le c + \epsilon \ \ \forall t \in \left(0, \frac{3}{4}\delta\right) \,,$$

it follows that

$$p(f \circ \eta(t,x)) = 1 \,.$$

Noticing that $\| \eta(t,x) - \eta(0,x) \| \leq t$, we have

$$\begin{aligned} \mathrm{dist}(\eta(t,x),(N')_{\frac{\delta}{8}}) &\geq \mathrm{dist}(\eta(0,x),(N')_{\frac{\delta}{8}}) - t \\ &> \left(\frac{7}{8} - \frac{1}{8} - \frac{3}{4}\right)\delta = 0 \,, \end{aligned}$$

so $g \circ \eta(t,x) = 1$. Now,

$$\begin{aligned} \frac{d}{dt} f \circ \eta(t,x) &= \langle f'(\eta(t,x)), \dot{\eta}(t,x)\rangle \\ &= -q(\| X(\eta(t,x)) \|)\langle f'(\eta(t,x)), X(\eta(t,x))\rangle \\ &\leq -q(\| X(\eta(t,x)) \|) \| f'(\eta(t,x)) \|^2 \ . \end{aligned}$$

If $\| X(\eta(t,x)) \| \leq 1$, then we have

$$\frac{d}{dt} f \circ \eta(t,x) \leq - \| f'(\eta(t,x)) \|^2 \leq -b^2 \,.$$

Otherwise,

$$\begin{aligned} \frac{d}{dt} f \circ \eta(t,x) &\leq - \| f'(\eta(t,x)) \|^2 / \| X(\eta(t,x)) \| \\ &\leq -\frac{1}{2} \| f'(\eta(t,x)) \| < -\frac{1}{2} b \,. \end{aligned}$$

In summary,

$$\frac{d}{dt} f \circ \eta(t,x) \leq - \min\left\{b^2, \frac{1}{2} b\right\} \,.$$

Thus,

$$\begin{aligned} f \circ \eta(t,x) &\leq f \circ \eta(0,x) - \frac{3}{4}\delta \min\left\{b^2, \frac{1}{2}b\right\} \\ &\leq c + \epsilon - 3\delta \min\left\{\frac{1}{4}b^2, \frac{1}{8}b\right\} < c - \epsilon \,. \end{aligned}$$

This is contradiction. □

We notice that in this theorem no (PS) condition is assumed, but the existence of N is attributed to the (PS) condition.

Corollary 5.2.9 *(First deformation lemma) Let M be a smooth Banach–Finsler manifold. Suppose that $f \in C^1(M, R^1)$ satisfies the $(PS)_c$ condition. Assume that N is a closed neighborhood of $K_c = K \cap f^{-1}(c)$. Then there exist a continuous map $\eta : [0,1] \times M \to M$ and constants $\overline{\epsilon} > \epsilon > 0$ such that (1)–(5) hold.*

With the aid of category, we introduce families of subsets of M. Let

$$\mathcal{F}_k = \{A |\ \text{closed subset of } M \text{ with } \mathrm{cat}_M(A) \geq k\}, \ \forall k \in N \,.$$

If $f \in C^1(M, R^1)$ is bounded from below, then $\mathcal{F}_k$ is $\Phi_a(f)$-invariant with $a < \inf_M f$, provided by the deformation nondecreasing property (4) of the category. According to minimax principle

$$c_k = \inf_{A\in\mathcal{F}_k} \sup_{x\in A} f(x) \quad k = 1, 2, \ldots, \tag{5.14}$$

are critical values whenever f satisfies the (PS) condition.

Since $\mathcal{F}_{k+1} \subset \mathcal{F}_k$, we have $c_k \le c_{k+1}$, $k = 1, 2, \ldots$. Generally, we do not have $c_k < c_{k+1}$. However, in order to get the multiplicity of critical points, we need:

Lemma 5.2.10 *Let M be a smooth Banach–Finsler manifold. Suppose that $f \in C^1(M, R^1)$ is bounded from below and satisfies the $(PS)_c$ condition, where $c = c_{k+1} = \cdots = c_{k+m}$, are defined by (5.14), then* $\mathrm{cat}_M(K_c) \ge m$.

Proof. Since K_c is compact, provided by the continuity of the category, we have a closed neighborhood N of K_c such that $\mathrm{cat}_M(N) = \mathrm{cat}_M(K_c)$.

From the definition of $c, \forall \epsilon > 0$ $\exists$ a closed subset $A_\epsilon \subset f_{c+\epsilon}$ such that $\mathrm{cat}_M(A_\epsilon) \ge k + m$. Applying first deformation lemma, $\exists \eta : [0, 1] \times M \to M$ satisfying (1)–(5) in Theorem 5.2.8. Let $\phi = \eta(1, \cdot)$, then

$$\phi(A_\epsilon \backslash \overset{\circ}{N}) \subset \phi(f_{c+\epsilon} \backslash \overset{\circ}{N}) \subset f_{c-\epsilon} \,.$$

It follows that

$$\begin{aligned} k + m \le \mathrm{cat}_M(A_\epsilon) &\le \mathrm{cat}_M(A \backslash \overset{\circ}{N}) + \mathrm{cat}_M(N) \\ &\le \mathrm{cat}_M(\overline{\phi(A \backslash \overset{\circ}{N})}) + \mathrm{cat}_M(K_c) \\ &\le \mathrm{cat}_M(f_{c-\epsilon}) + \mathrm{cat}_M(K_c) \\ &\le k + \mathrm{cat}_M(K_c) \,. \end{aligned}$$

Therefore

$$\mathrm{cat}_M(K_c) \ge m \,.$$

□

Proof. Proof of Theorem 5.2.7: Define $c_k, k = 1, 2, \cdots \mathrm{cat}_M(M)$; by (5.14), we have $c_1 \le c_2 \le \ldots \le c_k \le \ldots$. According to Lemma 5.2.10, the multiplicity of $c = c_k$ is dominated by $\mathrm{cat}_M(K_c)$, and then by the number of critical points in K_c. Combining with the fact that distinct critical values correspond to distinct critical points, the conclusion is proved. □

5.2.3 Cap Product

It is natural to ask: If we have two singular homology classes $[\tau_1], [\tau_2] \in H_*(f_b, f_a; G)$, both nontrivial, and if c_1, c_2 are defined in the same way as in (5.12), are there two distinct critical points? Generally speaking, no, but with the aid of the notion of the cap product, the answer is yes if $[\tau_1]$ and $[\tau_2]$ are related by a cap product.

Noticing that there is a ring structure on the cohomology groups. let X be a topological space and $C_q(X,G)$ be the set of all singular q-chains in X with coefficient-group G.

Let $C^q(X,G)$ be the set of all singular q-cochains, i.e., $\mathrm{Hom}(C_q(X,G),G)$. It is a module, and then the duality $[,]$ is a bilinear form over $C_q(X,G) \times C^q(X,G)$.

For a topological pair (X,Y), let

$$C^q(X,Y;G) = \{c \in C^q(X,G) | [\sigma, c] = 0 \ \forall \sigma \in C_q(Y,G)\} .$$

Under the duality, the coboundary operator $\delta_q : C^{q-1}(X,G) \to C^q(X,G)$ is defined to be the dual of the boundary operator $\partial_q : C_q(X,G) \to C_{q-1}(X,G)$:

$$[\partial_q \sigma, c] = [\sigma, \delta_q c] \quad \forall \sigma \in C_q(X,G), \forall c \in C^{q-1}(X,G) .$$

The *qth* singular cohomology group $H^q(X,G) = \ker \delta_{q+1}/\mathrm{Im}\delta_q$.

A cup product structure on the graded singular cochains $\oplus_{q=0}^{\infty} C^q(X,G)$ is defined as follows:

$$[\sigma, c \cup d] = [\sigma \circ \lambda_q, c] \cdot [\sigma \circ \mu_q, d]$$

$\forall c \in C^p(X,G), \forall d \in C^q(X,G)$ and $\forall \sigma \in C_{p+q}(X,G)$, where $\lambda_p : \triangle_p \to \triangle_{p+q}$ and $\mu_q : \triangle_q \to \triangle_{p+q}$ read as: $\lambda_p(\sum_{i=0}^p x_i e_i) = \sum_{i=0}^p x_i e_i$, and $\mu_q(\sum_{i=0}^q x_i e_i) = \sum_{i=0}^q x_i e_{i+p}$, and $\triangle_j$ is the standard j-simplex. It is well defined on $H^*(X,G)$, and makes the latter a graded algebra.

The cup product is bilinear, associative, and possesses the unit element, i.e., the 0-cochain 1, which is defined by $[x,1] = e$, the unit element of $G, \forall x \in X$.

The cap product is defined to be the dual operator of the cup product. For a topological pair (X,Y), one defines $\cap : C_{p+q}(X,Y;G) \times C^p(X,G) \to C_q(X,Y;G)$ by $[\sigma \cap c, d] = [\sigma, c \cup d]$ $\forall \sigma \in C_{p+q}(X,Y;G)$, $\forall c \in C^p(X;G)$, $\forall d \in C^q(X,Y;G)$. Since one has

$$\partial(\sigma \cap c) = (-1)^p (\partial\sigma \cap c - \sigma \cap \delta c) \quad \forall \sigma \in C_{p+q}(X,Y;G) \ \forall c \in C^p(X,G) ,$$

the cap product is also induced to relative singular homology groups:

$$\cap : H_{p+q}(X,Y;G) \times H^p(X,G) \to H_q(X,Y;G) .$$

Definition 5.2.11 *Let (X,Y) be a topological pair. $[\tau_1], [\tau_2] \in H_*(X,Y;G)$ are nontrivial classes. $[\tau_1]$ is called subordinate to $[\tau_2]$ if $\exists \omega \in H^*(X;G)$, dim $\omega > 0$ such that*

$$[\tau_1] = [\tau_2] \cap \omega ,$$

where $\cap$ is the cap product. In this case we write $[\tau_1] < [\tau_2]$.

Lemma 5.2.12 *Let $B \subset X$ be a contractible set, and $[\omega] \in H^q(X;G)$ with $q > 0$. Then $\exists \hat{\omega} \in [\omega]$ such that $|\hat{\omega}| \cap B = \emptyset$.*

Proof. Since B is contractible, there is a closed contractible neighborhood N of B (property (5) of the category). For all closed singular q-chains σ, by adding a $q+1$ singular chain τ with $|\tau| \subset N$ and $|\sigma| \cap N \subset |\partial\tau|$, after subdivision, we find a closed singular q-chain σ_N satisfying:

(1) $\sigma - \sigma_N = \partial\tau$, $\tau \in C_{q+1}(X,G)$ and $|\tau| \subset N$,
(2) $|\sigma_N| \cap B = \emptyset$.

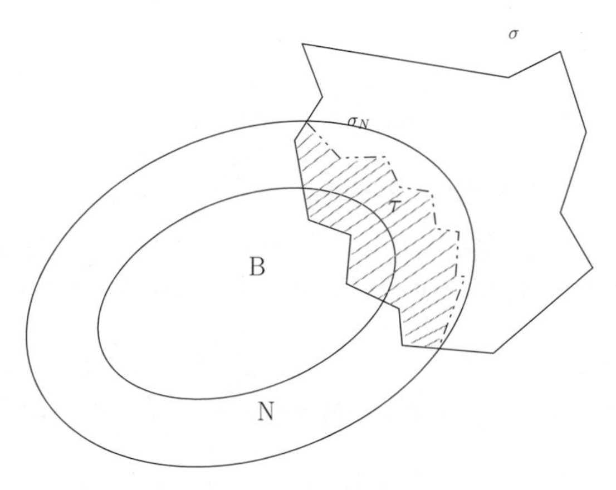

Fig. 5.4.

For any $\omega \in [\omega]$, define $\hat{\omega}$ by

$$[\sigma, \hat{\omega}] = [\sigma_N, \omega] \,,$$

then we have:

(1) $\hat{\omega} \in [\omega]$. Because

$$\begin{aligned} [\sigma, \hat{\omega}] &= [\sigma_N, \omega] \\ &= [\sigma, \omega] - [\partial\tau, \omega] \\ &= [\sigma, \omega] \quad \forall \sigma \in C_q(X,G) \,. \end{aligned}$$

(2) $|\hat{\omega}| \cap B = \emptyset$. Because $\forall \sigma \in C_q(X,G)$ with $|\sigma| \subset B$, we have $|\sigma_N| = \emptyset$. Therefore $[\sigma, \hat{\omega}] = [\sigma_N, \omega] = 0$. □

Theorem 5.2.13 *Let M be a smooth Banach–Finsler manifold. Suppose that $f \in C^1(M, R^1)$ with regular values $a < b$. Assume that $[\tau_1] < [\tau_2] < \cdots < [\tau_m]$ are nontrivial homology classes in $H_*(f_b, f_a; G)$. Let*

$$c_i = \inf_{\tau_i \in [\tau_i]} \sup_{x \in |\tau_i|} f(x), \ i = 1, 2, \ldots, m \, .$$

If $c = c_1 = c_2 = \cdots = c_m$, and if f satisfies the $(PS)_c$ condition, then we have

$$\mathrm{cat}_M(K_c) \geq m \, .$$

Proof. Since K_c is compact, we may choose neighborhoods $N' \subset \overline{N}' \subset N$ of K_c satisfying the following conditions:

(1) $\mathrm{cat}_M(N) = \mathrm{cat}_M(K_c)$;
(2) there exist constants $1 < \epsilon < \overline{\epsilon} < \min\{b - c, c - a\}$ and $\eta \in C(M, M)$ such that

$$\eta|_{f_{c-\overline{\epsilon}}} = \mathrm{id}_{f_{c-\epsilon}}$$
$$\eta \sim \mathrm{id} \, ,$$

and

$$\eta(f_{c+\epsilon} \backslash N') \subset f_{c-\epsilon} \, ,$$

according to the first deformation theorem.

We prove our theorem by contradiction. If $\mathrm{cat}_M(K_c) \leq m - 1$, then there are $(m-1)$ contractible closed sets $\{B_j\}_2^m$ that cover N: $\cup_{j=2}^m B_j \supset N$. Since $[\tau_1] < [\tau_2] < \cdots < [\tau_m]$, there exist $\omega_2, \omega_3, \ldots, \omega_m \in H^*(f_b, G)$, with dim $\omega_j > 0$ such that

$$[\tau_{j-1}] = [\tau_j] \cap \omega_j, \quad j = 2, 3, \ldots, m \, .$$

Provided by Lemma 5.2.12, we always may choose $\hat{\omega}_j \in \omega_j$, with $|\hat{\omega}_j| \cap B_j = \emptyset$, $j = 2, 3, \ldots, m$.

Again we choose $\tau \in [\tau_m]$ with $|\tau| \subset f_{c+\epsilon}$. Subdividing τ into $\tau = \sigma_1 + \sigma_2 + \cdots + \sigma_m$ such that

$$|\sigma_1| \subset f_{c+\epsilon} \backslash N', \text{ and } |\sigma_j| \subset B_j, \ j = 2, 3, \ldots, m \, ,$$

one has

$$\tau' = \tau \cap (\hat{\omega}_2 \cup \cdots \cup \hat{\omega}_m)$$
$$= \sigma_1 \cap (\hat{\omega}_2 \cup \cdots \cup \hat{\omega}_m) \, .$$

Hence $|\tau'| \subset f_{c+\epsilon} \backslash N'$ and $\eta(|\tau'|) \subset f_{c-\epsilon}$. However, $\tau' \in [\tau_1]$; therefore, $\eta(\tau') \in [\tau_1]$, which implies that $c_1 \leq c - \epsilon$. This is a contradiction. □

Definition 5.2.14 *For a pair of topological spaces (X, Y), let*

$$L(X, Y; G) = \sup\{l \in \mathbb{Z}_+ | \ \exists \text{ nontrivial classes } [\tau_1], [\tau_2], \ldots, [\tau_l] \in$$

$$H_*(X,Y;G) \text{ such that } [\tau_1] < \cdots < [\tau_l]\} .$$

$L(X,Y;G)$ measures the length of the chain of subordinate nontrivial singular homology classes. It is directly related to the notion of cup length:

$$CL(X,Y;G) = \sup\{l \in \mathbb{Z}_+ | \ \exists c_0 \in H^*(X,Y;G), \exists c_1,\ldots,c_l \in H^*(X;G) \\ \text{such that } \dim(c_i) > 0, i = 1,2,\ldots,l, \text{and } c_0 \cup \cdots \cup c_l \neq 0\} .$$

It is known from algebraic topology that $\text{cat}_X(X) \geq CL(X,\emptyset;G)$, i.e., the cup length provides a lower bound estimate for the category; we have $\text{cat}_{P^n}(P^n) = n+1$, $\text{cat}_{T^n}(T^n) = n+1$, where P^n is the n- dimensional real projective space, and T^n is the n-torus.

5.2.4 Index Theorem

Symmetric functions, or more generally, a function, which is invariant under some G-group action, may have more critical points. This is due to the fact that the underlying space quotient by the G-group action has more complicated topology. In this subsection we shall deal with this problem.

Let us recall the proof of the Ljusterik–Schnirelmann multiplicity theorem, in which only the fundamental properties (1)–(6) of the category were used. We are inspired to extend an abstract theory based on these properties for G-group action invariant functions.

Let M be a Banach–Finsler manifold with a compact group action G. Let Σ be the set of all G-invariant closed subsets of M, and $\mathcal{H}$ be the set of all G-equivariant continuous mappings from M into itself, i.e., $h \in \mathcal{H}$ if and only if $h \in C(M,M)$ and $h \circ g = g \circ h \ \forall g \in G$.

Definition 5.2.15 *An index $(\Sigma,\mathcal{H},i)$ with respect to G is defined by $i : \Sigma \to \mathbb{N} \cup \{+\infty\}$, which satisfies:*

(1) $i(A) = 0 \Leftrightarrow A = \emptyset, \ \forall A \in \Sigma$.
(2) (Monotonicity) $A \subset B \Rightarrow i(A) \leq i(B), \ \forall A,B \in \Sigma$.
(3) (Subadditivity) $i(A \cup B) \leq i(A) + i(B), \ \forall A,B \in \Sigma$.
(4) (Deformation nondecreasing) If $\eta : [0,1] \times M \to M$ satisfies $\eta(t,\cdot) \in \mathcal{H} \ \forall t \in [0,1]$, and $\eta(0,\cdot) = id_M$, then $i(A) \leq i(\overline{\eta(1,A)}), \ \forall A \in \Sigma$.
(5) (Continuity) If $A \in \Sigma$ is compact, then there is a neighborhood $N \in \Sigma$ of A such that $A \subset \text{int}(N)$ and $i(A) = i(N)$.
(6) (Normality) $i([p]) = 1 \ \forall p \notin \text{Fix}_G$ where $[p] = \{g \cdot p | \ g \in G\}$ and $\text{Fix}_G = \{x \in M | \ g \cdot x = x \ \forall g \in G\}$ is the fixed-point set of G, if $G \neq \{e\}$.

Example 1. If $G = \{e\}$, then category $(\Sigma,\mathcal{H},\text{cat}_M)$ is an index.

Example 2. Recall the genus defined in Sect. 3.3, where M is a Banach space, $G = \mathbb{Z}_2 = \{I,-I\}$, i.e., $Ix = x, (-I)x = -x \ \forall x \in M$,. Thus $\Sigma =$ the set of all closed symmetric subset, $\mathcal{H}$ is the set of odd continuous mappings.

Let

$$\gamma(A) = \begin{cases} 0 & \text{if } A = \emptyset \ , \\ \inf\{k \in \mathbb{N} | \ \exists \text{ an odd map } \phi \in C(A, R^k \backslash \{\theta\})\} \ , \\ +\infty & \text{if no such odd map} \ . \end{cases}$$

Then genus $(\Sigma, \mathcal{H}, \gamma)$ is an index with respect to $\mathbb{Z}_2$.
(Verifications) By definition, (1) and (2) are trivial.

(3) (Subadditivity) Set $\gamma(A) = n$, $\gamma(B) = m$; we may assume that both $n, m < \infty$. This means that $\exists \varphi : A \to R^n \backslash \{\theta\}$ $\exists \psi : B \to R^m \backslash \{\theta\}$ both are odd and continuous. By Tietze's theorem $\exists \widetilde{\varphi} \in C(M, R^n)$, $\exists \widetilde{\psi} \in C(M, R^m)$ such that $\widetilde{\varphi}|_A = \varphi$, $\widetilde{\psi}|_B = \psi$. Without loss of generality, we may assume that $\widetilde{\varphi}$ and $\widetilde{\psi}$ are odd. Define $f(x) = (\widetilde{\varphi}(x), \widetilde{\psi}(x))$, then $f \in C(M, R^{n+m})$ is odd and $f(A \cup B) \subset R^{n+m} \backslash \{\theta\}$. This implies that $\gamma(A \cup B) \leq \gamma(A) + \gamma(B)$.

(4) (Deformation nondecreasing) Suppose $\eta : [0, 1] \times M \to M$ is continuous and odd, and satisfies $\eta(0, \cdot) = \text{id}$ with $\gamma(\overline{\eta(1, A)}) = n$, i.e., $\exists \varphi : \overline{\eta(1, A)} \to R^n \backslash \{\theta\}$ is continuous and odd. Let $\psi = \varphi \circ \eta(1, \cdot)$ then $\psi : A \to R^n \backslash \{\theta\}$ is continuous and odd. Therefore $\gamma(A) \leq n$.

(5) (Continuity) Let $\gamma(A) = n$. From (2), with no loss of generality, we may assume $n < +\infty$. There exists an odd continuous $\varphi : A \to R^n \backslash \{\theta\}$. By Tietze's theorem, there exists an odd continuous mapping $\widetilde{\varphi} : M \to R^n$, with $\widetilde{\varphi}|_A = \varphi$.

Since A is compact and $\theta \notin \widetilde{\varphi}(A)$, $\exists \delta > 0$ such that $\theta \notin \widetilde{\varphi}(N)$, where $N = \overline{A_\delta}$, the closure of the δ-neighborhood of A, i.e.,

$$\widetilde{\varphi} : N \to R^n \backslash \{\theta\} \ .$$

Combining with the monotonicity, we obtain

$$n = \gamma(A) \leq \gamma(N) \leq n$$

(6) (Normality) Note that $\text{Fix}_G = \{\theta\}$, if $p \neq \theta$, then $[p] = \{p, -p\}$. We define $\phi(\pm p) = \pm 1$; it follows that $\gamma([p]) = 1$.

Moreover, applying the Borsuk–Ulam theorem in Sect. 3.2, we have computed $\gamma(S^{n-1}) = n$.

Example 3. (S^1-index) Let M be a Banach space, $S^1 = \{e^{i\theta} | \ \theta \in [0, 2\pi]\}$ be a compact Lie group. Let G be the isometry representation of S^1, i.e., the group consists of homomorphisms $T : S^1 \to L(M, M)$, i.e., $T(e^{i\theta})$ are isometric operators, satisfying $T(e^{i\theta}) \cdot T(e^{i\varphi}) = T(e^{i(\theta+\varphi)})$ $\forall \theta, \varphi \in [0, 2\pi]$. Set

$$\Sigma = \text{the set of all } G - \text{invariant closed subsets of } M \ .$$
$$\mathcal{H} = \{h \in C(M, M), \ T(e^{i\theta}) \circ h = h \circ T(e^{i\theta}) \ \forall \theta \in [0, 2\pi]\} \ .$$

$\forall A \in \Sigma$, define

$$\gamma(A) = \begin{cases} 0 & \text{if } A = \emptyset, \\ \inf\{k \in \mathbb{N} | \ \exists \phi \in C(A, \mathbb{C}^k \backslash \{\theta\}), \exists n \in \mathbb{N} \text{ satisfying} \\ \phi(T(e^{i\theta})x) = e^{in\theta}\phi(x) \ \forall (x, e^{i\theta}) \in A \times S^1\}, \\ +\infty & \text{if there is no such } \phi, \end{cases}$$

where $\mathbb{C}^k$ is k-dimensional unitary space (complex k space).

We shall verify that $(\Sigma, \mathcal{H}, \gamma)$ is an index with respect to S^1.

By definition, (1) and (2) are trivial. Before going on to verify other fundamental properties, we need:

Lemma 5.2.16 *For any $A \in \Sigma$, if $\phi \in C(A, \mathbb{C}^k \backslash \{\theta\})$ satisfies*

$$\phi(T(e^{i\theta})x) = e^{in\theta}\phi(x) \ \forall x \in A \text{ for some } n \in \mathbb{N} \, ,$$

then $\exists \widetilde{\phi} : M \to \mathbb{C}^k$ satisfying $\widetilde{\phi}|_A = \phi$ and

$$\widetilde{\phi}(T(e^{i\theta})x) = e^{in\theta}\widetilde{\phi}(x) \ \forall (x, e^{i\theta}) \in M \times S^1 \, .$$

Proof. By Tietze's theorem, $\exists \hat{\phi} \in C(M, \mathbb{C}^k)$ satisfying $\hat{\phi}|_A = \phi$. Setting

$$\widetilde{\phi}(x) = \frac{1}{2\pi} \int_0^{2\pi} e^{-in\varphi} \hat{\phi}(T(e^{i\varphi})x) d\varphi \, ,$$

we have

$$\widetilde{\phi}(T(e^{i\theta})x) = \frac{e^{in\theta}}{2\pi} \int_0^{2\pi} e^{-in(\theta+\varphi)} \hat{\phi}(T(e^{i(\theta+\varphi)})x) d\varphi = e^{in\theta}\widetilde{\phi}(x) \, ,$$

and $\forall x \in A$, by the assumption.

$$\widetilde{\phi}(x) = \frac{1}{2\pi} \int_0^{2\pi} \phi(x) dx = \phi(x) \, .$$

□

(Verifications) (3) (Subadditivity) $\forall A, B \in \Sigma$, let $\gamma(A) = n$, $\gamma(B) = m$. We may assume $n, m < \infty$. There exist $\varphi : A \to \mathbb{C}^n \backslash \{\theta\}$, $\psi : B \to \mathbb{C}^m \backslash \{\theta\}$ and $k, l \in \mathbb{N}$ such that $\varphi(T(e^{i\theta})x) = e^{ik\theta}\varphi(x)$ and $\psi(T(e^{i\theta})y) = e^{il\theta}\psi(y) \ \forall (x, y, e^{i\theta}) \in A \times B \times S^1$.

Define

$$f(x) = (\widetilde{\varphi}(x)^l, \widetilde{\psi}(x)^k) \, ,$$

where the power map $z \mapsto z^p$ is defined to be $z = (z_1, \ldots, z_k) \mapsto (z_1^p, \ldots, z_k^p)$. Thus $f : A \cup B \to \mathbb{C}^{n+m}$ is continuous and $f(z) \neq \theta$ as $z \in A \cup B$. Moreover,

$$\begin{aligned} f(T(e^{i\theta})x) &= (\widetilde{\varphi}(T(e^{i\theta})x)^l, \widetilde{\psi}(T(e^{i\theta})x)^k) \\ &= e^{ilk\theta}(\widetilde{\varphi}(x)^l, \widetilde{\psi}(x)^k) \\ &= e^{ilk\theta} f(x) \, . \end{aligned}$$

Therefore

$$\gamma(A \cup B) \leq \gamma(A) + \gamma(B) .$$

(4) (Deformation nondecreasing) Assume $A \in \Sigma$, $\eta \in [0,1] \times M \to M$ is continuous and satisfying $\eta(0,\cdot) = \mathrm{id}$, $\eta(t,\cdot) \in \mathcal{H}$, and $\gamma(\overline{\eta(1,A)}) = k$. By definition, $\exists \varphi : \overline{\eta(1,A)} \to \mathbb{C}^k \backslash \{\theta\}$ continuous and satisfying $\varphi(T(e^{i\theta})\eta(1,x)) = e^{in\theta}\varphi(\eta(1,x))$ for some $n \in \mathbb{N}$, $\forall (x, e^{i\theta}) \in A \times S^1$. Now let $\psi = \varphi \circ (\eta(1,\cdot))$. Then $\psi : A \to \mathbb{C}^k \backslash \{\theta\}$ and

$$\begin{aligned} \psi(T(e^{i\theta})x) &= \varphi(\eta(1, T(e^{i\theta})x)) \\ &= \varphi(T(e^{i\theta})\eta(1,x)) \\ &= e^{in\theta}\varphi(\eta(1,x)) = e^{in\theta}\psi(x) . \end{aligned}$$

Therefore $\gamma(A) \leq k$.

(5) (Continuity) Assume that $A \in \Sigma$ is compact and $\gamma(A) = k$. Provided by Lemma 5.2.16, $\exists$ a continuous $\widetilde{\phi} : M \to \mathbb{C}^k$ $\exists n \in \mathbb{N}$ satisfying $\widetilde{\phi}(T(e^{i\theta})x) = e^{in\theta}\widetilde{\phi}(x)$, $\widetilde{\phi}|_A = \phi$ and $\phi : A \to \mathbb{C}^k \backslash \{\theta\}$.

Setting $V = \widetilde{\phi}^{-1}(\mathbb{C}^k \backslash \{\theta\})$, then V is an open neighborhood of A. Since A is compact, $\exists \delta > 0$ such that the closure of the δ-neighborhood of A, $N \subset V$. Therefore

$$\widetilde{\phi} : N \to \mathbb{C}^k \backslash \{\theta\} .$$

It follows that $\gamma(N) \leq k$.

(6) (Normality) Let $p \notin \mathrm{Fix}_G$, i.e., $\exists \theta \in [0, 2\pi]$ such that $T(e^{i\theta})p \neq p$. Since $[p] = \{T(e^{i\theta})p \mid \theta \in [0, 2\pi]\} \in \Sigma$ and is not empty, $\gamma([p]) \geq 1$. There exists $\theta_0 > 0$, which is the minimal period of $T(e^{i\theta})p$, i.e., $T(e^{i\theta_0})p = p$ but $T(e^{i\theta})p \neq p$ $\forall \theta \in (0, \theta_0)$. $n = \frac{2\pi}{\theta_0}$ must be an integer. We define

$$\phi : [p] = \{T(e^{i\theta})p \mid \theta \in [0, \theta_0]\} \to \mathbb{C}^1 \backslash \{\theta\}$$

as follows:

$$\phi : T(e^{i\theta})p \mapsto e^{in\theta} .$$

It is continuous, and

$$\begin{aligned} \phi(T(e^{i\varphi})T(e^{i\theta})p) &= \phi(T(e^{i(\theta+\varphi)}p)) = e^{in(\theta+\varphi)} \\ &= e^{in\varphi}\phi(T(e^{i\theta})p) . \end{aligned}$$

Therefore $\gamma([p]) = 1$.

We have proved that γ so defined is an index.

In particular, if M is a Hilbert space, and G is an unitary representation of S^1, we have the following Borsuk–Ulam-type theorem for S^1 equivariant maps.

Theorem 5.2.17 *Under the above assumptions, suppose that V^{2k} is a $2k$-dimensional S^1-invariant subspace, which is isomorphic to $\mathbb{C}^k$. If $V^{2k} \cap \mathrm{Fix}_G = \{\theta\}$, then $\gamma(V^{2k} \cap \partial B_1(\theta)) = k$.*

Proof. According to Stone's representation theorem for one-parameter unitary groups, one may diagonalize $\{T(e^{i\theta})|\ \theta \in [0, 2\pi]\}$ simultaneously:

$$T(e^{i\theta}) = \text{diag}\{e^{i\lambda_1\theta}, \ldots, e^{i\lambda_k\theta}\} .$$

Since $T(e^{i\theta})$ is 2π-periodic, $\lambda_j \in \mathbb{Z},\ j = 1, 2, \ldots, k$. By the assumption that there is no fixed point of $\{T(e^{i\theta})|\ \theta \in [0, 2\pi]\}$ in $\mathbb{C}^k\backslash\{\theta\}$, therefore $\lambda_j \neq 0,\ j = 1, 2, \ldots, k$.

Firstly, we show $\gamma(V^{2k} \cap \partial B_1(\theta)) \leq k$ by constructing a continuous map $\phi : z = (z_1, \ldots, z_k) \mapsto (z_1^{\frac{\wedge}{\lambda_1}}, \ldots, z_k^{\frac{\wedge}{\lambda_k}})$, where $\wedge = |\lambda_1 \ldots \lambda_k|$. Obviously, $\phi : \mathbb{C}^k \cap \partial B_1(\theta) \to \mathbb{C}^k\backslash\{\theta\}$ continuously and $\phi(T(e^{i\theta})z) = e^{i\wedge\theta}\phi(z),\ \forall z \in \mathbb{C}^k \cap \partial B_1(\theta)$.

Next we are going to show that $\gamma(\mathbb{C}^k \cap \partial B_1(\theta)) \geq k$.

Suppose $\gamma(\mathbb{C}^k \cap \partial B_1(\theta)) = k_1 < k$, i.e., $\exists$ a continuous $\phi : \mathbb{C}^k \cap \partial B_1(\theta) \to \mathbb{C}^{k_1}\backslash\{\theta\}$, and $\exists n \in \mathbb{N}$ such that $\phi(T(e^{i\theta})z) = e^{in\theta}\phi(z)$. From Lemma 5.2.16, there is an continuous extension $\widetilde{\phi} : \mathbb{C}^k \to \mathbb{C}^{k_1}$ with $\widetilde{\phi}|_{\mathbb{C}^k \cap \partial B_1(\theta)} = \phi$. According to Theorem 3.3.10,

$$\deg(\widetilde{\phi}, B_1(\theta), \theta) = \deg(\phi, B_1(\theta), \theta) \neq 0 .$$

Thus for $y \in \mathbb{C}^k$ with small $\| y \|$, one has

$$\deg(\widetilde{\phi}, B_1(\theta), y) \neq 0 .$$

In particular, we take $y \in \mathbb{C}^k\backslash\mathbb{C}^{k_1}$, then $\widetilde{\phi}^{-1}(y) \cap B_1(\theta) \neq \emptyset$, which contradicts $\widetilde{\phi}(\mathbb{C}^k) \subset \mathbb{C}^{k_1}$. □

The Ljusternik–Schnirelmann multiplicity theorem is extended as follows:

Theorem 5.2.18 *(Multiplicity theorem) Let M be a smooth Finsher–Banach manifold with a compact Lie group action G. Let $(\Sigma, \mathcal{H}, i)$ be an index with respect to G. Suppose that $f \in C^1(M, \mathbb{R}^1)$ satisfies $(PS)_c$ conditions, where $c = c_{k+1} = \cdots = c_{k+m}$ is finite, and*

$$c_n = \inf_{i(A) \geq n} \sup_{x \in A} f(x) \quad n = 1, 2, \ldots .$$

Then $i(K_c) \geq m$.

Proof. The proof follows step by step that of Lemma 5.2.10. The only thing to be worked out is that the deformation constructed in Corollary 5.2.9 is G-equivariant. In fact, let $\eta : [0, 1] \times M \to M$ be the deformation obtained in Theorem 5.2.8. Then $\eta_g(t, z) = \int_G g^{-1}\eta(t, g \cdot z)d\mu$ is the G-equivariant deformation, where $d\mu$ is the right invariant Haar measure on compact continuous group G. We omit the verifications. □

5.2.5 Applications

Example 1. (An existence of three nontrivial solutions) Return to Sect. 5.1.5, Example 2, where we proved the existence of a nontrivial solution for the superlinear elliptic BVP both at infinity and at zero. In fact, it was also studied in Sect. 4.8, Example 2, where we showed the existence of a mountain pass solution p. After Corollary 5.2.5, we know that $C_1(f,p) \neq 0$; if further, the condition (Φ) in Theorem 5.1.20 is fulfilled, then we have $C_q(f,p) = \delta_{q1}G$.

We assume condition (Φ) at this moment, and conclude:

Statement 5.2.19 *Under assumptions (g1) (g2) and (g3) of Sect. 5.1.5 Example 2, equation (5.10) possesses at least three nontrivial solutions.*

Proof. We continue to use the notations in Sect. 5.1.5 Example 2.

1. From (g3),

$$J(u) = \frac{1}{2} \parallel u \parallel^2 + \circ(\parallel u \parallel^2) . \tag{5.15}$$

Therefore θ is a local minimum, and $C_q(J,\theta) = \delta_{q0}G$.

2. We claim the existence of two nontrivial solutions. Let us define

$$g_+(x,t) = \begin{cases} g(x,t) & t \geq 0 \\ 0 & t \leq 0 , \end{cases}$$

and

$$J_+(u) = \int_\Omega \left[\frac{1}{2}|\nabla u|^2 - G_+(x,u(x))\right] dx ,$$

where

$$G_+(x,t) = \int_0^t g_+(x,s)ds .$$

Similarly, one verifies that $J_+ \in C^2(H_0^1(\Omega),\mathbb{R}^1)$ satisfies (PS), and that

$$J_+(t\varphi_1) \to -\infty \text{ as } t \to +\infty ,$$

where $\varphi_1 > 0$ is the first eigenvector of $-\triangle$ with 0-Dirichlet data.

On the other hand, $\exists\delta > 0$ such that

$$J_+|_{\partial B_\delta(\theta)} \geq \frac{1}{4}\delta^2 ,$$

From (5.15). The mountain pass lemma is applied to obtain a critical point $u_+ \in H_0^1(\Omega)$, with critical value $c_+ > 0$, which satisfies

$$\begin{cases} -\Delta u_+ = g_+(x,u_+) \\ u_+|_{\partial\Omega} = 0 . \end{cases}$$

By using the maximum principle, $u_+ \geq 0$, so it is again a critical point of J.

Analogously, we define

$$g_-(x,t) = \begin{cases} g(x,t) & t \le 0 \\ 0 & t > 0 \,, \end{cases}$$

and obtain a critical point $u_- \le 0$, with critical value $c_- > 0$. Since (Φ) is assumed, $C_q(J_\pm, u_\pm) = \delta_{q1} G$. According to Theorem 5.1.16, we have

$$C_q(J_\pm, u_\pm) = C_q(\widetilde{J_\pm}, u_\pm) = C_q(\widetilde{J}, u_\pm) = C_q(J, u_\pm) \,,$$

where $\widetilde{J} = J|_{C_0^1}$. Therefore

$$C_q(J, u_\pm) = \delta_{q1} G \,.$$

3. Suppose that there were no other critical points of J. The Morse type numbers over the pair $(H_0^1(\Omega), J_a)$ would be

$$M_0 = 1, M_1 = 2, M_q = 0, q \ge 2 \,,$$

but the Betti numbers read as

$$\beta_q = 0 \quad \forall q = 0, 1, 2, \dots \,.$$

From Lemma 5.1.38: $H_q(H_0^1(\Omega), J_a) \simeq H_q(H_0^1(\Omega), S^\infty) \simeq 0$.

This contradicts the Morse relation. □

Now we return to verifying the condition (Φ): $J''(u) \ge 0$ and $0 \in \sigma(J''(u))$ imply $\dim \ker(J''(u)) = 1$.

It is based on the following:

Theorem 5.2.20 *(Manes–Micheletti) Suppose that $m \in C(\overline{\Omega})$ satisfies $\sup\{m(x) |\ x \in \Omega\} > 0$. Then the equation*

$$\begin{cases} -\triangle u = \lambda m u & \text{in } \Omega \\ u|_{\partial\Omega} = 0 \end{cases}$$

admits a smallest positive eigenvalue λ_1, associated with a positive eigenfunction. Moreover,

$$\dim \ker(-\triangle - \lambda_1 m \cdot) = 1 \,.$$

Continuation of the proof of Statement 5.2.19 (Verification of the condition (Φ))

Now

$$(J''(u)v, v) = \int_\Omega |\nabla v|^2 - m v^2 \quad \forall v \in H_0^1(\Omega) \,, \tag{5.16}$$

where $m(x) = g'(x, u(x))$.

Assume that the right-hand side of equation (5.16) is nonnegative and that $\exists \phi \in H_0^1(\Omega) \backslash \{\theta\}$ satisfies the equation

$$-\triangle v - m \cdot v = 0 \,. \tag{5.17}$$

We want to show that all solutions of equation (5.17) are multiples of ϕ. After Theorem 5.2.20, it is sufficient to show that $\lambda = 1$ is the smallest positive eigenvalue of

$$-\triangle v = \lambda m v \quad v \in H_0^1(\Omega) . \tag{5.18}$$

On the one hand,

$$\lambda_1 := \inf_{v \in H_0^1 \setminus \{\theta\}} \frac{\int |\nabla v|^2}{\int m v^2} \leq \frac{\int |\nabla \phi|^2}{\int m \phi^2} = 1 ,$$

from $\int_\Omega m\phi^2 = \int_\Omega |\nabla \phi|^2 \neq 0$. On the other hand, provided by the nonnegativeness of $J''(u)$,

$$\lambda = \frac{\int_\Omega |\nabla v|^2}{\int_\Omega m v^2} \geq 1$$

for every $\lambda \in \sigma(J''(u))$. Therefore the smallest positive eigenvalue λ_1 of equation (5.18) equals 1. Condition (Φ) is verified.

We turn to the proof of Theorem 5.2.20.

Proof. 1. Let $T : u \mapsto (-\triangle)^{-1} m \cdot u$ be the self-adjoint compact operator on $H_0^1(\Omega)$. According to Courant's max-min characterization,

$$\pm \lambda_{\pm n}^{-1} = \sup_{V \in \mathcal{F}_n} \inf_{u \in V} \frac{\pm \int_\Omega m u^2}{\int_\Omega |\nabla u|^2}, \quad n = 0, 1, 2, \ldots$$

are eigenvalues of T, where $\mathcal{F}_n$ is the family of all n-dimensional linear subspaces of $H_0^1(\Omega)$. Thus, $\pm\lambda_{\pm 1} \leq \pm\lambda_{\pm 2} \leq \cdots$, if they exist.

2. Under our assumption, $\lambda_1 > 0$. Let ω be an eigenfunction. We claim that ω does not change sign. Otherwise, let $\omega_+ = \omega \vee 0$ and $\omega_- = \omega \wedge 0$; they are not zero, then

$$\int_\Omega m\omega^2 = \int_\Omega m\omega_+^2 + \int_\Omega m\omega_-^2 = \alpha_+ + \alpha_- ,$$
$$\int_\Omega |\nabla \omega|^2 = \int_\Omega |\nabla \omega_+|^2 + \int_\Omega |\nabla \omega_-|^2 = \beta_+ + \beta_- ,$$

where $\alpha_\pm = \int_\Omega m\omega_\pm^2 > 0$ and $\beta_\pm = \int_\Omega |\nabla \omega_\pm|^2$.

Either $\frac{\alpha_+ + \alpha_-}{\beta_+ + \beta_-} = \frac{\alpha_+}{\beta_+} = \frac{\alpha_-}{\beta_-}$ or $\frac{\alpha_+ + \alpha_-}{\beta_+ + \beta_-} < \max\{\frac{\alpha_+}{\beta_+}, \frac{\alpha_-}{\beta_-}\}$. However, the latter case cannot occur, because $\lambda_1^{-1} = \frac{\alpha_+ + \alpha_-}{\beta_+ + \beta_-} = \sup_u \int_\Omega mu^2 / \int_\Omega |\nabla u|^2$. In the former case, both $\omega_\pm$ are eigenfunctions with respect to λ_1 and $\pm m \cdot \omega_\pm \geq 0$. From the maximum principle, we obtain $\pm\omega_\pm(x) > 0$ a.e. $x \in \Omega$. This is impossible.

3. We show that $\dim \ker(-\triangle - \lambda_1 m \cdot) = 1$.

Suppose that there are ω_1 and $\omega_2 \in H_0^1(\Omega)$ satisfying

$$-\triangle \omega_i = \lambda_1 m \omega_i \quad i = 1, 2 .$$

Then by step 2, $\forall\alpha \in R^1$, $\omega_1 + \alpha\omega_2$ does not change sign. Let $A_+ = \{\alpha \in R^1 \,|\, \omega_1 + \alpha\omega_2 \geq 0\}$ and $A_- = \{\alpha \in R^1 \,|\, \omega_1 + \alpha\omega_2 \leq 0\}$. Then both A_+ and A_- are nonempty closed subsets and $A_+ \cup A_- = R^1$. Let $\alpha_0 \in A_+ \cap A_-$, then $\omega_1 + \alpha_0\omega_2 = 0$, i.e., ω_1 and ω_2 are colinear. This proves $\dim\ker(-\triangle - \lambda_1 m\cdot) = 1$. □

Example 2. (Spectrum of the p-Laplacian)

Let $\Omega \subset R^n$ be a bounded open domain. $\forall p \in (1, \infty)$ we define the p-Laplacian operator to be

$$\triangle_p u = \mathrm{div}(|\nabla u|^{p-2}\nabla u) \quad \forall u \in W_0^{1,p}(\Omega)\ .$$

In particular, $\triangle_2$ is the Laplacian. λ is called an eigenvalue of $-\triangle_p$ if there exists a nontrivial weak solution $u \in W_0^{1,p}(\Omega)$ of the equation:

$$-\triangle_p u = \lambda|u|^{p-2}u \quad \text{in } \Omega, \tag{5.19}$$

i.e.,

$$\int_\Omega |\nabla u|^{p-2}\nabla u \nabla\varphi = \lambda \int_\Omega |u|^{p-2}u\cdot\varphi \quad \forall\varphi \in W_0^{1,p}(\Omega)\ ,$$

where $\frac{1}{p} + \frac{1}{p'} = 1$.

This is reduced to a constraint variational problem. Set

$$f(u) = \frac{1}{p}\int_\Omega |u|^p \text{ and } g(u) = \frac{1}{p}\int_\Omega |\nabla u|^p\ .$$

Let $M = g^{-1}(p^{-1})$. It is a nonempty closed submanifold of $W_0^{1,p}(\Omega)$. In fact, $\forall u \in M$, $\langle g'(u), u\rangle = \int_\Omega |\nabla u|^p = 1$, which implies that $g'(u) \neq \theta$.

Let $\widetilde{f} = f|_M$. We have

$$d\widetilde{f}(u) = |u|^{p-2}u + \left(\int_\Omega |u|^p\right)(\triangle_p u) \ \in W^{-1,p'}(\Omega)\ ,$$

which satisfies $\langle d\widetilde{f}(u), u\rangle = 0 \ \forall u \in M$.

Set $J_0 = (-\triangle_2)^{-1/2}$ and $J_1 w = |w|^{p'-2}w/\parallel w \parallel_{p'}^{p'-2}$, we have

$$\parallel J_0 w \parallel_{p'} = \parallel w \parallel_{-1,p'},\ \parallel J_0 w \parallel_{1,p} = \parallel w \parallel_p\ ,$$

and

$$\parallel J_1 w \parallel_p = \parallel w \parallel_{p'}\ .$$

Thus $J = J_0 \circ J_1 \circ J_0 : W^{-1,p'}(\Omega) \to W^{1,p}(\Omega)$ is well defined and satisfies:

(1) J is continuous,
(2) $\langle Jw, w\rangle = \parallel w \parallel_{-1,p'}^2 \quad \forall w \in W^{-1,p'}(\Omega)$,
(3) $\parallel Jw \parallel_{1,p} = \parallel w \parallel_{-1,p'} \quad \forall w \in W^{-1,p'}(\Omega)$.

An odd pseudo-gradient vector field for $\widetilde{f}$ is defined by

$$X(u) = J \cdot d\widetilde{f}(u) - \lambda u \in W^{1,p}(\Omega) ,$$

where $\lambda = \langle J \cdot d\widetilde{f}(u), -\triangle_p u\rangle$. We claim that $X(u) \in T_u M$.

In fact, let $s : (-1,+1) \to M$ be a C^1-curve with $s(0) = u$ and $\dot{s}(0) = X(u)$, then

$$\frac{d}{dt} g(s(t))|_{t=0} = \langle X(u), g'(u)\rangle = \langle X(u), -\triangle_p u\rangle = 0 .$$

But we have

$$\langle X(u), d\widetilde{f}(u)\rangle = \langle J \cdot d\widetilde{f}(u), d\widetilde{f}(u)\rangle = \| d\widetilde{f}(u) \|^2_{-1,p'} .$$

From

$$\| X(u) \|_{1,p} \leq \| d\widetilde{f}(u) \|_{-1,p'} + |\lambda| \, \| u \|_{1,p} ,$$

and

$$\begin{aligned} |\lambda| &\leq \| J \cdot d\widetilde{f}(u) \|_{1,p} \| -\triangle_p u \|_{-1,p'} \\ &\leq \| d\widetilde{f}(u) \|_{-1,p'} \| u \|_{1,p} = \| d\widetilde{f}(u) \|_{-1,p'} , \end{aligned}$$

it follows

$$\| X(u) \|_{1,p} \leq 2 \, \| d\widetilde{f}(u) \|_{-1,p'} .$$

Thus $X(u)$ is a pseudo-gradient vector field for $\widetilde{f}$. Now we verify the $(\mathrm{PS})_c$ condition for $\widetilde{f}$, as $c > 0$. Suppose that $\{u_j\}$ is a sequence on M such that $d\widetilde{f}(u_j) \to \theta$ and $\widetilde{f}(u_j) \to c > 0$. $\{u_j\}$ is bounded in $W_0^{1,p}(\Omega)$, and then after a subsequence $u_j \rightharpoonup u_0 \in W_0^{1,p}(\Omega)$. From $d\widetilde{f}(u_j) \to \theta$ $(W^{-1,p'}(\Omega))$, and $\widetilde{f}(u_j) \to c$ one has

$$-\triangle_p u_j \to (pc)^{-1} |u_0|^{p-2} u_0 \quad W^{-1,p'}(\Omega) .$$

From Sect. 2.6, Example 4, $-\triangle_p$ is a monotone operator with a continuous inverse: $(-\triangle_p)^{-1} : W^{-1,p'}(\Omega) \to W_0^{1,p}(\Omega)$, therefore u_j strongly converges in $W_0^{1,p}(\Omega)$.

We notice that M is the unit sphere in $W_0^{1,p}(\Omega)$, and that the function $\widetilde{f}$ is even. Quotient by a $\mathbb{Z}_2$-group, $M/\mathbb{Z}_2$ is homeomorphic to the infinite-dimensional projective space P^∞.

Applying the Ljusternik–Schnirelmann theorem, we have:

Statement 5.2.21 *For $1 < p < \infty$, the equation (5.19) has infinitely many pairs of solutions $(\lambda_m, \pm u_m) \in R^1_+ \times (W_0^{1,p}(\Omega) \backslash \{\theta\})$, $m = 1, 2, \ldots$. Moreover, let σ_p be the spectrum of the p-Laplacian, i.e., the set of all λ associated with nontrivial solutions. Then σ_p is unbounded.*

Proof. Since $\widetilde{f}$ satisfies $(\mathrm{PS})_c$, $c > 0$, we apply the Ljusternik–Schnirelmann theorem to $-\widetilde{f}$, i.e., let

$$c_k = \inf_{\gamma(A)\geq k} \sup_{u\in A} - \widetilde{f}(u) \quad k = 1, 2, \ldots$$

where A is a closed symmetric (with resp. to θ) subset of M and γ is the genus. All these c_k are negativet critical values with $c_k \leq c_{k+1}$. We have $\lambda_k = -\frac{1}{c_k}$, $k = 1, 2, \ldots$.

We shall prove that σ_p is unbounded by contradiction. If not, then there exists $\epsilon > 0$ such that $c_k < -\epsilon$. $\forall k$ large.

Let us choose a sequence of linear subspaces $\{E_j\}_1^\infty$, such that

$$E_j \subset E_{j+1}, \dim E_j = j, \text{and } \overline{\cup_{j=1}^\infty E_j} = W_0^{1,p}(\Omega) .$$

Set

$$d_k = \inf_{\gamma(A)\geq k} \sup_{u\in A\cap E_{k-1}^\perp} - \widetilde{f}(u) ,$$

where $E_k^\perp$ denotes a complementary linear subspace of E_k.

We need the following intersection result:

Lemma 5.2.22 *Let X be a Banach space and $E \subset X$ be a $j < \infty$ dimensional linear subspace. If A is a symmetric closed set with $\gamma(A) > j$ then $A \cap E^\perp \neq \emptyset$, where $E^\perp$ is any complementary subspace of E.*

Proof. We prove it by contradiction. If $A \cap E^\perp = \emptyset$, one defines the projection $P : X \to E$, according to the direct sum composition: $X = E \bigoplus E^\perp$. Define a function $\rho_\epsilon \in C(R_+^1, R_+^1)$ to be $\rho_\epsilon(t) = \epsilon$ as $t \leq \epsilon$, and $\rho_\epsilon(t) = t$ as $t \geq \epsilon$ for $\epsilon > 0$ and an odd deformation:

$$\eta(t, x) = (1 - t)x + t\frac{Px}{\rho_\epsilon(\| Px \|)} .$$

From $A \cap E^\perp = \emptyset$, we have $\epsilon > 0$ such that $P(A) \cap B_\epsilon(\theta) = \emptyset$. Since

$$\eta(1, x) = \frac{Px}{\rho_\epsilon(\| Px \|)}, \ \eta(1, A) \subset E \cap \partial B_1(\theta) .$$

Provided by the monotonicity and the deformation nondecreasing,

$$\gamma(A) \leq \gamma(\overline{\eta(1, A)}) \leq \gamma(E \cap \partial B_1(\theta)) = j .$$

This is a contradiction. □

(Continuing the proof of Statement 5.2.21) According to Lemma 5.2.22, if $\gamma(A) \geq k$, then $A \cap E_{k-1}^\perp \neq \emptyset$, therefore d_k is well defined, and $d_k \leq c_k$, $k = 1, 2, \ldots$.

We suppose $c_k < -\epsilon$ $\forall k$ large, then there exists a closed symmetric subset $A_k \in \sum$, with $\gamma(A_k) \geq k$, and $\exists u_k \in A_k \cap E_{k-1}^{\perp}$ such that $-\widetilde{f}(u_k) < -\epsilon$, i.e., $\widetilde{f}(u_k) > \epsilon$ $\forall k$ large.

However, $u_k \in M$ implies that after a subsequence $u_k \rightharpoonup v$ in $W_0^{1,p}(\Omega)$, and then $u_k \to v$ in $L^p(\Omega)$, in particular, $\widetilde{f}(u_k) \to f(v)$, as $k \to \infty$. But $u_k \in E_{k-1}^{\perp}$, which implies $u_k \rightharpoonup \theta$ in $W_0^{1,p}(\Omega)$, and $f(\theta) = 0$. This contradicts $\widetilde{f}(u_k) > \epsilon$. □

Example 3. Let $\Omega \subset \mathbb{R}^n$ be a bounded domain with smooth boundary. We consider the problem:

$$\begin{cases} -\Delta u(x) = \lambda g(x, u(x)) & \text{in } \Omega \\ u(x) = 0 & \text{on } \partial\Omega \end{cases} \tag{5.20}$$

where $g \in C(\overline{\Omega} \times R^1)$ is odd in t, and satisfies:

(1) $\exists t_0 > 0$ such that $g(x, t_0) \leq 0$ $\forall x \in \overline{\Omega}$
(2) $\exists a \in C(\overline{\Omega})$, $a > 0$, such that $g(x,t) - a(x)t = \circ(t)$ as $t \to 0$ uniformly in $x \in \overline{\Omega}$.

One has the following abstract result:

Theorem 5.2.23 *Let X be a Banach space, and $f \in C^1(X, R^1)$ be an even function satisfying the (PS) condition. Assume $a < b$ and either $f(\theta) < a$ or $f(\theta) > b$. If further,*

(1) there are an m-dimensional linear subspace E and $\rho > 0$ such that

$$\sup_{x \in E \cap \partial B_\rho(\theta)} f(x) \leq b\,,$$

(2) there is a j-dimensional linear subspace F such that

$$\inf_{x \in F^{\perp}} f(x) > a\,,$$

where $F^{\perp}$ is a complementary space of F,
(3) $m > j$,

then f has at least $m - j$ pairs of distinct critical points.

Proof. Let

$$c_n = \inf_{\gamma(A) \geq n} \sup_{x \in A} f(x) \quad n = 1, 2, \ldots$$

1. From condition (1) $E \cap \partial B_\rho(\theta) \subset f_b$, by the monotonicity of genus $\gamma(f_b) \geq \gamma(E \cap \partial B_\rho(\theta)) = m$. We obtain $c_m \leq b$.

2. We verify that $c_n \geq a$ as $n > j$. If it is not true, then $\exists n_0 > j$ such that $c_{n_0} < a$, i.e., $\exists$ a closed symmetric set A with $\gamma(A) \geq n_0$ such that

$f(x) < a\ \forall x \in A$. Condition (2) implies that $f(x) > a\ \forall x \in F^{\perp}$. But according to Lemma 5.2.22, $A \cap F^{\perp} \neq \emptyset$. This is a contradiction.

Combining steps 1 and 2, we obtain critical values in $[a, b]$

$$c_{j+1} \leq \cdots \leq c_m .$$

From the multiplicity theorem, there are at least $m - j$ pairs of critical points. □

Remark 5.2.24 *Another proof of Theorem 5.2.23 based on the estimation of cup length can be found in [Ch 5].*

Now we return to equation (5.20). First, we replace g by the function

$$\hat{g}(x,t) = \begin{cases} g(x,t) & t \leq t_0 \\ g(x,t_0) & t > t_0 . \end{cases}$$

From the maximum principle, we know that if $u \in H_0^1(\Omega)$ is a solution of the equation:

$$\begin{cases} -\triangle u(x) = \lambda \hat{g}(x, u(x)) \text{ in } \Omega \\ \qquad u(x) = 0 \qquad\qquad \text{on } \partial\Omega , \end{cases} \tag{5.21}$$

then $u(x) < t_0\ \forall x \in \overline{\Omega}$, and then it is a solution of equation (5.20).

Let us write

$$p(x,t) = \hat{g}(x,t) - a(x)t$$

and $P(x,t) = \int_0^t p(x,s)ds$. Consider the functional with a parameter λ:

$$J_\lambda(u) = \int_\Omega \left[\frac{1}{2}|\nabla u|^2 - \lambda \left(\frac{au^2}{2} + P(x, u(x)) \right) \right] dx .$$

Obviously, $\forall \lambda \in R^1$, J_λ is bounded from below.

Since $a > 0$, the eigenvalue problem:

$$-\triangle v = \mu a \cdot v \quad v \in H_0^1(\Omega)$$

has eigenvalues $\lambda_1 < \lambda_2 \leq \cdots \leq \lambda_k \leq \cdots$.

If $\lambda > \lambda_k$ then there exists $\rho > 0$ such that $J_\lambda|_{E_k \cap \partial B_\rho(\theta)} < 0$, where E_k is the subspace spanned by eigenfunctions with eigenvalues $\mu \leq \lambda_k$.

Omitting the verification of the (PS) condition, we apply Theorem 5.2.23 and obtain:

Theorem 5.2.25 *Under assumptions (1) and (2) of Example (3), (5.20) possesses at least k distinct pairs of solutions for $\lambda > \lambda_k$.*

5.3 Periodic Orbits for Hamiltonian System and Weinstein Conjecture

For $H \in C^1(R^n \times R^n, R^1)$, the following ODE system:

$$\begin{cases} \dot{x} = -H_p(x,p) \\ \dot{p} = H_x(x,p) \end{cases} \tag{5.22}$$

$\forall\,(x,p) \in R^n \times R^n$, is called a Hamiltonian system, and H is called a Hamiltonian function. Sometimes we use the notations:

$$z = (x,p)\ ,$$

and

$$J = \begin{pmatrix} 0 & -I \\ I & 0 \end{pmatrix}\ ,$$

where I is the $n \times n$ unit matrix. then (5.22) has a simple form:

$$-J\dot{z} = H'(z)\ ,$$

or

$$\dot{z} = JH'(z)\ . \tag{5.23}$$

J grad is called a symplectic gradient.

For the Hamiltonian system two types of periodic solution problems are often examined.

1. For a given period T, find a solution z satisfying (5.23) and $z(t) = z(t+T)$.
2. For a given energy surface, i.e., a real number c, find a periodic solution z of (5.23) lying on the given energy surface, i.e., $H(z(t)) = c$.

The reason why we can set up the type 2 problem is that Hamiltonian systems are conservative:

$$\begin{aligned} \frac{d}{dt}H(z(t)) &= (H'(z(t)), \dot{z}(t)) \\ &= (H'(z(t)), JH'(z(t))) = 0\ , \end{aligned} \tag{5.24}$$

i.e., along a trajectory $t \mapsto z(t)$, the Hamiltonian function is a constant.

Note: the periodic trajectory z on $\Sigma = H^{-1}(c)$ is virtually independent of the behavior of the function H outside Σ, that is, we have:

Lemma 5.3.1 *Let H, $H_1 \in C^1(R^{2n}, R^1), c \in R^1$. Assume that $\Sigma = H^{-1}(c)$ is a compact manifold, If $H|_\Sigma = H_1|_\Sigma$, then*

$$\dot{z} = JH'(z) \ \text{ and } \ \dot{z} = JH_1'(z)$$

has the same periodic trajectories on Σ.

Proof. By the hypotheses, $\exists\, \mu \in C(\Sigma, R^1)$ satisfying

$$H'(z) = \mu(z) H_1'(z), \quad \forall z \in \Sigma\,.$$

Since $\mu(z) \neq 0$ and Σ is compact, there are constants $m, M > 0$,

$$m \leqslant \mu(z) \leqslant M, \quad \forall z \in \Sigma\,. \tag{5.25}$$

Let z be an integral curve of H on Σ, define

$$z_1(t) = z \circ s(t)\,,$$

where $s : R^1 \to R^1$ will be determined later. First we solve the equation:

$$\begin{cases} \dot{\sigma}(t) = \frac{1}{(\mu \circ z)(\sigma(t))} \\ \sigma(0) = 0\,. \end{cases} \tag{5.26}$$

Since the functions μ, z are all bounded and continuous, (5.26) has a solution on every open interval in R^1. Let T be the period of z, since $\sigma(t) \to +\infty$ as $t \to +\infty$, there exists $\tau > 0$ such that $\sigma(\tau) = T$. Let

$$s(t) = \left[\frac{t}{\tau}\right] T + \sigma\left(t - \left[\frac{t}{\tau}\right]\tau\right),$$

where $[t]$ denotes the largest integer less than t, then $s \in C^1(R^1, R^1), s(0) = 0$.

It remains to verify the differentiability at $t = j\tau, j = 0, \pm 1, \pm 2, \ldots$. In fact, for $j = 1$,

$$\begin{aligned} s'(\tau + 0) = \dot{\sigma}(0) &= \frac{1}{(\mu \circ z)(\sigma(0))} = \frac{1}{(\mu \circ z)(0)} = \frac{1}{(\mu \circ z)(T)} \\ &= \dot{\sigma}(\tau - 0) = s'(\tau - 0) \end{aligned}$$

Similarly, we prove this for $j \neq 1$. Hence, $s : R^1 \to R^1$ is a diffeomorphism and satisfies

$$\begin{cases} \dot{s}(t) = \frac{1}{(\mu \circ z)(s(t))}\,, \\ s(0) = 0\,. \end{cases}$$

It follows that

$$\dot{z}_1(t) = \dot{s}(t)\dot{z}(t) \circ s(t) = J H_1'(z_1(t))\,,$$

and $z_1(0) = z(0) \in \Sigma$, hence z_1 is an integral curve of H_1 on Σ. The converse can be proved in the same way. □

Thus a given hypersurface Σ determines the orbits of every Hamiltonian vector field J grad H with Σ as regular energy surface $H^{-1}(c)$ for some constant c.

In this sense, one can ask the question: Which hypersurfaces carry a periodic orbit?

5.3.1 Hamiltonian Operator

Let S^1 be the unit circle $\{e^{it} \mid t \in [0, 2\pi]\}$ and $V = R^n \times R^n$. Define the real Hilbert space $L^2(S^1, V)$ as follows: $z \in L^2(S^1, V)$ means that z is defined on S^1 with range in V and $|z|^2$ is integrable, the inner product is defined by

$$\langle z, w \rangle = \frac{1}{2\pi} \int_{S^1} \sum_{j=1}^{2n} z_j(t) w_j(t) dt, \qquad \forall z, w \in L^2(S^1, V) .$$

Similarly define the Sobolev space $H^1(S^1, V)$ with the inner product

$$\langle z, w \rangle = \frac{1}{2\pi} \int_{S^1} \sum_{j=1}^{2n} [z_j(t) w_j(t) + \dot{z}_j(t) \dot{w}_j(t)] \, dt .$$

Let

$$J = \begin{pmatrix} 0 & -I_n \\ I_n & 0 \end{pmatrix} ,$$

where I_n is the $n \times n$ identity matrix. Consider the linear operator $A : z \mapsto -J \frac{d}{dt} z$ in $L^2(S^1, V)$ with domain $D(A) = H^1(S^1, V)$.

The real space V is isomorphic to the complex linear space $\mathbb{C}^n = R^n + iR^n$. The isomorphism is defined as follows: Let $e_1, \ldots, e_{2n}$ be an orthonormal basis in V, and let

$$\phi_j = e_j + i e_{j+n}, \quad j = 1, 2, \ldots, n .$$

They form a basis in $\mathbb{C}^n$. Let $z = \sum_{j=1}^{2n} z_j e_j$ be corresponding to $\hat{z} = \sum_{j=1}^{n} (z_j - i z_{j+n}) \phi_j$. Define the inner product

$$\begin{aligned} [\, \hat{z}, \hat{w} \,] &= \operatorname{Re} \sum_{j=1}^{n} (z_j - i z_{j+n}) \overline{(w_j - i w_{j+n})} \\ &= \sum_{j=1}^{n} (z_j w_j + z_{j+n} w_{j+n}) = \sum_{j=1}^{2n} z_j w_j . \end{aligned}$$

Then the correspondence $z \mapsto \hat{z}$ is linear and inner product preserving, and then $\{\phi_j \mid j = 1, \ldots, n\}$ is an orthogonal basis in $\mathbb{C}^n$.

The real Hilbert space $L^2(S^1, V)$ is isomorphic and homeomorphic to the complex Hilbert space $L^2(S^1, \mathbb{C}^n)$ with inner product defined by

$$\frac{1}{2\pi} \int_{S^1} [\, \hat{z}(t), \hat{w}(t) \,] dt = \langle z, w \rangle ,$$

then $z \mapsto \hat{z}$ is an inner product preserving linear isomorphism of $L^2(S^1, V) \to L^2(S^1, \mathbb{C}^n)$.

The space $L^2(S^1, \mathbb{C}^n)$ has an orthonormal basis:

$$\{ e^{-imt} \phi_j \mid j = 1, 2, \ldots, n; m = 0, \pm 1, \pm 2, \ldots \} .$$

Every $\hat{z} \in L^2(S^1, \mathbb{C}^n)$ has the Fourier expansion

$$\hat{z} = \sum_{j=1}^{n} \left(\sum_{m=-\infty}^{+\infty} c_{jm} e^{-imt} \right) \phi_j \, .$$

One has

$$\sum_{j=1}^{n} \sum_{m=-\infty}^{+\infty} |c_{jm}|^2 = \|z\|^2 = \frac{1}{2\pi} \int_{S^1} [\,\hat{z}, \hat{z}\,] dt \, .$$

Hence

$$\hat{z} \leftrightarrow c = \{c_{jm}\}$$

is an isometric isomorphism of $L^2(S^1, \mathbb{C}^n) \to (l^2)^n = \prod_{j=1}^{n} l^2$.

According to the Fourier expansion, $\hat{z} \in H^1(S^1, \mathbb{C}^n)$ if and only if

$$\sum_{j=1}^{n} \sum_{m=-\infty}^{+\infty} (1 + |m|)^2 |c_{jm}|^2 < +\infty \, .$$

Note that

$$-J \frac{d}{dt} (e^{-imt} \phi_j) = m e^{-imt} \phi_j \, ;$$

it is easy to verify that A is a self-adjoint operator with domain $D(A) = H^1(S^1, \mathbb{C}^n)$, and then $\{e^{-imt}\phi_j\}$ is just the diagonalized orthonormal basis. Let

$$M(m) = \mathrm{span}\{e^{-imt}\phi_1, \ldots, e^{-imt}\phi_n\}, \;\; m \in \mathbb{Z} \, ,$$

then

$$L^2(S^1, \mathbb{C}^n) = \oplus_{m \in \mathbb{Z}} M(m) \, ;$$

this is the spectral decomposition or A.

5.3.2 Periodic Solutions

Theorem 5.3.2 *Suppose that* $H \in C^2(R^n \times R^n, R^1)$ *satisfies:*

(1) $H(\theta) = 0, H'(\theta) = \theta, H''(\theta) = 0, H(z) \geqslant 0.$
(2) $H(z) = \frac{\gamma}{2}|z|^2, \quad$ *for* $\quad |z| \geqslant R, \; R > 0, \gamma \in (1, 2).$

Then the Hamiltonian system (5.22) possesses a nontrivial 2π*-periodic solution.*

Proof. Define $G(z) = H(z) - \frac{\gamma}{2}|z|^2$. Set $A = -J\frac{d}{dt}$,

$$\psi(z) = \int_{S^1} H(z(t)) dt \, ,$$

and

$$g(z) = \int_{S^1} G(z(t)) dt \, ;$$

then $\psi \in C^2(\mathcal{H}, R^1)$, $(\mathcal{H} = L^2(S^1, \mathbb{C}^n))$ possesses the following properties:

(1) $\psi(\theta) = 0, \psi'(\theta) = \theta, \psi''(\theta) = 0, \psi(x) \geqslant 0 \qquad \forall x \in \mathcal{H}$.
(2) $\exists\, b, M > 0$ such that $\forall x \in \mathcal{H}$

$$\begin{aligned} \|g'(x)\|_{L(\mathcal{H})} &\leqslant M \,, \\ \psi(x) &\geqslant \frac{\gamma}{2}\|x\|^2_{\mathcal{H}} - b \,, \\ \psi'(x) &= \gamma x + g'(x) \,. \end{aligned}$$

Denote the Sobolev space $H^{\frac{1}{2}}(S^1, \mathbb{C}^n) = D(|A|^{\frac{1}{2}})$ by E, and let E^* denote the subspace associated with the $*$ eigenvalues of E, where $* = +, 0, -$. Let P^* be the orthogonal projections onto E^*, $* = +, 0, -$. We define a bilinear form on E:

$$((z, w)) = (|A|^{\frac{1}{2}} z, |A|^{\frac{1}{2}} w) \,,$$

where $(\ ,\ )$ is the inner product on $\mathcal{H}$. The system (5.22) is rewritten as

$$Az = \psi'(z), \qquad z \in D(A) \,, \tag{5.27}$$

and equivalently,

$$(A - \gamma) z = g'(z) \,.$$

We introduce a functional on E:

$$\Phi(x) = \frac{1}{2}(((P^+ - P^-)x, x)) - \psi(x) \,.$$

Obviously, if $z \in D(A)$ is a critical point of Φ, then z solves (5.27).

We choose an arbitrary element e in $M(1)$ with norm one, and set

$$\begin{aligned} Q &= ((E^- \oplus E^0) \cap B_R(\theta)) \times [0, R]e \quad \text{for } R > 0 \text{ large, and} \\ S &= E^+ \cap \partial B_\varepsilon(\theta) \qquad \text{for } \varepsilon > 0 \text{ small} \,. \end{aligned}$$

Since $\psi(x) = o(\|x\|^2_{\mathcal{H}})$ as $\|x\|_{\mathcal{H}} \to 0$, we have $\beta > 0$ such that

$$\inf_S \Phi \geqslant \frac{\varepsilon^2}{2} + o(\varepsilon^2) \geqslant \beta \,.$$

We turn out to estimate the values of Φ on ∂Q:

$$\begin{aligned} \Phi(x) \leqslant -\psi(x) \leqslant 0 \qquad & \forall x = x^- + x^0 \in E^- \oplus E^0, \\ \Phi(x^- + x^0 + \mathrm{se}) \leqslant -\frac{1}{2}\|x^-\|^2 + \frac{1}{2}(1-\gamma)s^2 + b - \frac{\gamma}{2}\|x^0\|^2 \leqslant 0 \,, & \end{aligned}$$

as $x^- + x^0 + \mathrm{se} \in \partial Q, s > 0$ and R large enough. In summary $\Phi|_{\partial Q} \leqslant 0$.

Since both S and Q are infinite dimensional, and the intersection of ∂Q and S has not been discussed previously, we shall use the Galerkin method instead. Let $E_k = \oplus_{|j| \leqslant k} M(j)$, and $\Phi_k = \Phi|_{E_k}$. Then S_k and ∂Q_k link, where $S_k = S \cap E_k, Q_k = Q \cap E_k, \forall k$. We verify the (PS) condition for Φ_k. Denote by Π_k the orthogonal projection onto E_k.

Let $\{x_j\} \subset E_k$ be a sequence along which $\Phi_k'(x_j) \to \theta$. Then

$$(A-\gamma)x_j = \Pi_k g'(x_j) + o(1) ,$$

i.e.,

$$x_j = (A-\gamma)^{-1}(\Pi_k g'(x_j) + o(1)) .$$

Since $g'(x_j)$ is bounded, $\{x_j\}$ is bounded, and then $\{x_j\}$ is subconvergent. Applying the minimax theorem for linkings, we have a critical point $z_k \in E_k$ satisfying

$$\begin{cases} (A-\gamma)z_j = \Pi_k g'(z_k) , \\ \Phi(z_k) \geqslant \beta > 0 . \end{cases}$$

It remains to prove that $\{z_k\}$ is subconvergent in E.

Since we have

$$z_k = (A-\gamma)^{-1}\Pi_k g'(z_k) ,$$

and that the linear map $(A-\gamma)^{-1} : L^2 \to H^1 \to E$ is compact, $\{z_k\}$ has a convergent subsequence with limit $\hat{z}$. Thus $\Phi(\hat{z}) \geqslant \beta > 0$ and $\hat{z} = (A-\gamma)^{-1}g'(\hat{z})$, which implies $\hat{z} \in D(A)$ and that $\hat{z}$ is a nontrivial solution of (5.27). The proof is complete. □

5.3.3 Weinstein Conjecture

As to the given energy question, the breakthrough is due to Rabinowitz and Weinstein. In 1978, Rabinowitz [Ra 3] proved that a strongly star-shaped hypersurface (Weinstein [Wei 1] proved for a convex hypersurface) carries at least one periodic orbit. Nevertheless, the star-shapedness, and then the convexity, is not invariant under canonical diffeomorphism. Weinstein [Wei 2] called for a symplectic invariant version. He made the following conjecture:

Every compact smooth hypersurface of contact type carries at least a periodic orbit.

A hypersurface Σ is said to be of contact type if there exists a vector field X in a neighborhood of Σ, such that

(1) $X \pitchfork \Sigma$.
(2) The flow ϕ^t, generated by X (i.e., $\frac{d}{dt}\phi^t = X \circ \phi^t$ and $\phi^0 = \mathrm{id}$) satisfies $(d\phi^t)J(d\phi^t)^T = e^t J, \quad \forall t$.

A matrix A satisfying $AJA^T = \mu J$, where μ is a nonzero constant, is called a canonical matrix with multiple μ. A diffeomorphism $\phi : R^{2n} \to R^{2n}$ satisfying $(d\phi)J(d\phi)^T = \mu J$, is called a canonical diffeomorphism with multiple μ.

Remark 5.3.3 *In the definition of contact type, the condition (2) is equivalent to* $L_X\omega = \omega$, *where* ω *is the sympectic form* $\omega(u,v) = \langle Ju, v\rangle$, $\forall u, v \in R^{2n}$, $\langle\cdot\rangle$ *is the inner product in* R^{2n}, *and* L_X *is the Lie-derivative of the 2-form.*

Lemma 5.3.4 *Suppose that ϕ is a canonical diffeomorphism with multiple μ, and that $\widehat{z}$ is a solution of the Hamiltonian system*

$$\dot{z} = JH'(z) .$$

Let $\widehat{H}(u) = \mu H(z)$, where $u = \phi(z)$. Then $\hat{u} = \phi(\hat{z})$ is a solution of

$$\dot{u} = J\widehat{H}'(u) .$$

Proof. By changing variables:

$$\begin{aligned}\dot{u} = d\phi(z) \cdot \dot{z} &= (d\phi(z))\, J(\nabla_z H) \\ &= \frac{1}{\mu}\,(d\phi(z))\, J\,(d\phi(z))^T\, \nabla_u \widehat{H} \\ &= \frac{1}{\mu}\,(d\phi(z)) \cdot J \cdot (d\phi(z))^T \cdot \left(\nabla_u \widehat{H}\right) \\ &= J\widehat{H}'(u) .\end{aligned}$$

Example 5.3.5 *A hypersurface Σ is called strongly star-shaped, if there exists a point $x_0 \in R^{2n}$, such that*

$$\langle x - x_0, n(x)\rangle > 0 \qquad \forall x \in \Sigma ,$$

where $n(x)$ is the unit outward normal vector on Σ. A strongly star-shaped hypersurface is of contact type.

Indeed, without loss of generality, we may assume $x_0 = \theta$. Let the vector field

$$X(x) = \frac{1}{2}x .$$

The strong star-shapedness implies that $X \pitchfork \Sigma$; the flow ϕ^t generated by X is defined to be

$$\phi^t(x) = e^{t/2}x ,$$

i.e., $\phi^t(x) = e^{t/2}\mathrm{id}$. Therefore,

$$(d\phi^t)J(d\phi^t)^T = e^t J .$$

Now we turn to the Weinstein conjecture. By definition, for a contact-type hypersurface Σ, we have a family of hypersurfaces $\Sigma_\varepsilon = \phi^\varepsilon(\Sigma), \varepsilon \in (-1,1)$. Let $0 < \delta < 1$, then $R^{2n} \backslash \bigcup \{\Sigma_\varepsilon \,|\, |\varepsilon| < \delta\}$ has two components: The interior B, which with no with loss of generality contains the origin, and the exterior A, which is unbounded. Also one may assume $\bigcup \{\Sigma_\varepsilon \,|\, \varepsilon \in (-1,-\delta)\} \subset B$.

Let $d = \mathrm{diam} \bigcup \{\Sigma_\varepsilon \,|\, \varepsilon \in (-1,1)\}$. We define $d < r < 2d$, $\frac{\gamma}{2}r^2 < b < r^2, \gamma = \frac{3}{2}$, and functions $f \in C^\infty(-1,1)$, $g \in C^\infty(0,\infty)$ satisfying

$$f(s) = \begin{cases} 0, & s \in (-1, -\delta] , \\ b, & s \in [\delta, 1) , \end{cases}$$

$$g(s) = \begin{cases} b, \ s \leq r, \\ \frac{\gamma}{2} s^2, \ s \geq R, \text{ where } R > r , \end{cases}$$

$$g(s) \geq \frac{\gamma}{2} s^2 \text{ for } s > r ,$$

$$f'(s) > 0 \quad \text{for} \quad s \in (-\delta, \delta) \qquad 0 < g'(s) < \gamma s \qquad s \geqslant r .$$

Then we define a Hamiltonian function:

$$H(z) = \begin{cases} 0 & z \in B , \\ f(\varepsilon) & z \in \Sigma_\varepsilon \quad |\varepsilon| \leqslant \delta , \\ b & z \in A \quad \text{and} \quad |z| \leqslant r , \\ g(|z|) & |z| > r . \end{cases}$$

The function H satisfies assumptions (1) and (2) of Theorem 5.3.2.

According to Theorem 5.3.2, there is a nontrivial 2π-periodic solution $\hat{z}$ for this Hamiltonian system, with

$$\Phi(\hat{z}) = \frac{1}{2}\left(-J\frac{d}{dt}\hat{z}, \hat{z}\right) - \int_{S^1} H(\hat{z}(t))dt \geqslant \beta \text{ for some } \beta > 0 .$$

Lemma 5.3.6 *The solution $\hat{z}$ lies on Σ_ε for some $\varepsilon \in (-\delta, \delta)$.*

Proof. 1. If $|\hat{z}(0)| > r$, then $|\hat{z}(t)| = |\hat{z}(0)|$, because the energy surface of H at $\hat{z}(0)$ is a sphere. Thus

$$\begin{aligned} \Phi(\hat{z}) &= \frac{1}{2} g'(|\hat{z}(0)|)\, |\hat{z}(0)|^2 \cdot 2\pi - \int_0^{2\pi} g(|\hat{z}(t)|)dt \\ &= \gamma\pi|\hat{z}(0)|^2 - \gamma\pi|\hat{z}(0)|^2 = 0 . \end{aligned}$$

This is impossible.

2. If $|\hat{z}(0)| \leqslant r$ but $\hat{z}(0) \notin \bigcup_{|\varepsilon|<\delta} \Sigma_\varepsilon$, then $H(\hat{z}(0)) = b$ or 0. In either case, $\hat{z}(t) = \text{const}$. Thus

$$\Phi(\hat{z}) = -\int_0^{2\pi} H(|\hat{z}(t)|)dt \leqslant 0 .$$

Again, it is impossible.

In the remaining case, $\hat{z}(0) \in \Sigma_\varepsilon$ for some $\varepsilon \in (-\delta, \delta)$, then the whole orbit $\hat{z}(t)$ lies on Σ_ε, because Σ_ε is an energy surface of H.

The lemma is proved. □

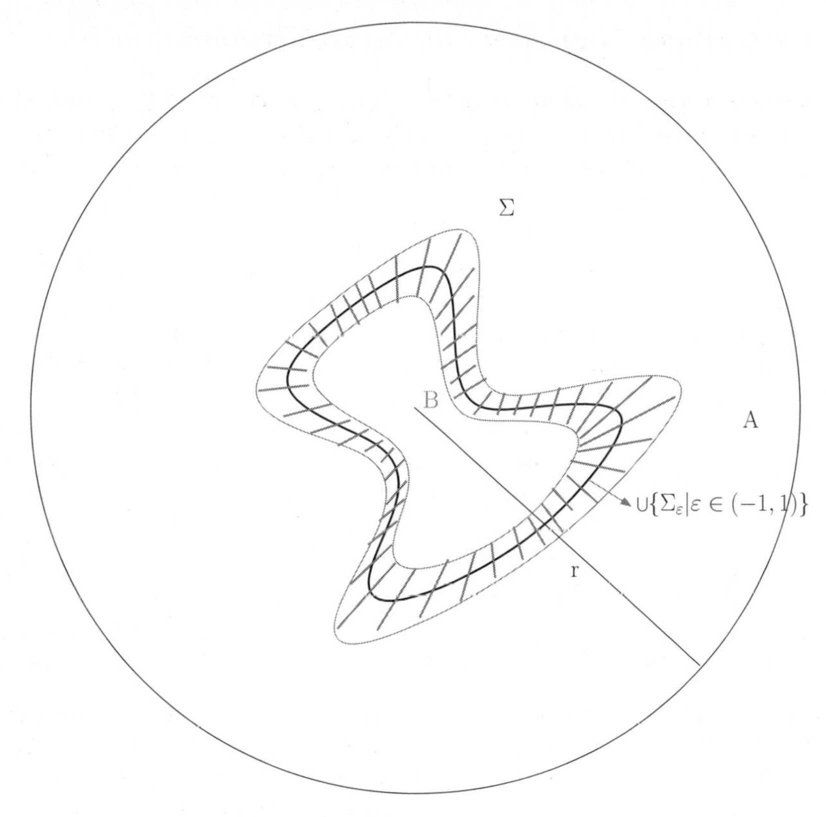

Fig. 5.5.

Theorem 5.3.7 *(Viterbo) Every compact smooth hypersurface of contact type carries at least one periodic orbit.*

Proof. Since $\phi^{-\varepsilon}$ is a canonical diffeomorphism, which maps Σ_ε onto Σ, the Hamiltonian $\overline{H} = H \circ \phi^{\varepsilon}$, i.e.,

$$\overline{H}(u) = H(z)$$

with $u - \phi^{-\varepsilon}(z)$, has Σ as an energy surface. According to Lemma 5.3.6, the periodic solution $\hat{z}$ on Σ_ε reduces to a periodic solution $\hat{u} = \phi^{-\varepsilon}(\hat{z})$ on Σ of the Hamiltonian system (5.22) with Hamiltonian function $\overline{H}$. However, $\overline{H}$ and H both have Σ as an energy surface; by Lemma 5.3.1, $\hat{u}$ is also a periodic solution of the Hamiltonian system (5.22) with Hamiltonian function H. The proof is complete.

The above proof is due to Hofer and Zehnder (see [HZ 1]).

5.4 Prescribing Gaussian Curvature Problem on S^2

We continue the study of the prescribing Gaussian curvature problem in Sect. 4.2 Example 4. Let g_0 be the canonical metric on S^2, where $k = 1$, and the problem is reduced to the following equations:

$$\triangle u = 1 - Ke^{2u} \text{ on } S^2, \tag{5.28}$$

where $K \in C^\infty(S^2)$ is given.

Again, $\wedge = \max K > 0$ is a necessary condition for the existence of a solution. However, in this section we always assume that $K(x) > 0 \; \forall x \in S^2$.

We introduce

$$S(u) = \frac{1}{4\pi} \int_{S^2} [|\nabla u|^2 + 2u] \, ,$$

and

$$J(u) = S(u) - \log \frac{1}{4\pi} \int_{S^2} Ke^{2u} \, .$$

The Euler–Lagrange equation for J reads as

$$\triangle u = 1 - \frac{4\pi Ke^{2u}}{\int_{S^2} Ke^{2u}} \text{ on } S^2 \, . \tag{5.29}$$

Noticing that $J(u+c) = J(u) \; \forall c \in \mathbb{R}^1$, u_0 is a critical point of J if and only if u solves (5.28), where $u = u_0 - c$, and $c = \frac{1}{2} \log \frac{1}{4\pi} \int_{S^2} Ke^{2u_0}$. Once we obtain a critical point of J, after adjusting a constant, we solve equation (5.29). In order to avoid the indetermination caused by the translation invariance of J, we add a constraint. Let

$$X = \left\{ u \in H^1(S^2) | \int_{S^2} e^{2u} = 4\pi \right\} ,$$

and consider the functional J on X. The Euler–Lagrange equation is again (5.29).

5.4.1 The Conformal Group and the Best Constant

In this section, we first analyze the special case $K \equiv 1$, and find out all the minimizers.

Now, we introduce the following stereographic projection:

$$\pi : S^2 \to \hat{\mathbb{C}} = \mathbb{C} \cup \{\infty\}$$
$$\pi(x_1, x_2, x_3) = z = \frac{x_1 + ix_2}{1 - x_3} \quad \forall (x_1, x_2, x_3) \in S^2 \subset R^3 \, .$$

This is a conformal diffeomorphism with

$$(\pi^{-1})^* g_0 = \frac{4}{(1 + |z|^2)^2} dz d\overline{z} \, .$$

Let $\mathrm{Conf}(S^2)$ be the conformal group consisting of all conformal diffeomorphisms from S^2 into itself. $\mathrm{Conf}(S^2)$ is isomorphic to the Möbius group $SL(2,\mathbb{C})$. A subgroup $\mathcal{D}$ of $\mathrm{Conf}(S^2)$ is defined as follows: for any $Q \in S^2$, we choose an orthonormal frame $\{e_1, e_2, e_3\}$, such that Q is the north pole, i.e., $Q = e_3$, by which we define a stereographic projection π_Q, $\forall t \in [1, \infty)$, let $\tau_t z = tz \ \forall z \in \hat{\mathbb{C}}, \phi_{Q,t} = \pi_Q^{-1} \circ \tau_t \circ \pi_Q$, and let

$$\mathcal{D} = \{\phi_{Q,t} | \ (Q,t) \in S^2 \times [1,\infty)\} \ .$$

Since

$$\phi_{Q,1} = \mathrm{id}, \ \text{and} \ \phi_{-Q,t} = \phi_{Q,t^{-1}} \ ,$$

$\mathcal{D}$ is a submanifold and is parameterized to be $B^3 \simeq (S^2 \times [1,\infty))/(S^2 \times \{1\})$ under the correspondence: $\phi_{Q,t} \to \frac{t-1}{t}Q$.

For any Riemannian metric g on S^2, $\forall \phi \in \mathrm{Conf}(S^2)$, the function

$$\psi = \frac{1}{2} \log \det(d\phi) \tag{5.30}$$

satisfies $\phi^* g_0 = e^{2\psi} g_0$. The conformal group action on the function v reads as:

$$\phi^* v := v_\phi = v \circ \phi + \psi \ . \tag{5.31}$$

We define the conformal Laplacian as follows:

$$L_g \varphi = -\triangle_g \varphi + K_g \quad \forall \varphi \in C^\infty(S^2) \ , \tag{5.32}$$

where $\triangle_g$ is the Laplacian–Beltrami operator and K_g is the Gaussian curvature with respect to g, respectively. Thus, for $g = e^{2u} g_0$, we have

$$\begin{aligned} L_g \varphi &= -\triangle_g \varphi + K_g \\ &= e^{2\varphi} K_{e^{2\varphi} g} \\ &= e^{-2u} \cdot e^{2(u+\varphi)} K_{e^{2(u+\varphi)} g_0} \\ &= e^{-2u} L_{g_0}(u + \varphi) \ . \end{aligned} \tag{5.33}$$

Lemma 5.4.1

$$L_{g_0} v_\phi = \det(d\phi)(L_{g_0} v) \circ \phi \tag{5.34}$$

Proof.

$$\begin{aligned} (L_{g_0} v) \circ \phi &= L_{e^{2\psi} g_0}(v \circ \phi) \\ &= e^{-2\psi} L_{g_0}(v \circ \phi + \psi) \\ &= \det(d\phi)^{-1} L_{g_0} v_\phi \ , \end{aligned}$$

provided by (5.33) and (5.31). □

Lemma 5.4.2 *All solutions of the equation:*

$$-\triangle\psi + 1 = e^{2\psi} \quad on\ S^2 \tag{5.35}$$

are of the form (5.30).

Proof. Since $L_{g_0}\psi = -\triangle\psi + 1$, plugging $\psi = o_\phi$ into (5.34), (5.35) follows from Lemma (5.4.1). Conversely, if ψ is a solution of (5.35), then $K_g = 1$ for $g = e^{2\psi}g_0$. According to the isometric theorem in Riemannian geometry, there exists an isometry $\phi : (S^2, g) \to (S^2, g_0)$ satisfying $\phi^* g_0 = g$, i.e., $\phi \in \text{Conf}(S^2)$. By (5.30), ψ has the form $\frac{1}{2}\log\det(d\phi)$. □

Lemma 5.4.3 *Both the functionals S and $T(u) = \frac{1}{4\pi}\int_{S^2} e^{2u}$ are conformal invariant. Furthermore, $S(\psi) = 0\ \forall\psi$ of form (5.30).*

Proof. 1. Obviously, $\forall\phi \in \text{Conf}(S^2), \forall v \in H^1(S^2)$,

$$T(v_\phi) = \frac{1}{4\pi}\int_{S^2} e^{2(v\circ\phi)}\det(d\phi) = \frac{1}{4\pi}\int_{S^2} e^{2v} = T(v)\,.$$

2. We prove that $S(\psi) = 0\ \forall\psi$ of the form (5.30).

For any $\phi_0 \in \text{Conf}(S^2)$, we introduce a differentiable path $\phi_t \in \text{Conf}(S^2)$, with $\phi_t|_{t=0} = \phi_0$. Let $\psi_t = \frac{1}{2}\log\det(d\phi_t)$, and $w = \frac{d}{dt}\psi_t|_{t=0}$. Then by step 1,

$$0 = \frac{d}{dt}T(o_{\phi_t})|_{t=0} = \frac{1}{2\pi}\int_{S^2} e^{2\psi_0}w\,, \tag{5.36}$$

and

$$\begin{aligned}\frac{d}{dt}S(\psi_t)|_{t=0} &= \frac{1}{2\pi}\int_{S^2}[\nabla\psi_0\cdot\nabla w + w]\\ &= \frac{1}{2\pi}\int_{S^2}[-\triangle\psi_0 + 1]w\\ &= \frac{1}{2\pi}\int_{S^2} e^{2\psi_0}w = 0\,.\end{aligned}$$

Since the path ϕ_t is arbitrary, we have $S(\psi) = \text{const}$. In particular, let $\phi_0 = \text{id}$, then $\psi_0 = 0$ and $S(\psi) = 0$.

3. Finally, we turn to proving the conformal invariant of S. $\forall\phi \in \text{Conf}(S^2)$,

$$\begin{aligned}S(v_\phi) &= \frac{1}{4\pi}\left(\int_{S^2}|\nabla(v\circ\phi)|^2 + 2\int_{S^2}[\nabla(v\circ\phi)\nabla\psi + v\circ\phi]\right) + S(\psi)\\ &= \frac{1}{4\pi}\left(\int_{S^2}|\nabla v|^2 + 2\int_{S^2}(L_{g_0}\psi)v\circ\phi\right)\\ &= \frac{1}{4\pi}\left(\int_{S^2}|\nabla v|^2 + 2\int_{S^2}v\circ\phi e^{2\psi}\right)\\ &= S(v)\,.\end{aligned}$$

□

In the sequel of this subsection we are going to prove that $S(u) \geq 0 \, \forall u \in X$, and then, all the minimizers of S on X are of the form (5.30).

Firstly, we give a necessary condition for the solvability of (5.28). Namely:

Theorem 5.4.4 *(Kazdan–Warner) If u is solution of (5.28) then*

$$\int_{S^2} \langle \nabla K, \nabla x_i \rangle e^{2u} = 0 \quad i = 1, 2, 3 \,. \tag{5.37}$$

Proof. By Lemma 5.4.3, $\forall \phi_{Q,t} \in \mathcal{D}$,

$$J(u_{\phi_{Q,t}}) = S(u) - \log \frac{1}{4\pi} \int_{S^2} K \circ \phi_{Q,t}^{-1} e^{2u} \,. \tag{5.38}$$

If u is a solution of (5.28), then u is a critical point of J, therefore

$$\frac{d}{dt} J(u_{\phi_{Q,t}})|_{t=1} = \left\langle J'(u), \frac{d}{dt} u_{\phi_{Q,t}}|_{t=1} \right\rangle = 0 \,.$$

By (5.38),

$$\frac{d}{dt} J(u_{\phi_{Q,t}})|_{t=1} = \frac{1}{\int_{S^2} K \circ \phi_{Q,t}^{-1} e^{2u}} \int_{S^2} \langle \nabla K, \nabla (x, Q) \rangle e^{2u} \,.$$

We choose $Q = e_1, e_2$, and e_3, respectively; (5.37) follows. □

For any $u \in H^1(S^2)$, we define the mass center of u to be

$$P(u) = \left(\int_{S^2} e^{2u} \right)^{-1} \cdot \left(\int_{S^2} x_1 e^{2u}, \int_{S^2} x_2 e^{2u}, \int_{S^2} x_3 e^{2u} \right) \,.$$

Obviously, $P(u) \in B^3$. We want to show that $\forall u \in H^1(S^2)$, after a conformal transform, the mass center can be moved to the origin, i.e., there exists $(Q, t) \in S^2 \times [1, \infty)$ such that $P(u \circ \phi_{Q,t}) = \theta$, or equivalently, the nonlinear system:

$$\int_{S^2} x_i e^{2u_{\phi_{\theta,t}}} = 0, \quad i = 1, 2, 3 \,, \tag{5.39}$$

is solvable.

We solve the nonlinear system by IFT. Let $F : H^1(S^2) \times B^3 \to B^3$ be the map

$$F(u, \xi) = \int_{S^2} x \circ \phi_{-Q,t} e^{2u}, \tag{5.40}$$

where $\xi = s(t)Q$, and

$$s(t) = \begin{cases} 1 - t^{-1} & t \in [1, 2] \,, \\ 1 - t^{-2} \ln t & \text{for } t \text{ large} \,, \end{cases}$$

is a diffeomorphism: $[1, \infty) \mapsto [0, 1)$.

We need some estimate for $\partial_\xi F$.

Lemma 5.4.5 *Assume* $\{Q(t)\} \subset S^2,\ Q(t) \to Q_0$ *as* $t \to +\infty$. *Then*

$$\int_{S^2} x \circ \phi_{Q(t),t} e^{2u} \to Q_0 \int_{S^2} e^{2u} \ \ as\ t \to +\infty \tag{5.41}$$

uniformly in $\{u \in H^1(S^2) |\ \parallel u \parallel \leq c\}\ \forall c > 0$. *Moreover,*

$$e^{2\psi_{Q,t}} \to 4\pi\delta(-Q) \ \ as\ t \to +\infty\ . \tag{5.42}$$

Proof. Since the mapping $H^1(S^2) \to L^1(S^2) : w \mapsto e^{2w}$ is compact, it is sufficient to prove (5.41) for fixed u. Noticing that

$$x \circ \phi_{Q,t} = D^{-1}\{2t(x - (Q \cdot x)Q) + [t^2(1 + Q \cdot x) - (1 - Q \cdot x)]Q\}\ ,$$

where

$$D = t^2(1 + Q \cdot x) + (1 - Q \cdot x)\ ,$$

we have

$$x \circ \phi_{Q(t),t} \to Q_0 \text{ except } x = -Q_0\ .$$

Our conclusion follows from the Lebesgue dominance theorem. □

Lemma 5.4.6 *For any compact subset* $C \subset H^1(S^2)$ *and any constants* $b > 1$, *the matrix* $\partial_\xi F(u, \xi)$ *has a uniformly bounded inverse* $\forall u \in C,\ t \leq b$, *where* $\xi = s(t)Q$ *solves the equation* $F(u, \xi) = \theta$.

Proof. We write $\widetilde{x} = x \circ \phi_{Q,t}$, and compute

$$\partial_t \widetilde{x} = \frac{1}{t}(Q - (\widetilde{x} \cdot Q)\widetilde{x}),$$
$$\partial_e \widetilde{x} = \frac{t^2 - 1}{2t}(e - (\widetilde{x} \cdot e)\widetilde{x}) + \frac{(t^2 - 1)^2}{2t}[(\widetilde{x} \cdot e)Q - (\widetilde{x} \cdot Q)e]\ ,$$

$\forall e \in T_Q(S^2)$. One chooses an orthonormal frame $\{e_1, e_2\}$ in $T_Q(S^2)$; the matrix $\partial_\xi F(u, \xi)$ is then an array of the three vectors $(\frac{t^2-1}{2t} g_1, \frac{t^2-1}{2t} g_2, \frac{1}{ts'(t)} g_3)$, where

$$g_i = \frac{1}{4\pi} \int_{S^2} (e_j - (\hat{x} \cdot e_j)\hat{x}) e^{2u} \ \ j = 1, 2, 3\ ,$$

$\hat{x} = x \circ \phi_{-Q,t}$ and $e_3 = Q$, because ξ is a solution.

Let A be the matrix consisting of (g_1, g_2, g_3). It is symmetric and positive definite. Indeed, $\forall \lambda = \sum_{i=1}^3 \lambda_i e_i$ with $|\lambda| = 1$,

$$\begin{aligned}(A\lambda) \cdot \lambda &= \left(\sum_{i=1}^3 \lambda_i g_i\right) \cdot \lambda \\ &= 1 - \frac{1}{4\pi} \int_{S^2} (\hat{x} \cdot \lambda)^2 e^{2u}\ .\end{aligned}$$

Fixing u, $\exists\delta = \delta(u) > 0$ such that

$$\frac{1}{4\pi}\int_{S^2}(\hat{x}\cdot\lambda)^2 e^{2u} \le 1-\delta$$

uniformly in $(Q,t) \in S^2 \times [1,b]$. Since C is compact, $\exists\delta_0 > 0$ such that

$$\frac{1}{4\pi}\int_{S^2}(\hat{x}\cdot\lambda)^2 e^{2u} \le 1-\delta_0, \quad \forall (u,\xi) \in C \times (S^2 \times [1,b]) .$$

i.e.,

$$(A\lambda)\cdot\lambda \ge \delta_0 .$$

Thus $\exists\delta_1 > 0$ such that $|\det(A)| \ge \delta_1$, and then

$$|\det(\partial_\xi F(u,\xi))| \ge \frac{(t^2-1)^2\delta_1}{4t^3 s'(t)} .$$

The lemma is proved. □

We are ready to solve (5.40) by the continuity method. According to Lemma 5.4.6, equation (5.39) is locally solvable for any solution (u_0,ξ_0) of (5.39) from the implicit function theorem. One chooses a starting point (u_0,ξ_0), satisfying $F(u_0,\xi_0) = \theta$. For any $u \in H^1(S^2)$, we consider a path $u(\tau) = \frac{1}{2}\log(\tau e^{2u} + (1-\tau)e^{2u_0})$ connecting u_0 and u, and study the equation $F(u(\tau),\xi(\tau)) = \theta$ for $\tau \in [0,1]$.

Let $[0,\tau_0)$ be the maximal existence interval. If $\tau_0 = 1$, then (5.39) is solvable for such a u. Otherwise, there must be a sequence $\tau_j \to \tau_0$ such that $t_j = t(\tau_j) \to +\infty$. Because, if not, $t(\tau)$ is bounded on $[0,\tau_0)$, and then by Lemma 5.4.6, $(\partial_\xi F(u(\tau),\xi(\tau)))^{-1}$ is uniformly bounded. The solution is extendible beyond τ_0. This contradicts the maximality of τ_0.

After subtracting a subsequence from $\{\tau_j\}$, again denoted by $\{\tau_j\}$, we have $Q_j = Q(\tau_j) \to Q_0 \in S^2$. By Lemma 5.4.5,

$$\begin{aligned}\theta &= F(u(\tau_j),\xi(\tau_j)) \\ &= \int_{S^2} x\circ\phi_{-Q_j,t_j} e^{2u(\tau_j)} \to -Q_0\int_{S^2} e^{2u(\tau_0)} .\end{aligned}$$

But this is impossible. Thus we have completed the proof.

Theorem 5.4.7 *$\forall u \in H^1(S^2)$, there exists $(Q,t) \in S^2 \times [1,\infty)$ satisfying (5.39).*

Now let us define

$$I(u) = S(u) - \log\frac{1}{4\pi}\int_{S^2} e^{2u} .$$

Obviously, $I(u) = S(u)$ $\forall u \in X$. We shall prove:

Theorem 5.4.8 *$I(u) \geq 0\ \forall u \in H^1(S^2)$, and the equality holds if and only if $u = \frac{1}{2}\log\det(d\phi) + c,\ \forall \phi \in \mathrm{Conf}(S^2)\ \forall c \in R^1$. Thus all the minimizers of S on X are of the form (5.30).*

Proof. First we prove the existence of a minimizer of I on $Y = \{u \in H^1(S^2)|\ P(u) = \theta,\ \int_{S^2} u = 0\}$. According to Aubin's inequality Sect. 4.2, $\forall u \in Y$.

$$\begin{aligned} I(u) &= \frac{1}{4\pi}\int_{S^2}[|\nabla u|^2 + 2u] - \log\frac{1}{4\pi}\int_{S^2} e^{2u} \\ &\geq \left(\frac{1}{4\pi} - \frac{1}{8\pi - \epsilon}\right) \| u \|^2 - \log C_\epsilon\ , \end{aligned}$$

i.e., I is coercive on Y. Since Y is weakly closed in $H^1(S^2)$, we conclude that there exists $u_0 \in Y$ satisfying $I(u_0) = \min\{I(u)|\ u \in Y\}$. The Euler–Lagrange equation reads as

$$-\triangle u_0 + 1 - \left(\frac{1}{4\pi}\int_{S^2} e^{2u_0}\right)^{-1} e^{2u_0} = \mu \cdot x e^{2u_0},\ \text{for some } \mu \in R^3\ .$$

Let $c = \frac{1}{2}\log\frac{1}{4\pi}\int e^{2u_0}$ and $v = u_0 - c$, then

$$-\triangle v + 1 = (1 + \mu \cdot x e^{2c})e^{2v}\ .$$

Applying the necessary condition in Theorem 5.4.4, we have

$$\int_{S^2} \mu e^{2u_0} = \theta\ ,$$

i.e., $\mu = \theta$. Hence v is a solution of (5.35), it must be of form (5.30). However, $P(v) = \theta$; it follows that $v = \theta$, or $u_0 =$ const. Therefore $I(u_0) = 0$ and then $I(u) \geq 0\ \forall u \in Y$.

Since I is translation invariant: $I(u+c) = I(u)\ \forall c \in R^1$, and also conformal invariant: $I(u_\phi) = I(u)\ \forall \phi \in \mathrm{Conf}(S^2)$, our conclusion follows from Theorem 5.4.7. □

Returning to equation (5.28), in which $K \equiv 1$, all solutions are also the minimizers of the functional I. We may even write down these solutions explicitly:

$$\psi = \frac{1}{2}\log\frac{1 - \alpha^2}{(1 + \alpha Q \cdot x)^2} \tag{5.43}$$

where $\alpha = \frac{1-t^2}{1+t^2}$, for $(Q, t) \in S^2 \times [1, \infty)$. Although $\mathrm{Conf}(S^2)$ is a group with six parameters, the representation $\psi_{Q,t} = \frac{1}{2}\log\det(d\phi_{Q,t})$ only has three, i.e., it is totally determined by the subgroup $\mathcal{D}$.

Corollary 5.4.9 *(Onofri) $\forall u \in H^1(S^2)$, we have*

$$\frac{1}{4\pi}\int_{S^2} e^{2u} \leq \exp\left[\frac{1}{4\pi}\int_{S^2}(|\nabla u|^2 + 2u)\right]\ .$$

see also Hong [Hon 1].

5.4.2 The Palais–Smale Sequence

From Theorem 5.4.8, we see that J is bounded below. We claim that if $K \neq$ const, then the infimum of J is not achieved. Indeed, let $P_0 \in S^2$ be the maximum of K. On one hand, we have $J(u) \geq -\log K(P_0)$, provided by Corollary 5.4.9. On the other hand,

$$J(\psi_{-P_0,t}) = S(\psi_{-P_0,t}) - \log \frac{1}{4\pi} \int_{S^2} K e^{2\psi_{-P_0,t}} = -\log \frac{1}{4\pi} \int_{S^2} K e^{2\psi_{-P_0,t}} .$$

Applying (5.42), we obtain $J(\psi_{-P_0,t}) \to -\log K(P_0)$, i.e., $\inf J = -\log K(P_0)$.

If u_0 were a minimum of J, then we would have $J(u_0) = -\log K(P_0)$. Define

$$\hat{J}(u) = S(u) - \log \frac{1}{4\pi} \int_{S^2} K(P_0) e^{2u} ,$$

then $\min \hat{J} = -\log K(P_0)$, which is achieved by $u = \psi_{Q,t} + c \forall c \in R^1$, $\forall (Q,t) \in S^2 \times [1, \infty)$. Then

$$-\log K(P_0) \leq \hat{J}(u_0) \leq J(u_0) = -\log K(P_0) ,$$

which implies that u_0 is a minimum of $\hat{J}$. Therefore, $u_0 = \psi_{Q,t} + c$ for some $(Q,t) \in S^2 \times [1, \infty)$ and $c \in R^1$. But this is impossible.

The above argument proved indirectly that the PS condition does not hold for J (Ekeland variational principle). Now, we shall analyze the (PS) sequence carefully, and then reveal the reason why the (PS) condition breaks down.

Firstly, we continue the proof in Theorem 5.4.7, and obtain:

Theorem 5.4.10 *Let $P(u)$ be the mass center of u, and let*

$$X_0 = \{u \in X |\ P(u) = \theta\} ,$$

then

$$X \cong X_0 \times \overset{\circ}{B^3} ,$$

where $\cong$ means diffeomorphism.

Proof. After Theorem 5.4.7, it remains to prove that the solution of equation (5.39) is unique.

Suppose not, then we may assume that $\xi_i = s(t_i) Q_i$ satisfies (5.39), $i = 0, 1$. Define $w_i = u_{\phi_{Q_i,t_i}}$, $i = 0, 1$; we pick a path $(Q(\tau), t(\tau))$ connecting (Q_0, t_0) and (Q_1, t_1), and let $w(\tau) = \frac{1}{2} \log(\tau e^{2w_1} + (1-\tau) e^{2w_0}) \forall \tau \in [0,1]$. Set

$$u(\tau) = w(\tau)_{\phi^{-1}_{Q(\tau),t(\tau)}} ,$$

then

$$F(u(\tau), \xi(\tau)) = \int_{S^2} x \circ \phi_{-Q(t),t(\tau)} e^{2u(\tau)} = \theta .$$

We introduce a new parameter σ, and define for $(\sigma, \tau) \in [0,1]^2$,

$$u(\sigma, \tau) = \frac{1}{2} \log[(1-\sigma)e^{2u(\tau)} + \sigma e^{2u}]$$

and we solve the equation:

$$F(u(\sigma, \tau), \xi(\sigma, \tau)) = \theta \, .$$

By the same argument as used in Theorem 5.4.8, we have solutions $(Q(\sigma, \tau), t(\sigma, \tau))$ depending on σ, τ continuously, and satisfying

$$\begin{aligned} &Q(0,\tau) = Q(\tau), \ t(0,\tau) = t(\tau) \, , \\ &Q(\sigma, i) = Q_i, \ t(\sigma, i) = t_i, \ i = 0, 1, \ \text{and} \\ &F(u, \xi(1, \tau)) = \theta \, . \end{aligned}$$

This contradicts the local uniqueness of the solution of (5.39), which follows from the implicit function theorem. The smoothness of (Q, t) depending on u follows directly. □

Theorem 5.4.10 provides a parameterization of $X : X \to X_0 \times (S^2 \times [1, \infty)/S^2 \times \{1\})$, $u \mapsto (w, Q, t)$.

Theorem 5.4.11 *Suppose that $\{u_j = (w_j, Q_j, t_j)\}$ is a Palais–Smale sequence of J on X, then either it is subconvergent or $t_j \to +\infty$.*

Proof. Since

$$J(u_j) = S(w_j) - \log \int K \circ \phi_{Q_j, t_j} e^{2w_j} \to c \, ,$$

and

$$-\Delta w_j + 1 = \frac{4\pi K(\widetilde{x}_j) e^{2w_j}}{\int K(\widetilde{x}_j) e^{2w_j}} + \lambda_j e^{2w_j} + (\mu_j \cdot x) e^{2w_j} + \circ(1) \, , \tag{5.44}$$

where $(\lambda_j, \mu_j) \in \mathbb{R}^1 \times \mathbb{R}^3$, if $\{t_j\}$ is bounded, then after a subsequence, $\widetilde{x}_j = x \circ \phi_{Q_j, t_j} \to \widetilde{x} = x \circ \phi_{Q,t}$ for some $(Q, t) \in S^2 \times [1, \infty)$.

It follows that $S(w_j)$ and λ_j are both bounded. Since

$$\wedge(w_j) \mu_j = \frac{-\int K(\widetilde{x}_j) x e^{2w_j}}{\int K(\widetilde{x}_j) e^{2w_j}} + \circ(1) \, ,$$

where

$$\wedge(w_j) = \left(\frac{1}{4\pi} \int_{S^2} x_i x_k e^{2w_j} \right)_{3 \times 3} ,$$

and $\wedge(w_j)$ is uniformly positive definite, $\{\mu_j\}$ is bounded. Moreover, the embedding $w \mapsto e^{2w}$ from H^1 to L^p, $p > 1$, is compact; we may use (5.44) to conclude that $\{w_j\}$ is subconvergent. □

5.4.3 Morse Theory for the Prescribing Gaussian Curvature Equation on $\mathbf{S^2}$

From Theorem 5.4.11, we know that the Palais–Smale Condition for J breaks down at $t = +\infty$. We are inspired with courage to compactify the open ball $\overset{\circ}{B}{}^3$. By simple estimates, we obtain the limit of the integral:

$$\frac{1}{4\pi}\int_{S^2} K \circ \phi_{Q,t} e^{2w} = K(Q) + \circ(1) \text{ as } t \to \infty\,.$$

Let us extend J to be $\widetilde{J}$ defined on the manifold with boundary $M = X_0 \times \overline{B}^3$ where $\partial M = X_0 \times S^2$:

$$\widetilde{J}(u) = \begin{cases} J(u) & u \in X \\ S(w) - \log K(Q) & (w, Q) \in X_0 \times S^2\,. \end{cases}$$

If $\widetilde{J}$ were C^1 on M, then Remark 5.1.31 would be applicable directly to this problem. Unfortunately, this is not true. We are forced to figure out the asymptotic expansions for the partial derivatives for J. Namely we have

Lemma 5.4.12 *$\exists b > 0, c_b > 0$ and $\exists \widetilde{v} \in T_w(X_0)$ such that*

$$\langle S'(w), \widetilde{v}\rangle \geq c_b S^{\frac{1}{2}}(w) \parallel \widetilde{v} \parallel\,,$$

if $S(w) \leq b$. Furthermore, $\forall\ \mu, \delta > 0,\ \exists N = N(\delta, b, \mu),\ \exists T = T(\delta, b, \mu)$ such that

$$\begin{aligned} |\partial_e J + \frac{\partial_e K(Q)}{K(Q)}| &\leq N t^{-1+\mu} \log t \quad \forall e \in T_Q(S^2)\,, \\ |\partial_s J - \frac{2\triangle K(Q)}{K(Q)}| &\leq N(|\nabla K(Q)| t \log^{-\frac{1}{2}} t S^{\frac{1}{2}}(w) \\ &\quad + (\log t)^{-1} + t^2 (\log t)^{-1} S^2(w))\,. \end{aligned}$$

if $K(Q) \geq \delta$ and $t \geq T$.

The proof is referred to Chang and Liu [CL 2].

Combining Lemma 5.4.12 with the partition of unity, we obtain:

Theorem 5.4.13 *Assume $K > 0$. If $\triangle K \neq 0$ whenever $\nabla K(x) = \theta$, then there exists a function f defined on $X_0 \times \overline{B}^3$ possessing the same critical set as J, and $\exists$ a neighborhood U of $X_0 \times S^2$, such that on U*

$$f(w, sQ) = S(w) - \log K(Q) - \frac{2\triangle K(Q)}{K(Q)}(1 - s)\,,$$

where $s : [1, \infty) \to [0, 1)$ is a diffeomorphism with $s(t) = 1 - t^{-2}\log t$ for t large.

Furthermore, f satisfies the (PS) condition.

Now we are ready to apply the Morse theory under general boundary conditions to f on $X_0 \times \overline{B}^3$. For given $a < b$ such that e^{-a}, e^{-b} are regular values of K, we assume:

(I) K has only a finite number of critical points with values in the interval $[e^{-b}, e^{-a}]$.

Set

$$\begin{aligned} &\Omega = \{x \in S^2 | \; \triangle K(x) < 0\}, \\ &CR_0(a,b) = \{x \in \Omega | \; K(x) \in (e^{-b}, e^{-a}), \; x \text{ is a local maximum of K}\}, \\ &CR_1(a,b) = \{x \in \Omega | \; K(x) \in (e^{-b}, e^{-a}), \; x \text{ is a saddle point of K}\} \,. \end{aligned}$$

Theorem 5.4.14 *Under assumption (I), assume*
(II)

$$\triangle K(x) \neq 0 \text{ whenever } \nabla K(x) = \theta \,.$$

If J has only isolated critical points, then the following Morse inequalities hold:

$$\sum (m_q + \mu_q - \beta_q) t^q = (1+t) Q(t) \,,$$

where Q is a formal series with nonnegative coefficients, and

$$\begin{aligned} m_q &= q\text{th Morse type number of } J \text{ in } J^{-1}[a,b] \,, \\ \beta_q &= \text{rank } H_q(J_b, J_a) \,, \\ \mu_g &= \begin{cases} 0 & q \geq 2 \\ \sharp CR_q(a,b) & q = 0,1 \,. \end{cases} \end{aligned}$$

Proof. It is equivalent to study the Morse inequalities for f.We notice that f is of separate variables, and that $S(w)$ has no critical point except the minimum θ on X_0, from Lemma 5.4.12. Moreover, f possesses the same critical set as J and equals J in a neighborhood of the critical set.

Now the restriction $\hat{f}$ on the boundary $X_0 \times S^2$ reads as $S(w) - \log K(Q)$, which has only isolated critical points. Since

$$f(w, sQ) = S(w) - \log K(Q) - 2\frac{\triangle K(Q)}{K(Q)}(1-s)$$

near the boundary, and the normal

$$n(w,Q) = (\theta, Q) \in T_w(X_0) \times R^3 \,,$$

we see that

$$\begin{aligned} \Sigma_- &= \{(w,Q) \in X_0 \times S^2 | \; \langle f'(u), n(u) \rangle \leq 0\} \\ &= \{(w,Q) \in X_0 \times S^2 | \; Q \in \Omega \cap K^{-1}[e^{-b}, e^{-a}]\} \,. \end{aligned}$$

It remains to compute the critical groups of the critical points of $-\log K$ on Ω.

Now $\forall Q \in CR_0(a,b)$,

$$\text{rank}C_q(-\log K, Q) = \delta_{q0} \; ,$$

and $\forall Q \in CR_1(a,b)$

$$\text{rank}C_q(-\log K, Q) = \delta_{q1} \; ,$$

from (II).

We claim that $\text{rank}C_2(-\log K, Q) = 0 \; \forall Q \in \Omega$. If not, Q must be a local minimum of K, and then $\triangle K(Q) \geq 0$. This is a contradiction. Obviously we have

$$\text{rank}C_q(-\log K, Q) = 0 \;\; \forall q > 2 \; .$$

On the other hand, θ is the minimum of S on X_0; it follows that

$$\text{rank}C_q(S, \theta) = \delta_{q0} \;\; \forall q \; .$$

Thus the Morse type numbers of $\hat{f}$ read as

$$\mu_q = \begin{cases} 0 & \forall q \geq 2 \\ \sharp CR_q(a,b) & q = 0, 1 \; . \end{cases}$$

The proof is complete. □

Corollary 5.4.15 *Under assumptions (I) and (II). Let*

$$p = \sharp \textit{local maxima of } K \textit{ in } \Omega \; ,$$
$$q = \sharp \textit{saddle points of } K \textit{ in } \Omega \; .$$

If $p \neq q + 1$, then (5.28) admits a solution.

Proof. By taking $e^{-b} < \min K$ and $e^{-a} > \max K$. Since $X_0 \times \bar{B}^3$ is contractible, the conclusion follows from the Morse inequalities. □

Even assumption (I) can be dropped. Namely,

Theorem 5.4.16 *Assume (II) and if* $\deg(\Omega, \nabla K, \theta) \neq 1$, *then (5.28) admits a solution.*

Proof. First, we perturb K to $\widetilde{K}$ satisfying:

(1) $\widetilde{K} > 0$ possesses only nondegenerate critical points.
(2) For a given neighborhood U of the critical set V of K, $K(x) = \widetilde{K}(x) \; \forall x \notin U$.
(3) $\| K - \widetilde{K} \|_{C^2}$ is small.

$\widetilde{K}$ is constructed as follows: Find a cover $\cup_{i=1}^{l} B_\epsilon(x_i) \subset U$ of V, $\forall i$, we pull back the linear function on $B_\epsilon(x_i)$, and obtain a function $\phi_{a,i}(x)$ such that $\nabla\phi_{a,i}(x) = a\ \forall x \in B_\epsilon(x_i)$, where a is any given vector in R^2. Let $\{\rho_i\}_1^l$ be the partition of unity of V with supp$\rho_i \subset B_\epsilon(x_i)$, and let

$$\widetilde{K}(x) = K(x) - \sum \phi_{a,i}(x)\rho_i(x)\ .$$

By Sard's theorem, $\| a \|$ can be chosen as small as we wish such that (1) and (3) hold.

Second, as in Theorem 5.4.13, we can construct a functional $\widetilde{f}$ similar to f satisfying:

(1) In a neighborhood O of $X_0 \times S^2$, $\widetilde{f}$ is similar to f with respect to $\widetilde{K}$, and there are no critical points there.
(2) Outside O, $\widetilde{f} = f = J$.

Now, let p, q be the numbers of local maxima and saddle points of $\widetilde{K}$ in $\widetilde{\Omega} = \{x \in S^2 \mid \triangle\widetilde{K}(x) < 0\}$, respectively. By homotopy invariance of the Brouwer degree:

$$1 \neq \det(\Omega, \nabla K, \theta) = \deg(\widetilde{\Omega}, \nabla\widetilde{K}, \theta) = p - q\ .$$

This proves the existence of a critical point of $\widetilde{f}$ so does for J. □

Remark 5.4.17 *Other sufficient conditions were obtained by Chen and Ding [CD]. Chang and Yang [CY 1, 2], Han [Han], and Hong [Hon 2]. All these results are contained in Theorem 5.4.14, see [CL 2].*

Remark 5.4.18 *The prescribing scalar curvature problem on S^n is a similar problem. For $n = 3$, it was due to Bahri and Coron [BC] and later, Schoen and Zhang [SZ]. For general n, see Li [Li], and Chen and Lin [ChL].*

5.5 Conley Index Theory

The Conley index is another topological tool in nonlinear analysis. It can be seen as an extension of both the Leray–Schauder degree and the critical groups. In fact, the Conley index theory is an extension of the Morse theory. It is based on flows without a variational structure.

In contrast with the degree theory and the Morse theory, Conley's theory studies the invariant sets of a flow rather than either the fixed points for a compact vector field or the critical points for differentiable functions.

Let (X, d) be a metric space, and let $\varphi : R^1 \times X \to X$ be a flow, i.e., it is continuous and satisfies:

(1) $\varphi(0, x) = x \quad \forall x \in X$,
(2) $\varphi(t, \varphi(s, x)) = \varphi(t + s, x) \quad x \in X,\ \forall t, s \in R^1$.

Definition 5.5.1 *A set $S \subset X$ is called invariant with respect to the flow φ, if $\varphi(R^1, S) = S$, i.e., $\forall x \in S$, $\forall t \in R^1$, $\varphi(t, x) \in S$.*

Example 1. Let X be a Banach space, and let $K : X \to X$ be compact. Let φ be the flow derived by the compact vector field $f = \mathrm{id} - K$:

$$\begin{cases} \dot{\varphi}(t,x) = f(\varphi(t,x)) & \forall (t,x) \in R^1 \times X \, , \\ \varphi(0,x) = x \, , \end{cases}$$

then any subset of the fixed-point set of K is an invariant set for the flow φ.

Example 2. Let M be a smooth Banach Finsler manifold and let $f \in C^1(M, R^1)$ satisfy the (PS) condition. If V is a pseudo-gradient vector field of f and φ is the associated flow:

$$\dot{\varphi}(t,x) = -g(\varphi(t,x)) \frac{V(\varphi(t,x))}{\|V(\varphi(t,x))\|} \text{ with } \varphi(0,x) = x \, , \tag{5.45}$$

where $g(x) = \mathrm{Min}\{\mathrm{d(x,K)}, 1\}$, and K is the critical set of f.

Then any subset of the critical set of f is an invariant set of the flow φ.

Definition 5.5.2 *We call the triple (M, f, φ) a pseudo-gradient flow if M is a smooth Banach–Finsler manifold, $f \in C^1(M, R^1)$ satisfies the (PS) condition, and φ is defined by equation (5.45.)*

5.5.1 Isolated Invariant Set

Invariant sets can be extremely complicated. It is known from dynamical systems that there are chaotic dynamics and fractal structure. As we have learned in Chap. 3 and in Sects. 5.1 and 5.2, we single out isolated invariant sets as the object of our study. Before going on, we introduce some notions from dynamical systems.

$\forall T \in R^1_+ \cup \{+\infty\}$, for any subset A of X, we write

$$G^T(A) = \bigcap_{|t| \le T} \varphi(t, \overline{A}), \text{ and } \Gamma^T(A) = \{x \in G^T(A) | \; \varphi([0,T], x) \cap \partial A \neq \emptyset\} \, .$$

In particular, we write $I(A) = G^\infty(A)$. It is the maximal invariant subset of $\overline{A}$.

By definition, we have the following properties:

(1) $G^T(A) = G^T(\overline{A})$,
$G^{T_1}(A) \subset G^{T_2}(A)$, if $T_1 \ge T_2$,
$A_1 \subset A_2 \Rightarrow G^T(A_1) \subset G^T(A_2)$,
$G^T(A)$ is closed,
$G^{T_1+T_2}(A) = G^{T_2}(G^{T_1}(A))$.

(2) If $G^T(A) \subset \mathrm{int}(A)$, then $G^{2T}(A) \subset \mathrm{int}(G^T(A))$.

Proof. Suppose not, then $\exists y \in G^{2T}(A) \cap \partial G^T(A)$, i.e., $\exists y_n \to y$ with $\varphi([-T,T],y_n) \not\subset \overline{A}$, or $\exists t_n \in [-T,T]$, $\varphi(t_n,y_n) \notin \overline{A}$. There exists a subsequence $\{t'_n\}$ of $\{t_n\}$ such that $t'_n \to \bar{t} \in [-T,T]$ and $\varphi(\bar{t},y) \in \partial A$. But $y \in G^{2T}(A)$, i.e., $\varphi([-2T,2T],y) \in \overline{A}$, which implies that $\varphi(\bar{t},y) \in G^T(\overline{A}) = G^T(A) \subset \mathrm{int}(A)$, a contradiction. □

Now, we introduce a family of closed sets:

$$\Sigma = \Sigma(\varphi) = \{A \subset X|\ A \text{ is closed, and } \exists\, T > 0 \text{ such that } G^T(A) \subset \mathrm{int}(A)\}\,.$$

Definition 5.5.3 *A neighborhood U of an invariant set S for the flow φ is called isolating if $U \in \Sigma$ and $I(U) = S$. An invariant set S is called an isolated invariant set if there exists an isolating neighborhood.*

Remark 5.5.4 *If U is compact and if $I(U) = S \subset \mathrm{int}(U)$, then $U \in \Sigma$, i.e., $\exists T > 0$ such that $G^T(U) \subset \mathrm{int}(U)$.*

In fact, if not, i.e., $G^n(U) \not\subset \mathrm{int}(U)\ \forall n \in \mathbb{N}$, we have $y_n \in G^n(U)\backslash\mathrm{int}(U)$, i.e., $\varphi([-n,n],y_n) \subset U$ but $y_n \notin \mathrm{int}(U)$. Since U is compact, there is a subsequence $y_{n_i} \to y_0 \in I(U)$, but $y_0 \notin \mathrm{int}(U)$. A contradiction.

Moreover, we have:

(3) If $U \in \Sigma$, then $\exists T_1 > 0$ such that $G^{T_1}(U) \in \Sigma$, and $\forall t \in R^1$, $\varphi(t,U) \in \Sigma$.

Proof. It is known that both $G^{T_1}(U)$ and $\varphi(t,U)$ are closed. Since $\exists T > 0$ such that $G^T(U) \subset \mathrm{int}(U)$. Taking $T_1 = T$, from (1) and (2), we have $G^T(G^{T_1}(U)) = G^{2T}(U) \subset \mathrm{int}(G^T(U)) = \mathrm{int}(G^{T_1}(U))$. Also

$$G^T(\varphi(t,U)) = \bigcap_{|s|\le T}\varphi(t+s,U) = \varphi(t,G^T(U)) \subset \varphi(t,\mathrm{int}(U)) = \mathrm{int}(\varphi(t,U))\,.$$

□

(4) If $U \in \Sigma$, then $\Gamma^T(U) \subset \partial G^T(U)$ is closed.

Proof. If $\{x_n\} \in \Gamma^T(U)$ and $x_n \to x$, then $\exists t_n \in [0,T]$ such that $\varphi(t_n,x_n) \in \partial U$, after a subsequence, we have $t_n \to \bar{t} \in [0,T]$ and $\varphi(\bar{t},x) \in \partial U$. Thus $x \in \Gamma^T(U)$. This shows that $\Gamma^T(U)$ is closed.

Next, $\forall x \in \Gamma^T(U)$, we verify $x \in \partial G^T(U)$. By definition, $\exists \bar{t} \in [0,T], \exists y_n \notin U$ such that $y_n \to \varphi(\bar{t},x)$. Let $x_n = \varphi(-\bar{t},y_n)$, we have $x_n \to x$ and $x_n \notin G^T(U)$. Since $G^T(U)$ is closed, we have $x \notin \mathrm{int}(\mathrm{G^T(U)})$, therefore $x \in \partial G^T(U)$. □

The following terminologies are taken from dynamical systems:

Definition 5.5.5 $\forall x \in X$, *the set*

$$\omega(x) = \bigcap_{t>0}\overline{\varphi([t,\infty),x)}$$

is called the ω-limit set of x, and the set

$$\omega^*(x) = \bigcap_{t>0} \overline{\varphi((-\infty, -t], x)}$$

is called the ω^*-limit set of x. Given a subset $S \subset X$, the set

$$[S] = \{x \in X | \ \omega(x) \cup \ \omega^*(x) \subset S\}$$

is called the invariant hull of S.

By definition, $\omega(x) = \omega(\varphi(t,x))$, $\omega^* = \omega^*(\varphi(t,x))$ $\forall t \in R^1$, and then $[S]$ is invariant. Moreover, if S is invariant, then $S \subset [S]$.

Lemma 5.5.6 *Let (M, f, φ) be a pseudo-gradient flow. Then for any $x \in M$ the set $\omega(x)$ is compact and is a subset of K_c for some critical value c, where $K_c = K \cap f^{-1}(c)$. The same holds true for the set $\omega^*(x)$.*

Proof. Since the function is nonincreasing along the pseudo-gradient flow, first we show that $\omega(x)$ is on one level, say $\omega(x) \subset f^{-1}(c)$ for some $c \in R^1$. Indeed, if not, then there exist t_n, $t'_n \uparrow \infty$ such that $\varphi(t_n, x) \to y$ and $\varphi(t'_n, x) \to y'$ with $f(y) < f(y')$. We may always assume that $t'_n > t_n$, which means that

$$f(y') = \lim_n f(\varphi(t'_n, x)) \le \lim_n f(\varphi(t_n, x)) = f(y) \ .$$

This is a contradiction.

Next, we prove that $\omega(x) \subset K$. Indeed, if $\exists y \in \omega(x) \backslash K$, then we choose regular values $a < b$ such that both $f(x)$ and $f(y)$ are in (a, b). Since $K_a^b = K \cap f^{-1}[a,b]$ is compact, there exists $r > 0$ such that $B_r(y) \cap (K_a^b)_r = \emptyset$. By definition, there exists $t_n \to +\infty$ such that $x_n = \varphi(t_n, x) \to y$. From the (PS) condition, there exists $\delta > 0$ such that $\| \ f'(x) \ \| \ge \delta$ for all $x \in f^{-1}[a,b] \backslash (K_a^b)_r$,

We claim that there exists $t'_n \to +\infty$ such that $x'_n = \varphi(t'_n, x) \in \partial (K_a^b)_r$. If not, $\exists T > 0$ such that $\varphi([T, \infty), x) \cap (K_a^b)_r = \emptyset$, then we would have

$$f(y) = \lim_n f(x_n) = \liminf_{t \to +\infty} f(\varphi(t, x)) \le a \ ,$$

This is impossible. Now we choose $t''_n \to +\infty$ with $t'_n < t''_n$ such that

$$x''_n = \varphi(t''_n, x) \in B_r(y), \ \ \varphi([t'_n, t''_n], x) \cap (K_a^b)_r = \emptyset \ .$$

It follows that

$$f(x'_n) - f(x''_n) \ge \delta \| x''_n - x'_n \| \ge \delta \text{dist}(B_r(y), (K_a^b)_r) \ .$$

Again, this is impossible. The lemma is proved. □

Definition 5.5.7 *A subset W of X is said to have a mean value property (MVP) with respect to the flow φ, if $\forall x \in X$, $\forall t_1 < t_2$, $\varphi(t_i, x) \in W$, $i = 1, 2$ implies that $\varphi([t_1, t_2], x) \subset W$.*

We now introduce the following key concept:

Definition 5.5.8 *(Dynamically isolated critical set) Let (M, f, φ) be a pseudo-gradient flow. A subset S of the critical set K is said to be a dynamically isolated set if there exist a closed neighborhood $\mathcal{O}$ of S and regular values $\alpha < \beta$ of f such that*

$$\mathcal{O} \subset f^{-1}[\alpha, \beta] ,$$

and

$$\mathrm{cl}(\widetilde{\mathcal{O}}) \cap K \cap f^{-1}[\alpha, \beta] = S ,$$

where $\widetilde{\mathcal{O}} = \bigcup_{t \in R^1} \varphi(t, \mathcal{O})$. We shall then say that $(\mathcal{O}, \alpha, \beta)$ is an isolating triplet for S.

Lemma 5.5.9 *Let (M, f, φ) be a pseudo-gradient flow and K be the critical set of f. If U is a closed (MVP) neighborhood of S satisfying $U \cap K = S$, then $[S] = I(U)$. If further, there exist real numbers $\alpha < \beta$ and a (MVP) closed set W, satisfying $U \subset W \subset f^{-1}[\alpha, \beta]$ and $W \cap K = S$, then $\exists T > 0$ such that $G^T(W) \subset \mathrm{int(U)}$.*

Proof. 1. $[S] = I(U)$

"$\subset$" $\forall x \in [S]$, by definition $\omega(x) \cup \omega^*(x) \subset S$, then $\exists t_n^{\pm} \to \pm\infty$ such that $\varphi(t_n^{\pm}, x) \in U$, and then $\varphi([t_n^-, t_n^+], x) \subset U$. Since n is arbitrary, it follows that $\varphi(t, x) \in U \ \forall t \in R^1$, i.e., $x \in I(U)$.

"$\supset$" $\forall x \in I(U)$, $\varphi(t, x) \in U \ \forall t \in R^1$. Since U is closed, $\omega(x) \cup \omega^*(x) \subset U$. From Lemma 5.5.6, $\omega(x) \cup \omega^*(x) \subset K$, therefore $\omega(x) \cup \omega^*(x) \subset U \cap K = S$, i.e., $x \in [S]$.

2. $G^T(W) \subset \mathrm{int}(U)$.

From the (PS) condition, $\exists \delta \in (0, 1)$ such that $\mathrm{dist}(x, K) \geq \delta$ and $\| f'(x) \| \geq \delta \ \forall x \in W \backslash \mathrm{int}(U)$. Set $T > \delta^{-2}(\beta - \alpha)$. We shall prove that $\forall x \notin \mathrm{int}(U) \ \exists t \in [-T, T]$ such that $\varphi(t, x) \notin W$.

It is divided into three cases:

(a) $x \notin W$. By taking $t = 0$, it is done.

(b) $x \in W \backslash \mathrm{int}(U)$. If $\varphi([-T, T], x) \subset W$, then

$$\begin{aligned} f(\varphi(-T, x)) - f(\varphi(T, x)) &= \int_{-T}^{T} \langle f'(\varphi(s, x)), \dot{\varphi}(s, x) \rangle ds \\ &\geq 2T\delta^2 > 2(\beta - \alpha) . \end{aligned}$$

The contradiction shows $\varphi([-T, T], x) \not\subset W$.

(c) $x \in (\widetilde{\mathrm{int}(U)} \backslash \mathrm{int}(U)) \cap W = \widetilde{\mathrm{int}(U)} \cap W \backslash \mathrm{int}(U)$. Since either $x \in \bigcup_{t > 0} \varphi(t, \mathrm{int}(U))$ or $x \in \bigcup_{t < 0} \varphi(t, \mathrm{int}(U))$.

In the first case, by the use of (MVP) of U, we have $t_1 \leq 0 \leq t_2$ such that

$$\varphi([t_1, t_2], x) \subset (\widetilde{\mathrm{int}(U)} \cap W) \backslash \mathrm{int}(U) \text{ and } \varphi(t_1 - \epsilon, x) \in U, \ \varphi(t_2 + \epsilon, x) \notin W$$

for all $\epsilon > 0$ small. Again, we would have $\beta - \alpha \geq \delta^2(t_2 - t_1)$ so that $t_2 < T$. Therefore $\varphi([-T, T], x) \not\subset W$.

Similarly for the second case. □

Theorem 5.5.10 *Let (M, f, φ) be a pseudo-gradient flow. If $(\mathcal{O}, \alpha, \beta)$ is an isolating triplet for a dynamically isolated critical set S for f, then $[S]$ is an isolated invariant set.*

Moreover, any closed MVP neighborhood U of $[S]$, satisfying $U \subset cl(\widetilde{\mathcal{O}}) \cap f^{-1}[\alpha, \beta]$, is an isolating neighborhood for $[S]$, and $U \in \Sigma$.

Proof. 1. Applying Lemma 5.5.9, $W = \text{cl}(\widetilde{\mathcal{O}}) \cap f^{-1}[\alpha, \beta] \in \Sigma$, is an isolated invariant neighborhood of $[S]$.

2. To prove that $U \in \Sigma$ is an isolating neighborhood for $[S]$, it is sufficient to show that $[S] = I(U) \subset G^T(U) \subset \text{int}(U)$ for some $T > 0$. Since

$$[S] = I([S]) \subset I(U) \subset I(W) = [S] ,$$

where $W = \text{cl}(\widetilde{\mathcal{O}}) \cap f^{-1}[\alpha, \beta]$, we obtain $[S] = I(U)$.

Again by Lemma 5.5.9, we have $G^T(U) \subset G^T(W) \subset \text{int}(U)$. □

Example 1. If c is an isolated critical value, i.e., $K_c = K \cap f^{-1}(c) \neq \emptyset$, and there is no critical point on the levels in $[c - \epsilon, c + \epsilon] \backslash \{c\}$ for some $\epsilon > 0$, then the set K_c is a dynamically isolated critical set.

Example 2. If x_0 is an isolated critical point of f, then $S = \{x_0\}$ is a dynamically isolated critical set.

5.5.2 Index Pair and Conley Index

As we have seen in the Morse theory, isolated neighborhoods only are not enough in characterizing the isolated critical points, the local dynamic behavior provides the necessary information. This leads us to:

Definition 5.5.11 *Let (N, L) be a pair of subspaces of X. A subset L of N is called positively invariant in N with respect to the flow φ, if $x \in L$ and $\varphi([0, t], x) \subset N$ imply $\varphi([0, t], x) \subset L$. It is called an exit of N, if $\forall x \in N, \exists t_1 > 0$ such that $\varphi(t_1, x) \notin N$, implies $\exists t_0 \in [0, t_1)$ such that $\varphi([0, t_0], x) \subset N$ and $\varphi(t_0, x) \in L$.*

Example. Let (M, f, ϕ) be a pseudo-gradient flow, and let $\alpha < \beta < \gamma$. Let $N = f^{-1}[\alpha, \gamma]$ and $L = f^{-1}[\alpha, \beta]$. L is positively invariant in N, and also an exit set of N.

To an isolated invariant set S, we introduce:

Definition 5.5.12 *For $U \in \Sigma$, let (N, L) be a pair of closed subsets of U with $L \subset N$. It is called an index pair relative to U if:*

(1) $\overline{N \backslash L} \in \Sigma$,

(2) L is positively invariant in N,
(3) L is an exit set of N,
(4) $\overline{N\backslash L} \subset U$ and $\exists T > 0$ such that $G^T(U) \subset \overline{N\backslash L}$.

According to the definition $S := I(U) = I(\overline{N\backslash L})$ is an isolated invariant set, and both U and $\overline{N\backslash L}$ are isolating neighborhoods of S.

Example. Let (M, f, φ) be a pseudo-gradient flow. If $(\mathcal{O}, \alpha, \beta)$ and $(\mathcal{O}', \alpha', \beta')$ are two isolating triplets for a dynamically isolated critical set S for f, with $\mathcal{O}' \subset \mathcal{O}$, $[\alpha', \beta'] \subset [\alpha, \beta]$, then $\exists T > 0$ such that $N = G^T(W), L = \varphi(-T, W_-)$ is an index pair relative to $U = \text{cl}(\widetilde{\mathcal{O}'}) \cap f^{-1}[\alpha', \beta']$, where $W = \text{cl}(\widetilde{\mathcal{O}}) \cap f^{-1}[\alpha, \beta]$ and $W_- = \text{cl}(\widetilde{\mathcal{O}}) \cap f^{-1}(\alpha)$. Moreover, both U and N are isolating neighborhoods of $[S]$.

Note that $\overline{N\backslash L} = N$ and $\text{int}(N\backslash L) = \text{int}(N)$. In order to verify the conclusion, we need:

1. $\forall T > 0, N = G^T(W)$ is a closed MVP neighborhood of $[S]$.

It is sufficient to verify that $[S] \subset \text{int}(N)$. From Lemma 5.5.9, we have $[S] \subset N$. If the conclusion is not true, then $\exists x \in [S] \cap \partial N$, i.e., $\exists x_n \notin G^T(W)$ with $x_n \to x$. This means that there are $t_n \in [-T, T]$, such that $\varphi(t_n, x_n) \notin W$. After a subsequence, we have $t_n' \to t \in [-T, T]$ and then $\varphi(t, x) \notin \text{int}(W)$. But $x \in [S]$ implies that $\varphi(t, x) \in [S] \subset \text{int}(W)$, provided by Lemma 5.5.9. This is a contradiction.

Now, we are going to verify conditions (1)–(4).

2. Applying Lemma 5.5.9 to $U = N$, $\exists T_0 > 0$ such that

$$G^{T+T_0}(W) \subset G^{T_0}(W) \subset \text{int}(N) .$$

Since $G^{T+T_0}(W) = G^{T_0}(N)$. This shows that $\overline{N\backslash L} \in \Sigma$, (1) is verified.

3. Again applying Lemma 5.5.9, $\exists T > 0$ such that $N = G^T(W) \subset \text{int}(U) \subset U$. Moreover,

$$G^T(U) \subset G^T(W) = N = \overline{N\backslash L} .$$

Poperty (4) is verified.

4. Since W_- is an exit set of W, L is an exit set of N. Obviously, L is positively invariant in N.

This completes the verification. From Lemma 5.5.9, both U and N are isolating neighborhoods of $[S]$.

For a system without variational structure, does there exist an index pair relative to any set $U \in \Sigma$? We have:

Theorem 5.5.13 *(Existence of an index pair) Let φ be a flow on a metric space X. $\forall U \in \Sigma$, $(G^T(U), \Gamma^T(U))$ is an index pair relative to U, where $T > 0$ is assumed such that $G^T(U) \subset \text{int}(U)$.*

Proof. From the properties (1) and (4), both $G^T(U)$ and $\Gamma^T(U)$ are closed. We shall verify the four conditions in Definition 5.5.11 successively.

(1) By property (4), $\mathrm{int}(G^T(U)\backslash\Gamma^T(U)) = \mathrm{int}(G^T(U))$. Applying property (2),

$$G^T(G^T(U)\backslash\Gamma^T(U)) \subset G^{2T}(U) \subset \mathrm{int}(G^T(U)) .$$

Thus $\overline{G^T(U)\backslash\Gamma^T(U)} \in \Sigma$.

(2) $\Gamma^T(U)$ is positively invariant in $G^T(U)$, i.e., if $x \in \Gamma^T(U)$ and $\varphi([0,T_1],x) \in G^T(U)$, then $\varphi([0,T_1],x) \subset \Gamma^T(U)$.

Suppose not, then $\exists t \in [0,T_1]$ such that $\varphi(t,x) \notin \Gamma^T(U)$. Let $t^* = \inf\{s \in [0,T_1]|\ \varphi(s,x) \notin \Gamma^T(U)\}$. Since $\Gamma^T(U)$ is closed, $y = \varphi(t^*,x) \in \Gamma^T(U)$ and $\exists \epsilon_n \to +0$ such that $\varphi(t^* + \epsilon_n, x) \notin \Gamma^T(U)$. Thus $\varphi([0,T],y) \cap \partial U \neq \emptyset$ and $\varphi([\epsilon_n,T],y) \cap \partial U = \emptyset$; it follows that $y \in \partial U$. But $y \in G^T(U) \subset \mathrm{int}(U)$. This is a contradiction.

(3) $\Gamma^T(U)$ is an exit set of $G^T(U)$, i.e., if $x \in G^T(U)$ and if $\exists t_1 > 0$ such that $\varphi(t_1,x) \notin G^T(U)$, then $\exists t_0 \in [0,t_1)$ such that $\varphi([0,t_0],x) \subset G^T(U)$ and $y = \varphi(t_0,x) \in \Gamma^T(U)$.

Let us define

$$\begin{aligned} t_0 &= \inf\{s > 0|\ \varphi([s-T,s+T],x) \not\subset U\} \\ &= \inf\{s > 0|\ \varphi(s,x) \notin G^T(U)\} ; \end{aligned}$$

we have $t_0 \in [0,t_1]$. Since $G^T(U)$ is closed, $\varphi([0,t_0],x) \subset G^T(U)$, therefore $t_0 < t_1$. Defining $y = \varphi(t_0,x)$, we have

$$\varphi(T,y) = \varphi(t_0 + T, x) \in \partial U ,$$

therefore $y \in \Gamma^T(U)$.

(4) From property (4), $\overline{G^T(U)\backslash\Gamma^T(U)} = G^T(U) \subset U$. For $T_1 > T$, we obtain

$$G^{T_1}(U) \subset G^T(U) = \overline{G^T(U)\backslash\Gamma^T(u)} .$$

□

A topological invariant is introduced to describe the index pair (N,L) relative to an isolating neighborhood U. Conley called the homotopy type $h(U) = [N/L]$ the invariant. In comparing with that in the Morse theory, we prefer to replace it by the relative homology groups. However, in order to match Conley's definition, Alexander–Spanier cohomology is more suitable, because it possesses a special excision property not shared by singular cohomology theory.

For a topological pair (X,A) and a coeficient field F, $\overline{H}^*(X,A;F)$ stands for Alexander–Spanier cohomology. The following excision property holds.

Suppose that X and Y are paracompact Hausdorff spaces, and that A and B are closed in X and Y respectively. If $X\backslash A$ is homeomorphic to $Y\backslash B$. Then $\overline{H}^*(X,A;F) \cong \overline{H}^*(Y,B;F)$.

Definition 5.5.14 *(Conley index) For a given flow φ and a given $U \in \Sigma$, let (N,L) be an index pair with respect to (U,φ). We call*

$$h(U) = h(U, \varphi) = \overline{H}^*(N, L; F)$$

the Conley index for U (or for the isolated invariant set $S = I(U)$) with respect to the flow φ.

In contrast with equation (5.3), we also write

$$h(t, U) = \sum_{q=0}^{\infty} t^q \bar{H}^q(N, L; F) .$$

In the following we omit φ in the notation h, if there is no ambiguity.

In order to show that the Conley index is well-defined, one has to verify that if (N_1, L_1) and (N_2, L_2) are two index pairs relative to U, then $\overline{H}^*(N_1, L_1; F) \cong \overline{H}^*(N_2, L_2; F)$.

In fact by definition, $\exists T > 0$ such that

$$G^T(N_1 \backslash L_1) \subset G^T(U) \subset \overline{N_2 \backslash L_2} \text{ and } G^T(N_2 \backslash L_2) \subset G^T(U) \subset \overline{N_1 \backslash L_1} .$$

Because $\overline{N_i \backslash L_i} \in \Sigma$, $i = 1, 2$, for sufficient large $T > 0$, we have

$$G^T(N_1 \backslash L_1) \subset \text{int}(\overline{N_2 \backslash L_2}) \text{ and } G^T(N_2 \backslash L_2) \subset \text{int}(\overline{N_1 \backslash L_1}) . \tag{5.46}$$

Lemma 5.5.15 *If (N_1, L_1) and (N_2, L_2) are two index pairs relative to $U \in \Sigma$, and if $T > 0$ such that (5.46) holds, then the map $f : [T, \infty) \times N_1/L_1 \to N_2/L_2$ defined by*

$$f(t, [x]) = \begin{cases} \varphi(3t, x) & \textit{if } \varphi([0, 2t], x) \subset N_1 \backslash L_1 \textit{ and } \varphi([t, 3t], x) \subset N_2 \backslash L_2 , \\ [L_2] & \textit{otherwise} \end{cases}$$

is continuous.

Proof. There are three cases in logic: (1) $\varphi([t, 3t], x) \not\subset \overline{N_2 \backslash L_2}$, (2) $\varphi([0, 2t], x) \not\subset \overline{N_1 \backslash L_1}$, (3) $\varphi([t, 3t], x) \subset \overline{N_2 \backslash L_2}$ and $\varphi([0, 2t], x) \subset \overline{N_1 \backslash L_1}$.

In case (1), by the continuity of the flow, there exists a neighborhood W of (t, x) such that $\varphi([s, 3s], y) \not\subset \overline{N_2 \backslash L_2}$ $\forall (s, y) \in W$ then $f(s, [y]) = [L_2] = f(t, [x])$, i.e., f is continuous at $(t, [x])$.

Similarly for case (2).

In case (3), we consider two possibilities:

(a) $\varphi([t, 3t], x) \cap L_2 = \emptyset$. Since $\varphi([t, 3t], x) \subset N_2 \backslash L_2$, it follows that $\varphi(2t, x) \in G^t(N_2 \backslash L_2) \subset \text{int}(\overline{N_1 \backslash L_1}) \subset N_1 \backslash L_1$. Again from $\varphi([0, 2t], x) \subset \overline{N_1 \backslash L_1}$ and the positive invariance of L_1 in N_1, it follows that $\varphi([0, 2t], x) \subset N_1 \backslash L_1$. Therefore $f(t, [x]) = \varphi(3t, x) \in N_2 \backslash L_2$. For any open neighborhood U of $\varphi(3t, x)$ in $N_2 \backslash L_2$, by the continuity of the flow φ, $\exists$ a neighborhood W in N_1/L_1 of (t, x) such that $\varphi([0, 2s], y) \subset N_1 \backslash L_1, \varphi([s, 3s], y) \subset N_2 \backslash L_2$ and $\varphi(3s, y) \in U \, \forall (s, y) \in W$. Thus $f(s, [y]) = \varphi(3s, y) \in U$ as $(s, y) \in W$ with $y \in N_1$, i.e., f is continuous at this point (t, x).

(b) $\varphi([t,3t],x) \cap L_2 \neq \emptyset$. Since L_2 is an exit set of N_2, $\varphi(3t,x) \in L_2$. Let $[U]$ be any neighborhood of $[L_2]$ in $N_2 \backslash L_2$, define $V = \{x \in N_2 \backslash L_2 \,|\, x \in [U]\} \cup (X \backslash N_2) \cup L_2$, then V is a neighborhood of L_2 in X, and $[U] = (V \cap (N_2 \backslash L_2)) \cup [L_2]$.

By the continuity of the flow φ, $\exists$ a neighborhood W of (t,x) such that $\varphi(3s,y) \in V$ $\forall (s,y) \in W$. Thus

$$f(s,[y]) \in \{[\varphi(3s,y)],[L_2]\} \subset (V \cap (N_2 \backslash L_2)) \cup [L_2] \subset [U]\,,$$

$\forall (s,y) \in W$ with $y \in N_1$. Again f is continuous at this point. The proof is complete. □

Theorem 5.5.16 *If (N_1,L_1) and (N_2,L_2) are two index pairs relative to $U \in \Sigma$, then $\overline{H}^*(N_1,L_1;F) \cong \overline{H}^*(N_2,L_2;F)$.*

Proof. According to Lemma 5.5.15, there are continuous functions: $f : [T,\infty) \times N_1/L_1 \to N_2/L_2$ and $g : [T,\infty) \times N_2/L_2 \to N_1/L_1$ defined by

$$f(t,[x]) = \begin{cases} \varphi(3t,x) & \text{if } \varphi([0,2t],x) \subset N_1 \backslash L_1 \text{ and } \varphi([t,3t],x) \subset N_2 \backslash L_2\,, \\ [L_2] & \text{otherwise}\,, \end{cases}$$

and

$$g(t,[x]) = \begin{cases} \varphi(3t,x) & \text{if } \varphi([0,2t],x) \subset N_2 \backslash L_2 \text{ and } \varphi([t,3t],x) \subset N_1 \backslash L_1\,, \\ [L_1] & \text{otherwise}\,, \end{cases}$$

respectively.

One defines $\eta_i : [0,T] \times N_i/L_i \to N_i/L_i$

$$\eta_i(t,[x]) = \begin{cases} \varphi(6t,x) & \text{if } \varphi([0,6t],x) \subset N_i \backslash L_i\,, \\ [L_i] & \text{otherwise}\,, \end{cases}$$

$i = 1,2$. Again they are continuous and satisfy $\eta_1(T,[x]) = g(T,f(T,[x]))$, $\eta_2(T,[x]) = f(T,g(T,[x]))$ and $\eta_i(0,[x]) = \mathrm{id}_{N_i/L_i}$ $i = 1,2$. This shows that N_1/L_1 and N_2/L_2 have the same homotopy type. It follows that

$$\overline{H}^*(N_1/L_1,[L_1];F) \cong \overline{H}^*(N_2/L_2,[L_2];F)\,.$$

By the special excision property, we have

$$\overline{H}^*(N_i,L_i;F) \cong \overline{H}^*(N_i/L_i,[L_i];F), \quad i = 1,2\,.$$

Combining them together, we obtain

$$\overline{H}^*(N_1,L_1;F) \cong \overline{H}^*(N_2,L_2;F)\,.$$

□

Example 1. (Hyperbolic fixed point) Recall the hyperbolic system and the Hartman–Grobman theorem in Chap. 1. Let $f : R^n \to R^n$ be C^1. Assume that θ is a hyperbolic fixed point with k-dimensional unstable manifold, i.e., $A = e^L$ with $L = f'(\theta)$ satisfying $\sigma(A) \cap S^1 = \emptyset$. Then we have the decomposition $R^n = E^u \oplus E^s$ where E^u and E^s are invariant subspaces of A, on which the eigenvalues of $A^u = A|_{E^u}$ lie outside the unit circle, and those of $A^s = A|_{E^s}$ lie inside the unit circle with $\dim E^u = k$. Thus $L = L^u \oplus L^s$, where $L^u = L|_{E^u}$ and $L^s = L|_{E^s}$. The real parts of all eigenvalue of L^u are greater than zero, while those of L^s are less than zero. According to the Hartman–Grobman theorem the flow in a neighborhood of θ for the ODE $\dot{x} = f(x)$ is topologically equivalent to the flow in a neighborhood of $\dot{x} = Lx$, i.e., $\varphi(t,x) = e^{Lt}x$. An isolating neighborhood of the origin is given by $B^k \times B^{n-k}$ and the exit set by $S^{k-1} \times B^{n-k}$. Let $U = N = B^k \times B^{n-k}$ and $L = S^{k-1} \times B^{n-k}$. It follows that

$$h(t, U) = t^k .$$

Example 2. An invariant manifold S is an invariant set with manifold structure. It is called normally hyperbolic, if the tangent bundle over R^n restricted on S has the bundle decomposition: $(\mathrm{TR}^n)_S = E^u \oplus E^s$ with

$$E^u_{\varphi(t,x)} = T\varphi(t,\cdot)E^u_x ,\ E^s_{\varphi(t,x)} = T\varphi(t,\cdot)E^s_x \quad \forall x \in S,\ \forall t \in R^1 .$$

According to Thom's isomorphism theorem, see [BT],[Mi 3],[Sp], we have the following result: Let S be a normally hyperbolic invariant manifold. Let E^u be the vector bundle over S defined by the local unstable manifold of S. If E^u is a rank k orientable bundle, then

$$h(U) = \overline{H}^{*+k}(S; F) ,$$

for $U \in \Sigma$ with $I(U) = S$.

Thus, if S is a hyperbolic periodic orbit with an oriented unstable manifold of dimension $k + 1$, then

$$h(t, U) = t^k + t^{k+1} .$$

Example 3. (Critical groups) Let (M, f, ϕ) be a pseudo-gradient flow, and let $\{\mathcal{O}, \alpha, \beta\}$ be an isolating triplet for a dynamically isolated critical set S of f. Then

$$h(\mathcal{O}) = \overline{H}^*(\widetilde{\mathcal{O}} \cap f_\beta, \widetilde{\mathcal{O}} \cap f_\alpha; F) .$$

In particular, if S is an isolated critical point p, then

$$h(\mathcal{O}) = C_*(f, p) ,$$

where $C_*(f, p)$ is the critical groups of f at S.

The Conley index possesses many important properties: Wazewski's principle, the continuation and the Morse–Smale decomposition, which are the counterparts of the Kronecker existence, the homotopy invariance and the (sub) additivity of the degree theory.

First, we introduce a type of compactness condition on $U \in \Sigma$.

Definition 5.5.17 *(**Condition(B)**) $U \in \Sigma$ is said to satisfy Condition (B) if $\forall$ closed neighborhoods W of $I(U)$ there exists $T > 0$ such that $G^T(U) \subset W$.*

In particular, if $U \in \Sigma$ satisfies Condition (B) and $I(U) = \emptyset$, then $\exists T > 0$ such that $G^T(U) = \emptyset$.

It is known that if $U \in \Sigma$ is compact, then U satisfies Condition (B) (see Remark 5.5.4).

We investigate other sufficient conditions of Condition (B): Rubakowski [Ry] introduced:

Condition (A) $U \in \Sigma$ is said to satisfy Condition (A) if $\forall x_n \in U$, $\forall t_n \to +\infty$, $\varphi([0, t_n], x_n) \subset U$ implies that $\varphi(t_n, x_n)$ is subconvergent.

Condition (A) $\Rightarrow$ Condition (B).

Indeed, if not, $\exists W$, a closed neighborhood of $I(U)$ such that $G^n(U) \not\subset W$ $\forall n$, i.e., $\exists y_n \in G^n(U) \backslash W$. Let $x_n = \varphi(-n, y_n)$, then $\varphi([0, n], x_n) \subset U$. By Condition (A), $y_n = \varphi(n, x_n)$ subconvergent to $y_0 \in U$, and then $y_0 \in I(U)$. But W is closed, therefore $y_0 \notin \text{int}(W)$. This is a contradiction.

The isolating neighborhoods of the examples of dynamically isolated critical sets after Theorem 5.5.10 satisfy Condition (B). More precisely, let (M, f, φ) be a pseudo-gradient flow, and let c be an isolated critical value. If $K_c = K \cap f^{-1}(c)$ is isolated (or, $p \in K_c$ is isolated), then the set $W = cl(\tilde{\mathcal{O}}) \cap f^{-1}[c - \epsilon, c + \epsilon]$ is an isolating neighborhood of K_c (or p resp.) satisfying Condition (B), if $\epsilon > 0$ is small so that $f^{-1}[c - \epsilon, c + \epsilon] \cap K = K_c$, and $\mathcal{O}$ is a neighborhood of K_c (or p resp.) containing only critical points in K_c (or p resp.).

From Lemma 5.5.9, it is sufficient to show that any neighborhood V of p, containing p as the only critical point, contains a (MVP) neighborhood U of p.

In fact, let $B_R(p) \subset V$ for some $R > 0$, we choose $r = \frac{R}{2}, \mathcal{O} = B_r(p)$, and $U = cl(\tilde{\mathcal{O}}) \cap f^{-1}(c - \delta, c + \delta)$ for small $\delta > 0$. By standard estimates, $U \subset B_R(p) \subset V$.

The following Wazewski principle holds:

Theorem 5.5.18 *Let $U \in \Sigma$ satisfy Condition (B). If $h(U) \neq 0$ then $I(U) \neq \emptyset$.*

Proof. We prove that $I(U) = \emptyset$ implies $\overline{H}^*(N, L; F) = 0$ for all index pairs (N, L) relative to U. According to Theorem 5.5.13, it is sufficient to verify that for a special index pair $(N, L) = (G^T(U), \Gamma^T(U))$. Condition (B) yields the existence of large $T > 0$ such that $G^T(U) = \emptyset$. The conclusion follows trivially. □

Next, we turn to the continuation. A special form of the homotopy invariance of the topological degree asserts that the degree $\deg(f, \Omega, p)$ is continuous with respect to the vector field f. In Conley index theory, the vector field f is replaced by the flow φ, so we should study: Under what condition on $U \in \Sigma$, and in which sense of the variance between flows φ and ψ, does one have

$$h(U, \varphi) = h(U, \psi)?$$

First we notice:

Lemma 5.5.19 *Assume that $U \in \Sigma(\varphi)$ satisfies Condition (B) and that $I(U)$ is compact. Then U satisfies the condition:*

$$\Sigma_0(\varphi) : \exists T, \delta > 0 \text{ such that } N_\delta(G^T(U)) \subset \operatorname{int}(U) ,$$

where $N_\delta(A)$ is the δ-neighborhood of the set A. Moreover, $\exists T, \delta > 0$ such that

$$N_\delta(G^T(U)) \subset G^{T/2}(U), \; N_\delta(G^{T/2}(U)) \subset \operatorname{int}(U) . \tag{5.47}$$

Proof. 1. Let $\epsilon = d(\partial U, I(U))$ where $d(\cdot, \cdot)$ is the distance between two sets induced by the metric d. Since $I(U)$ is compact, $\epsilon > 0$. Let $V = N_{\epsilon/2}(I(U))$; by Condition (B), one has $T > 0$ such that $G^T(U) \subset V$. Let $\delta = \epsilon/2$; the first conclusion follows.

2. To prove the second conclusion, we claim that $\partial G^T(U) \cap I(U) = \emptyset$. Indeed, if not, $\exists x_0 \in \partial G^T(U) \cap I(U)$. From $x_0 \in \partial G^T(U)$, it implies that $\exists t_0 \in [-T, T]$ such that $\varphi(t_0, x_0) \in \partial U$. But from $x_0 \in I(U)$, we have $\varphi([t_0 - T, t_0 + T], x_0) \subset U$, then $\varphi(t_0, x_0) \in G^T(U) \subset \operatorname{int}(\mathrm{U})$. This is a contradiction.

Then $\epsilon_1 := d(\partial G^T(U), I(U)) > 0$; we repeat the proof in step 1, and obtain $T_1 > 0$ such that $G^{T_1}(U) \subset V := N_{\frac{\epsilon_1}{2}}(I(U))$. Thus, $N_{\frac{\epsilon_1}{2}}(G^{T_1}) \subset G^T(U)$. □

One defines the Hausdorff metric between two sets U and V as follows:

$$d_H(U, V) = \sup_{x \in U} d(x, V) + \sup_{x \in V} d(U, x) .$$

Lemma 5.5.20 *For $U \in \Sigma_0(\varphi)$, there exist $T, \epsilon > 0$ such that if a flow ψ satisfies*

$$d(\varphi(t, x), \psi(t, x)) \leq \epsilon \;\forall (t, x) \in [-T, T] \times N_\delta(U) \tag{5.48}$$

and a closed set V satisfies

$$d_H(U, V) \leq \epsilon , \tag{5.49}$$

then $U, V \in \Sigma_0(\psi)$ and $V \in \Sigma_0(\varphi)$.

Proof. By definition, one has $T, \delta > 0$ such that

$$N_\delta(G^T_\varphi(U)) \subset \text{int}(U) .$$

Letting $\epsilon \le \delta/4$, we have

$$N_\epsilon(G^T_\psi(U)) \subset N_{2\epsilon}(G^T_\varphi(U)) \subset \text{int}(U) ,$$

i.e., $U \in \Sigma_0(\psi)$, and

$$N_{2\epsilon}(G^T_\psi(V)) \subset N_{4\epsilon}(G^T_\varphi(U)) \subset \text{int}(U) .$$

Thus $d_H(N_\epsilon(G^T_\psi(V)), \partial V) \ge \epsilon$; it follows that $N_\epsilon(G^T_\psi(V)) \subset \text{int}(V)$ i.e., $V \in \Sigma_0(\psi)$. Similarly, we prove that $V \in \Sigma_0(\varphi)$. □

Theorem 5.5.21 *(Continuation) For $U \in \Sigma_0(\varphi)$, $\exists T, \epsilon > 0$ such that*

$$h(U, \varphi) = h(V, \psi) ,$$

whenever (ψ, V) satisfies (5.48) and (5.49), and that both φ and ψ are uniformly continuous on $[-T, T] \times N_{2\delta}(U)$, namely,

$$d(\phi(t,x), \phi(t,y)) < \frac{\delta}{3}, d(\psi(t,x), \psi(t,y)) < \frac{\delta}{3}, \text{ as } d(x,y) < \epsilon .$$

Proof. The proof is fairly long, we divide it into several steps.

First we assume $U = V$, and use the following simplified notations:

$$\widetilde{\varphi} = \psi,\ G^T = G^T_\varphi(U), \Gamma^T = \Gamma^T_\varphi(U),$$
$$\widetilde{G}^T = G^T_{\widetilde{\varphi}}(V), \widetilde{\Gamma}^T = \Gamma^T_{\widetilde{\varphi}}(V) .$$

According to (5.47), we may assume

$$N_\delta(G^T) \subset G^{T/2},\ N_\delta(G^{T/2}) \subset \text{int}(U) .$$

Setting $\epsilon \in (0, \delta/3)$ in (5.48), we have

$$N_{\frac{2\delta}{3}}(\widetilde{G}^T) \subset G^{T/2},\ N_{\frac{2\delta}{3}}(\widetilde{G}^{T/2}) \subset \text{int}(U),\ N_{\frac{2\delta}{3}}(G^T) \subset \widetilde{G}^{T/2} ,$$

and

$$N_{\frac{\delta}{3}}(\widetilde{G}^T) \subset \widetilde{G}^{T/2} .$$

1. $\forall \lambda \in [-1, +1]$, let $\xi_\lambda(T, x) = \widetilde{\varphi}(\lambda T, \varphi(-\lambda T, x))$, then $\xi_\lambda(T, \cdot) : G^T \to \text{int}(G^{T/2}) \cap \text{int}(\widetilde{G}^{T/2})$ and $\Gamma^T \to N_\epsilon(\Gamma^T)$ is continuous.

In fact, $\forall x \in G^T, y = \varphi(-\lambda T, x) \in U$, we have

$$d(\xi_\lambda(T,x), x) = d(\widetilde{\varphi}(\lambda T, y), \varphi(\lambda T, y)) < \epsilon ,$$

and $N_\epsilon(G^T) \subset G^{T/2}$, $N_\epsilon(G^T) \subset \widetilde{G}^{T/2}$.

2. The map $h_1 : \mathrm{int}(G^{T/2}) \to G^T/\Gamma^T$ defined by

$$h_1(x) = \begin{cases} [\varphi(T,x)] & \varphi(T,x) \in G^T \\ [\Gamma^T] & \text{otherwise} \end{cases}$$

is continuous.

In fact, it is sufficient to verify that $\varphi(T,x) \in \partial G^T \Rightarrow \varphi(T,x) \in \Gamma^T$. Since

$$\varphi(T,x) \in G^T \Leftrightarrow \varphi\left(\left[\frac{T}{2}, \frac{3}{2}T\right], x\right) \subset G^{T/2} \subset \mathrm{int}(U)\,,$$

in combination with $x \in \mathrm{int}(G^{T/2})$, i.e., $\varphi([-\frac{T}{2}, \frac{T}{2}], x) \subset \mathrm{int}(U)$, we have $\varphi([-\frac{T}{2}, \frac{3T}{2}], x) \subset \mathrm{int(U)}$, i.e., $\varphi([-\frac{T}{2}, \frac{3}{2}T], x) \cap \partial U = \emptyset$. Since $\varphi(T,x) \in \partial G^T$ implies that $\varphi([0,2T],x) \cap \partial U \neq \emptyset$, so $\exists \bar{t} \in [\frac{3}{2}T, 2T]$ such that $\varphi(\bar{t},x) \in \partial U$, and then $\varphi(T,x) \in \Gamma^T$. Also the map $\widetilde{h_1} : \mathrm{int}(\widetilde{G}^{T/2}) \to \widetilde{G}^T/\widetilde{\Gamma}^T$ defined by

$$\widetilde{h_1}(x) = \begin{cases} [\widetilde{\varphi}(T,x)] & \widetilde{\varphi}(T,x) \in \widetilde{G}^T \\ [\widetilde{\Gamma}^T] & \text{otherwise} \end{cases}$$

is continuous.

Note: On the right-hand side of the definitions of h_1 and $\widetilde{h}_1$, it is equivalent to write $\varphi([\frac{T}{2}, T], x) \subset G^T$ and $\widetilde{\varphi}([\frac{T}{2}, T], x) \subset \widetilde{G}^T$ instead, respectively.

3. The composition map $g_1 = \widetilde{h}_1 \circ \xi_1(T, \cdot) : G^T/\Gamma^T \to \widetilde{G}^T/\widetilde{\Gamma}^T$ is well defined and continuous:

$$g_1(x) = \begin{cases} [\widetilde{\varphi}(2T, \varphi(-T,x))] & \widetilde{\varphi}(2T, \varphi(-T,x)) \in \widetilde{G}^T \\ [\widetilde{\Gamma}^T] & \text{otherwise}\,. \end{cases}$$

It is sufficient to verify that $x \in \Gamma^T \Rightarrow g_1(x) = [\widetilde{\Gamma}^T]$.

In fact, $x \in G^T$ implies that $\varphi(\frac{T}{2}, x) \in G^{\frac{T}{2}} \subset \mathrm{int}(U)$. Therefore $x \in \Gamma^T \Rightarrow \exists \bar{t} \in (\frac{T}{2}, T]$ such that $\varphi(\bar{t}, x) \in \partial U$. From step 1, the assumptions on the uniform continuity of $\widetilde{\varphi}$ and (5.48), we have

$$\begin{aligned} d(\widetilde{\varphi}(\bar{t}, \xi_1(T,x)), \partial U) &\leq d(\widetilde{\varphi}(\bar{t}, \xi_1(T,x)), \varphi(\bar{t},x)) \\ &\leq d(\widetilde{\varphi}(\bar{t}, \xi_1(T,x)), \widetilde{\varphi}(\bar{t},x)) + d(\widetilde{\varphi}(\bar{t},x), \varphi(\bar{t},x)) \\ &\leq \delta/3 + \epsilon\,. \end{aligned}$$

It follows that $\widetilde{\varphi}(\bar{t}, \xi_1(T,x)) \notin \widetilde{G}^T$, and then $\widetilde{\varphi}([\frac{T}{2}, T], \xi_1(T,x)) \not\subset \widetilde{G}^T$. But $\widetilde{\varphi}(2T, \varphi(-T,x)) \in \widetilde{G}^T$ is equivalent to $\widetilde{\varphi}([0,T], \xi_1(T,x)) \subset \widetilde{G}^T$. The conclusion follows.

4. Similarly, we define $\widetilde{h}_2 : \mathrm{int}(\widetilde{G}^{T/2}) \to G^T/\Gamma^T$ by

$$\widetilde{h}_2(x) = \begin{cases} [\varphi(T,x)] & \varphi(T,x) \in \widetilde{G}^T \\ [\Gamma^T] & \text{otherwise}\,, \end{cases}$$

and $\widetilde{\xi_\lambda}(T,x) = \varphi(\lambda T, \widetilde{\varphi}(-\lambda T, x)) : \widetilde{G}^T \to \mathrm{int}(\widetilde{G}^{T/2})$. Also

$$g_2 = \widetilde{h}_2 \circ \widetilde{\xi}_1(T, \cdot) : \widetilde{G}^T / \widetilde{\Gamma}^T \to G^T / \Gamma^T$$

is well defined and continuous.

5. In order to verify that $h(U,\varphi) = h(U,\widetilde{\varphi})$, it is sufficiently to take $(N,L) = (G^T, \Gamma^T)$, $(\widetilde{N},\widetilde{L}) = (\widetilde{G}^T, \widetilde{\Gamma}^T)$ for sufficiently large $T > 0$, and show that $g_2 \circ g_1$ and $g_1 \circ g_2$ are homotopic to identity maps on G^T/Γ^T and $\widetilde{G}^T/\widetilde{\Gamma}^T$, respectively.

Let us define for $\lambda \in [0,1]$ the map

$$H(\lambda, [x]) = \begin{cases} [\varphi(2T, \xi_{-\lambda}(T,x))] & \varphi([0,2T], \xi_{-\lambda}(T,x)) \subset G^T \\ [\Gamma^T] & \text{otherwise} . \end{cases}$$

Note: It is equivalent to replace $\varphi([0,2T], \xi_{-\lambda}(T,x)) \subset G^T$ by $\varphi(2T, \xi_{-\lambda}(T, x)) \in G^T$ on the right-hand side of the definition.

Since $H(1,[x]) = g_2 \circ g_1$ and

$$H(0, [x]) = \begin{cases} [\varphi(2T, x)] & \varphi([0,2T], x) \subset G^T \\ [\Gamma^T] & \text{otherwise} , \end{cases}$$

the latter is homotopic to $\mathrm{id}|_{G^T/\Gamma^T}$ as shown in Theorem 5.5.16.

It remains to verify that the definition is well-defined and that $H(\lambda, [x])$ is continuous.

We shall show that $x \in \Gamma^T \Rightarrow H(\lambda, x) = [\Gamma^T]$ i.e., $\varphi([0,2T], \xi_{-\lambda}(T,x)) \not\subset G^T$.

From $x \in \Gamma^T$, by step 1, $\xi_{-\lambda}(T,x) \in N_\epsilon(\Gamma^T)$, i.e., $\exists \overline{x} \in \Gamma^T$ such that $d(\xi_{-\lambda}(T,x), \overline{x}) < \epsilon$. Following the argument in 3, $\varphi([\frac{1}{2}T, \frac{3}{2}T], \overline{x}) \cap \partial U \neq \emptyset$, we have $\varphi([0,2T], \xi_{-\lambda}(T,x)) \not\subset G^T$.

As to the continuity, we define $h_2 : G^T/\Gamma^T \to G^T/\Gamma^T$ by

$$h_2([x]) = \begin{cases} [\varphi(T,x)] & \varphi(T,x) \in G^T \\ [\Gamma^T] & \text{otherwise} . \end{cases}$$

Since $\varphi(T,x) \in G^T$ can be replaced by $\varphi([0,T],x) \subset G^T$, the continuity of h_2 has been proved in Theorem 5.5.16, and from $\varphi([0,T],x) \subset G^T \subset \mathrm{int}(U)$, it implies $x \notin \Gamma^T$, h_2 is well defined.

By definition

$$H(\lambda, [x]) = h_2 \circ h_1 \circ \xi_{-\lambda}(T,x) ,$$

therefore H is continuous on $[0,T] \times G^T/\Gamma^T$.

6. Finally, we consider the case $U \neq V$, but with $\widetilde{\varphi} = \varphi$, and write $G^T(U) = G^T_\varphi(U)$, $G^T(V) = G^T_\varphi(V)$. Since $N_\delta(G^T(U)) \subset \mathrm{int}(U)$, $N_\delta(G^T(V)) \subset \mathrm{int}(V)$, if one takes $0 < \epsilon < \delta/3$ and $d_H(U,V) < \epsilon$, we have $N_\epsilon(G^T(U)) \subset V$ and $N_\epsilon(G^T(V)) \subset U$.

Let (N, L) and $(\widetilde{N}, \widetilde{L})$ be index pairs related to U and V resp., then $\exists T_1, T_2 > 0$ such that $G^{T_1}(\overline{N \backslash L}) \subset G^{T_1}(U) \subset V$, and $G^{T_2}(V) \subset \overline{\widetilde{N} \backslash \widetilde{L}}$. Thus $G^{T_1+T_2}(\overline{N \backslash L}) \subset \overline{\widetilde{N} \backslash \widetilde{L}}$. Similarly one has $G^{T_1+T_2}(\overline{\widetilde{N} \backslash \widetilde{L}}) \subset \overline{N \backslash L}$. By the proof of Theorem 5.5.16, we obtain $\overline{H}^*(N, L; F) \cong \overline{H}^*(\widetilde{N}, \widetilde{L}; F)$, i.e., $h(U, \varphi) = h(V, \varphi)$. Combining steps 5 and 6, we have

$$h(U, \varphi) = h(V, \varphi) = h(V, \psi) .$$

The proof is complete. □

5.5.3 Morse Decomposition on Compact Invariant Sets and Its Extension

In the study of the dynamics on invariant sets the Morse–Smale decomposition is an extension of the Morse relation. Before going on, let us introduce the notion of attractor–repeller pairs.

On a metric space (X, d) with flow φ, $\forall Y \subset X$, define

$$\omega(Y) = \bigcap_{t \geq 0} \overline{\varphi([t, \infty), Y)} \text{ and } \omega^*(Y) = \bigcap_{t > 0} \overline{\varphi((-\infty, -t], Y)} .$$

By definition, $\omega(Y)$ is an invariant closed set.

Definition 5.5.22 *Let S be a compact invariant set for φ; a subset $A \subset S$ is called an attractor in S, if there exists a neighborhood U of A such that $\omega(U \cap S) = A$. The dual repeller of A in S is defined by*

$$A^* = \{x \in S |\ \omega(x) \cap A = \emptyset\} .$$

The pair (A, A^*) is called an attractor–repeller pair.

The set

$$C(A^*, A, S) = \{x \in S |\ \omega(x) \subset A,\ \omega^*(x) \subset A^*\}$$

is called the set of connecting orbits from A^* to A in S.

The following properties hold.

(1) A and A^* are disjoint compact invariant sets.

Proof. In the following, we use the notation U as in Definition 5.5.22. In fact, $A = \omega(U \cap S)$ is invariant. Both $\omega(x)$ and A are invariant, so is A^*. If $\exists x \in A \cap A^*$, then $\omega(x) \subset A$. But by definition of A^*, $\omega(x) \cap A = \emptyset$. Since S is compact, $\omega(x) \neq \emptyset$. This is a contradiction. Therefore $A \cap A^* = \emptyset$.

By definition, A is closed.

It remains to verify the closedness of A^*.

If $\{x_n\} \subset A^*$ with $x_n \to x \in S$, and if $\omega(x) \cap A \neq \emptyset$, then $\varphi(t, x) \in U$ for some $t > 0$. Consequently $\varphi(t, x_n) \in U$ for n large. Therefore $\omega(x_n) \subset \omega(U \cap S) = A$. This is impossible. Therefore $\omega(x) \cap A = \emptyset$, i.e., $x \in A^*$. □

(2) For any open neighborhood V of A, $\exists T = T(V)$ such that $\bigcup_{t \geq T} \varphi(t, U \cap S) \subset V$.

Proof. If not, $\exists V \supset A$ open, $\exists x_n \in U \cap S$ $\exists t_n \to +\infty$ such that $\varphi(t_n, x_n) \notin V$. Since S is a compact invariant set, $\varphi(t_n, x_n) \to y \in S \backslash V$. But $y \in \omega(U \cap S) = A$. This is a contradiction. □

(3) If B is a closed set disjoint from A, then $\forall \epsilon > 0$ $\exists T = T_\epsilon > 0$ such that $d(x, A^*) < \epsilon$, whenever $x \in S$ and $t \geq T$ such that $\varphi(t, x) \in B$.

Proof. If not, $\exists \epsilon > 0$ $\exists t_n \to \infty$ $\exists x_n \in S$ such that $\varphi(t_n, x_n) \in B$ but $d(x_n, A^*) \geq \epsilon$.

Since S is compact and invariant, $\exists x \in S$ such that $x_n \to x$. Then $d(x, A^*) \geq \epsilon$, which implies that $x \notin A^*$, i.e., $\omega(x) \cap A \neq \emptyset$. Thus for the neighborhood U of A; we have $t_1 \in R^1$ such that $\varphi(t_1, x) \in U \cap S$ and then $\varphi(t_1, x_n) \in U \cap S$ for n large. Let $V = X \backslash B$, which is an open neighborhood of A, from (2), $\exists T > 0$ such that $\varphi(t, x_n) \in V$ $\forall t \geq T$. This contradicts $\varphi(t_n, x_n) \in B$ $\forall n$. □

For any $x \in X$, we denote the orbit passing through x by $o(x) = \{\varphi(t, x) |\ t \in R^1\}$.

(4) $\omega(y) \cap A^* \neq \emptyset \Rightarrow o(y) \subset A^*$.

Proof. Let B be a closed neighborhood of A^* such that $A \cap B = \emptyset$. Since $\omega(y) \cap A^* \neq \emptyset$, $\exists t_n \to +\infty$ such that $\varphi(t_n, y) \in B$. $\forall t \in R^1$, let $z = \varphi(t, y)$; we have $\varphi(t_n - t, z) = \varphi(t_n, y) \in B$, thus from (3), $\forall \epsilon > 0$, we have $d(z, A^*) < \epsilon$. Since $\epsilon > 0$ is arbitrarily small, $o(y) \subset A^*$. □

(5) $\omega^*(y) \cap A \neq \emptyset \Rightarrow o(y) \subset A$.

Proof. By the assumption that $\exists t_n \to +\infty$ such that $\varphi(-t_n, y) \in U \cap S$, then $\forall t \in R^1$, $t + t_n \geq 0$ for n large, from $\varphi(t, y) = \varphi(t_n + t, \varphi(-t_n, y))$ we have $\varphi(t, y) \in \omega(U \cap S)$, i.e., $o(y) \subset A$. □

Lemma 5.5.23 *Let (A, A^*) be an attractor–repeller pair of a compact invariant set S, then*

$$S = A \cup A^* \cup C(A^*, A, S)\ .$$

Proof. It is sufficient to prove that $\forall x \in S \backslash (A \cup A^*), \omega(x) \subset A$, and $\omega^*(x) \subset A^*$.

1. $\omega(x) \subset A$: $\forall y \in \omega(x)$, i.e., $\exists t_n \to +\infty$ such that $\varphi(t_n, x) \to y$. Let $B = S \backslash U$, where U is as in Definition 5.5.22, then either $\exists n_0 \in \mathbb{N}$ such that $z = \varphi(t_{n_0}, x) \in U$ or $\varphi(t_n, x) \in B$, $\forall n$. But the latter case is impossible, because from (3) we would have $x \in A^*$. In the former case, $y \in \omega(S \cap U) = A$.

2. $\omega^*(x) \subset A^*$: $\forall y \in \omega^*(x)$, $\exists t_n \to +\infty$ such that $z_n = \varphi(-t_n, x) \to y$. Let $B = \{x\}$. Since $\{x\} \cap A = \emptyset$, we have $d(z_n, A^*) \to 0$, provided by (3). Thus $y \in A^*$. □

Combining the conclusions of Lemma 5.5.23 and properties (4) and (5), we see that $\forall x \in C(A^*, A, S)$, $\omega(x) \subset A$ and $\omega^*(x) \subset A^*$, and there are connecting orbits from A to A^* but none from A^* to A.

Let us define the Morse decomposition for a compact invariant set.

Definition 5.5.24 *Let S be a compact invariant set of X with respect to the flow φ. An ordered collection $(M_1, \ldots, M_n)$ of invariant subsets $M_j \subset S$ is called a Morse decomposition of S, if there exists an increasing sequence of attractors in S:*

$$\emptyset = A_0 \subset A_1 \subset \cdots \subset A_n = S$$

such that $M_j = A_j \cap A^*_{j-1},\ 1 \le j \le n$.

Example 1. An attractor–repeller pair (A, A^*) of a compact invariant set S is a Morse decomposition, where $A_0 = \emptyset,\ A_1 = A,\ A_2 = S$.

Example 2. Suppose $f \in C^1(M, \mathbb{R}^1)$, where M is a compact manifold. Assume that $f^{-1}[a,b] \cap K = \{p_1, \ldots, p_n\}$, where a, b are regular values of f with $f(p_i) \le f(p_{i+1})$, $i = 1, \ldots, n-1$. Then $(\{p_1\}, \ldots, \{p_n\})$ is a Morse decomposition of $S = I(f^{-1}([a,b]))$.

In fact, by setting $A_0 = \emptyset$, and $A_i = \{x \in S|\ \omega^*(x) \subset \{p_1, \ldots, p_i\}\}$, $i = 1, 2, \ldots, n$. We shall verify that this is an increasing sequence of attractors in S with $A_i^* = \{x \in S|\ \omega(x) \subset \{p_{i+1}, \ldots, p_n\}\}$, and then $A_i \cap A^*_{i-1} = \{p_i\}$.

It is proved by induction. Let $S_k = I(f^{-1}[a, a_k])$, where $a_k \in (f(p_k), f(p_{k+1}))$, $k = 1, \ldots, n-1$, and $a_n = b$. Thus $S_n = S$. We verify that $A_i = \{x \in S_k\,|\,\omega^*(x) \subset \{p_1, p_2, \ldots, p_i\}\}$, $i = 1, 2, \ldots, k$, is an increasing sequence of attractors in S_k, and $A_i^* = \{x \in S_k\,|\,\omega(x) \subset \{p_{i+1}, \ldots, p_k\}\}$.

For $n = 1$. Obviously, $A_1 = \{p_1\}$ is an attractor in $S_1 = \{p_1\}$, with $A_1^* = \emptyset$ and $A_0^* = S_1$. Thus $A_1 \cap A_0^* = \{p_1\}$.

If the conclusion holds for $n = k$, i.e., $A_i = \{x \in S_k|\ \omega^*(x) \subset \{p_1, \ldots, p_i\}\}$, $i = 1, \ldots, k$, is an increasing sequence of attractors in S_k.

For $n = k+1$. It is easily seen that S_k is an attractor in S_{k+1}. Indeed, let $U_k = f^{-1}(a, a_k)$, then $\omega(S_{k+1} \cap U_k) = S_k$ with $S_k^* = \{p_{k+1}\}$. Then

$$A_k = \{x \in S_{k+1}|\ \omega^*(x) \subset \{p_1, \ldots, p_k\}\} = S_k$$

is an attractor in S_{k+1} with $A_k^* = S_k^* = \{p_{k+1}\}$. In this case

$$\emptyset = A_0 \subset A_1 \subset \ldots \subset A_k = S_k \subset A_{k+1} = S_{k+1}$$

is an increasing sequence of attractors in $S = S_{k+1}$. By easy computation,

$$A_i^* = \{x \in S_{k+1}|\ \omega(x) \subset \{p_{i+1}, \ldots, p_{k+1}\}\} \text{ and then } A_i \cap A^*_{i-1} = \{p_i\}\,.$$

Our conclusion follows from the mathematical induction.

By the definition of the Morse decomposition $(M_1, \ldots, M_n)$ of S, we have the following conclusions:

(1) $\{M_i|\ 1 \le i \le n\}$ are pairwise disjoint.

Indeed if $i < j$, then $M_i \cap M_j = A_i \cap A^*_{i-1} \cap A_j \cap A^*_{j-1} \subset A_i \cap A^*_{j-1} \subset A_{j-1} \cap A^*_{j-1} = \emptyset$, from property (1).

(2) $\forall x \in S \backslash (\bigcup_{1 \le i \le n} M_i)$, $\exists i < j+1$ such that $\omega(x) \subset M_i$ and $\omega^*(x) \subset M_{j+1}$.

Indeed, set

$$i = \min\{k \in \mathbb{N} |\ \omega(x) \subset A_k\} \text{ and } j = \max\{k \in \mathbb{N} |\ \omega^*(x) \subset A^*_k\}\ .$$

From $A_0 = \emptyset$ and $A_n = S$, we see $0 < i$ and $j < n$. Since $\omega(x) \not\subset A_{i-1}$, provided by the decomposition lemma, $x \in A^*_{i-1}$, and then by property (4), $o(x) \subset A^*_{i-1}$, in particular, $\omega(x) \subset A^*_{i-1}$. By the same reason, $\omega^*(x) \not\subset A_{j+1}$, we have $o(x) \subset A_{j+1}$ and $\omega^*(x) \subset A_{j+1}$.

We claim that $i \le j+1$. For otherwise, $j+1 \le i-1$, then $o(x) \subset A_{j+1} \cap A^*_{i-1} \subset A_{i-1} \cap A^*_{i-1} = \emptyset$, a contradiction. For $i = j+1$, we have $o(x) \subset A_i \cap A^*_{i-1} = M_i$, it contradicts $x \notin M_i$. Therefore the only possibility is $i < j+1$. In this case we have $\omega(x) \subset A_i \cap A^*_{i-1} = M_i$ and $\omega^*(x) \in A_{j+1} \cap A^*_j = M_{j+1}$.

Comparing with Example 2, the following conclusion is easily verified:

(3) $A_j = \{x \in S |\ \omega(x) \cup \omega^*(x) \subset M_1 \cup \ldots \cup M_j\}$ and $A^*_j = \{x \in S |\ \omega(x) \cup \omega^*(x) \subset M_{j+1} \cup \ldots \cup M_n\}$, $j = 1, 2, \ldots, n$.

(4) If $(M_1, \ldots, M_n)$ is a Morse decomposition of S, and if $\emptyset = A_0 \subset A_1 \subset \ldots \subset A_n = S$ is the associated increasing sequence of attractors, then for $0 < i \le j < n$, $(M_1, \ldots, M_{i-1}, M_{ji}, M_{j+1}, \ldots, M_n)$ is again a Morse decomposition of S, where $M_{ji} = A_j \cap A^*_{i-1} = \{x \in S |\ \omega(x) \cup \omega^*(x) \subset M_i \cup \ldots \cup M_j\}$.

(5) If S is an isolating invariant compact set, and if $\{M_1, \ldots, M_n\}$ is a Morse decomposition of S, then M_i, $1 \le i \le n$ are isolating.

Proof. We assume that there exists a compact neighborhood U of S such that $I(U) = S \subset \text{int}(U)$. According to (1), M_i and M_j are disjoint, $\forall i \ne j$, we have a neighborhood $U_i \subset U$ of M_i such that $U_i \cap M_j = \emptyset\ \forall j \ne i$. We claim that $I(U_i) = M_i$.

"$\subset$" $\forall x \in I(U_i)$, $o(x) \subset I(U_i) \subset U_i \subset U$. Therefore $\omega(x) \cup \omega^*(x) \subset U_i \cap S$. According to (2), $\omega(x) \cup \omega^*(x) \subset M_i$ and then $x \in M_i$.

"$\supset$" Since M_i is an invariant set in U_i, $M_i \subset I(U_i)$. □

For a compact isolating invariant set S, let $U \in \Sigma$ be its compact isolated neighborhood, i.e., $I(U) = S \subset \text{int}(U)$. We define

$$I^+_j = \{x \in U |\ \varphi(R^1_+, x) \subset U, \omega(x) \subset M_j \cup \ldots \cup M_n\} \text{ and}$$

$$I^-_j = \{x \in U |\ \varphi(R^1_-, x) \subset U, \omega^*(x) \subset M_1 \cup \ldots \cup M_j\},\ 1 \le j \le n\ .$$

(6)

$$I^+_1 \supset I^+_2 \supset \ldots \supset I^+_n,\ \varphi(R^1_+, I^+_j) = I^+_j,\ 1 \le j \le n\ ,$$

$$I^-_1 \supset I^-_2 \supset \ldots \supset I^-_n,\ \varphi(R^1_-, I^-_j) = I^-_j,\ 1 \le j \le n\ .$$

$$A_j \subset I_j^-, \; A_{j-1}^* \subset I_j^+ ,$$

and then

$$M_{ji} = A_j \cap A_{i-1}^* = I_i^+ \cap I_j^- ,$$

for $i \leq j$.

(7) $I_j^{\pm}$, $1 \leq j \leq n$ are compact.

Proof. It is sufficient to show that they are closed. For I_1^+ and I_n^- it follows from the continuity of the flow and the closedness of U. In the case where $n = 2$, it suffices to verify the closedness of I_2^+ and I_1^-. Let (M_1, M_2) be the Morse decomposition, and let $\{x_n\} \subset I_2^+$ and $x_n \to x$; we want to show that $x \in I_2^+$. Since $x \in I_1^+$, we have $\omega(x) \subset M_1 \cup M_2$. It is sufficient to show that $\omega(x) \not\subset M_1$.

Suppose not, i.e., $\omega(x) \subset M_1$. We choose open neighborhoods V_i of M_i, $i = 1, 2$ such that $\overline{V}_1 \cap \overline{V}_2 = \emptyset$. Since $\omega(x_n) \subset M_2$ and $x_n \to x$, we have $t_n' \geq 0$ and $t_n'' \geq 0$ such that $\varphi([t_n', \infty), x_n) \subset V_2$ and $\varphi(t_n'', x_n) \in V_1$. Thus $\exists t_n \in [t_n'', t_n']$ such that $\varphi([t_n, \infty), x_n) \subset U \backslash V_1$ and $\varphi(t_n, x_n) \in U \backslash (V_1 \cup V_2)$. From the compactness of U, after a subsequence $\varphi(t_n, x_n) \to y \notin M_1 \cup M_2$ and $\varphi([0, +\infty), y) \subset U \backslash V_1$. From $x_n \in I_1^+$, we have $\varphi(t_n, x_n) \in I_1^+$; it follows that $y \in I_1^+$, and then $\omega(y) \subset M_2$. If $\{t_n\}$ is bounded, then after a subsequence, $t_n \to s$; we have $\omega(\varphi(-s, y)) = \omega(y) \subset M_2$. But $\varphi(-s, y) = x$. This contradicts $\omega(x) \subset M_1$. If $\{t_n\}$ is unbounded, without loss of generality may assume, $t_n \to +\infty$. $\forall t > 0$ $\exists n$ large such that $\varphi([-t, 0], \varphi(t_n, x_n)) = \varphi([t_n - t, t_n], x_n) \subset \varphi([0, \infty), x) \subset U$; it follows that $\varphi([-t, 0], y) \subset U$. Since $t > 0$ is arbitrary, we have $\varphi(R_-^1, y) \subset U$. Combining with $\varphi(R_+^1, y) \subset U \backslash V_1$, it follows that $o(y) \subset U$, i.e., $y \in I(U) = S$. Since (M_1, M_2) is a Morse decomposition of S, and $\omega(y) \subset M_2$, it follows that $y \in M_2$, again a contradiction. Similarly, I_1^- is closed.

For $n > 2$, it can be reduced to the case where $n = 2$ by defining $M_1' = M_{j-1,1}$ and $M_2' = M_{n,j}$. Thus (M_1', M_2') is again a Morse decomposition of S from (4), since $I_j^+ = I_2^{+'}$, where $I_2^{+'}$ is that in (M_1', M_2'). Thus I_j^+ is closed. Similarly $I_j^- = I_j^{-''}$, where $M_1'' = M_{j1}$ and $M_2'' = M_{n,j+1}$. The proof is complete. □

(8) Let $U \in \Sigma$ be compact. If $Z \subset U$ is closed, and if $\forall x \in Z$ $\exists t > 0$ such that $\varphi(t, x) \notin U$, then the minimal positively invariant set in U which contains Z, defined by

$$P(Z, U) = \{x \in U \,|\, \exists t \geq 0 \,\text{such that}\; \varphi([-t, 0], x) \subset U \,\text{and}\; \varphi(-t, x) \in Z\}$$

is closed.

It follows directly from the continuity of the flow and the closedness of U.

Lemma 5.5.25 *Suppose that $U \in \Sigma$ is a compact isolating neighborhood with $S = I(U)$, and that $(M_1, \ldots, M_n)$ is a Morse decomposition of S. Then for every open neighborhood V of I_j^-, there exists a compact neighborhood N_j such that $N_j \subset V$ and N_j is positively invariant in U.*

Proof. Since $I_{j+1}^+ \cap I_j^- = \emptyset$, we choose a compact neighborhood $W \subset U$ of $I_{j+1}^+ \cap I_n^-$ such that $W \cap I_j^- = \emptyset$.

1. $\exists t^* > 0$ such that $\varphi([-t^*, 0], x) \subset \overline{U \backslash W} \Rightarrow x \in V \cap (U \backslash W)$. If not, $\exists \{x_n\} \subset U$ and $\{t_n\} \subset R_+^1$ with $t_n \to +\infty$ such that $\varphi([-t_n, 0], x_n) \subset \overline{U \backslash W}$, but $x_n \notin V \cap (U \backslash W)$. Let x be a limit point of x_n, then $x \notin V \cap (U \backslash W)$ and $\varphi(R_-^1, x) \subset \overline{U \backslash W}$. Thus $\omega^*(x) \subset M_1 \cup \ldots \cup M_j$. Therefore $x \in I_j^- \subset V \cap (U \backslash W)$. A contradiction.

2. Define two disjoint subsets of I_j^-

$$A = \{x \in I_j^- \mid \varphi([0, t^*], x) \subset U\} \text{ and } B = \{x \in I_j^- \mid \varphi([0, t^*], x) \not\subset U\},$$

then $\forall x \in A \, \exists \delta = \delta(x) > 0$ such that $\varphi([0, t^*], B_\delta(x)) \subset V \cap (X \backslash W)$, from step 1. $\forall x \in B \, \exists t = t(x) > 0$ such that $\varphi([0, t], x) \subset V \cap (X \backslash W)$ and $\varphi(t, x) \notin U$. This enables us to choose $\delta = \delta(x) > 0$ such that $\varphi([0, t(x)], B_\delta(x)) \subset V \cap (X \backslash W)$ and $\varphi(t(x), B_{\delta(x)}(x)) \cap U = \emptyset$. Since I_j^- is compact, $\exists x_1, \ldots, x_k \in I_j^-$ such that $I_j^- \subset \bigcup_{i=1}^k B_{\delta(x_i)}(x_i)$, and then we choose a compact neighborhood of I_j^-, $Z \subset \bigcup_{i=1}^k B_{\delta(x_i)}(x_i)$.

3. We claim that $P(Z, U) \subset V \cap (U \backslash W)$.

$\forall x \in P(Z, U)$, by definition $\exists t \geq 0$ such that $\varphi([-t, 0], x) \subset U$, and $\varphi(-t, x) \in Z$. Then by the definition of Z, $\exists i \in [1, k]$, such that $\varphi(-t, x) \in B_{\delta(x_i)}(x_i)$. If $x \notin V \cap (U \backslash W)$, there are two cases:

Either $x_i \in A$, then $\varphi([-t, t^* - t], x) \subset V \cap (X \backslash W)$ which implies $t > t^*$, and then $\exists t_1 \in [0, t^* - t]$ such that $\varphi([-t, -t_1], x) \subset V \cap (\overline{X \backslash W})$ and $y = \varphi(-t_1, x) \notin V \cap (X \backslash W)$. It implies that $\varphi([-t^*, 0], y) \subset \overline{U \backslash W}$, but $y \notin V \cap (X \backslash W)$. This contradicts step 1.

Or $x_i \in B$, then $\varphi([-t, t(x_i) - t], x) \subset V \cap (X \backslash W)$ and $\varphi(t(x_i) - t, x) \notin U$. From $\varphi([-t, 0], x) \subset U$, we see that $t(x_i) > t$. But if $x \notin V \cap (X \backslash W)$, we would have $t(x_i) < t$. Again this is a contradiction. We have proved that $x \in V \cap (X \backslash W) \cap U = V \cap (U \backslash W)$.

4. $P(Z, U)$ is compact.

In fact, if $\{x_n\} \subset P(Z, U)$ with $x_n \to x$, i.e., $\exists t_n \geq 0$ such that $\varphi([-t_n, 0], x_n) \subset U$ and $\varphi(-t_n, x_n) \in Z$, then $\varphi([-t_n, 0], x_n) \subset P(Z, U) \subset V \cap (U \backslash W)$. If $\{t_n\}$ is unbounded, then $\varphi(R_-^1, x) \subset \overline{V \cap (U \backslash W)}$. It follows that $\omega^*(x) \in M_i$ for some $i \leq j$, and then $x \in I_j^- \subset P(Z, U)$. Otherwise, $\{t_n\}$ is bounded; after a subsequence, $t_n \to s \geq 0$, then $\varphi([-s, 0], x) \subset U$ and $\varphi(-s, x) \in Z$; again $x \in P(Z, U)$.

Now we set $N_j = P(Z, U)$. Combining steps 1 and 4 and (8), the conclusion follows. □

Inductively applying Lemma 5.5.25, we have the following:

Theorem 5.5.26 *Let S be an isolating invariant set with a compact isolated neighborhood U_0. If $(M_1, \ldots, M_n)$ is a Morse decomposition of S, then there exists an increasing sequence of compact sets: $N_0 \subset N_1 \subset \ldots \subset N_n$ such that*

(N_j, N_{i-1}) is an index pair for M_{ji} $\forall i \le j$. In particular, (N_n, N_0) is an index pair for S and (N_j, N_{j-1}) is an index pair for M_j, $\forall j$.

Proof. According to Theorem 5.5.13, in combination with Remark 5.5.4, there exists an index pair (N_n, N_0) for S. Setting $U = \overline{N_n \backslash N_0}$, we define compact sets $I_j^\pm$ $j = 1, 2, \ldots, n$ see (7), and $U_n = U \cap N_n$. Applying Lemma 5.5.25, we have a compact neighborhood $U_{n-1} \subset U$ of I_{n-1}^-, which is positively invariant in N_n, and $U_{n-1} \cap I_n^+ = \emptyset$. Define $N_{n-1} = U_{n-1} \cup N_0$. Inductively, one defines a compact neighborhood $U_j \subset U$ of I_j^-, which is positively invariant in N_{j+1} and $U_j \cap I_{j+1}^+ = \emptyset$, and $N_j = U_j \cup N_0$, $j = n-1, \ldots, 1$. □

Recalling (5.2) and (5.3), as a consequence of Theorem 5.5.26, we have the Morse relations for the Morse decomposition $(M_1, \ldots, M_n)$, i.e.,

$$\sum_{j=1}^{n} P(t; N_j, N_{j-1}) = P(t; N_n, N_0) + (1+t)Q(t) ,$$

where Q is a formal series with nonnegative coefficients, and

$$P(t; N_j, N_{j-1}) = \sum_{q=0}^{\infty} t^q \quad \text{rank } \overline{H}^q(N_j, N_{j-1}; F) ,$$

or, in the spirit of Theorem 5.1.29,

$$\sum_{j=1}^{n} h(t; M_j) = h(t; U_0) + (1+t)Q(t) .$$

Let us return to Example 2, and assume that f is a Morse function, i.e., $\forall p \in K$, $f''(p)$ is invertible, or in other words, the gradient system $\dot{x} = -f'(x)$ is hyperbolic.

$\forall p \in K$, let

$$W^s(p) = \{x \in M |\ \omega(x) = p\} \text{ and } W^u(p) = \{x \in M |\ \omega^*(x) = p\}$$

be the stable and unstable manifolds at p, respectively. One has

$$\dim W^s(p) = \text{ind}(f, p) = \text{codim} W^u(p) .$$

According to the Sard–Smale theorem, one may choose a generic Riemannian metric on M such that $W^s(p) \pitchfork W^u(q)$ $\forall p, q \in K$. Such Morse functions are said to be of Morse–Smale type.

Thus, if $\text{ind}(f, q) - \text{ind}(f, p) = k$, and let

$$M(q, p) = W^u(q) \cap W^s(p),\ \dim M(q, p) = k .$$

We have seen that if $K = \{p_1, \ldots, p_n\}$, then $(\{p_1\}, \ldots, \{p_n\})$ is a Morse decomposition of M, where $p_1, \ldots, p_n$ are ordered by their values $\{f(p_i) |\ i \le i \le$

$n\}$. However, there is another way to make a Morse decomposition according to their Morse indices. Setting

$$S_k = \{p \in K |\ \mathrm{ind}(f,p) = k\},\ 0 \le k \le m = \dim M\ ,$$

and letting

$$A_j = \{x \in M |\ \omega(x) \cup \omega^*(x) \subset S_0 \cup \ldots \cup S_j\}\ 0 \le j \le m\ ,$$

then $(S_0, S_1, \ldots, S_m)$ is a Morse decomposition of M.

According to Theorem 5.5.26, there exists an increasing sequence of compact sets: $\varnothing = N_{-1} \subset N_0 \subset \ldots \subset N_m$ such that (N_j, N_{j-1}) is an index pair of $S_j, 0 \le j \le m$.

Noticing the following exact sequences, we have: (1) j is injective, (2) $\partial'_{k-1} \circ j = 0$, and (3) $\partial'_k = j \circ i$. Thus $\partial'_{k-1} \circ \partial'_k = 0$.

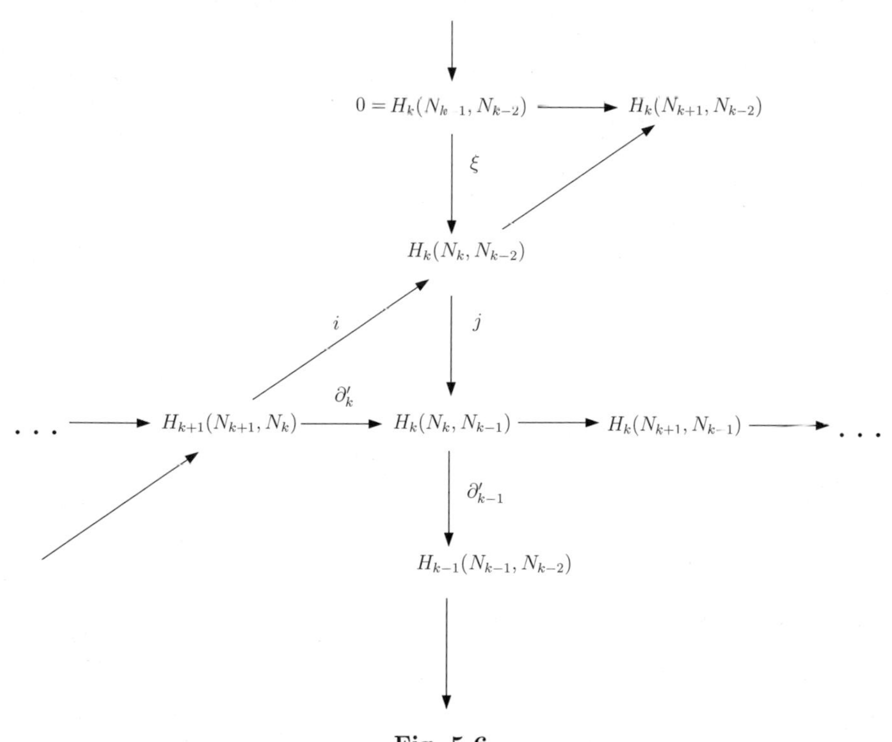

Fig. 5.6.

With suitable geometric representation of $\{H_k(N_k, N_{k-1}; G) |\ k = 0, 1, \ldots, m\}$, Floer used ∂' to establish the Floer homology.

In order to extend the notion of Morse decomposition to noncompact isolated invariant sets, let us keep in mind the Morse relations for (PS) functions on Banach–Finsler manifolds, and introduce the following:

Definition 5.5.27 *Let $U_1, U_2 \in \Sigma$ with $\mathrm{int}(U_1) \cap \mathrm{int}(U_2) = \emptyset$. One defines an order $U_2 > U_1$, if $\exists T > 0$ such that $U_1 \cap G^T(U_1 \cup U_2)$ is positively invariant with respect to $G^T(U_1 \cup U_2)$.*

If $U_2 > U_1$ and $U_1 > U_2$, then we say that U_1 and U_2 are φ-disconnected. Otherwise, we say that they are φ-connected.

Lemma 5.5.28 *Let $U_1, U_2 \in \Sigma$ with $U_2 > U_1$, and let $U = U_1 \cup U_2$. Then there exists closed subsets $N_0 \subset N_1 \subset N_2$ such that (N_2, N_0), (N_2, N_1) and (N_1, N_0) are index pairs for U, U_2 and U_1, respectively.*

Proof. Taking $T > 0$ large enough such that $(G^T(U), \Gamma^T(U))$ is an index pair for U, $G^T(U_1) \subset \mathrm{int}(U_1)$ and $U_1 \cap G^T(U)$ is positively invariant with respect to $G^T(U)$. Setting $N_0 = \Gamma^T(U)$, $N_1 = (U_1 \cap G^T(U)) \cup \Gamma^T(U)$ and $N_2 = G^T(U)$. Obviously (N_2, N_0) is an index pair for U (Theorem 5.5.13).

(1) In order to verify that (N_1, N_0) is an index pair for U_1, it is sufficient to verify the four conditions in Definition 5.5.12 as follows:

1.

$$G^T(N_1 \backslash N_0) = G^T(U_1 \cap G^T(U)) \subset G^T(U_1) \cap G^{2T}(U) \subset \mathrm{int}(U_1 \cap G^T(U))$$
$$= \mathrm{int}(N_1 \backslash N_0) .$$

2. Since $\Gamma^T(U)$ is positively invariant in $G^T(U)$, and $U_1 \cap G^T(U)$ is positively invariant in $G^T(U)$, $\Gamma^T(U)$ is positively invariant in N_1.
3. By the same reasoning, $\Gamma^T(U)$ is the exit set of N_1.
4. $\overline{N_1 \backslash N_0} = (U_1 \cap G^T(U)) \subset U_1$,
   $G^T(U_1) \subset U_1 \cap G^T(U) = N_1 \backslash N_0 \subset \overline{N_1 \backslash N_0}$.

(2) Similarly, we verify that (N_2, N_1) is an index pair for U_2. Since $\overline{N_2 \backslash N_1} = \overline{G^T(U) \backslash U_1} = G^T(U) \cap U_2$, the verification of $G^T(N_2 \backslash N_1) \subset \mathrm{int}(N_2 \backslash N_1)$ follows similarly to that of (N_1, N_0). Moreover

$$G^T(U_2) \subset G^T(U) \cap U_2 = \overline{N_2 \backslash N_1} .$$

Other verifications are omitted. □

Definition 5.5.29 *Let $U \in \Sigma$, a family of subsets of U, $\{U_1, \ldots, U_n\}$ is called a Morse decomposition of U, if*

(1) $U_i \in \Sigma$, $i = 1, \ldots, n$,
(2) $U = \bigcup_{1 \le i \le n} U_i$,
(3) $\mathrm{int}(U_i) \cap \mathrm{int}(U_j) = \emptyset$ as $i \neq j$,
(4) $U_{i+1} > \bigcup_{1 \le j \le i} U_j$, for $i = 1, 2, \ldots, n-1$.

By observing the exactness of sequences

$$\to \overline{H}^q(N_2, N_1; G) \to \overline{H}^q(N_2, N_0; G) \to \overline{H}^q(N_1, N_0; G) \to \ldots .$$

and the proof of (5.3), we arrive at:

Theorem 5.5.30 *If* $\{U_1, \ldots, U_n\}$ *is a Morse decomposition of* $U \in \Sigma$, *then there is a formal series* Q *with nonnegative coefficients such that*

$$\sum_{j=1}^{n} h(t; U_j) = h(t; U) + (1 + t)Q(t) .$$

Notes

Chapter 1

Section 1.1 contains the basic differential calculus on Banach spaces; the material can be found in any nonlinear functional analysis book, for instance, Schwartz [Scw], Nirenberg [Ni 1], Deimling [De], Zeidler [Zei], Berger [Ber 1] etc. The discussion on Nemytscki operator can be found in Vainberg [Va], but the proof is much simpler than that in the reference.

The presentations of the implicit function theorem and the inverse function theorem are standard. Interesting applications are scattered in the literature, for instance, Nirenberg [Ni 4], Kazdan [Ka], Chow, Hale [CH], Mawhin [Maw 3] etc. The continuity method is extensively used in the existence proof of differential equations. The global implicit function theorem is due to Hadamard [Ha], Caccioppoli [Cac 1]; extensions can be found in Browder [Bd 1] and Plastock [Pl]. However, the proof presented here is very different. The application of the continuity method to an a priori bound as a simpler proof for semi-linear elliptic equations with quadratic growth was given by Amann and Crandall [AC]; the result for quasi-linear elliptic equations can be found in Ladyzenskaya and Ural'zeva [LU].

The Lyapunov–Schmidt reduction is extensively used in nonlinear problems. The application to the study of bifurcation problems introduced here is due to Crandall and Rabinowitz [CR 1], [CR 2]. The references on bifurcation theory are recommended to Chow and Hale [CH].

Sections 1.3.3–1.3.4 provide examples on gluing, which is an important technique in symplectic geometry. The material here is taken from Floer and Weinstein [FW] and Oh [Oh]. Further references are Floer [Fl 1] [Fl 2], Hofer and Zehnder [HZ 2]. Parallel to the implicit function theorem method, a gluing technique via the variational method was developed by Sere [Se]; it has been applied to the homoclinic orbits in Hamiltonian systems, multi-bump solutions, and multi-peak solutions for elliptic differential equations, see also Coti Zelati and Rabinowitz [CR], Li [Li 1], Gui [Gu] etc. Section 1.3.5 is another important technique in applying the implicit function theorem. The notion of transversality is taken from differential geometry. Combining with the Sard theorem, it provides a method for proving various generic type results. The finite-dimensional form of the transversality theorem can be found in Guillemin and Podollak [GP]. The Sard–Smale theorem is taken from Smale [Sm 3]. The simplicity of eigenvalues of the Laplacian on generic domains is due to Uhlenbeck [Uh].

Section 1.4 is on the Nash–Moser technique. See Nash [Na 2] and Moser [Mos 1] [Mos 2]. KAM theory is due to Kolmogorov [Ko], Arnold [Ar 2] and Moser [Mo 3]. The presentation here can be found in Hormander [Hor 2]. Applications in differential geometry can be found in Hamilton [Ham]; a version setting up on Frechet spaces is given therein. Other versions can be found in Nirenberg [Ni 1], see also Zehnder [Ze]. For further reading on recent developments of KAM theory to partial differential equations, Bourgain [Bou], Kuksin [Kuk], Wayne [Way], and Poschel [Po] are recommended.

Chapter 2

The order method is very different from other methods in this book. In concrete problems, once the assumptions are met, the method is simple and powerful. Our discussions start with the Bourbaki–Kneser principle [Bo], [Kn], see also Tarski [Ta]. The Amann theorem [Am 1] is a version that is easy to apply. The sub- and super-solutions method is extensively applied in ODE and PDE whenever the maximum principle is applicable. However, the constructions of sub- and super-solutions require special knowledge and techniques. We are satisfied introducing the method by an example. The Caristi fixed-point theorem [Ca] is among a few fixed-point theorems without assuming the continuity of the nonlinear mappings, applications can be found in [Lie]. Also, an equivalent version of this theorem is the very important Ekeland variational principle [Ek 1]; various applications in different branches of analysis can be found in [Ek 1] and de Figueiredo [dF].

There are lots of books on convex analysis; in Sect. 2.2, we only present very briefly the necessary material for the sequel discussions. References can be found in Ekeland and Temam [ET], Aubin and Ekeland [AE] etc.

Fixed points for nonexpansive mappings have been studied by Browder [Bd 2], Goebel [GK], etc. The importance of this class of mappings is that one of the fixed points can be figured out by iteration methods. Many algorithms for finding feasible solutions for convex programming have been studied in recent years, see [B], [BB].

There are many ways to introduce the Schauder fixed-point theorem and related topics. We use the KKM theorem [KKM] and Ky Fan's inequality [FK 3] as the starting point. The Nash equilibrium [Na 3], the Von Neumann–Sion minimax theorem [VN] [Si], the Schauder fixed-point theorem [Sc], the Schauder–Techonoff theorem, the Ky Fan–Glicksberg theorem [FK 1], [FK 2], and the existence result of Hartman and Stampacchia on variational inequalitiy [HS] are direct consequences. The variational inequality [LS], [Bd 3], [Stm] is another direction in convex analysis with many applications in free boundary problems from mathematical physics, see Duvaut and Lions [DL], Kinderlehrer and Stampacchia [KS], Friedman [Fr] etc. However, the approach based on the fixed points of set-valued mappings is due to the author, see [Ch 1], [Ch 2].

The theory of monotone operators and pseudo-monotone operators attracted much attention in the 1960s and 70s. The works of Minty [Min], Browder [Bd 4], Hartman and Stampacchia [HS], H. Brezis [Br 1], etc. constitute the basic content of the theory. Again we use the version of the Ky Fan inequality due to Brezis, Nirenberg and Stampacchia [BNS] to derive the most important results on this topics. For applications to quasi-linear elliptic equations see Leray and Lions [LL], to boundary value problems in nonlinear partial differential equations see Lions [LJ 1], and to nonlinear semigroups of operators see Brezis [Br 1], Crandall and Ligget [CLi].

Chapter 3

The Brouwer degree is a topological invariant; roughly speaking, there are two approaches: algebraic (see for instance Spanier [Sp], Greenberg [GH]) and differential (Milnor [Mi 2]). The Leray–Schauder degree is its extension to compact vector fields [LS] on Banach space.

The analytic presentation can be found in many books and lecture notes, for instance: J. Schwartz [Scw], Nirenberg [Ni 1], Rabinowitz [Ra 2], Guillemin and Podollak [GP], Zeidler [Zei] etc. The materials of Sects. 3.1–3.4 are taken from these references. An application of the transversality theorem to the proof for the Borsuk–Ulam theorem and to the computation of S^1-invariant degree are due to Nirenberg [Ni 3].

Sections 3.5 and 3.6.3 are adapted from Rabinowitz [Ra 1], [Ra 2]. Section 3.6.2 is based on Dancer [Dan 1]. The material of Sects. 3.6.4 and 3.6.5 is taken from Amann [Am 2], Krasnosel'ski [Kr 2], de Figueiredo, Lions and Nussbaum [FLN], Chang [Ch 3], and Dancer [Dan 2]. The blowing up method in a priori estimates is a useful technique, see for instance, Gidas and Spruck [GS].

Section 7 contains various extensions of the Leray–Schauder degree; for α-set contraction mappings, see Darbo [Dar], Stuart and Toland [ST]; for condensing mappings, see Nussbaum [Nu], Sadovskii [Sa], set-valued mappings see Browder [Bd 5], Cellina [Ce], Ma [Ma], Chang [Ch 2]. There are other directions: Fredholm operators, see Elworthy and Tromba [ElT 1] [ElT 2], Nirenberg [Ni 4], Llyord [Ll], Fitzpatrick, Pejsachowicz and Rabier [FP], [FPR], [PR 1, PR 2]; coincidence degree theory, Mawhin [Maw 1], etc.

Chapter 4

Section 4.1 is an introduction to the calculus of variations. The derivations of the Euler–Lagrange equation, the Legendre–Hadamard condition and the Ljusternik theorem on constraint variational problems can be found in any standard textbook. For the dual variational principle, i.e., the Legendre–Fenchel transformation, see for instance, Arnold [Ar 1], Ekeland and Temam [ET] or Aubin and Ekeland [AE]. For the Hamiltonian systems, the second version of the dual variational principle, i.e., the Legendre transform to the Hamiltonian function of all variables if the latter is strictly convex, see Clarke and Ekeland [CE].

In Sect. 4.2, the direct method is a general principle in the calculus of variations. We introduce a few interesting examples showing how the principle works.

Harmonic maps were introduced by Eells and Sampson [ES] in representing homotopy classes of mappings between two manifolds. Here we only touch on the existence of a weak solution. For $m = 2$ the harmonic map with minimal energy is smooth, see Morrey [Mo 2]; for $m > 2$, Schoen and Uhlenbeck [ScU 1], [ScU 2] proved that the singularity has at most a codimension 3 finite Hausforff measure. A nonsmooth minimal energy harmonic map was given by Lin [Lin]. As to nonminimal energy harmonic maps, the smoothness for $m = 2$ was proved by Helein [Hel 1], and the existence of a nowhere continuous harmonic for $m = 3$ constructed by Riviere [Ri].

The example on constant mean curvature surface is taken from Hildebrandt [Hi 1]; a systematic introduction can be found in Struwe [St 2].

The prescribing scalar curvature problem can be found in Kazdan [Ka], Kazdan and Warner [KW 1], [KW 2]. The result for $\chi(M) = 0$ is due to Berger [Ber 2], that for real projective space P^2 is due to Moser [Mos 4]; however, we present an easy proof with the aid of Aubin's inequality [Au 3].

Section 4.3 is from Morrey [Mo 1] [Mo 2], Dacorogna [Dac], Acerbi and Fusco [AF], Marcellini [Mar], Muller [Mul 1] [Mul 2] Sverak [Sv 1] [Sv 2], Zhang [Zh 1] and Ball [Bal 1].

The relaxation method is from Buttazzo [Bu] and Dacorogna [Dac]. For the two-well problem, Zhang [Zh 2] has obtained an explicit expression of the quasi-convex envelope for the square distance function.

The Young measure was introduced by Young [Yo]; the presentation here is due to Ball [Bal 2], see also [Mul 1], Kinderlehrer and Pedregal [KP 1, KP 2]. In Chipot [Chi], there are examples on the computation of Young measures.

There are few books in dealing with the BV space, see for instance, Giusti [Gi], Evans and Gariepy [EG], Zimer [Zi], Ambrosio [Amb] etc. Section 4.5.1 is adapted from these books.

The Hardy space and BMO space are important parts in harmonic analysis, see Stein [Ste 1, Ste 2] and Stein–Weiss [SW]. They are applied to PDEs in compensated compactness (as seen in Sect. 4.5.2) and in regularity (for harmonic maps see Helein [Hel 2] Evans [Ev 2]. For the background material we recommend Brezis and Nirenberg [BN 3], and Semmes [Se] as references. The connection between Hardy space and the compensated compactness is revealed by Coifman, Lions, Meyer and Semmes [CLMS]. In the applications to the calculus of variations, the bitting theorem due to Chacon is required, see Brooks and Chacon [BC]. Compensated compactness is also an important tool in nonlinear analysis; we refer the reader to Tartar [Tar 1, Tar 2], DiPerna [Di] and Evans [Ev 2].

We briefly introduce the Γ-convergence in Sect. 4.6 by a result of Modica and Mortola [MM], see also Alberti [Al]. Readers who want to know more can read Dal Maso [D].

The Munford–Shah model [MS] in the segmentation of the image processing is an interesting subject; the existence proof was given by Ambrisio [Amb], De Giorgi and [DG 2], and De Giorgi and Ambrosio [GA]. More details can be found in Ambrosio Fusco and Pallara [AFP], Dal Maso, Morel and Solimini [DMS] and Morel and Solimini [MS]. As to the regularity of the edge curve, see Morel [Mor]. However, all of these results require deep knowledge on BV functions. A much simplified model presented here is from Nordström [No].

Concentration compactness [LP] in combination with blowing up analysis [SU] is one of the most important techniques in nonlinear problems without compactness. Since interesting concrete problems require more background knowledge, they are out of the scope of this book. We present in Sect. 4.7 just an introduction of the idea: The concentration phenomenon and the role of the bubble (or best constant). Two examples on semilinear elliptic equations are studied: The subcritical exponent on R^n, see Coti Zelati [CZ], and the critical exponent on bounded domain, see Brezis and Nirenberg [BN 1].

Section 4.8 is the preliminary of the minimax method. The Palais–Smale condition [PS] and the mountain pass lemma due to Ambrosetti and Rabinowitz [AR] are introduced in accordance with the Ekeland minimizing principle. The proof was given by Shi [Shi] and Aubin and Ekeland [AE]. Examples are taken from Mawhin [Maw 2] and Ambrosetti and Rabinowitz [AR].

Chapter 5

Section 5.1: The most recommended book in introducing Morse theory on compact manifolds is Milnor [Mi 1]. The classics are Morse [Mo] and Morse and Cairns [MC]. The extension to infinite-dimensional space can be found in Chang [Ch 4], [Ch 5]. See also Mawhin and Willem [MW].

The background material in this section can be found in Palais [Pa 1], Palais and Smale [PS], Rothe [Ro 1], [Ro 2], [Ro 3], Marino and Prodi [MP 1], [MP 2], Chang [Ch 7], [Ch 8], Wang [Wa 1], [Wa 2], Castro and Lazer [CaL].

The critical point theory has been extended to non-differentiable functionals, see Chang [Ch 6] for locally Lipschitzian functionals on Banach space, and Kartiel [Ka], Ioffe, Schwartzmann [IS], and Corvellec, De Giovanni, Marzocchi [CGM] for continuous functions on metric spaces.

Section 5.2: There are several books on minimax methods, see for instance Rabinowitz [Ra 4], Struwe [St 1], Ghoussoub [GH], and Willem [Wi] etc.

The background material can be found in Rabinowitz [Ra 4], [Ra 5], Palais [Pa 2], [Pa 3], Ni [Ni], Chang [Ch 5], Tian [Ti], Liu [Liu 1], [Liu 2], Viterbo [Vi 2], Solimini [So], Chang, Long and Zehnder [CYZ], Fournier and Willem [FoW].

The notion of category was discovered by Ljusternik and Schnirelmenn [LjS 1]. The genus was introduced by Krasnosel'ski in [Kr 1], while for general index theory see Fadell and Rabinowitz [Fr]. The geometric index for S^1 is taken from Benci [Be 1] and Nirenberg [Ni 3]; other extensions of the index theory can be found in Benci and Rabinowitz [BR 2].

Section 5.3: There are many papers on periodic solutions for Hamiltonian systems via variational methods. The results related to this section are Rabinowitz [Ra 3], Weinstein [We 1], [We 2], Viterbo [Vi 1], Hofer and Zehnder [HZ 2], Struwe [St 1]. Further development on the Weinstein conjecture can be found in Hofer and Viterbo [HV], Hofer [Ho 2] and Liu and Tian [LT 1].

The following books are recommended: Hofer and Zehnder [HZ 1], Ekeland [Ek 2], Mawhin and Willem [MW].

Other important topics on the periodic solutions include:

- For Arnold conjecture on the symplectic fixed points and Lagrangian intersections, see Conley and Zehnder [CZ 1], Floer [Fl 1], [Fl 2], [Fl 3], Liu and Tian, [LT 2], Fukaya Ono [FO].
- The number of the periodic orbits on compact convex hypersurface has been estimated by Long and Zhu [LZ].
- For the N-body problem, see Bahri and Rabinowitz [Br].

Section 5.4: The main result of this section is taken from Chang and Yang [CY 1], [CY 2], and Chang and Liu [CL 2]. See also Han [Han], Chen and Li [ChL 1], [ChL 2]. The related prescribing scalar curvature problem on S^n has received extensive attention. For $n = 3$ see Bahri and Coron [BC], and for high dimension, see Schoen and Zhang [SZ], Bahri [Bar 1], [Bar 2], Li [Li 1], and Chen and Lin [ChL].

Section 5.5: Conley's index theory was introduced by Conley [Co], in which isolating neighborhoods for isolated invariant sets are compact. The theory is completed by Conley and Zehnder [CZ 2], Salamon [Sal], Salamon and Zehnder [SZ]. There are many ways to extend the theory to infinite-dimensional spaces, see Rybakowski [Ry], Rybakowski and Zehnder [RZ], Benci [Be 2] etc. The relationship between the Conley theory and the Morse theory can be found in Chang and Ghoussoub [CG]. For further reading see Michaikov [Mic] and Smoller [Smo].

References

[AF] Acerbi E., Fusco N., Semicontinuity problems in the calculus of variations, Arch. Rat. Mech. Anal. 86 (1984), 125–145.

[Ad] Adams R. A., Sobolev spaces, Acad. Press (1975).

[Al] Alberti G., Variational models for phase transition, an approach via Γ-convergence calculus of variations and partial differential equations. Topics on geometrical evolution problems and degree theory (ed. by Ambrossio, Dancer), Springer-Verlag, (2000) 95–114.

[Am 1] Amann H., Order structures and fixed points, Ruhr-Universität, Bochum (1977).

[Am 2] Amann H., Fixed point equations and elliptic eigenvalue problems in ordered Banach spaces, SIAM Rev. 18 (1976), 620–709.

[Am 3] Amann H., On the number of solutions of nonlinear equations in ordered Banach spaces, J. Funct. Anal. 14 (1973), 346–384.

[Am 4] Amann H., Saddle points and multiple solutions of differential equations, Math. Zeit. 169 (1979), 122–166.

[AC] Amann H., Crandall M., On some existence theorems for semi-linear elliptic equations, Indiana Univ. Math. J. 27 (1978), 779–790.

[AR] Ambrosetti A., Rabinowitz P., Dual variational methods in the critical point theory and applications, J. Funct. Anal. 14 (1973), 349–381.

[Amb] Ambrosio L., Variational problems in SBV and image segmentation, Acta Appl. Math. 17 (1989), 1–40.

[AFP] Ambrosio L., Fusco N., Pallara D., Functions of bounded variation and free discontinuity problem, Oxford Sci Publ. (2000).

[Ar 1] Arnold V. I., Mathematical methods of classical mechanics, Springer-Verlag (1978).

[Ar 2] Arnold V. I., Proof of a theorem of A. N. Kolmogorov on the conservation of quasi-periodic motions under a small change of the Hamiltonian function, Uspekhi Mat. Nauk 18:5 (1963), 9–36.

[AE] Aubin J. P., Ekeland I., Applied nonlinear analysis, John Wiley and Sons (1984).

[Au 1] Aubin Th., Nonlinear analysis on manifolds, Monge Ampere equations, Grundlehren 252 (1982), Springer-Verlag.

[Au 2] Aubin Th., Equations differentielles nonlinéaires et probléme de Yamabe concernant la courbure scalaire, J. Math. Pure Appl. 55 (1976), 269–293.

[Au 3] Aubin Th., Meilleures constants dans le théorème d'inclusion de Sobolev et un théorème de Fredholm nonlinéaire pour la transformation conforme de la courbure scalaire, J. Funct. Anal. 32 (1979), 148–174.

[Ba 1] Bahri A., Critical points at infinity in some variational problems, Pitman Research Notes in Math. V. 182, Longman (1989).

[Ba 2] Bahri A., The scalar curvature problem on sphere of dimension larger or equal to 7, Preprint (1994).

[BC] Bahri A., Coron J. M., The scalar curvature problem on the standard three dimensional sphere, J. Funct. Anal. 95 (1991), 106–172.

[Bal 1] Ball J., Convexity conditions and existence theorems in nonlinear elasticity, Arch. Rat. Mech. Anal. 63 (1977), 337–403.

[Bal 2] Ball J., A version of fundamental theorem for Young measures, PDE's and continuum models of phase transitions (Rascle M. Serre D. Slemrod M. eds.), Lecture Notes in Physics, 344, Springer-Verlag (1989), 207–215.

[Bm] Ballmann W., Der Satz von Lyusternik und Schnirelmann, Bonn. Math. Schr. Nr. 102 (1978).

[B] Bauschke H. H., The approximation of fixed points of compositions of nonexpansive mappings in Hilbert space, JMAA 202 (1996) 150–159.

[BB] Bauschke H. H., Borwein J. M., On projection algorithms for solving convex feasible problems, SIAM Revi. 38 (1996), 367–426.

[Be 1] Benci V., A geometric index for the group S^1 and some applications to the study of periodic solutions of ordinary differential equations. Comm. Pure Appl. Math. 274 (1981) 393–432.

[Be 2] Benci V., A new approach to the Morse Conley theory and some applications. Ann. Mat. Pura Appl. (IV) 158 (1991), 231–305.

[BR] Benci V., Rabinowitz P., Critical point theorems for indefinite functionals, Invent. Math. 52 (1979), 241–273.

[Ber 1] Berger M. S., Nonlinearity and functional analysis, Acad. Press (1977).

[Ber 2] Berger M. S., On Riemannian structures of prescribing Gaussian curvature for compact 2-manifolds, J. Diff. Geom. 5 (1971), 325–332.

[Bi] Birkhoff G. D., Dynamical systems with two degrees of freedom, Trans. AMS 18 (1917), 199–300.

[BZS] Borisovich Y. G., Zvyagin V. G., Sapronov Y. I., Nonlinear Fredholm maps and the Leray Schauder degree theory, Russian Math. Surveys 324 (1977) 1–54.

[Bo 1] Bott R., Nondegenerate critical manifolds, Ann. Math. 60 (1954), 248–261.

[Bo 2] Bott R., Lectures on Morse theory, old and new, Bull. AMS 7 (1982), 331–358.

[BT] Bott R., Tu L. W., Differential forms in algebraic topology, Springer-Verlag (1982).

[Bo] Bourbaki N., Topologie générale, Hermann, Paris (1940).

[Bou] Bourgain J., Global solutions of nonlinear Schrodinger equations, AMS Colloquium Publ. Vol. 46 (1999).

[Br 1] Brezis H., Opérateurs maximaux monotones, Lecture Notes, Vol. 5, North Holland Amsterdam (1973).

[BrC] Brezis H., Coron J. M., Multiple solutions of H-systems and Rellich's conjecture, Comm. Pure Appl. Math. 37 (1984), 149–187.

[BL] Brezis H., Lieb E. H., A relation between pointwise convergence of functions and convergence of functionals, Proc. AMS 88 (1983), 486–490.

[BN 1] Brezis H., Nirenberg L., Positive solutions of nonlinear elliptic equations involving critical Sobolev exponents, Comm. Pure Appl. Math. 36 (1983), 437–477.

[BN 2] Brezis H., Nirenberg L., Remarks on finding critical points, Comm. Pure Appl. Math. 64 (1991), 939–963.

[BN 3] Brezis H., Nirenberg L., Degree theory and BMO, Selecta Math. 1 (1995), 197–263.

[BNS] Brezis H., Nirenberg L., Stampacchia G., A remark on Ky Fan's minimax principle, Boll. Un. Mat. Ital. (4) 6, (1973), 293–300.

[BC] Brooks J. K., Chacon R. V., Continuity and compactness of measures, Adv. Math. 37 (1980), 16–26.

[Br] Brouwer L. E. J., Über Abbildung von Mannigfaltikeiten, Math. Ann. 71 (1912), 97–115.

[Bd 1] Browder F., Covering spaces, fibre spaces and local homeomorphisms, Duke Math. J. 21 (1974), 329–336.

[Bd 2] Browder F., Non-expansive nonlinear operators in a Banach space, Proc. Nat. Acad. Sci. USA 54 (1965), 1041–1044.

[Bd 3] Browder F., On the unification of the calculus of variations and the theory of monotone nonlinear operators in Banach space, Proc. Nat. Acad. USA 56 (1966), 419–425.

[Bd 4] Browder F., Nonlinear elliptic boundary value problems, Bull. AMS 69 (1963), 862–874.

[Bd 5] Browder F., The fixed point theory of multi-valued mappings in topological vector spaces, Math. Ann. 117 (1968), 283–301.

[Bu] Buttazzo G., Semicontinuity, relaxation and integral representation in the calculus of variations, Pitman Res. Notes Math. Ser., 207, Longman (1989).

[Cac 1] Caccioppoli R., Sugli elementi uniti delle transformazioni funzionali, Rend. Sem. Mat. Padova 3 (1932), 1–15.

[Cac 2] Caccioppoli R., Sulle corrispondenze funzionali inverse diramate, teoria generale e applicazioni ad alcune equazioni funzionali nonlineari al problema di Plateau, I, II, Rend. Acad. Naz. Lincei (6) 24 (1936), 258–263, 416–421.

[CGS] Caffarelli L., Gidas B., Spruck J., Asymptotic symmetry and local behavior of semilinear elliptic equations with critical Sobolev growth, Comm. Pure Appl. Math. 42 (1989), 271–297.

[Ca] Caristi J., Fixed point theorems for mappings satisfying inwardness conditions, Tran. AMS 215 (1976), 241–251.

[CC] Carleson L., Chang A. S. Y., On the existence of an extremal function for an inequality of J. Moser, Bull. Sc. Math. 2 110 (1986), 113–127.

[CaL] Castro A., Lazer A. C., Critical point theory and the number of solutions of a nonlinear Dirichlet problem, Ann. Pura Appl. 70 (1979), 113–137.

[Ce] Cellina A., Approximation of set-valued functions and fixed point theorems, Ann. Pura Appl. Mat. 4 (1969), 17–24.

[CY 1] Chang A. S. Y., Yang P., Prescribing Gaussian curvature on S^2, Acta Math. 159 (1987), 215–259.

[CY 2] Chang A. S. Y., Yang P., Conformal deformation of metric on S^2, J. Diff. Geom. 23 (1988), 259–296.

[Ch 1] Chang K. C., The obstacle problem and partial differential equations with discontinuous nonlinear terms, Comm. Pure Appl. Math., 33 (1980), 117–146.

[Ch 2] Chang K. C., Free boundary problems and set valued mappings, J. Diff. Eq. 49 (1981), 1–28.

[Ch 3] Chang K. C., Remarks on some free boundary problems for equilibrium equation of plasmas, Comm. PDE 5 (1980), 741–751

[Ch 4] Chang K. C., Infinite dimensional Morse theory and its applications, les presses de l' univ. de Montreal SMS 97 (1985).

[Ch 5] Chang K. C., Infinite dimensional Morse theory and multiple solution problems, Birkhauser (1993).

[Ch 6] Chang K. C., Variational methods for non-differentiable functionals and their applications to PDE, J. Math. Anal. Appl. 80 (1981), 102–129.

[Ch 7] Chang K. C., Solutions of asymptotic linear operator equations via Morse theory, Comm. Pure Appl. Math. 34 (1981), 693–712.

[Ch 8] Chang K. C., H^1 versus C^1 isolated critical points, C. R. Acad. Sci. Paris 319 (1994), 441–446.

[CG] Chang K. C., Ghoussoub N., The Conley index and the critical groups via an extension of Gromoll-Meyer theory, Topol. Methods in Nonlinear Analysis 7 (1996), 77–93.

[CL 1] Chang K. C., Liu J. Q., Morse theory under general boundary conditions, J. System Sci. and Math. Sci. 4 (1991), 78–83.

[CL 2] Chang K. C., Liu J. Q., On Nirenberg's problem, International J. Math. 4 (1993) 35–58.

[CYZ] Chang K. C., Long Y., Zehnder E., Forced oscillations for the triple pendulum, Analysis et cetera (P. Rabinowitz, E. Zehnder eds.) Academic Press, (1990).

[ChL] Chen C. C., Lin C. S., Blowing up with infinite energy of conformal metric on S^n, Comm. PDE 24 (1999), 785–799.

[CD] Chen W. X., Ding W. Y., Scalar curvatures on S^2. Tran. Amer. Math. Soc. 303 (1987), 365–382.

[ChL 1] Chen W., Li C., A priori estimates for prescribing scalar curvature equations, Ann. Math. 145 (1997), 547–564.

[ChL 2] Chen W., Li C., A necessary and sufficient condition for the Nirenberg problem, Comm. Pure Appl. Math. 48 (1995), 657–667.

[Chi] Chipot M., Elements in nonlinear analysis, Birkhauser, (2000).

[CH] Chow S. N., Hale J., Methods of bifurcation theory, Springer-Verlag, (1982).

[Cl] Clarke F. H., Optimization and nonsmooth analysis, Wiley Interscience (1983).

[CE] Clarke F. H., Ekeland I., Hamiltonian trajectories having prescribed minimal period, Comm. Pure Appl. Math. 33 (1980), 103–116.

[CLMS] Coifman R., Lions, P. L., Meyer Y., Semmes S., Compensated compactness and Hardy spaces, J. Math. Pure Appl. (9), 72 (1993), 247–286.

[Co] Conley, C., Isolated invariant sets and the Morse index. CBMS 38 (1978), AMS.

[CZ 1] Conley C., Zehnder E., The Birkhoff-Lewis fixed point theorem and a conjecture of V. I. Arnold, Inv. Math. 73 (1983) 33–49.

[CZ 2] Conley C., Zehnder E., A Morse type index theory for flows and periodic solutions to Hamiltonian systems, Comm. Pure Appl. Math. 37 (1984), 207–253.

[CGM] Corvellec J. N., De Giovanni M., Marzocchi M., Deformation properties for continuous functions and critical point theory, preprint Univ. di Pisa (1992).

[CZ] Coti Zelati V., Critical point theory and applications to elliptic equations in R^n, Nonlinear Functional Analysis and Appl. to Diff. Eqs. (A. Ambrosetti, K. C. Chang, I. Ekeland eds.) World Sci. (1998), 102–121.

[CR] Coti Zelati V., Rabinowitz P., Homoclinic type solutions for a semilinear elliptic partial differential equation on R^n, Comm. Pure Appl. Math. 45, (1992), 1217–1269.

[CLi] Crandall M., Ligget T., Generations of semigroups of nonlinear transformations on general Banach spaces, Amer. J. Math. 93 (1971), 265–298.

[CR 1] Crandall M., Rabinowitz P., Bifurcation, perturbations of simple eigenvalue and linearized stability, Arch. Rat. Mech. Anal. 52 (1971), 161–180.

[CR 2] Crandall M., Rabinowitz P., The Hopf bifurcation theorem, Arch. Rat. Math. Anal. 67 (1977), 53–72.

[Dac] Dacorogna B., Direct methods in the calculus of variations, Springer-Verlag (1989).

[DM] Dacorogna B., Marcellini, P. A counter example in the vectorial calculus of variations, Material Instabilities in Continuum Mechanics, Proc. (ed. J. Ball), Oxford Sci. Publ. (1988), 77–83.

[D] Dal Maso G., An introduction to Γ -convergence, Birkhauser (1993).

[DMS] Dal Maso G., Morel J. M., Solimini S., A variational method in image segmentation: existence and approximation results, Acta Math. 168 (1992), 89–151.

[Dan 1] Dancer N., Global solution branches for positive maps, Arch. Rat. Math. Anal. 55 (1974), 207–213.

[Dan 2] Dancer N., The effect of domain shape on the number of positive solutions of certain nonlinear equations, J. Diff. Equs. 87 (1990), 316–339.

[Dar] Darbo G., Punti uniti in transformazioni a codominio non compatto, Rend. Sem. Univ. Padua 24 (1955), 84–92.

[dF] de Figueiredo D. G., Lectures on Ekeland variational principle with applications and detours, Lect. Notes, College on variational problems in analysis, ICTP Trieste (1988).

[FLN] de Figueiredo D. G., Lions P. L., Nussbaum R., J. A priori estimates and existence of positive solutions of semilinear elliptic equations, Math. Pures et Appl. 61 (1982), 41–63.

[DG 1] De Giorgi E., Frontiere orientate di misura minima, Sem. Mat. Scuola Norm. Sup. Pisa (1961).

[DG 2] De Giorgi E., Free discontinuity problems in the calculus of variations, Frontiers in pure and applied Mathematics, a collection of papers dedicated to J. L. Lions (Dautray R., ed.) North Holland (1991), 55–62.

[DA] De Giorgi E., Ambrosio L., Un nuovo funzionale del calcolo delle variazioni, Atti Accad. Naz. Lincei Cl. Sci. Fis. Mat. Natur. Rend. Lincei (8), Mat. Appl. 82 (1988), 199–210.

[De] Deimling K., Nonlinear functional analysis, Springer-Verlag (1985).

[DT] DeTurck D., Existence of metrics with prescribing Ricci curvature: local theory, Invent. Math. 65 (1981), 179–208.

[DiT] Ding W. Y., Tian G., Energy identity for a class of approximate harmonic maps from surfaces, Comm Anal. Geom. 3 (1995), 543–544.

[Di] DiPerna R. J., Compensated compactness and general systems of conservation laws, Trans. AMS 292 (1985), 383–420.

[DM] DiPerna R. J., Majda A. J. Oscillations and concentration in weak solutions of the incompressible fluid equations, Comm. Math. Phys. 108, (1987), 667–689.

[Du 1] Dugundji J., An extension of Tietze's theorem, Pac. J. Math. 1 (1951), 353–367.

[Du 2] Dugundji J., Topology, Allyn and Bacon (1966).

[DG] Dugundji J., Granas A., Fixed point theory, Monografie Matematyczne 16, Polish Sci. Publ. (1982).

[DS] Dunford N., Schwartz, J. T., Linear operators, Vol. I., Interscience (1958).

[DL] Duvaut G., Lions J. L., Les inéquations en mécanique et en physique, Dunod (1972).

[ES] Eells J., Sampson J. H., Harmonic mappings of Riemannian manifolds, J. AMS 86 (1964), 109–160.

[Ek 1] Ekeland I., Nonconvex minimization problems, Bull. AMS 1 (1979), 443–474.

[Ek 2] Ekeland I., Convexity methods in Hamiltonian mechanics, Ergebnisse d. Math. (Ser. III) 19, Springer-Verlag (1990).

[ET] Ekeland I., Temam R., Convex analysis and variational problems, North Holland (1976).

[ElT 1] Elworthy K. D., Tromba A. J., Degree theory on Banach manifolds, Nonlinear Functional Analysis (F. Browder ed.) Proc. Symp. Pure Math. Part I 18 (1970), 86–94.

[ElT 2] Elworthy K. D., Tromba A. J., Differential structures and Fredholm maps on Banach manifolds, Global analysis (S. S. Chern and S. Smale eds.) Proc. Symp. Pure Math. 15 (1970), 45–94.

[Ev 1] Evans L. C., Weak convergence methods for nonlinear partial differential equations, AMS, Providence (1990).

[Ev 2] Evans L. C., Quasi-convexity and partial regularity in the calculus of variations, Arch. Rat. Mech. Anal. 95 (1986), 227–252.

[EG] Evans L. C., Gariepy R., Measure theory and fine properties of functions, CRC Press (1992).

[EM] Evans L. C., Muller S., Hardy spaces and the two dimensional Euler equations with nonnegative vorticity, J. AMS (1994), 199–219.

[Fa] Faber C., Beweiss, dass unter allen homogenen Membrane von gleicher Flache und gleicher Spannung die Kriesformige die tiefsten Grundton gibt, Sitzungsber-Bayer Acad. Wiss. Math. Phys. Munich (1923), 169–172.

[FR] Fadell E. R., Rabinowitz P., Generalized cohomological index theories for Lie group actions with applications to bifurcation questions for Hamiltonian systems, Invent. Math. 45 (1978), 139–174.

[FK 1] Fan K., Fixed point and minimax theorems in locally convex topological spaces, Proc. Nat. Acad. Sci. USA 38 (1952), 121–126.

[FK 2] Fan K,. A generalization of Tychnoff's fixed point theorem, J. Math. Anal. Appl. 142 (1961), 305–310.

[FK 3] Fan K., A minimax inequality and applications, Inequalities-III (Shisha O. ed.) Acad. Press (1972).

[Fe] Federer H., Geometric measure theory, Springer-verlag (1969).

[FP] Fitzpatrick P. M., Pejsachowicz J., Parity and generalized multiplicity, TAMS 326 (1991), 281–305.

[FPR] Fitzpatrick P. M., Pejsachowicz J., Rabier P. J., Orientability of Fredholm families and topological degree for orientable Fredholm mappings, J. Funct. Anal. 124 (1994), 1–39.

[Fl 1] Floer A., Morse theory for fixed points of symplectic diffeomorphisms, Bull. AMS 16 (1987), 279–281.

[Fl 2] Floer A., Witten complex and infinite dimensional Morse theory, J. Diff. Geom. 18 (1988), 207–221.

[Fl 3] Floer A., Symplectic fixed points and holomorphic spheres, Comm. Math. Phys. 120 (1989), 576–611.

[FW] Floer A., Weinstein A., Nonspreading wave packets for cubic Schrodinger equations with a bounded potential, J. Funct. Anal. 69 (1986), 397–408.

[FoW] Fournier G., Willem M., Relative category and the calculus of variations, Preprint.

[Fr] Friedman A., Variational principle and free boundary problems, John Wiley and Sons (1982).

[FO] Fukaya K., Ono K., Arnold conjecture and Gromov-Witten invariants for general symplectic manifolds, The Arnoldfest, Fields Inst. Commun. 24, (173–190).

[Gh] Ghoussoub N., Duality and perturbation method in critical point theory, Cambridge Univ. Press (1993).

[GMS] Giaquinta M., Modica G., Soucek J., Cartesian currents in the calculus of variations, Part I, Cartesian currents, Part II, Variational integrals, Springer-Verlag (1998).

[GNN] Gidas B., Ni W. M., Nirenberg L., Comm. Math. Phys. 68 (1979), 209.

[GS] Gidas B., Spruck J., A priori bounds for positive solutions of nonlinear elliptic equations, Comm. PDE 6 (1981), 883–901.

[GT] Gilbarg D., Trudinger N. S., Elliptic partial differential equations of second order, 2nd edition, Grundlehren der Mathematik, 224, Springer-Verlag (1983).

[Gi] Giusti E., Minimal surfaces and functions of bounded variations, Monographs in Mathematics 80, Birkhauser (1984).

[GK] Goebel K., Kirk W., Iteration process for nonexpansive mappings, Comtemp. Math. 21 (1983), 115–123.

[GH] Greenberg M. J., Harper J. R., Algebraic topology, A first course, The Benjamin/Cummings Publ. Co. (1981).

[GM] Gromoll D., Meyer W., On differential functions with isolated critical points, Topology 8 (1969), 361–369.

[Gu] Gui C., Existence of multi-bump solutions for nonlinear Schrodinger equations via variational method, Comm. Part. Diff. Eq. 21 (1996), 787–820.

[GP] Guillemin V., Pollack A., Differential topology, Prentice-Hall (1974).

[Ha] Hadamard J., Sur les transformations ponctuelles, Bull. Soc. Math. France, 34, (1906), 7–84.

[Ham] Hamilton, R., The inverse function theorem of Nash and Moser, Bull. AMS 7 (1982), 65–222.

[Han] Han Z. C. Prescribing Gaussian curvature on S^2, Duke Math. J. 61 (1990), 679–703.

[Hann] Hanner, Some theorems on absolute neighborhood retracts, Arkiv Math. 1 (1951), 389–408.

[HS] Hartman P., Stampacchia G., On some nonlinear elliptic differential equations, Acta Math. 115 (1966), 271–310.

[Hel 1] Helein F., Régularité des applications faiblement harmoniques entre une surface et ue variété riemannienne, CR Acad. Sci. Paris 312 (1991), 591–596.

[Hel 2] Helein F., Constant mean curvature surfaces, harmonic maps and integrable systems, Birkhauser (2001).

[Hi 1] Hildebrandt S., Nonlinear elliptic systems and harmonic mappings, Proc. Beijing Symp. Diff Geom. and Diff Eqs., Gordon and Breach (1983), 481–615.

[Hi 2] Hildebrandt S., Calculus of variations, I, II. Grudlehrer Mathematik Wiss. 310 311, Springer-Verlag (1996).

[Ho 1] Hofer H., A note on the topological degree at a critical point of mountain pass type, Proc. AMS 90 (1984), 309–315.

[Ho 2] Hofer H., Pseudo holomorphic curves in symplectisations with applications to the Weinstein conjecture in dimension three, Invent. Math. 114 (1993), 515–563.

[HV] Hofer H., Viterbo C., The Weinstein conjecture in the presence of holomorphic curves, Comm. Pure Appl. Math. 45 (1992), 583–622.

[HZ 1] Hofer H., Zehnder E., Periodic solutions on hyper-surfaces and a result by C. Viterbo, Invent. Math. 90 (1987), 1–9.

[HZ 2] Hofer H., Zehnder E., Symplectic invariants and Hamiltonian dynamics, Birkhauser (1994).

[Hon 1] Hong C. W., A best Constant and the Gaussian curvature, Proc. AMS 97 (1986), 737–747.

[Hon 2] Hong C. W., A note on prescribed Gaussian curvature on S^n, Partial Diff. Eqs, (in Chinese), 1 (1987), 13–20.

[Hor 1] Hormander L., The analysis of linear partial differential operators, Vol. III, Grundlehren der Math. Wiss. 274, Springer-Verlag (1984).

[Hor 2] Hormander L., The boundary value problems of physical geodesy, Arch. Rat. Math. Anal. 62 (1976), 1–52.

[IS] Ioffe A., Schwartzmann E., Metric critical point theory 1, Morse regularity and homology stability of a minimum, J. Math. Pure Appl. 75 (1996), 125–153.

[Iz] Ize J., Bifurcation theory for Fredholm operators, Memoirs, AMS, 174, Providence (1976).

[JLS] Jeanjean H., Lucia M., Stuart C. A., Branches of solutions to semilinear elliptic equations on R^n, Math. Z. 230 (1999), 79–105.

[Kak] Kakutani S., A generalization of Brouwer's fixed point theorem, Duke Math. J. 8 (1941), 457–459.

[Kar] Kartiel G., Mountain pass theorems and global homeomorphism theorems, Analyse Nonlineaire 11 (1994), 189–209.

[Ka] Kazdan J. L., Prescribing the curvature of a Riemannian manifold, CBMS no. 57 (1987).

[KW 1] Kazdan J., Warner F., Curvature functions for compact 2-manifolds, Ann. of Math. 99 (1974), 14–47.

[KW 2] Kazdan J., Warner F., Existence and conformal deformation of metrics with prescribing Gaussian and scalar curvatures, Ann. of Math. 101 (1975), 317–331.

[KP 1] Kinderlehrer D., Pedregal P., Characterization of Young measure generated by gradients, Arch. Rat. Mech. Anal. 115 (1991), 329–365.

[KP 2] Kinderlehrer D., Pedregal P., Gradient Young measure generated by sequences in Sobolev spaces, J. Geometric Analysis 4 (1994), 59–90.

[KS] Kinderlehrer D., Stampacchia G., An introduction of variational inequalities and their applications, Acad. Press (1980).

[Kl] Klingenberg W., Lecture on closed geodesics, Grundlehren der Math. 230, Springer-Verlag (1978).

[Ko] Kolmogorov A. N., On the conservation of conditionally periodic motions for a small change in Hamilton's function (Russian), Dokl. Acad. Nauk SSSR 98 (1954), 525–530.

[KKM] Knaster B., Kuratowski K., Mazurkiewicz S., Ein Beweis des Fixpunktsatze für n-dimensionale Simplexe, Fundamenta Mathematica 14 (1929), 132–137.

[Kn] Kneser H., Eine direkte Ableitung des Zornsche Lemmas aus dem Auswahlaxiom, Math. Zeit. 53 (1950), 110–113.

[Krh] Krahn E., Über eine von Rayleigh formulierte Minimaleigenschaft des Kreises, Math. Ann. 94 (1925), 97–100.

[Kr 1] Krasnosel 'ski M. A., Topological methods in the theory of nonlinear integral equations, MacMillan (1964).

[Kr 2] Krasnosel 'ski M. A., Positive solutions of operator equations, Groningen, Noordhoff (1964).

[Kui] Kuiper N. H., The homotopy type of the unitary group of Hilbert space, Topology 3 (1965) 19–30.

[Kuk] Kuksin S. B., Nearly integrable infinite dimensional Hamiltonian systems, Lecture Notes in Math. 1556, Springer-Verlag (1993).

[LU] Ladyzhenskaya O. A., Ural'ceva N. N., Linear and quasilinear elliptic equations, Acad. Press (1968).

[La] Larsen R., Functional analysis, Marcel Dekker (1973).

[Le] Leray J., La théorie des points fixes et ses applications en analyse, Proc. ICM, Cambridge 2 (1950), 202–208.

[LL] Leray J., Lions J. L., Quelques résultats de Vishik sur les problémes elliptiques nonlinéaires par les méthodes de Minty Browder, Bull. Soc. Math. France 93 (1965), 97–107.

[LS] Leray J., Schauder J., Topologie et équations fonctionelles, Ann. Sci. Ecole Norm. Sup. 51 (1934), 45–78.

[Li 1] Li Y. Y. Liouville-type theorems and Harnack-type inequalities for semilinear elliptic equations, J. d'analyse mathématique 90 (2003), 27–87.

[Li 2] Li Y. Y. Prescribing scalar curvature on S^n and related problems, Part I, J. Diff. Eqs. 120, (1995), 319–410, Part II, Comm. Pure Appl. Math. 49 (1996), 541–597.

[Lieb] Lieb E. H. Sharp constants in the Hardy-Littlewood-Sobolev and related inequalities, Ann. of Math. 118 (1983), 349–374.

[Lie] Lieberman G. M., 2nd order parabolic differential equations, World Sci. (1996).

[Lin] Lin F. H., A remark on the map $\frac{x}{|x|}$, C. R. Sci. Paris 305 (1987), 529–531.

[Lid] Lindqvist P., On the equation $div(|\nabla u|^{p-2}\nabla u)+\lambda|u|^{p-2}u=0$, Proc. AMS 109 (1992), 157–164.

[LJ 1] Lions J. L., Quelques méthodes de résolution des problémes aux limites non linéaires, Dunod (1969).

[LS] Lions J. L., Stampacchia G., Variational inequality, Comm. Pure Appl. Math. 20 (1967), 493–519.

[LP 1] Lions P. L., The concentration compactness principle in the calculus of variations, Part I and II, Analyse Nonlinéaire 1 (1984), 109–145, 223–283.
[LT 1] Liu G., Tian G., Weinstein conjecture and GW-invariants,
[LT 2] Liu G., Tian G., Floer homology and the Arnold conjecture, J. Diff. Geom. 49 (1998), 1–74.
[Liu 1] Liu J. Q., A Morse index of saddle points, System Sci. and Math. Sci. 2 (1989), 32–39.
[Liu 2] Liu J. Q., A generalized saddle point theorem, J. Diff. Eqs. (1989) 372–385.
[Lj] Ljusternik L., The topology of function spaces and the calculus of variations in the large, Trudy Mat. Inst. Steklov 19 (1947).
[LjS 1] Ljusternik L., Schnirelmann L., Méthodes topologiques dans les problémes variationnelles, Actualités Sci. Industr. 188 (1934).
[LjS 2] Ljusternik L., Schnirelmann L., Sur le probléme de trois géodesiques fermées sur les surfaces de genre 0, C. R. Acad. Sci Paris 189 (1929), 269–271.
[Ll] Lloyd N. G., Degree theory, Cambridge Univ. Press (1977).
[Lo] Long Y. M., Index theory for symplectic paths with applications, Progress in Math. 207, Birkhauser (2002).
[LZ] Long Y. M., Zhu C. Closed characteristics on compact hypersurfaces in R^{2n}, Ann. of Math. 155 (2002), 317–368.
[Ma] Ma T. W., Topological degrees of set-valued compact fields in locally convex spaces, Dissertationes Mathematicae, 92 Warzawa (1972).
[Man] Mann W. R., Mean value methods in iterations, Proc. AMS 4 (1953), 506–510.
[Mar] Marcellini P., Approximation of quasiconvex functions and lower semicontinuity of multiple integrals, Manuscripta Math. 51 (1985), 1–28.
[MP 1] Marino A., Prodi G., La teoria di Morse per spazi di Hilbert, Rend. Sem. Mat. Univ. Padova, 41 (1968), 43–68.
[MP 2] Marino A., Prodi G., Metodi perturbativi nella teoria di Morse, Boll. Un. Mat. Ital. Suppl. 3 (1975), 1–32.
[Maw 1] Mawhin J., Topological degree methods in nonlinear boundary value problems, AMS (1979).
[Maw 2] Mawhin J., Problémes de Dirichlet, variationnels nonlinéaires, Univ. de Montréal 104 (1987)
[Maw 3] Mawhin J. Functional analysis and BVP, Studies in Math. 14 (J. Hale ed.) (1977), 128–168.
[MW] Mawhin J., Willem M., Critical point theory and Hamiltonian systems, Appl. Math. Sci. 74, Springer-Verlag (1989).
[Mic] Michaikov K., Conley index theory, Lect. Notes, 1609 Springer-Verlag (1996), 119–207.
[Mi 1] Milnor J., Morse theory, Princeton Univ. Press (1963).
[Mi 2] Milnor J., Topology from the differential viewpoint, Univ. Press of Virginia, Charlottesville (1969).
[Mi 3] Milnor J., Lecture on the h-cobordism theorem, Princeton Univ. Press (1965).
[Min] Minty G., On a monotonicity method for the solutions of nonlinear equations in Banach spaces, Proc. Nat. Acad. Sci. USA, 50 (1963), 1038–1041.
[MM] Modica L., Mortola S., Un esempio di Γ-convergenza, Boll. U. M. I. 14-B (1977), 285–299.

[Mor] Morel J. M., The Mumford Shah conjecture in image processing, Asterisque, 241, Expose 813 (1997), 221–242.

[MS] Morel J. M., Solimini S., Variational models in image segmentation, Birkhauser (1994).

[Mo 1] Morrey C. B., Quasi-convexity and the lower semicontinuity of multiple integrals, Pacific J. Math. 2 (1952), 25–53.

[Mo 2] Morrey C. B., Multiple integrals in the calculus of variations, Springer (1966).

[Mo] Morse M., The calculus of variations in the large, AMS Coll. Publ. 18 (1934).

[MC] Morse M., Cairns S. S., Critical point theory in global analysis and differential topology, Acad. Press (1969).

[MT] Morse M., Tompkins C. B., The existence of minimal surfaces of general critical types, Ann. of Math. 40 (1939), 443–472.

[Mos 1] Moser J., A new technique for the construction of solutions of nonlinear differential equations, Proc. Nat. Acad. Sci. USA 47 (1961), 1824–1831.

[Mos 2] Moser J., A rapidly convergent iteration method and nonlinear partial differential equations, I, II, Ann. Scuola Norm. Sup. Pisa 20 (1966), 226–315, 449–535.

[Mos 3] Moser J., Convergent series expansions for quasi-periodic motions, Math. Ann. 169 (1967), 136–176.

[Mos 4] Moser J., On a nonlinear problem in differential geometry, Dynamical systems, Acad. Press (1973).

[Mos 5] Moser J., A sharp form of an inequality by N. Trudinger, Indiana U. Math. J. 20 (1971), 1077–1092.

[Mul 1] Muller S., Variational models for microstructure and phase transitions, Lecture Notes in Math, 1713 Calculus of variations and geometric evolution problems (Hildebrandt S., Struwe M., eds.) Springer-Verlag (1999), 85–210.

[Mul 2] Muller S., A sharp version of Zhang's theorem on truncating sequences of gradients, Trans. AMS,

[MS] Mumford D., Shah J., Boundary detection by minimizing functionals, IEEE Conference on computer vision and pattern recognition, San Francisco (1985).

[Na 1] Nash J., The embedding problem for Riemann manifolds, Ann. of Math. 63 (1956), 20–63.

[Na 2] Nash J., Continuity of solutions of parabolic and elliptic equations, Amer. J. Math. 80 (1958), 931–954.

[Na 3] Nash J., Non-cooperative games, Ann. of Math. 54 (1951), 286–295.

[NN] Newlander A., Nirenberg L., Complex coordinates in almost complex manifolds, Ann. Math. 65 (1957), 391–404.

[Ni] Ni W. M., Some minimax principles and their applications in nonlinear elliptic equations, J. d'analyse Math. 37 (1980), 248–275.

[NT] Ni W. M., Takagi I., On the shape of least energy solutions to a semilinear Neumann problem, Comm. Pure Appl. Math. 45 (1991), 819–851.

[Ni 1] Nirenberg L., Topics in nonlinear functional analysis, Courant Institute Lecture Notes, New York (1974).

[Ni 2] Nirenberg L., Linear partial differential equations, CBMS, no.17 (19720.

[Ni 3] Nirenberg L., Comments on nonlinear problems, Le Mathematische 16 (1981).

[Ni 4] Nirenberg L., Variational and topological methods in nonlinear problems, Bull. AMS 3 (1981), 267–302.

[No] Nordstrom N., Minimization of energy functional with curve-represented edges (J. J. Koenderink ed.), 155–168.

[Nu] Nussbaum R. D., The fixed point index for local condensing maps, Ann. Mat. Pura Appl. 37 (1972), 741–766.

[Ob] Obata M., The conjectures on conformal transformations of Riemannian manifolds, J. Diff. Geom. 6 (1971), 247–258.

[Oh] Oh Y. G., On positive multi-lump bound states of nonlinear Schrodinger equations under multiple well potential, Comm Math. Phys. 131 (1990), 223–253.

[Pa 1] Palais R. S., Morse theory on Hilbert manifolds, Topology (1963), 299–340.

[Pa 2] Palais R. S., Ljusternik Schnirelmann theory on Banach manifolds, Topology 5 (1966), 115–132.

[Pa 3] Palais R. S., Homotopy theory of infinite dimensional manifolds, Topology 5 (1966), 1–16.

[PS] Palais R. S., Smale S., A generalized Morse theory, Bull. AMS 70 (1964), 165–171.

[PR 1] Pejsachowicz J., Rabier P. J., Degree theory for C^1 Fredholm mappings of index 0, J'Analyse mathematique 76 (1998), 289–319.

[PR 2] Pejsachowicz J., Rabier P. J., A substitute for the Sard-Smale theorem in the C^1 case, ibid., 76 (1998), 265–288.

[Pl] Plastock R., Homeomorphisms between Banach spaces, TAMS 200 (1974), 169–183.

[Po] Poschel J., Small divisors with spatial structure in infinite dimensional Hamiltonian systems, Comm. Math. Phys. 127 (1990), 351–393.

[Pw] Protter M. H., Weinbergeer H. F, Maximum principles in differential equations, Prentice–Hall, (1967).

[QS] Quinn F., Sard A., Hausdorff conullity of critical images of Fredholm maps, Amer. J. Math. 94 (1972), 1101–1110.

[Ra 1] Rabinowitz P., A global theorem for nonlinear eigenvalue problems and applications, Contributions to Nonlinear Functional Analysis, Academic Press (1971), 11–30.

[Ra 2] Rabinowitz P., Théorie du degré topologique et applications á des problémes aux limites non linéaires, Univ. laborat. anal. num. Paris (1975).

[Ra 3] Rabinowitz P., Periodic solutions for Hamiltonian systems, Comm. Pure Appl. Math. 31 (1978), 157–184.

[Ra 4] Rabinowitz P., Minimax methods in critical point theory with applications to differential equations, CBMS Regional Conference Series Math. 65 (1986), AMS Providence.

[Ra 5] Rabinowitz P., Variational method for nonlinear eienvalue problems (ed. LG. Prodi), Cremonese, Roma (1974), 141–195.

[Re 1] Reshetnyak Yu. G., On the stability of conformal mappings in multidimensional spaces, Sib. Math. J. 8 (1967), 69–85.

[Re 2] Reshetnyak Yu. G., Space mapping with bounded distortion, AMS (1989).

[Ri] Riviere T., Everywhere discontinuous harmonic maps into spheres, Acta Math. 175 (1995), 197–226.

[Ro 1] Rothe E., Critical points theory in Hilbert space under regular boundary conditions, J. Math. Anal. Appl. 36 (1971), 377–431.

[Ro 2] Rothe E., Morse theory in Hilbert space, Rocky Mountain J. Math. 3 (1973), 251–274.

[Ro 3] Rothe E., On the connection between critical point theory and Leray Schauder degree, J. Math. Anal. Appl. 88 (1982), 265–269.

[Ry] Rybakowski K. P., The homotopy index and partial differential equations, Springer-Verlag (1987).

[RZ] Rybakowski K. P., Zehnder E., On the Morse equation in Conley's index theory for semiflows on metric spaces, Ergodic Theory Dyn, Syst. 5 (1985), 123–143.

[SU] Sacks J., Uhlenbeck K., The existence of minimal immersions of 2 spheres, Ann. Math. 113 (1981), 1–24.

[Sa] Sadovskii B., Limit compact and condensing operators (Russian), Uspehi Mat. Nauk 27(1) (1972), 81–146.

[Sal] Salamon D., Connected simple systems and the Conley index of isolated invariant sets, Trans. Amer. Math. Soc. 291 (1985), 1–41.

[SZ] Salamon D., Zehnder, E., Morse theory for periodic solutions of Hamiltonian systems and the Morse index, Comm. Pure Appl. Math. 45 (1992), 1303–1360.

[Sca] Schaefer H., Über die Methode der a-priori Schranken, Math. Ann. 129 (1955), 415–416.

[Sc] Schauder J., Der Fixpunktsatz in Funktionalräumen, Studia Math. 2 (1930), 171–180.

[Sch] Schiffman M., The Plateau problem for non-relative minima, Ann. of Math. 40 (1939), 834–854.

[Sco] Schoen R., Conformal deformation of a Riemannian metric to a constant scalar curvature, J. Diff. Geom. 20 (19084), 479–495.

[ScU 1] Schoen R., Uhlenbeck K., A regularity for harmonic maps, J. Diff. Geom. 17 (1982), 307–335.

[ScU 2] Schoen R, Uhlenbeck K., Boundary regularity and the Dirichlet problem for harmonic maps, J. Diff. Geom. 18 (1983), 253–268.

[SZ] Schoen R., Zhang D., Prescribing scalar curvature on the n-sphere, Calc. Var. PDE 4 (1996), 1–25.

[Scw] Schwartz J. T., Nonlinear functional analysis, Golden Beach Publ. (1969).

[Se] Semmes S., A primer on Hardy spaces and some remarks on a theorem of Evans and Muller, Comm. PDE, 19 (1994), 277–319.

[Se] Séré E., Existence of infinitely many homoclinic orbits in Hamiltonian systems, Math. Zeit. 209 (1992), 27–42.

[Shi] Shi S.Z., Ekeland's variational principle and the mountain pass lemma, Acta Math. Sinica (NS) 1 (1985) 348–355.

[Si] Sion M., On general minimax theorems, Pacific J. Math. 8 (1958), 171–176.

[Sm 1] Smale S., Generalized Poincaré's conjecture in dimension great than four, Ann. of Math. 74 (1961), 391–406.

[Sm 2] Smale S., Morse theory and a nonlinear generalization of the Dirichlet problem, Ann. of Math. 80 (1964), 382–396.

[Sm 3] Smale S., An infinite dimensional version of Sard's theorem, Amer. J. Math. 87 (1965), 861–867.

[Sma] Smart D. R., Fixed point theorems, Cambridge Univ. Press (1974).

[Smo] Smoller J., Shock waves and reaction diffusion equations, Grundlehren der Mathematischen Wissenschaften 258, Springer-Verlag (1982).

[So] Solimini S., Morse index estimates in Min-Max theorems, Manusc. Math. 32 (1989), 421–454.

[Sp] Spanier E. H., Algebraic topology, Springer-Verlag (1966).

[Ste 1] Stein E., Singular integrals and differentiability properties of functions, Princeton Univ. Press (1970).

[Ste 2] Stein E., Harmonic analysis: real variable methods, orthogonality and oscillatory integrals, Princeton Univ. Press (1993).

[SW] Stein E., Weiss G., Introduction to Fourier analysis on Euclidean spaces, Princeton Univ. Press (1971).

[St 1] Struwe M., Variational Methods, Applications to nonlinear partial differential equations and Hamiltonian systems, 2nd ed., Springer-Verlag (1996).

[St 2] Struwe M., Plateau problem and the calculus of variations, Princeton Univ. Press (1988).

[ST] Stuart C. A., Toland J. F., The fixed point index of a linear k-set contraction, J. London Math. Soc. 6 (1973), 317–320.

[Sv 1] Šverák V., Rank one convexity does not imply quasiconvexity, Proc. Royal Soc. Edinburgh 120 (1992), 185–189.

[Sv 2] Šverák V., Lower semicontinuity of variational integrals and compensated compactness, Proc. ICM 1994, Vol 2, Birkhauser (1995), 1153–1158.

[Tal] Talenti G., Best constant in Sobolev inequality, Ann. Mat. Pure Appl. 110 (1976), 353–372.

[Tar 1] Tartar L., The compensated compactness method applied to systems of conservations laws, Systems of Nonlinear Partial Differential Equations(Ball J., ed.), NATO ASI series, Vol CIII, Reidel (1983), 263–285.

[Tar 2] Tartar L., H-measures, a new approach for studying homogenization, oscillations and concentation effects in partial differential equations, Proc. Roy. Soc. Edinburgh A 115 (1990), 193–230,

[Ta] Tarski A., A lattice theoretical fixed point theorem and its applications, Pac. J. Math. 5 (1955), 285–309.

[Tau] Taubes C. H. Self-dual connections on non-self-dual 4-manifolds, JDG 17 (1982), 139–170.

[Ti] Tian G., On the mountain pass lemma, Kexue Tongbao 14 (1983), 833–835.

[Tr] Trudinger N. S., On imbeddings into Orlicz spaces and some applicatons, J. Math Mech. 17 (1967), 473–483.

[Tu] Turner R., Positive solutions of nonlinear eigenvalue problems (ed. LG. Prodi) Eigenvalues of Nonlinear Problems, Crmenese, Roma (1974), 213–239.

[Uh] Uhlenbeck K., Generic properties of eigenfunctions, Amer. J. Math. 98 (1976), 1059–1078.

[Va] Vainberg M., Variational method for the study of nonlinear operators (Russian), Gostehizdt (1956).

[Vi 1] Viterbo C., A proof of the Weinstein conjecture on R^{2n}, Ann. Inst. H. Poincaré, Annlyse non lineaire 4 (1987), 337–356.

[Vi 2] Viterbo C., Indices de Morse des points critiques obtenus par minimax, Ann. Inst. H. Poincaré, Analyse nonlinéaire 5 (1988), 221–225.

[VN] Von Neumann J., Zur Theorie der Gesellschaftsspiele, Math. Ann. 100 (1927), 295–320.

[Wa 1] Wang Z. Q., A note on the deformation theorem, Acta Math. Sinica 30 (1987), 106–110.

[Wa 2] Wang Z. Q., On a superlinear elliptic equation, Analyse Nonlineaire 8 (1991), 43–58.

[Way] Wayne C. E., Periodic and quasi-periodic solutions of nonlinear wave equations via KAM theory, Comm. Math. Phys. 127 (1990), 479–528.

[Wei 1] Weinstein A., Periodic orbits for convex Hamiltonian systems, Ann Math. 108 (1978), 507–518.

[Wei 2] Weinstein A., On the hypotheses of Rabinowitz's periodic orbit theorem, J. Diff. Eq. 33 (1979), 353–358.

[Wi] Willem M., Minimax methods, Birkhauser (1994).

[Ya] Yamabe H., On the deformation of Riemannian structures on compact manifolds, Osaka Math. J. 12 (1960), 21–37.

[Yo] Young L. C., Lectures on the calculus of variations and optimal control theory, Saunders (1969).

[Ze] Zehnder E., Generalized implicit function theorem with applications to some small divisor problems, I, II, Comm. Pure Appl. Math. 28 (1975), 91–140 (1976), 49–111.

[Zei] Zeidler, E., Nonlinear functional analysis and its applications I-IV, Springer-Verlag, New York (1988).

[Zh 1] Zhang K. W., A construction of quasiconvex function with linear growth at infinity, Ann. S. N. S. Pisa 19 (1992), 313–326.

[Zh 2] Zhang K. W., Two well structure and intrinsic mountain pass points, preprint.

[Zi] Ziemer W. P., Weakly differentiable functions, Springer-Verlag (1989).

Springer Monographs in Mathematics

This series publishes advanced monographs giving well-written presentations of the "state-of-the-art" in fields of mathematical research that have acquired the maturity needed for such a treatment. They are sufficiently self-contained to be accessible to more than just the intimate specialists of the subject, and sufficiently comprehensive to remain valuable references for many years. Besides the current state of knowledge in its field, an SMM volume should also describe its relevance to and interaction with neighbouring fields of mathematics, and give pointers to future directions of research.

Abhyankar, S.S. **Resolution of Singularities of Embedded Algebraic Surfaces** 2nd enlarged ed. 1998
Alexandrov, A.D. **Convex Polyhedra** 2005
Andrievskii, V.V.; Blatt, H.-P. **Discrepancy of Signed Measures and Polynomial Approximation** 2002
Angell, T. S.; Kirsch, A. **Optimization Methods in Electromagnetic Radiation** 2004
Ara, P.; Mathieu, M. **Local Multipliers of C*-Algebras** 2003
Armitage, D.H.; Gardiner, S.J. **Classical Potential Theory** 2001
Arnold, L. **Random Dynamical Systems** corr. 2nd printing 2003 (1st ed. 1998)
Arveson, W. **Noncommutative Dynamics and E-Semigroups** 2003
Aubin, T. **Some Nonlinear Problems in Riemannian Geometry** 1998
Auslender, A.; Teboulle M. **Asymptotic Cones and Functions in Optimization and Variational Inequalities** 2003
Bang-Jensen, J.; Gutin, G. **Digraphs** 2001
Baues, H.-J. **Combinatorial Foundation of Homology and Homotopy** 1999
Brown, K.S. **Buildings** 3rd printing 2000 (1st ed. 1998)
Chang, K.-C. **Methods in Nonlinear Analysis** 2005
Cherry, W.; Ye, Z. **Nevanlinna's Theory of Value Distribution** 2001
Ching, W.K. **Iterative Methods for Queuing and Manufacturing Systems** 2001
Crabb, M.C.; James, I.M. **Fibrewise Homotopy Theory** 1998
Chudinovich, I. **Variational and Potential Methods for a Class of Linear Hyperbolic Evolutionary Processes** 2005
Dineen, S. **Complex Analysis on Infinite Dimensional Spaces** 1999
Dugundji, J.; Granas, A. **Fixed Point Theory** 2003
Elstrodt, J.; Grunewald, F. Mennicke, J. **Groups Acting on Hyperbolic Space** 1998
Edmunds, D.E.; Evans, W.D. **Hardy Operators, Function Spaces and Embeddings** 2004
Engler, A.J.; Prestel, A. **Valued Fields** 2005
Fadell, E.R.; Husseini, S.Y. **Geometry and Topology of Configuration Spaces** 2001
Fedorov, Y.N.; Kozlov, V.V. **A Memoir on Integrable Systems** 2001
Flenner, H.; O'Carroll, L. Vogel, W. **Joins and Intersections** 1999
Gelfand, S.I.; Manin, Y.I. **Methods of Homological Algebra** 2nd ed. 2003
Griess, R.L. Jr. **Twelve Sporadic Groups** 1998
Gras, G. **Class Field Theory** corr. 2nd printing 2005
Hida, H. ***p*-Adic Automorphic Forms on Shimura Varieties** 2004
Ischebeck, F.; Rao, R.A. **Ideals and Reality** 2005
Ivrii, V. **Microlocal Analysis and Precise Spectral Asymptotics** 1998
Jech, T. **Set Theory** (3rd revised edition 2002)
Jorgenson, J.; Lang, S. **Spherical Inversion on SLn (R)** 2001
Kanamori, A. **The Higher Infinite** corr. 2nd printing 2005 (2nd ed. 2003)
Kanovei, V. **Nonstandard Analysis, Axiomatically** 2005
Khoshnevisan, D. **Multiparameter Processes** 2002
Koch, H. **Galois Theory of *p*-Extensions** 2002

Komornik, V. **Fourier Series in Control Theory** 2005
Kozlov, V.; Maz'ya, V. **Differential Equations with Operator Coefficients** 1999
Landsman, N.P. **Mathematical Topics between Classical & Quantum Mechanics** 1998
Leach, J.A.; Needham, D.J. **Matched Asymptotic Expansions in Reaction-Diffusion Theory** 2004
Lebedev, L.P.; Vorovich, I.I. **Functional Analysis in Mechanics** 2002
Lemmermeyer, F. **Reciprocity Laws: From Euler to Eisenstein** 2000
Malle, G.; Matzat, B.H. **Inverse Galois Theory** 1999
Mardesic, S. **Strong Shape and Homology** 2000
Margulis, G.A. **On Some Aspects of the Theory of Anosov Systems** 2004
Murdock, J. **Normal Forms and Unfoldings for Local Dynamical Systems** 2002
Narkiewicz, W. **Elementary and Analytic Theory of Algebraic Numbers** 3rd ed. 2004
Narkiewicz, W. **The Development of Prime Number Theory** 2000
Parker, C.; Rowley, P. **Symplectic Amalgams** 2002
Peller, V. (Ed.) **Hankel Operators and Their Applications** 2003
Prestel, A.; Delzell, C.N. **Positive Polynomials** 2001
Puig, L. **Blocks of Finite Groups** 2002
Ranicki, A. **High-dimensional Knot Theory** 1998
Ribenboim, P. **The Theory of Classical Valuations** 1999
Rowe, E.G.P. **Geometrical Physics in Minkowski Spacetime** 2001
Rudyak, Y.B. **On Thom Spectra, Orientability and Cobordism** 1998
Ryan, R.A. **Introduction to Tensor Products of Banach Spaces** 2002
Saranen, J.; Vainikko, G. **Periodic Integral and Pseudodifferential Equations with Numerical Approximation** 2002
Schneider, P. **Nonarchimedean Functional Analysis** 2002
Serre, J-P. **Complex Semisimple Lie Algebras** 2001 (reprint of first ed. 1987)
Serre, J-P. **Galois Cohomology** corr. 2nd printing 2002 (1st ed. 1997)
Serre, J-P. **Local Algebra** 2000
Serre, J-P. **Trees** corr. 2nd printing 2003 (1st ed. 1980)
Smirnov, E. **Hausdorff Spectra in Functional Analysis** 2002
Springer, T.A. Veldkamp, F.D. **Octonions, Jordan Algebras, and Exceptional Groups** 2000
Sznitman, A.-S. **Brownian Motion, Obstacles and Random Media** 1998
Taira, K. **Semigroups, Boundary Value Problems and Markov Processes** 2003
Talagrand, M. **The Generic Chaining** 2005
Tauvel, P.; Yu, R.W.T. **Lie Algebras and Algebraic Groups** 2005
Tits, J.; Weiss, R.M. **Moufang Polygons** 2002
Uchiyama, A. **Hardy Spaces on the Euclidean Space** 2001
Üstünel, A.-S.; Zakai, M. **Transformation of Measure on Wiener Space** 2000
Vasconcelos, W. **Integral Closure. Rees Algebras, Multiplicities, Algorithms** 2005
Yang, Y. **Solitons in Field Theory and Nonlinear Analysis** 2001
Zieschang P.-H. **Theory of Association Schemes** 2005